Compliance Handbook for Pharmaceuticals, Medical Devices, and Biologics

DRUGS AND THE PHARMACEUTICAL SCIENCES

DRUGS AND THE PHARMACEUTICAL SCIENCES

A Series of Textbooks and Monographs

1. Pharmacokinetics, *Milo Gibaldi and Donald Perrier*
2. Good Manufacturing Practices for Pharmaceuticals: A Plan for Total Quality Control, *Sidney H. Willig, Murray M. Tuckerman, and William S. Hitchings IV*
3. Microencapsulation, *edited by J. R. Nixon*
4. Drug Metabolism: Chemical and Biochemical Aspects, *Bernard Testa and Peter Jenner*
5. New Drugs: Discovery and Development, *edited by Alan A. Rubin*
6. Sustained and Controlled Release Drug Delivery Systems, *edited by Joseph R. Robinson*
7. Modern Pharmaceutics, *edited by Gilbert S. Banker and Christopher T. Rhodes*
8. Prescription Drugs in Short Supply: Case Histories, *Michael A. Schwartz*
9. Activated Charcoal: Antidotal and Other Medical Uses, *David O. Cooney*
10. Concepts in Drug Metabolism (in two parts), *edited by Peter Jenner and Bernard Testa*
11. Pharmaceutical Analysis: Modern Methods (in two parts), *edited by James W. Munson*
12. Techniques of Solubilization of Drugs, *edited by Samuel H. Yalkowsky*
13. Orphan Drugs, *edited by Fred E. Karch*
14. Novel Drug Delivery Systems: Fundamentals, Developmental Concepts, Biomedical Assessments, *Yie W. Chien*
15. Pharmacokinetics: Second Edition, Revised and Expanded, *Milo Gibaldi and Donald Perrier*
16. Good Manufacturing Practices for Pharmaceuticals: A Plan for Total Quality Control, Second Edition, Revised and Expanded, *Sidney H. Willig, Murray M. Tuckerman, and William S. Hitchings IV*
17. Formulation of Veterinary Dosage Forms, *edited by Jack Blodinger*
18. Dermatological Formulations: Percutaneous Absorption, *Brian W. Barry*
19. The Clinical Research Process in the Pharmaceutical Industry, *edited by Gary M. Matoren*
20. Microencapsulation and Related Drug Processes, *Patrick B. Deasy*
21. Drugs and Nutrients: The Interactive Effects, *edited by Daphne A. Roe and T. Colin Campbell*
22. Biotechnology of Industrial Antibiotics, *Erick J. Vandamme*

23. Pharmaceutical Process Validation, *edited by Bernard T. Loftus and Robert A. Nash*
24. Anticancer and Interferon Agents: Synthesis and Properties, *edited by Raphael M. Ottenbrite and George B. Butler*
25. Pharmaceutical Statistics: Practical and Clinical Applications, *Sanford Bolton*
26. Drug Dynamics for Analytical, Clinical, and Biological Chemists, *Benjamin J. Gudzinowicz, Burrows T. Younkin, Jr., and Michael J. Gudzinowicz*
27. Modern Analysis of Antibiotics, *edited by Adjoran Aszalos*
28. Solubility and Related Properties, *Kenneth C. James*
29. Controlled Drug Delivery: Fundamentals and Applications, Second Edition, Revised and Expanded, *edited by Joseph R. Robinson and Vincent H. Lee*
30. New Drug Approval Process: Clinical and Regulatory Management, *edited by Richard A. Guarino*
31. Transdermal Controlled Systemic Medications, *edited by Yie W. Chien*
32. Drug Delivery Devices: Fundamentals and Applications, *edited by Praveen Tyle*
33. Pharmacokinetics: Regulatory • Industrial • Academic Perspectives, *edited by Peter G. Welling and Francis L. S. Tse*
34. Clinical Drug Trials and Tribulations, *edited by Allen E. Cato*
35. Transdermal Drug Delivery: Developmental Issues and Research Initiatives, *edited by Jonathan Hadgraft and Richard H. Guy*
36. Aqueous Polymeric Coatings for Pharmaceutical Dosage Forms, *edited by James W. McGinity*
37. Pharmaceutical Pelletization Technology, *edited by Isaac Ghebre-Sellassie*
38. Good Laboratory Practice Regulations, *edited by Allen F. Hirsch*
39. Nasal Systemic Drug Delivery, *Yie W. Chien, Kenneth S. E. Su, and Shyi-Feu Chang*
40. Modern Pharmaceutics: Second Edition, Revised and Expanded, *edited by Gilbert S. Banker and Christopher T. Rhodes*
41. Specialized Drug Delivery Systems: Manufacturing and Production Technology, *edited by Praveen Tyle*
42. Topical Drug Delivery Formulations, *edited by David W. Osborne and Anton H. Amann*
43. Drug Stability: Principles and Practices, *Jens T. Carstensen*
44. Pharmaceutical Statistics: Practical and Clinical Applications, Second Edition, Revised and Expanded, *Sanford Bolton*
45. Biodegradable Polymers as Drug Delivery Systems, *edited by Mark Chasin and Robert Langer*
46. Preclinical Drug Disposition: A Laboratory Handbook, *Francis L. S. Tse and James J. Jaffe*
47. HPLC in the Pharmaceutical Industry, *edited by Godwin W. Fong and Stanley K. Lam*
48. Pharmaceutical Bioequivalence, *edited by Peter G. Welling, Francis L. S. Tse, and Shrikant V. Dinghe*

49. Pharmaceutical Dissolution Testing, *Umesh V. Banakar*
50. Novel Drug Delivery Systems: Second Edition, Revised and Expanded, *Yie W. Chien*
51. Managing the Clinical Drug Development Process, *David M. Cocchetto and Ronald V. Nardi*
52. Good Manufacturing Practices for Pharmaceuticals: A Plan for Total Quality Control, Third Edition, *edited by Sidney H. Willig and James R. Stoker*
53. Prodrugs: Topical and Ocular Drug Delivery, *edited by Kenneth B. Sloan*
54. Pharmaceutical Inhalation Aerosol Technology, *edited by Anthony J. Hickey*
55. Radiopharmaceuticals: Chemistry and Pharmacology, *edited by Adrian D. Nunn*
56. New Drug Approval Process: Second Edition, Revised and Expanded, *edited by Richard A. Guarino*
57. Pharmaceutical Process Validation: Second Edition, Revised and Expanded, *edited by Ira R. Berry and Robert A. Nash*
58. Ophthalmic Drug Delivery Systems, *edited by Ashim K. Mitra*
59. Pharmaceutical Skin Penetration Enhancement, *edited by Kenneth A. Walters and Jonathan Hadgraft*
60. Colonic Drug Absorption and Metabolism, *edited by Peter R. Bieck*
61. Pharmaceutical Particulate Carriers: Therapeutic Applications, *edited by Alain Rolland*
62. Drug Permeation Enhancement: Theory and Applications, *edited by Dean S. Hsieh*
63. Glycopeptide Antibiotics, *edited by Ramakrishnan Nagarajan*
64. Achieving Sterility in Medical and Pharmaceutical Products, *Nigel A. Halls*
65. Multiparticulate Oral Drug Delivery, *edited by Isaac Ghebre-Sellassie*
66. Colloidal Drug Delivery Systems, *edited by Jörg Kreuter*
67. Pharmacokinetics: Regulatory • Industrial • Academic Perspectives, Second Edition, *edited by Peter G. Welling and Francis L. S. Tse*
68. Drug Stability: Principles and Practices, Second Edition, Revised and Expanded, *Jens T. Carstensen*
69. Good Laboratory Practice Regulations: Second Edition, Revised and Expanded, *edited by Sandy Weinberg*
70. Physical Characterization of Pharmaceutical Solids, *edited by Harry G. Brittain*
71. Pharmaceutical Powder Compaction Technology, *edited by Göran Alderborn and Christer Nyström*
72. Modern Pharmaceutics: Third Edition, Revised and Expanded, *edited by Gilbert S. Banker and Christopher T. Rhodes*
73. Microencapsulation: Methods and Industrial Applications, *edited by Simon Benita*
74. Oral Mucosal Drug Delivery, *edited by Michael J. Rathbone*
75. Clinical Research in Pharmaceutical Development, *edited by Barry Bleidt and Michael Montagne*

76. The Drug Development Process: Increasing Efficiency and Cost Effectiveness, *edited by Peter G. Welling, Louis Lasagna, and Umesh V. Banakar*
77. Microparticulate Systems for the Delivery of Proteins and Vaccines, *edited by Smadar Cohen and Howard Bernstein*
78. Good Manufacturing Practices for Pharmaceuticals: A Plan for Total Quality Control, Fourth Edition, Revised and Expanded, *Sidney H. Willig and James R. Stoker*
79. Aqueous Polymeric Coatings for Pharmaceutical Dosage Forms: Second Edition, Revised and Expanded, *edited by James W. McGinity*
80. Pharmaceutical Statistics: Practical and Clinical Applications, Third Edition, *Sanford Bolton*
81. Handbook of Pharmaceutical Granulation Technology, *edited by Dilip M. Parikh*
82. Biotechnology of Antibiotics: Second Edition, Revised and Expanded, *edited by William R. Strohl*
83. Mechanisms of Transdermal Drug Delivery, *edited by Russell O. Potts and Richard H. Guy*
84. Pharmaceutical Enzymes, *edited by Albert Lauwers and Simon Scharpé*
85. Development of Biopharmaceutical Parenteral Dosage Forms, *edited by John A. Bontempo*
86. Pharmaceutical Project Management, *edited by Tony Kennedy*
87. Drug Products for Clinical Trials: An International Guide to Formulation • Production • Quality Control, *edited by Donald C. Monkhouse and Christopher T. Rhodes*
88. Development and Formulation of Veterinary Dosage Forms: Second Edition, Revised and Expanded, *edited by Gregory E. Hardee and J. Desmond Baggot*
89. Receptor-Based Drug Design, *edited by Paul Leff*
90. Automation and Validation of Information in Pharmaceutical Processing, *edited by Joseph F. deSpautz*
91. Dermal Absorption and Toxicity Assessment, *edited by Michael S. Roberts and Kenneth A. Walters*
92. Pharmaceutical Experimental Design, *Gareth A. Lewis, Didier Mathieu, and Roger Phan-Tan-Luu*
93. Preparing for FDA Pre-Approval Inspections, *edited by Martin D. Hynes III*
94. Pharmaceutical Excipients: Characterization by IR, Raman, and NMR Spectroscopy, *David E. Bugay and W. Paul Findlay*
95. Polymorphism in Pharmaceutical Solids, *edited by Harry G. Brittain*
96. Freeze-Drying/Lyophilization of Pharmaceutical and Biological Products, *edited by Louis Rey and Joan C. May*
97. Percutaneous Absorption: Drugs–Cosmetics–Mechanisms–Methodology, Third Edition, Revised and Expanded, *edited by Robert L. Bronaugh and Howard I. Maibach*

98. Bioadhesive Drug Delivery Systems: Fundamentals, Novel Approaches, and Development, *edited by Edith Mathiowitz, Donald E. Chickering III, and Claus-Michael Lehr*
99. Protein Formulation and Delivery, *edited by Eugene J. McNally*
100. New Drug Approval Process: Third Edition, The Global Challenge, *edited by Richard A. Guarino*
101. Peptide and Protein Drug Analysis, *edited by Ronald E. Reid*
102. Transport Processes in Pharmaceutical Systems, *edited by Gordon L. Amidon, Ping I. Lee, and Elizabeth M. Topp*
103. Excipient Toxicity and Safety, *edited by Myra L. Weiner and Lois A. Kotkoskie*
104. The Clinical Audit in Pharmaceutical Development, *edited by Michael R. Hamrell*
105. Pharmaceutical Emulsions and Suspensions, *edited by Francoise Nielloud and Gilberte Marti-Mestres*
106. Oral Drug Absorption: Prediction and Assessment, *edited by Jennifer B. Dressman and Hans Lennernäs*
107. Drug Stability: Principles and Practices, Third Edition, Revised and Expanded, *edited by Jens T. Carstensen and C. T. Rhodes*
108. Containment in the Pharmaceutical Industry, *edited by James P. Wood*
109. Good Manufacturing Practices for Pharmaceuticals: A Plan for Total Quality Control from Manufacturer to Consumer, Fifth Edition, Revised and Expanded, *Sidney H. Willig*
110. Advanced Pharmaceutical Solids, *Jens T. Carstensen*
111. Endotoxins: Pyrogens, LAL Testing, and Depyrogenation, Second Edition, Revised and Expanded, *Kevin L. Williams*
112. Pharmaceutical Process Engineering, *Anthony J. Hickey and David Ganderton*
113. Pharmacogenomics, *edited by Werner Kalow, Urs A. Meyer, and Rachel F. Tyndale*
114. Handbook of Drug Screening, *edited by Ramakrishna Seethala and Prabhavathi B. Fernandes*
115. Drug Targeting Technology: Physical • Chemical • Biological Methods, *edited by Hans Schreier*
116. Drug–Drug Interactions, *edited by A. David Rodrigues*
117. Handbook of Pharmaceutical Analysis, *edited by Lena Ohannesian and Anthony J. Streeter*
118. Pharmaceutical Process Scale-Up, *edited by Michael Levin*
119. Dermatological and Transdermal Formulations, *edited by Kenneth A. Walters*
120. Clinical Drug Trials and Tribulations: Second Edition, Revised and Expanded, *edited by Allen Cato, Lynda Sutton, and Allen Cato III*
121. Modern Pharmaceutics: Fourth Edition, Revised and Expanded, *edited by Gilbert S. Banker and Christopher T. Rhodes*
122. Surfactants and Polymers in Drug Delivery, *Martin Malmsten*
123. Transdermal Drug Delivery: Second Edition, Revised and Expanded, *edited by Richard H. Guy and Jonathan Hadgraft*

124. Good Laboratory Practice Regulations: Second Edition, Revised and Expanded, *edited by Sandy Weinberg*
125. Parenteral Quality Control: Sterility, Pyrogen, Particulate, and Package Integrity Testing: Third Edition, Revised and Expanded, *Michael J. Akers, Daniel S. Larrimore, and Dana Morton Guazzo*
126. Modified-Release Drug Delivery Technology, *edited by Michael J. Rathbone, Jonathan Hadgraft, and Michael S. Roberts*
127. Simulation for Designing Clinical Trials: A Pharmacokinetic-Pharmacodynamic Modeling Perspective, *edited by Hui C. Kimko and Stephen B. Duffull*
128. Affinity Capillary Electrophoresis in Pharmaceutics and Biopharmaceutics, *edited by Reinhard H. H. Neubert and Hans-Hermann Rüttinger*
129. Pharmaceutical Process Validation: An International Third Edition, Revised and Expanded, *edited by Robert A. Nash and Alfred H. Wachter*
130. Ophthalmic Drug Delivery Systems: Second Edition, Revised and Expanded, *edited by Ashim K. Mitra*
131. Pharmaceutical Gene Delivery Systems, *edited by Alain Rolland and Sean M. Sullivan*
132. Biomarkers in Clinical Drug Development, *edited by John C. Bloom and Robert A. Dean*
133. Pharmaceutical Extrusion Technology, *edited by Isaac Ghebre-Sellassie and Charles Martin*
134. Pharmaceutical Inhalation Aerosol Technology: Second Edition, Revised and Expanded, *edited by Anthony J. Hickey*
135. Pharmaceutical Statistics: Practical and Clinical Applications, Fourth Edition, *Sanford Bolton and Charles Bon*
136. Compliance Handbook for Pharmaceuticals, Medical Devices, and Biologics, *edited by Carmen Medina*

ADDITIONAL VOLUMES IN PREPARATION

New Drug Approval Process: Fourth Edition, Accelerating Global Registrations, *edited by Richard A. Guarino*

Compliance Handbook for Pharmaceuticals, Medical Devices, and Biologics

edited by
Carmen Medina
Precision Consultants
Coronado, California, U.S.A.

CRC Press is an imprint of the
Taylor & Francis Group, an **informa** business

CRC Press
Taylor & Francis Group
6000 Broken Sound Parkway NW, Suite 300
Boca Raton, FL 33487-2742

First issued in paperback 2019

ISBN-13: 978-0-8247-4078-8 (hbk)
ISBN-13: 978-0-367-39476-9 (pbk)

A CIP record for this book is available from the British Library.

Library of Congress Cataloging-in-Publication Data available on application

Visit the Taylor & Francis Web site at
http://www.taylorandfrancis.com

and the CRC Press Web site at
http://www.crcpress.com

To the 80 million strong GenXers, many of whom
will comprise the future of this marvelous industry. May you take the
accumulated wisdom offered in this book and make our industry
the very best it can be.
Your destiny, our future—it is all in your hands.

Preface

The art of compliance is a craft much like that of a carpenter who learns to work with various kinds of wood and designs. Ours is an industry wrought with differences. While the regulations that govern FDA-regulated industries are proscribed, how the regulations are interpreted and applied really depends on experience and how well one has mastered the craft of compliance. There are no proscribed procedures, but there are guideposts common to all FDA-regulated industries, whether a firm manufactures medical devices, pharmaceuticals, or biological products.

This book is about how to hone the craft of compliance, and how to take basic guideposts and apply them to your specific company culture and needs. This book is not about how to adopt completely new methods and systems, but rather how to adapt proven ideas and strategies to your current practices.

Furthermore, this book is about providing a bridge for the Generation X entrepreneurs and employees who are such a vital part of our workforce and this industry's destiny, so that they may benefit from the decades of accumulated wisdom and experience that the various contributors have poured into this text.

The contributors have been carefully selected to provide readers with the state-of-the-art thinking on 17 of the most critical compliance and quality systems required by the FDA and practiced throughout the industry. These contributors are not offering their opinions or theory. Instead, they bridge the gap between the theoretical and the practical by sharing their understanding of what they have experienced from within the industry and by reflecting on what they know works and does not work relative to the critical compliance categories delineated in the table of contents.

Read it in any order, skip what you don't need and enjoy what you can!

ACKNOWLEDGMENTS

I am forever indebted to the many contributors who put their professional commitments aside long enough to write their respective chapters and who in doing so have allowed themselves to be used as bridges that readers have been invited to cross. They gave generously of their time, knowledge, and expertise to help the future of this marvelous industry.

I am also eternally grateful to two women who gave me their best efforts during a time when I was frenetic with professional commitments of my own and facing a looming deadline: Christine Simmonds and Dawn Silva. Without their continual support, organizational skills, patience, and word processing expertise this book would never have been born.

To my friends who never let me give up and always provided gentle prodding to "*Get the book done!*" I send out one million thank yous.

The idea for this book would never have occurred to me were it not for the mentoring and example of one man—Dr. David A. Kessler, the former Commissioner of the U.S. Food and Drug Administration. His longstanding example during my tenure as an FDA investigator inspired me beyond words and gave me the courage to embark on a project of this magnitude. He forever changed the complexion of the agency and provided me with the impetus to always give a little more than I expected to get when dealing with FDA-regulated industries. This book is my way of saying thank you to an industry I love and that over the course of 17 years has rewarded me beyond my imagination.

I extend my greatest thanks to Ms. Sandra Beberman for her remarkable patience, belief in my abilities, and, most of all, her confidence in the value and timeliness of this text. Her steadfastness never let me give up. To all of her staff who have supported this effort through the years of its making—thank you!

Finally, I will attempt to put into words the gratitude I feel for my family. Brian and Quinn were my rocks, my eternal shoulders to lean on when I felt overwhelmed and discouraged. Words cannot convey my appreciation, but I can say that without their unyielding support, encouragement, love, and belief in my abilities this book would surely not have been completed.

Thank you one and all.

Carmen Medina

Introduction

The current regulations were not meant to fully provide answers to the myriad of compliance quandaries facing FDA-regulated industries. Recognizing and accepting this fact clears the way for evaluating critical compliance and quality challenges within the context of FDA's expectations, industry standards, and, most importantly, emerging trends around the globe.

The 21 CFR regulations, particularly the sections related to cGMPs for medical devices, pharmaceuticals, and biologics, offer these industries the minimum threshold of compliance requirements, and only begin to address issues related to product quality. It is up to the experts, such as quality assurance personnel and seasoned consultants, to help interpret those regulations and apply them in today's rapidly changing and increasingly demanding world of FDA-regulated products. Legislation such as the FDA Modernization Act of 1997, which amends the FD&C Act and the biological products provisions in the Public Health Service Act, aids in giving particular categories of regulated products (e.g., new drugs, biological products, and medical devices) more regulatory latitude, while increasing regulatory demands on regulated activities such as clinical trials and other categories of regulated products such as OTCs and cosmetics.

The *Compliance Handbook for Pharmaceuticals, Medical Devices, and Biologics* provides cutting-edge guidance on several critical compliance challenges that frequently are not adequately addressed within new and established firms and that often result in irreconcilable regulatory situations. The FDA, along with other regulatory and educational organizations, has never concisely or comprehensively addressed many of these compliance challenges. The strategies and programs offered in this book can significantly decrease a company's compliance vulnerability and regulatory liability and increase its overall quality.

This book is a compilation of regulatory interpretation, technical support, extensive practical expertise, and assorted experiences from seasoned individuals at the top of their game. Some of the contributors once worked for the FDA, and others have spent decades evaluating and installing programs and systems for enhanced compliance and quality in such areas as method development, computer and process validation, change management, internal auditing, personnel training, annual product reviews, and strategic planning.

There are many areas for which the cGMPs and FDA Guidelines and compliance programs simply do not offer enough clarity and direction. Change control—an essential subject—is addressed in the 21 CFR one time, under 211.100(b) with a mere 11 words. Yet any FDA-regulated industry is aware that the management of change within the facility, or change related to contract services, is one of the most critical compliance and quality systems demanded during any phase of product development and commercialization. Cleaning validation is cited in the 21 CFR Part 211 over 30 times, yet the industry is repeatedly cited for poor housekeeping and lack of validated cleaning procedures and analytical methods. This book will not solve all the industry's problems, but it will certainly put into perspective the compliance and regulatory topics that merit closer attention and enhanced resource allocation. The guidance in this book will assist management in prioritizing and ultimately developing an overall compliance upgrade master plan.

The book provides readers with the edge needed to compete globally and meet FDA regulatory and compliance demands. The cGMPs are only a platform from which to launch total quality performance and forward quality principles. This book's models for success in areas related to personnel training, SOP compliance, handling out-of-specification results in the laboratory and in manufacturing operations, and vendor qualification move beyond the cGMPs and allow for an integrated approach to compliance and total quality management. Several authors also cover topics of international scope and interest that have not received adequate coverage elsewhere.

Another topic on which the agency does not provide specific guidance is how to manage and control the FDA inspection process. Chapter 13 guides the reader through the preparations, execution, and follow-up required for this arduous process. Whether it is a preapproval, for-cause, routine, or foreign inspection—it all boils down to inspection readiness and overall compliance. Throughout the book, the most critical compliance categories have been identified and elaborated on, eliminating any guesswork relative to the investigators' expectations during their visit to a domestic or foreign facility. This book is guaranteed to increase your company's overall compliance readiness, whether for an FDA inspection, overseas preapproval inspection, internal audit, ISO audit, or vendor qualification audit, or to simply prepare a regulatory submission that will pass muster with the agency and its critical reviewers.

The science of compliance lies entirely in knowing how to interpret the regulations and apply them against the current backdrop of the FDA's shifting expectations and the industry's constantly evolving standards. This book is a useful tool toward increasing a company's overall compliance and quality status and decreasing its regulatory vulnerability, particularly in areas for which there is currently limited or no guidance.

The subject matter covered herein applies to the medical device, pharmaceutical, and biologics industries; however, a distinction is made when a topic has a unique or specific application to any one particular industry. For example, the new medical device regulations require management review; hence, the specifics of a comprehensive management review program are delineated in the chapters covering annual product reviews and consumer complaint handling. The information provided in this book will prove beneficial to readers concerned and involved with compliance and quality systems in U.S. firms or overseas facilities seeking U.S. distribution of their products.

Contents

Contributors

James Agalloco Agalloco & Associates, Belle Mead, New Jersey, U.S.A.

Graham Bunn Astra Pharmaceuticals LP, Wayne, Pennsylvania, U.S.A.

Timothy Carey Wyeth BioPharma, Andover, Massachusetts, U.S.A.

Anisa Dhalla UCB Pharma, Inc., Smyrna, Georgia, U.S.A.

Susan Freeman Antioch, Illinois, U.S.A.

Patricia Fritz UCB Pharma, Inc., Smyrna, Georgia, U.S.A.

Troy Fugate Compliance Insight, Inc., Hamilton, Ohio, U.S.A.

Maria A. Geigel MAG Associates, LLC, Washington Crossing, Pennsylvania, U.S.A.

William E. Hall Hall and Pharmaceutical Associates, Inc., Kure Beach, North Carolina, U.S.A.

Anne P. Hoppe Serologicals, Inc., Clarkston, Georgia, U.S.A.

Timothy Horgan Wyeth BioPharma, Andover, Massachusetts, U.S.A.

Karen L. Hughes Guilford Pharmaceuticals, Inc., Baltimore, Maryland, U.S.A.

Martin D. Hynes III Pharmaceutical Projects Management, Lilly Research Laboratories, Eli Lilly and Company, Indianapolis, Indiana, U.S.A.

Ron Johnson Quintiles Consulting, Rockville, Maryland, U.S.A.

Robert B. Kirsch R. B. Kirsch Consulting, Arlington Heights, Illinois, U.S.A.

Carmen Medina Precision Consultants, Inc., Coronado, California, U.S.A.

Richard F. Meyer Quantitative Technologies, Inc., Whitehouse, New Jersey, U.S.A.

Alan G. Minsk Arnall Golden & Gregory, Atlanta, Georgia, U.S.A.

Curtis L. Scribner Biomedicines, Inc., Emeryville, California, U.S.A.

Elizabeth M. Troll Chesapeake Biological Laboratories, Inc., Baltimore, Maryland, U.S.A.

Paul A. Winslow Quantitative Technologies, Inc., Whitehouse, New Jersey, U.S.A.

Compliance Handbook for Pharmaceuticals, Medical Devices, and Biologics

1

Regulatory Submissions

Patricia Fritz and Anisa Dhalla
UCB Pharma, Inc., Smyrna, Georgia, U.S.A.

Regulatory submissions refers to applications providing data or information to the Food and Drug Administration (FDA) related to the development, approval, or postapproval reporting for prescription drugs, biologics, and medical device products. Regulatory submissions are the primary means by which the pharmaceutical and the medical device industry communicate product-specific information to the FDA. Submissions of applications for either premarket investigations or market authorization are generally a series of submissions reflecting product or applicant information from the development stage throughout the marketing life cycle of a product.

The FDA is responsible for the review and market approval of new drugs, biologics, and medical devices in the United States under the authority of the Federal Food Drug and Cosmetic Act (the Act) and Section 351 of the Public Health Service Act (the PHS Act). The FDA defines "premarket review" as the examination of data and information in an application as described in Sections 505, 510(k), 513(f), 515, or 520(g) or 520(I) of the Act or Section 351 of the PHS Act. This refers to the premarket review of data and information contained in any Investigational New Drug application (IND), Investigational Device

Exemption (IDE), New Drug Application (NDA), Abbreviated New Drug Application (ANDA), Biologics License Application (BLA), device premarket notification [510(k)], or device Premarket Approval Application (PMA).

The review of these applications is performed by three major program centers in the FDA organization: the Center for Drug Evaluation and Research (CDER), with responsibility for drug products, the Center for Biologics Evaluation and Research (CBER), with responsibility for biologic products, and the Center for Devices and Radiological Health (CDRH), with responsibility for medical devices. Each center is organized into scientific review divisions and also includes divisions responsible for quality and compliance aspects.

Applications to gain marketing authorization for new prescription drugs, biologics, or medical devices generally fall within one of the following: NDAs for new prescription drug products, BLAs for new biologic products, and PMAs for certain new medical device products. These marketing applications require the inclusion of the results of human testing to support the safe and effective use of the new product. In order to ship investigational products to the investigators for the required clinical testing, applicants must submit either an IND for a new drug or a biologic product or an IDE for new medical device products.

Abbreviated new drug applications are submitted to gain approval of generic versions of already approved drug products. Premarket notification [510(k)] applications are the mechanisms for marketing medical devices that are substantially equivalent to already marketed device products. Both of these applications are based on approved similar product information.

There are numerous other regulatory submissions, including product listings and establishment registrations and modifications or changes to already submitted applications, as well as routine periodic reporting. While each of these applications is not discussed in detail, the overall philosophical approach for the preparation of regulatory submissions across these product categories is similar. A tabular summary of the major applications is provided in Appendix A.

The submission of an application conveys an acceptance of certain responsibilities, including the accuracy and the quality of the data as well as the required subsequent reporting and technical commitments for the product and its intended use. To assure the accuracy and quality of the data and information provided in applications, the Act gives the FDA broad authority to inspect pharmaceutical and medical device establishments, including manufacturers and other research testing facilities from which data are derived. Applicants therefore must have documented systems in place for all processes from which data are derived and included in regulatory submissions. These systems should also assure the required reporting, including the reporting of changes after the submission of an application.

1 CRITICAL ELEMENTS OF REGULATORY SUBMISSIONS

Regulatory citations in Title 21 of the Code of Federal Regulations (CFR) and FDA guidance documents provide the content requirements for all required applications. While these references discuss the content and format of regulatory submissions, they are not rich in detail and often do not specifically address an applicant's specific product concerns. Outlines of the required elements of INDs, NDAs or BLAs, ANDAs, IDEs, PMAs, and 510(k)s are included in Appendix B. The data and information required in these applications encompass a wide range of disciplines, including medical, pharmacology, toxicology, microbiology, biopharmaceutics, statistics and chemistry, manufacturing, and controls.

There are some common administrative elements for all applications. For example, a cover letter should address the purpose of the submission and make reference to any relevant previous submissions or previous agreements with FDA. To the extent possible, a summary of the information provided in the submission should be included. Cover letters should also include, if applicable, the assigned application or serial number, user fee information, contact persons, a description of the sections being submitted on paper and being submitted electronically, and an antivirus statement for electronic submissions.

Nearly every application has a required FDA transmittal form. Transmittal forms are primarily screening tools and are used to identify the information required and also provide administrative information for the application. Electronically generated forms may be used, provided the FDA approves the form prior to its initial use. A copy of FDA's approval letter should accompany the form the first time it is submitted. In some instances in which there are no transmittal forms, specific cover sheets or checklists are recommended. These are found within the guidance documents specific to the application type.

A table of contents should accompany all submissions and specify the volume and page number for each item included. This is a critical navigation tool to orient the reviewer to the information provided in the application.

Most applications require that certifications be included with the submission, either in the administrative section of the application or within reports of specific types of data included in the application. These include field copy certifications, debarment certifications, current good manufacturing practice (cGMP) certifications, current good laboratory practice (cGLP) certifications, current good clinical practice (cGCP) certifications, patent certifications, and financial disclosure certifications.

Field-copy certification—a certification, as stated in 21 CFR 314.94, that a true third field copy of the technical sections (chemistry, manufacturing, and controls) of the application has been submitted to the appropriate FDA district office. The district office should also receive a certification with the submission that it is a true copy of the information submitted to the reviewing division.

Debarment certification—a certification that the applicant did not and will not use in any capacity the services of any debarred persons in connection with a drug product application. If the application is an ANDA, it must also include a list of all convictions described under the Act that occurred within the previous 5 years and were committed by the applicant or affiliated persons responsible for the development or submission of the ANDA.

cGMP certification—a statement that facilities and controls used in the manufacture of the product comply with the applicable GMP regulations.

cGLP certification—a statement for each nonclinical laboratory study. A statement that the study was conducted in compliance with the requirements set forth in 21 CFR Part 58, or, if the study was not conducted in compliance, it should include a brief statement justifying the noncompliance.

cGCP certification—a statement with regard to each clinical investigation involving human subjects that it either was conducted in compliance with the requirements in 21 CFR 56 or was not subject to such requirements in accordance with 21 CFR 56.104 and 56.105 and was conducted in compliance with the requirements for informed consent in 21 CFR Part 50.

Patent certification—a certification regarding "any patents that claim the listed drug or that claim any other drugs on which investigations relied on by the applicant for approval of the application were conducted, or that claim a use for the listed or other drug."

Financial disclosure certification—(1) A financial statement for any clinical investigator conducting any clinical study submitted in a marketing application that the applicant or FDA relies on to establish that the product is effective; (2) additionally, any study in which a single investigator makes a significant contribution to the demonstration of safety. This applies to marketing applications for drugs, biologics, and medical devices. The applicant is required to submit as part of this certification a list of investigators who conducted

TABLE 1 Some of the More Common Certifications Included in Applications and Supplements

Certification	IND	IDE	NDA	ANDA	BLA	PMA	510k
Field copy	X	X	X				
cGMP	X	X	X	X	X	X	X
cGLP	X	X	X	X	X	X	X
cGCP[a]	X	X	X	X	X	X	X
Patent	X	X	X				
Financial disclosure	X	X	X	X	X		
Debarment	X	X	X				

[a]Only applications that contain clinical studies.

applicable clinical studies and certify or disclose certain financial arrangements as follows:

Certification that no financial arrangements have been made with an investigator where by the study outcome could affect compensation, that the investigator has no proprietary interest in the product of interest, that the investigator does not have a significant equity interest in the study; and that there were no significant payments.

Disclosure of financial arrangements and steps to minimize the potential bias on the study of interest.

Failure to include the required certifications in submissions can result in an application being refused for filing. In addition, NDAs and BLAs require the submission of a user fee form 3397 indicating that the fee has been submitted in the amount as required. Table 1 provides a matrix of the certifications required for drug, biologic, and medical device applications.

2 SUBMISSIONS PLANNING—THE KEY TO SUCCESS

The planning and preparation of either a new product or investigational product application requires a multidisciplinary approach. Most companies developing new products utilize formal project management systems to facilitate the collaboration between the technical disciplines, which may include personnel from research and development, manufacturing, formulation development, regulatory affairs, quality assurance, and other disciplines, as required. In addition to planning for the required elements, most FDA review divisions have specific preferences on how data should be analyzed and presented. When complex applications are planned, such as INDs, IDEs, PMAs, NDAs or BLAs it is critical for communications with the review division to be held prior to the actual submission of the application. The purpose and the complexity of the application will determine the form and manner of these interactions. It is important to build your submission team to match the disciplines of the FDA review team and to keep it intact throughout the review process.

2.1 MEMBERSHIP AND FUNCTIONS OF SUBMISSION TEAM

Regulatory affairs: Coordination of application planning, preparation, follow-up, and communication with FDA; identification of all applicable regulatory requirements.

Scientific disciplines (e.g., Toxicology, pharmacology, pharmacokinetics, clinical, manufacturing): The areas that provide necessary scientific support.

Marketing: Review of proposed labeling and the potential affect on product marketing; for example, labeling changes.

Legal: Review of label claims and product indications in comparison to scope of clinical studies.

Quality control: Assurance of the quality aspects of all submissions and facilities in which data were generated.

The submission project plan should identify the critical path of the application and the major milestones influencing the timeline. An index should be prepared identifying the critical sections and associated documents and the targeted availability of the documents for inclusion in the application. Systems should be in place for the controlled physical assembly of the application, including appropriate procedures for assembling the application and performing adequate quality control. Checklists are very helpful in ensuring that all necessary elements are included.

Some basic tips to remember in the planning of a complex submission are to:

1. Plan early
2. Build a submission team that includes all the required expertise to evaluate the data and to support the preparation as well as the review process
3. Plan for effective and meaningful communication with the review division, including prefiling meetings
4. Rely on the most effective written and illustrative communication tools, including graphical or tabular presentations when communicating large amounts of data
5. Use well-defined, documented systems to manage change control

6. Select electronic filing strategies appropriate to the application
7. Ensure quality and consistency of the individual reports, individual summaries, and integrated summary documents
8. Evaluate regulatory compliance status of all facilities providing clinical and nonclinical testing as well as manufacturing facilities

The quality of a submission depends not only on the content of the submission but the tools provided to facilitate the review process. Now more than ever, FDA reviewers are receptive to electronic review aids and the use of electronic files. Simple and relatively inexpensive approaches are available. Important issues to consider during planning include the need for electronic review aids. These are electronic tools requested by the reviewer outside the electronic submission policy. Examples of these include bookmarked, searchable versions of the integrated summaries and final clinical study reports for the core studies on CD-ROM or Microsoft Word versions of summaries that reviewers can cut and paste for their reviews. Since the preparation of the review summary can often be the longest part of the review, providing these tools can be useful in significantly reducing the total review time.

Submission planning must include systems for assurance that the contents of any application are accurate and tailored to the audience and that they disclose the required information. One of the most important compliance aspects of submission planning is assurance that the application reflects accurately the data and the processes in which the data were collected and evaluated.

2.2 Communication with FDA

Effective communication, both within the applicant's organization and with the FDA review team, is an integral part of planning any submission. It has been demonstrated consistently that the success of a new product application is dependent on the effectiveness of the organization's submission team and the working relationship developed between this group and the agency review team.

Early communications with your prospective review division will improve the quality, clarify the expectation of the contents of the application, and improve the timeliness of the review process. It is important to identify early whether the information available is adequate for filing the application, thus preventing refusal-to-file actions.

Communications with the FDA will take the form of meetings, telephone contacts, regulatory submissions, general correspondence, and faxes. Regardless of the form, each communication should be considered as an opportunity to build on the working relationship necessary for long-term success. All communications should be documented in a log or database and the relevant information should be disseminated appropriately. Valuable FDA comments and agreements are frequently lost because companies are not diligent about adequate documentation and dissemination of these less formalized interactions.

2.2.1 Meetings with FDA

Through early and interactive meetings from the multidisciplinary review staff, applicants can obtain valuable regulatory feedback on development plans or clinical trial design. FDA meetings are a valuable opportunity to resolve issues related to product development, application contents and review processes, compliance actions, and policy development. Meetings are routinely held at critical milestones of development and at all other times as needed. Those related to premarket and marketing applications can be generalized in the following categories:

Pre-IND—usually held for products for the treatment of life-threatening or severely debilitating diseases (or in some cases orphan drugs). This is the first introduction of the product to the FDA.

End of phase I—usually held for products for the treatment of life-threatening or severely debilitating diseases. Discussions focus on the design of phase II studies.

End of phase II—held to review planning and ensure well-designed confirmatory studies that will most efficiently confirm a drug's effectiveness and also determine what additional information will be necessary to demonstrate the safety and efficacy of the product for the intended claim.

Pre-NDA, pre-PMA, pre-BLA—held to determine the requirements for producing a high-quality marketing application that will be accepted for filing. The meeting provides reviewers an overview of the data that will be presented in the marketing application.

There are other types of meetings, and in general, an applicant can request a meeting at any time to discuss development, scientific, medical, or other concerns regarding product.

Planning for FDA prefiling meetings can be thought of in the following four phases:

1. The request (who, what, when, and why)
2. The preparation (meeting package, selection of participants, and rehearsals)
3. The actual meeting
4. The follow-up (preparation of the meeting minutes and the follow-up agreements)

Meeting requests can initially be communicated over the telephone to the FDA regulatory project manager (RPM), although it is necessary to follow up with a written request. The RPM is the primary contact with the FDA review division and is pivotal to facilitating communication between the applicant and the review team. Requests should clearly state the objective of the meeting, a tentative list of agenda items, who will be attending the meeting, and some proposed dates for the meeting. Meeting requests need to be submitted 30 to 75 days prior to the meeting, depending on the type of meeting requested.

Generally about 4 weeks prior to the meeting the applicant must submit a meeting package. It should include a proposed agenda, some background material to support the topics for discussion, and the proposed planning and other information necessary to provide sufficient details to enable the FDA reviewers to provide comments and to reach agreement. Be transparent and factual with your concerns, and seek discussion on key issues or problems identified in the evaluation of the data. Meeting packages can become voluminous and should only provide the information necessary for the objective. For example, protocol synopses with the critical information are more manageable than the entire protocols with template information. Rehearsals before the meeting with the meeting participants are important. There should be at least one rehearsal before the submission of the meeting package. This is one of the most important rehearsals, providing a forum to identify any weaknesses or errors prior to submission of the meeting package.

Those submission team members most familiar with the data and capable of responding to questions should attend the meeting. It is not recommended to take large groups to these meetings. The rehearsals should be attended by an expanded team and should include some peer review or external experts to challenge the proposed approaches to ensure that the applicants are well prepared. The goals of a prefiling meeting for an application seeking market approval of a new product should include the following:

- Identify and provide an overview of the most important studies in the application.
- Acquaint the reviewers with the data and discuss issues or concerns.
- Discuss and mutually agree upon the methods of statistical analysis.
- Mutually agree upon the data presentation and formating of data tables.
- Identify any additional analyses or studies that maybe required.
- Determine if the other technical sections as presented appear adequate for filing.

- Discuss and mutually agree upon the electronic filing and electronic review aids.

Meeting etiquette with FDA is as with any other professional meeting. If applicants have articulated clearly their points in the briefing package and brought the appropriate personnel, the intended purpose for the meeting will be achieved. It is recommended to keep presentations to a minimum and use them for clarifying data or issues for discussion. During the meeting someone from the applicant's team should be assigned the task of taking minutes, noting all of FDA's concerns and resolutions, as well as agreements reached. In addition, it is important to follow up on issues that need to be resolved as soon after the meeting as possible. The review division will issue the official minutes, and if agreements are made generally these are outlined and signed by the division director.

2.3 Key Points Contributing to Success of a Prefiling Meeting

- Identify clearly the objectives of the meeting.
- Identify the issues and develop proposed solutions.
- Prepare a quality meeting package.
- Know your data and be well prepared to discuss the topics.
- Rehearse, rehearse, and rehearse.
- Anticipate questions.
- Include peer review through the meeting preparation process.
- Take only the most necessary team members who can speak to the data and the overall project.
- Present the data dispassionately, objectively, and accurately.
- Do not spend a lot of time on marketing the project—stick to substantive topics.
- Keep accurate minutes for the purpose of sharing with FDA and the applicant's regulatory project manager.

3 PREPARATION OF THE SUBMISSION

3.1 Application Assembly

Assembly of the application should ideally begin when supporting documentation is available. Enough time should be planned in the submission time line to allow for physical compilation of the submission and preparation of any review aids and quality control.

Generally, for a marketing application, the time needed for this process is 4 to 6 weeks, although this may be reduced with the introduction of electronic submission publishing tools. The submission plan should address the logistical aspects of submission preparation: supplies, space, resources, and transportation for the submission to the agency.

3.1.1 Required Copies

Archival Copy. This is a complete copy of the entire submission and serves as a reference for information not included in specific review copies and also as a repository for case report tabulations and case report forms. This copy is maintained on file at FDA after the review of the submission is completed. In accordance with recent guidances, certain sections (Sections 11 and 12) of a marketing application archival copy may be submitted in electronic format.

Review Copy(ies). These copies are bound separately to allow concurrent review of the technical sections.

INDs: Two review copies are required for all submissions.
NDAs: A single copy is required for each section. Two copies are required for Section 4 (Chemistry, Manufacturing, and Controls) and Section 6 (Human Pharmacokinetics and Bioavailability). A review copy of the application summary is required for each reviewer.

Field Copy. This is a certified copy of the chemistry, manufacturing, and controls section that is sent to the FDA district office with jurisdiction over the manufacturing site. For international sites, these copies are sent to the Division of Emergency and Investigational Operations in Washington, D.C. The field copy must be accompanied by a certification that it is a true and accurate copy of the application (for NDAs, BLAs, and ANDAs).

3.1.2 Format and Assembly of the Application

Jackets. The copies should be bound in color-coded jackets and appropriately labeled on each cover. The volumes should be bound with fasteners rather than three-ring binders. Jackets are available from FDA's forms and consolidated publications division or can be ordered from commercial sources, provided they meet the requirements outlined in FDA's guidelines.

3.1.3 Media Specifications

Paper Submissions. Applications must be submitted on paginated, three-hole-punched $8\frac{1}{2} \times 11''$ paper. The left margin should be at least $3/8''$ from the edge of the paper. Pages should be printed on one side only. Volumes should be no more than 2 inches thick and must be numbered consecutively through the submission.

Electronic Submissions.

Less than 10 MB—3.5″ DOS-formatted floppy disks (one to 10 disks)

Less than 3.25 GB—CD-ROM ISO 9660 (one to five CDs)

Greater than 3.25 GB—Digital tape, Digital Equipment Corp. DLT 20/40 and 10/20 GB format using OPENVMS with VMS backup or NT server 4.0 with NT backup or backup exec.

The print area for pages should have a margin of at least 1″ on all sides. Pages should be correctly oriented.

3.1.4 Pagination

All pages of paper submissions must be numbered using any numbering system. Paging and indexing must provide rapid access to the entire submission. For electronic submissions, pagination should be provided only for individual documents.

3.1.5 Packaging

Large submissions should be packed in boxes measuring 14″ × 12″ × 9½″. All electronic media and any reviewer desk copies should be clearly marked and included in the first box of the shipment. The shipping containers should be identified with application number, product name, volume numbers, review copy or archival copy, and applicant name and address. Specific instructions for marking the mailing package (e.g., safety reports) should be followed.

4 SUGGESTED NAVIGATIONAL TOOLS FOR PAPER PORTIONS OF REGULATORY SUBMISSIONS

Tables of contents that increase in detail at each level.

Submission tables of contents at the beginning of the submission.

Section tables of contents at the beginning of each section.

Volume tables of contents at the beginning of each volume.

Study report tables of contents at the beginning of study reports listing all of the sections and appendices of the study may be helpful in extremely large submissions.

Labeled divider tabs for major sections and for marking important information; for example, appendices of study reports.

Cover sheets for submission sections and documents in the submission are useful.

Colored divider sheets used between sections of documents can help the reviewer(s) navigate the submission.

Pagination should be provided for the submission. Pagination can be by volume or section or for the entire submission. (Refer to the preferences of the center responsible for review.)

For legacy documents that may not meet current standards for document authoring, provide as many navigational tools as possible, including explanatory cover pages or notes.

4.1 Quality Control

Quality checks of submissions are critical to ensure that methods, processes, equipment, and facilities have been accurately reported and that the documents included in the submission represent the appropriate scope of information. In addition, the actual data presented in the submission must be checked for integrity and accuracy to the extent possible. The submission of inaccurate or fraudulent documents could result in the invocation of the fraud policy or the application integrity policy (AIP) that covers the "failure to have and implement systems or procedures to ensure the quality and accuracy of submissions." It is therefore clear that ensuring the quality of submissions and authenticating all data is critical to maintaining the good compliance standing of the applicant.

It is recommended that the regulatory status be reviewed by examining the most recent inspection reports on the facilities from which all data were derived for inclusion in the application. This includes manufacturing, nonclinical laboratories, and clinical sites.

5 QUALITY CHECKS FOR REGULATORY SUBMISSIONS

5.1 Before Preparation of the Submission

Check all source documents for document quality. Make sure there are no missing pages, cropped pages, or text that is not legible. For poor-quality documents that cannot be improved, mark each page "best copy available."

Check to ensure translations are provided for all foreign language text.

Authenticate all data against all source documents.

5.2 After Preparation of the Submission

Check printed documents for quality.

Check pagination to make sure that all pages are in order and are clearly paginated.

Check tables of contents to ensure that page references correspond.

5.3 Quality Control (QC) Checks

Check organization of submission to ensure that all relevant documents are included and that the most recent version of all documents (including amendments) is presented.

Check data presented in submission to ensure completeness and accuracy.

Ensure that all QC checks used during the document compilation process were performed.

6 THE REVIEW PROCESS

Factors influencing review times are not limited to agency review practices. Some factors creating delays in the review reflect back on the applicant. The quality of the application, the applicant's response time, the comprehensiveness of responses to the FDA reviewer's questions, and requests for additional data or analysis as a result of inadequate submissions as well as the applicant's ability to provide new data during the course of the review all influence the timeliness of the application review.

6.1 Refusal to File

The FDA has taken seriously the need to address the quality of the applications accepted by clarifying the refusal-to-file (RTF) regulations. Applications that are poorly organized or provide inadequate data are difficult to review, therefore in 1993 the FDA introduced a new guidance entitled refusal to file. Although existing regulations did provide circumstances in which FDA could refuse to file an application, the intent of the guideline was to clarify its practices regarding this policy. Prior to the guidance, decisions to refuse to file were generally based on extreme deficiencies (e.g., total omission of a section or lack of the controlled studies to support the intended claim).

Recently applications have been refused when less extreme deficiencies existed but when the deficiencies meant reviewability or approvability was impossible without major modifications to the file. The practice of submitting incomplete or inadequate applications and then repairing them during the course of the review is a waste of FDA resources. It does not benefit the applicant or the FDA when deficient applications are filed. With the introduction of the user fee program for marketing applications of new drugs and biologics, the RTF policy has become even more important. As a result, the quality of applications submitted has improved. Before 1993, 25–30% of original NDAs, were refused for filing. In recent years the rate has dropped to approximately 4%.

The RTF policies are applicable to drugs, biologics, and certain medical device applications, and the basis for refusal of an application are similar across the products. Below are some examples of reasons why the FDA has invoked the RTF policy for a new drug product.

6.2 Examples of Refusal-to-File Issues Across Products

The application does not contain a completed application form.

The application is not submitted properly in terms of the content and format requirements outlined in the applicable regulations.

The applicant did not complete the environmental assessment per 21 CFR 25.40 or fails to establish exclusion under 21 CFR 25.30 or 21 CFR 25.31.

The application does not include an accurate and complete English translation of each part of the application that is not in English.

Use of a study design is clearly inappropriate for the intended use of the product.

Total patient exposure (numbers and duration) at relevant doses is clearly inadequate to evaluate safety.

There is no comprehensive analysis of safety data.

There is no assessment of the carcinogenic potential for a chronically administered product and no explanation of why such an assessment is not applicable.

Not all nonclinical laboratory studies include a statement that the study was conducted in compliance with the requirements set forth in 21 CFR Part 58 (or if the study was not conducted in compliance with such regulations, a brief statement justifying the noncompliance).

Not all clinical investigations involving human subjects include a statement that they either were conducted in compliance with the requirements of 21 CFR 56 or were not subject to such requirements are were conducted in compliance with the requirements for informed consent in 21 CFR Part 50.

Data are missing establishing the stability of the product through the dating period and a stability protocol describing the test methods used and time intervals for the product stability assessment in accordance with the *FDA Guideline for Submitting Documentation for the Stability of Drugs and Biologics.*

Failure to describe all manufacturing sites (including contract sites).

Failure to describe all major production equipment and support systems.

Failure to submit complete production flow diagrams.

Lack of validation protocols and data summaries, including environmental monitoring.

Even if an application is accepted, incomplete or substandard submissions will result in a fragmented and subsequently extended review period.

When submissions are poorly organized and difficult to review, the reviewers have to contact the applicant for assistance. This leads to significant delays since reviewers will often start reviewing another submission while waiting for a response and will not restart their review until they are done with the other submission. It is therefore extremely important that submissions include all the information that is necessary for the review and that it is well organized so that the information can be located quickly.

6.3 Applications or Supplements Requiring Compliance Status Check Prior to Approval

Compliance status will be determined for applications submitted for PMAs and 510(k)s, NDAs, BLAs, and ANDAs. This includes original applications and supplements to these applications. Supplements for the approval of new or expanded indications of use, a new production site, an increase in the scale of the production, an extensive modification of the production process, an extensive modification to a production area, or a process change that uses a new production area require the FDA district office for the manufacturing site to issue a satisfactory rating for the site. This may involve conducting a cGMP inspection of the facility. If the site is in good compliance standing for the type of product being submitted, the inspection may be waived. The level of manufacturing changes typically reported in annual reports will generally not trigger a compliance status check.

In addition, the Office of Scientific Investigations (CDER) conducts biomedical audits of clinical investigator's sites. These investigations focus primarily on sites involved in pivotal clinical trials, but may also involve other sites. Investigators should be informed that they should contact the applicant when the FDA notifies them of an upcoming inspection. The sponsor/applicant is generally not permitted to be involved in an FDA inspection of an investigator's site. Refer to Chap. 3 for an extensive overview of the FDA's expectations of clinical trial activities.

Prior to using these facilities, the sponsor/applicant should evaluate the compliance backdrop of the clinical investigator(s), the investigational product's manufacturing operations, and any contract laboratories involved in the clinical trial. This process may involve reviewing FDA's listing of disqualified clinical investigators, facilities subject to the AIP or listings of debarred individuals. Recent FDA-483s and inspection reports should also be requested through the Freedom of Information Act to determine the current inspection status and any potential impact on the use of the facility to perform testing to generate data for applications.

6.4 Application Review Process—Example 1 (NDA)

A schematic of the NDA review process is presented in Appendix C. CDER's central document room (CDR) initially handles administrative processing of the application, including stamping the application with a date, which starts the review timeline. A determination of the user fee status is made and a copy of the user fee cover sheet (FDA form 3397) is sent to the regulatory information management staff. The CD is responsible for distribution of the copies of the NDA to the various divisions for evaluation. An acknowledgment letter is sent to the applicant and a project manager is assigned to coordinate the NDA review process. The project manager performs an initial screening of the application, and seriously deficient applications are refused for filing. The technical sections are then distributed to reviewers for a more thorough screening of each section. The FDA review team will convene at a "45-day meeting" to determine whether the application should be filed or refused. Oftentimes these meetings can be used as a review planning session in which internal review milestones are projected. If not already done, a priority of either a "priority review" or a "standard review" will be assigned to the application.

Once the acceptability of the application is established, the "primary" review begins. Reviewers communicate with other reviewers and with the applicant regarding issues or questions that arise during the review. During the review process, the FDA reviewer may contact the application sponsor to discuss issues and obtain clarifications. Interactions between the review team and the applicant team can range from telephone calls to letters. If the reviewer requests assistance in finding information, it is important to respond quickly. A submission response team that is familiar with the information in the submission and its organization should be available to address questions as they are received.

Upon completion of the review, a written evaluation of the product is prepared in a review document and the comments of the various reviewers are reconciled and reviewed by the division director. The results of the decisions are communicated to the applicant in an approvable or not-approvable letter. In some cases, if the questions have been satisfactorily addressed during the review process, the agency may proceed directly to an approval letter. The scientific review divisions are independent from the district offices that conduct field facility inspections. The review division will wait for assurance that the preapproval inspections are completed satisfactorily prior to issuing an approval letter. In certain cases, the inspection of a facility may be waived based on accepted compliance standing and history of the company.

6.5 Application Review Process—Example 2 (IND)

After review by the CDER or CBER CDR, the application is sent to the document control center within the division responsible for the review of the application. An acknowledgment letter is sent to the applicant, and a project manager is assigned to coordinate the IND review process. The project manager may perform an initial screening of the application. The technical sections are then distributed to reviewers, each of whom undertakes a more thorough screening of the application. If there are no concerns with the safety profile or the risks anticipated in the proposed clinical trial, the 30-day review period is allowed to expire, thereby permitting the sponsor to initiate clinical trials.

If concerns are found, a deficiency letter will be sent, and if the deficiencies are serious enough to delay clinical trials the agency will impose a clinical hold on the product that can be lifted after the deficiencies are corrected. Clinical holds are classified as complete or partial, depending on whether the issues relate to the product or its manufacture or are specifically related to protocol concerns. Application sponsors should respond to clinical hold notifications promptly. Additionally, FDA is required to respond to completed responses within 30 days of receipt. Examples of reasons for clinical holds are

The product presents unreasonable health risks to the subjects in the initial IND trials (C. F., a product made with unknown or impure components).

The product possesses chemical structures of unknown or high toxicity.

The Product cannot remain chemically stable throughout the proposed testing program.

The Product presents an impurity profile indicative of a potential health hazard.

The impurity profile is insufficiently defined to assess a potential health hazard

A master or working cell bank is poorly characterized.

In addition, inspections of clinical sites during the clinical investigation phase have risen and therefore the numbers of FDA-483s and warning letter for these issues have also increased. This means that compliance issues in relation to clinical trial activities could also delay the reviewer's final determination of acceptability. Typical compliance issues in relation to clinical trial activities include

- Inadequate drug reconciliation and accountability
- Nonconformance with cGMPs during the manufacture of clinical trial materials
- Clinical investigator noncompliance with the protocol (investigational plan)
- Inadequate clinical trial material labeling
- Lack of change control
- Lack of quality assurance throughout the course of the trial
- Lack of adequate trial monitoring by the sponsor or contract research organization (CRO)*

Frequent internal and third party audits are critical to identify possible issues and institute corrective actions promptly.

7 LEGISLATION AFFECTING FDA REVIEW PROCESS

The FDA's performance relative to the review of new product applications is always under the microscope. The review process is the area of most concern for application sponsors and has been the source of re-engineering within FDA's centers governing the regulation of drugs, biologics, and medical devices. The increasing complexity of science and technology and the political pressures on the government to improve the efficiency and effectiveness of new product review has led to several legislative changes in the last decade.

The Prescription Drug User Fee Act of 1992 (PDUFA I) was one of the first major legislative efforts to attempt to address the inefficiencies in the review process. PDUFA I authorized the agency to levy fees on new prescription drug and biologics applications in an effort to provide additional resources for the review process. The agency was authorized to collect three different user fees: annual fees on drug manufacturing establishments, annual fees on prescription drug products, and application fees. The amount of the fee is dependent on whether or not clinical data are provided in support of safety or effectiveness. Applications with no clinical data and supplements with clinical data are assessed half the established user fee. The amount for fees is adjusted annually based on inflation and FDA's review workload. In conjunction with the user fees from industry, CDER, the CBER, and the

---

*A CRO is an organization contracted by the sponsor to be responsible for some or all aspects of the clinical trial.

Office of Regulatory Affairs (ORA) were held to stringent performance goals. These included completing priority reviews in 6 months or less and standard reviews in 12 months or less. The agency successfully used these revenues to increase existing staffing to improve its new drug and biologics review process, resulting in reduced review times without compromising the quality of the review. The median approval time for new drugs has been substantially reduced from 20 months in 1993 to around 12 months in 1999. Additionally, these fees have gone a long way in assisting the agency with expediting its preapproval inspection process.

The current evolution in policy changes within the agency can be attributed to the FDA's Modernization Act of 1997 (FDAMA). This legislation was part of REGO—Vice-President Gore's Reorganization of Government initiative that attempted to reform and bring into the twenty-first century the regulation of food, medical products, and cosmetics. This act reauthorized, until September 2002, the Prescription Drug User Fee Act of 1992 (PDUFA 5-year plan, FY 1999 revision, Health and Human Services (HHS), FDA, Office of Management and Systems (OMS), July 1999). The goal of PDUFA II was to continue to increase the efficiency and quality of the review of new drug and biologic applications and established new goals. It established new goals for industry-sponsored meetings, dispute resolution, and the electronic receipt and electronic review of submissions by 2002.

Section 406(b) of FDAMA provides the following requirements [1]:

> "Establishing mechanism, by July 1, 1999, for meeting the time periods specified in this Act [the FFD&C act] for the review of all applications and submissions submitted after the date of enactment of the FDAMA."

PDUFA II also focuses on reducing the application review time during new product development and enhancing the quality of the review process. Essentially, FDA is investing review resources early in the process, resulting in a productive and ongoing review during the development phase for new products. Performance goals are provided that address meetings between the agency and industry and dispute resolution. PDUFA II describes meetings as type A, B, or C, each with a defined time frame for scheduling. Type A meetings are considered critical path meetings and have to be scheduled within 30 days of the request. Type B refers to regulatory meetings, such as pre-NDA meetings. These have to be scheduled within 60 days of the request. Type C meetings cover the rest and have to be scheduled within 75 days of the request. With the defined and short time frames, applicants must assure their readiness for such meetings at the time of the request.

The Prescription Drug User Fee Act (PDUFA) was renewed in 2002 (PDUFA III) and included similar performance goals targeting process improvements in the agencies review practices of new products and changes for already marketed products and in their interactions with the product sponsors, the implementation of risk management activities and also improvements in information technology (http://www.fda.gov/oc/pdufa/PDUFAIIIGoals.html). Table 2 provides an overview and comparison of the major goals in PDUFA I, II, and III that are intended to enhance both the review time and the communication between application sponsors and the FDA.

The Medical Device User Fee and Modernization Act of 2002 (MDUFMA) amended the Federal Food, Drug and Cosmetic Act to provide FDA new responsibilities, resources, and challenges. The purpose was to provide FDA with additional resources for "the process for the review of devices and the assurance of device safety and effectiveness so that statutorily mandated deadlines may be met." MDUFMA has three significant provisions: (1) user fees for premarket reviews; (2) establishment inspection may be confucted by accredited person; and (3) established new regulatory requirements for reprocessed single use devices. MDUFMA includes several additional provisions that are less complex and have a narrower scope. The collection of fees will add $25.1 million to the FDA's medical device budget authority during FY 2003, rising to $35 million in FY 2007. As with PDUFA, the revenues from the fees are intended to add appropriations for infrastructure and allow FDA to pursue ambitious performance goals. The performance goals can be reviewed at www.fda.gov/cdrh/mdufma.

The initiatives codified in the FDAMA and MDUFMA legislation outline innovative approaches to meet the increasingly complex and technological challenges of health care in the twenty-first century. It has become increasingly evident that to succeed there has to be collaboration between FDA and the various industries it regulates. A successful implementation depends on commitment of resources by both FDA and the industry, most of which are directed toward enhancing both the quality and the timeliness of the application review process [2].

8 IMPACT OF LEGISLATION ON SUBMISSION STRATEGIES

This new legislation has resulted in a re-engineering of the regulatory review process at the agency to both improve the quality and reduce the time required for application review. While these changes have provided the industry with substantial opportunity, they have also brought some challenges. The implications include the increased need for submissions to include concise and comprehensive high-quality documents with review tools. In the past, industry would send submissions with the hopes

TABLE 2 Comparison of Goals at the End of PDUFA I, PDUFA II, and PDUFA III

Goal	PDUFA I	PDUFA II	PDUFA III
Complete review of priority original new drug applications and efficacy supplements	90% in 6 months	90% in 6 months	90% in 6 months
Complete review of standard original new drug applications and efficacy supplements	90% in 12 months	90% in 10 months	90% in 10 months
Complete review of manufacturing if supplements prior approval needed	90% in 6 months	90% in 4 months	90% in 4 months
Complete review of resubmitted new drug applications	90% in 6 months	90% of class 1 in 2 months and 90% of class 2 in 6 months	90% of class 1 in 2 months and 90% of class 2 in 6 months
Respond to industry requests for meetings	No goal	90% within 14 days	90% within 14 days
Meet with industry within set times	No goal	90% within 30, 60, or 75 days, depending on type of meeting	90% within 30, 60, or 75 days, depending on type of meeting
Provide industry with meeting minutes	No goal	90% within 30 days	90% within 30 days
Communicate results of review of complete industry responses to FDA clinical holds	No goal	90% within 30 days	90% within 30 days
Resolve major disputes appealed by industry	No goal	90% within 30 days	90% within 30 days
Complete review of special protocols	No goal	90% within 45 days	90% within 45 days
Electronic application receipt and review	No goal	In place by 2002	IT 5-year plan will be developed within 6 months of authorization (Oct. 2002)

Source: Ref. 20.

that FDA would function as the submission reviewer and as a consultant. Since there was no penalty for sending incomplete data and information, the industry relied upon FDA's feedback to bring its submissions to completion. The accelerated review process has forced the industry to take a critical look at the quality of its applications before submission. Greater emphasis needs to be placed on the preparation of a product application and for the provision of support during the review and evaluation process.

There has been a major impact on the need for firms to be ready for the preapproval inspection (PAI) earlier than ever because of the increased number of FDA reviewers and the greatly reduced review times. Firms do not have the lag time that they had become accustomed to and typically used to focus on and prepare their facilities for inspection. Prior to PDUFA, the industry was accustomed to submitting its applications and subsequently having a long lead time before FDA initiated the pre-approval inspection. Lack of planning results in lengthy FDA-483 at the PAI because firms are simply not ready for the inspection at the time of the submission of the marketing application.

Overall, firms have had to coordinate planning submissions across cross-functional teams working together to ensure that information presented in the application is reflective of company practice at the time of the submission. This has created the need for professionals with both regulatory and technical skills. These professionals need to be ready for change, have an awareness of future direction of legislation affecting products, and have the capability to work proactively with the FDA.

9 ACCELERATED DRUG APPROVAL AND ACCESSIBILITY PROGRAMS

In response to the need to provide expedited access to new therapies for patients FDA has developed the following programs:

Treatment IND: Mechanism to provide patients with experimental products for serious or life-threatening diseases.

Parallel track: Mechanism to provide patients with AIDS or related diseases early access to experimental therapies.

Accelerated drug development program: Mechanism to accelerate development of products designed to treat life-threatening or seriously debilitating diseases.

Accelerated drug approval program: Mechanism to accelerate approval of products designed to treat life-threatening or seriously debilitating diseases based on modified criteria for marketing approval; for example, the use of surrogate end points.

Oncology initiative: Several reforms have been initiated to reduce development times and approval times for products for the treatment of cancer.

Fast track program: This program was added under the FDA Modernization Act of 1997 as an extension of the accelerated drug and biologic product approval process. It was designed to facilitate development and expedite review for products that demonstrate potential or unmet medical needs in the treatment of serious or life-threatening conditions.

When considering any of these mechanisms for expedited approval, it is important to keep in mind that manufacturing facilities and supporting processes should be in place early during the review process to allow for more aggressive time lines with the PAI.

10 SUBMISSIONS MAINTENANCE

Submissions should be treated as living documents and must be continually updated in order to keep the submission active, up-to-date, and reflective of current company practices. In addition, it is very important to maintain archival files of all submissions and related documentation, including meeting minutes, contact reports, and correspondence. These documents are generally maintained for the life cycle of the product. It is essential to maintain the records from the development phase through commercialization for the purpose of adequate historical accountability as well as for providing new personnel with the full scope of the project. These records are often relied upon to acquaint new personnel to the product team or to review previous regulatory agency agreements.

It is important to maintain control of any changes to submission commitments. Any proposed changes should be reviewed and evaluated through a formal change control mechanism that includes a review of the impact of any changes to processes that are currently in place and that have been validated, such as manufacturing controls and methods.

10.1 Preapproval Maintenance Requirements

10.1.1 IND Maintenance

The following reporting mechanisms are available for changes that may occur postsubmission of the IND. Investigational new drug applications are submitted for the purpose of shipping clinical trial material intended solely for investigational use. The FDA does not approve INDs.

Protocol amendments: Submitted to report changes in previously submitted protocols or to add protocols not previously submitted.

Information amendments: Submitted to report new information that would not be included as a protocol amendment or safety report. Examples include the results of animal testing, chemistry, manufacturing, and controls data, reports of completed or discontinued clinical trials, or changes in administrative information.

Safety reports: Applicants are required to submit reports of any adverse experiences associated with the use of the product. Safety reports should also bring to the agency's attention any trends resulting from product use, even if they are expected and not very serious. Any correlation between manufacturing and quality problems and these trends should be presented in the safety and annual reports as well.

Annual reports: Annual reports should be submitted within 60 days of the anniversary date on which the IND went into effect and should include an overview of the information collected during the previous year.

10.1.2 Investigational Device Exemption Maintenance

The FDA is required to approve investigational device exemptions.

Investigational device exceptors are submitted to request authorization for shipment of devices intended solely for investigational use. Investigational device exceptors are submitted for individual clinical studies, and FDA approval is required prior to the initiation of the clinical study.

Safety reports: Applicants are required to submit reports of adverse experiences associated with the use of the product within 10 days of becoming aware of the event. Safety reports should also bring to the agency's attention any trends resulting from product use, even if they are expected and not very serious. Any correlation between manufacturing and quality problems and these trends should be presented in the safety and annual progress reports as well.

Annual progress reports: These must be submitted to the institutional review boards (IRB) and should be submitted to FDA for significant risk devices only.

Final reports: Final reports on the clinical study should be submitted within 6 months of the completion of the study.

10.1.3 Maintenance of Pending Marketing Applications

Amendments may be submitted either at an applicant's own initiative or in response to an FDA request. Amendments are usually intended to clarify and supplement information provided in applications during the review. Depending on the information being submitted and the timing of the submission, amendments to pending applications may cause an extension in FDA's time line for review of the application.

Updates of safety information are required for marketing applications. These should be submitted at intervals after the initial submission as required for the type of product, immediately prior to approval of the product (unless not requested by the reviewing division), and upon request during the review process.

10.2 Postapproval Reporting Requirements

The fundamentals of postapproval responsibilities are very similar to premarketing responsibilities. The basic responsibilities are as follows:

- To ensure that the product is produced according to accepted manufacturing standards
- To report postmarketing data or information that might cause the FDA to reassess the safety and effectiveness of the product
- To comply with the conditions of use detailed in the approved application and subsequent supplemental applications

10.2.1 General Reporting Requirements

Field alert reports (FARs for drugs and biologics): Applicants are required to report within 3 days any information "concerning any incident that causes the drug product or its labeling to be mistaken for another article" or "concerning any bacteriological contamination, or any significant chemical, physical, or other change or deterioration in the distributed drug product, or any failure of one or more distributed batches of the drug product to meet specifications established in the application."

Annual reports (drugs and biologics): Applicants are required to submit annual reports within 60 days of the anniversary date of the approval. These contain a summary of new research data, distribution information, and labeling information.

Advertising and promotional labeling: At both the time of the initial dissemination of the labeling and the time of the initial publication of the advertisement for a prescription drug product applicants must submit specimens of mailing pieces and any other labeling or advertising devised for promoting it.

Product listing and establishment listing: Applicants are required to submit product listing and establishment listing information for approved products. For new establishments, the facility should be registered within 5 days of submission of the marketing application. Approved products should be listed no later than the first biannual update after the product is introduced for commercial distribution.

10.2.2 Adverse Drug Experience Reporting (AER) Requirements

After approval, applicants should continue to collect, analyze, and submit data on adverse drug experiences so that the product can continue to be assessed within the larger population. Currently AER reporting requirements are in transition, and there are several pending initiatives for safety reporting. To meet the safety reporting requirements, formalized systems should be in place to gather safety information reported worldwide and to submit those reports in accordance with global and FDA regulations. The agency has initiated several compliance actions against companies in recent years for failure to comply with safety reporting requirements.

It is also important to establish postmarketing surveillance for safety signals that may result in labeling changes. The FDA's recent position on safety information provided in labeling is to minimize the lists of adverse events reported to be more reflective of the adverse events that may actually be expected with use of the product within a larger population. This will require applications to report many adverse events as "unlabeled" and will allow FDA epidemiologists to develop a more realistic impression of the true adverse event profiles associated with use of a drug.

There are three types of postmarketing AERs for drugs and biologics.

Fifteen-day alert reports: Applicants must report AERs that are "both serious and unexpected, whether foreign or domestic, as soon as possible, but in no case later than 15 calendar days of initial receipt of the information by the applicant."

Fifteen-day alert follow-up reports: Applicants must report follow-up information on 15-day reports as soon as possible, but in no case later than 15 calendar days of initial receipt of the information by the applicant.

Periodic adverse experiences reports: These reports must be submitted quarterly for the first 3 years after approval of the product (within 30 days of the end of the quarter) and annually (within 60 days of the anniversary of the approval date) thereafter. Periodic adverse drug experience reports should present a narrative overview and discussion of the safety information received during the reporting period.

There are five types of postmarketing medical device reports (MDRs) for medical devices.

Thirty-day reports: Applicants must report deaths, serious injuries, and malfunctions within 30 days of becoming aware of the event.

Five-day reports: Applicants must report events that require remedial action to prevent an unreasonable risk of substantial harm to the public health and other types of events designated by FDA within 5 working days of becoming aware of an event.

Baseline reports: Applicants are required to submit baseline reports to identify and provide basic data on each device that is the subject of an MDR report when the device or device family is reported for the first time. Interim and annual updates are also required if the baseline information changes after the initial submission.

Supplemental reports: Applicants must report follow-up information on MDR reports as soon as possible, but in no case later than 30 calendar days of initial receipt of the information by the applicant.

Annual certification: Applicants must submit an annual certification that reports were filed for all reportable events and include a numerical summary of all reports submitted. This report should be submitted at the same time that the firm's annual registration is required.

10.3 Current Good Manufacturing Practice (cGMP)

Applicants must be sure that the manufacturing sites for their products maintain satisfactory cGMP inspection status. The FDA assigns profile class codes to help manage the cGMP inspection process, evaluate the findings and follow-up needed, and to communicate the results of the inspections. Profile class codes relate to the manufacturer of particular dosage forms, types of drug substances, or specific functions performed. Maintaining satisfactory cGMP status allows companies flexibility in making changes to some product manufacturing conditions without prior agency approval.

10.4 Phase IV Commitments

Phase IV commitments are agreements made between the agency and sponsors to conduct postapproval studies for the purpose of gathering further safety and efficacy information. Under the FDA Modernization Act of 1997, applicants are required to submit annual reports on the status of postmarketing commitments. Additionally, under FDAMA, marketing an approved product for off-label claims would be allowed, providing one or more clinical study corrobates safety and efficacy.

10.5 Postapproval Changes

After approval of the application, applicants can supplement an approved application to provide for authorization to market variations in the product beyond those approved in the application. Changes to the product can include chemistry, manufacturing, and controls changes (dosage form, route of administration, manufacturing process, ingredients, strength, container-closure system) and labeling changes (indication, patient population, and other labeling changes, such as safety changes in response to accumulated safety reporting data).

Supplemental applications vary in complexity but should include all the traditional elements of a submission and should be formatted like the original submission with the omission of sections that are not affected. Postapproval changes are also classified into various classes—changes that require FDA approval before they are implemented, changes that should be submitted prior to implementation, and changes that are described in the annual report.

With the recent efforts to improve efficiency at the agency, several initiatives have been undertaken to simplify the requirements for postapproval reporting of changes. These initiatives are intended to reduce the regulatory burden of the change mechanism, not reduce the body of evidence needed to support the change. Since 1995, FDA issued several guidances on scale-up and postapproval changes (SUPAC), which classify postapproval manufacturing changes into three levels and establish postmarketing reporting requirements for changes within each level.

The SUPAC guidances describe various changes relating to the chemistry, manufacturing, and controls sections of applications. The guidance allows many of these changes to be submitted as annual reports or changes being effected (CBE) supplements. This allows application sponsors greater control in planning manufacturing changes since in many cases they do not have to wait for FDA approval. The SUPAC procedures also reduce the number of batches required for stability testing in support of these changes.

The challenge that arises with the new regulations is the risk of releasing unapproved product to the marketplace based on a CBE supplement that may be rejected. Adhering to such compliance systems as change management, validation, personnel training, quality assurance, and enhanced documentation practices can offset the risk. In certain circumstances, however, it may be prudent to submit changes more conservatively than required by the SUPAC guidances and await FDA approval prior to implementation of the changes.

Table 3 provides examples of SUPAC-IR (immediate release) changes and the regulatory requirements.

Table 3 Examples of SUPAC IR Changes and Regulatory Requirements

Level	Type of change	Compliance documents	Regulatory filings	Compliance challenges
1	Change to operating targets within validated range	Master and batch record revisions Addendum to validation study	Annual report	Change management Tie in with development report Personnel training SOP revisions
2	Change to operating range outside validated parameters	Amend and expand validation protocol Stability protocol revisions for expanded long-term stability Review methods validation for possible changes Master and batch record revisions	Changes being effected (CBE) with new data submitted in annual report	Change management Tie in with development report Personnel training SOP revisions Equipment qualification

3	Site change (maintaining same specifications)	New validation protocol Stability protocol revisions for expanded accelerated and long-term stability Update methods validation Master and batch record revisions Equipment comparability study	Changes being effected supplement (CBE) with new data submitted in annual report	Change management Tie in with development report Personnel qualifications and training SOP revisions Equipment and site qualification Process validation Methods validation Prior equipment and site comparability
3	Manufacturing process change	New validation protocol Stability protocol revisions for expanded accelerated and long-term stability Update methods validation Master and batch record revisions Methods and specifications revisions	Prior Approval Supplement (PAS) with new data submitted in annual report	Change management Tie in with development report Personnel qualifications and training SOP revisions Equipment and site qualification Process validation Methods validation Prior equipment and site comparability

11 ELECTRONIC SUBMISSIONS

In recent years the agency has been working to develop standards for electronic submissions. This started with the publication of the Electronic Records; Electronic Signatures regulations (21 CFR Part 11) in March 1997. This regulation provided for the voluntary submission of parts or all of regulatory records in electronic format without an accompanying paper copy. This allowed the agency to develop guidance on the format and requirements for these electronic submissions. In 1999, CDER and CBER published an important guidance governing the electronic submission process that describes the requirements for electronic submissions and the conditions under which they would be accepted by the agency.

The new publications concerning electronic submissions have moved away from the CANDA guidance published in the 1980s that provided for applicants to develop electronic review tools for their submissions in agreement with the review division. This meant that each electronic review tool was different and often required companies to provide their hardware, software, and training to the FDA reviewers in order to facilitate the review process. The new guidance provides for a much more standardized submission format that will allow the development of more consistent submissions that can be reviewed utilizing tools currently available at FDA. Although the new guidance allows for the development of specialized review aids in certain instances, these are not encouraged and require prior approval from the specific division. The CDER guidance states that "a review aid should only be requested or agreed to if (1) it will add functionality not found in a submission provided in accordance with guidance and (2) we agree that the review aid will contribute significantly to the review of the application."

12 SUMMARY OF ELECTRONIC SUBMISSIONS REQUIREMENTS

File format: All files should be submitted in portable document format (PDF). The version of Acrobat Reader to be used for review should be confirmed with the agency. Electronic data sets should be provided in SAS System XPORT transport format (version 5 SAS transport file).

Fonts: Limit the number of fonts used in each document, use only True Type or Adobe Type I fonts. FDA recommends Times New Roman, 12 point (fonts smaller than 12 point should be avoided wherever possible). Black font is recommended; blue can be used for hyperlinks.

Page orientation: Page orientation should be set correctly so documents can be viewed on the screen.

Page size and margin: The print area should be $8\frac{1}{2}'' \times 11''$ with a margin of at least 1″ on all sides.

Source of electronic documents: Electronic source documents should be used for creation of PDF documents instead of scanned documents wherever possible.

Hypertext linking and bookmarks: Bookmarks and hypertext links should be provided for each item listed in the table of contents, including tables, figures, publications, other references, and appendices. Hypertext links should be used throughout the document for supporting annotations, related sections, references, appendices, tables, figures.

Pagination: Pagination should be provided for individual documents only.

Document information fields: Used for searching. Requirements are specified for each document type.

Naming PDF files: Files should be named in accordance with FDA recommendations.

Indexing PDF documents: Full text indices are used to help find specific documents or search for text within a document. For scanned documents, this indexing is not possible.

Electronic signatures: At the present time, hard copies of documents requiring signatures are required.

Both CDER and CBER have indicated that they will stop accepting paper submissions in the near future, although the actual date for these mandates is not clearly defined. Under PDUFA II commitments, FDA agreed to develop a paperless electronic submission program for all applications by 2002. This means that companies planning submissions should develop standards and procedures to ensure that the electronic submission requirements can be met.

Several software development companies have developed software to meet FDA's extensive electronic submission requirements. As a quick solution to the electronic submission requirements, some firms have purchased these software programs. Other companies have elected to develop an in-house solution to this challenge. A very important factor in the development of electronic submissions systems is the development of company-specific user requirements that describe the current procedures for handling documents at the company and the needs for any electronic system. These needs vary from company to company and the solution should be designed to accommodate all authoring groups at a company. Important issues in this process are the development of a house standard for documents and the agreement of all contributing departments to these standards. It is very useful to have document templates developed to assist in the standardization of document preparation. Most companies use an electronic file management system as the basis of their development of electronic document processes.

As with any computerized system, it is important that the implementation of the electronic submissions system be documented and validated. Changes to the system must be controlled in order to maintain the system's state of validation. Refer to Chap. 7 for an extensive overview of computer validation.

Recent FDA trends reflect an increasing desire to implement electronic tools and standards with the goal of increasing the efficiency and the quality of the review process. For applicants to be prepared to meet the emerging standards, it is important that appropriate technology is put into operation and procedures be developed for electronic document management with the end goal of creating electronic submissions. Electronic submissions are becoming the standard because they make the review process easy for both the agency and the industry.

13 CASE STUDY IN REGULATORY SUBMISSIONS

Recently an NDA for a new chemical entity for adjunctive treatment in adult epilepsy patients was submitted and approved in approximately 10 months. The contents of the application presented more than 15 years of research and development activities conducted in Europe and the United States. The planning and preparation of the NDA was a challenge for both the company and the FDA review team. Because of the long and complex development history there were voluminous amounts of data available that had to be evaluated in the application. Planning involved the review and organization of data recorded in multiple languages and varying quality. Negotiations in prefiling meetings—both in person and by telephone conference—on the contents and presentation of the data spanned nearly 18 months.

One of the first challenges was organizing the data in a manner that could be included in a meeting package for the first of several prefiling meeting. Several topics were discussed, specifically determining the readiness of the submission for filing. During a series of meetings, specific statistical analyses were discussed and agreed upon for inclusion in the final submission.

The "submission team" from the company and the FDA review team worked together to find the optimal solution to present the data in the most efficient manner, including abbreviated reporting strategies, electronic review aids, and the inclusion of comprehensive tabular and narrative summaries for each technical section. Each discipline (chemistry and manufacturing, pharmacology and toxicology, clinical/medical biopharmaceutics, and statistics) was reviewed and discussed prior to the finalization of the sections of the NDA. To facilitate the review process, every effort was made to eliminate all redundancies and provide very detailed index features throughout the paper volumes of the NDA.

Electronic documents included bookmarking and hyperlinking to assist the reviewers in navigating through multivolume reports and sections. Case report forms and case report tabulations were provided electronically, not only facilitating the review but also saving on application preparation time. The electronic files were tested by the FDA reviewers prior to submission of the NDA to determine if the files were as specified during prefiling discussions. The testing was invaluable for early identification of some minor formatting problems that were resolved prior to submission.

While the documents were being reviewed and summaries were being prepared, the facilities were readied for inspection. Independent experts evaluated the facility and assisted in the final preparations to ensure readiness for the inspections.

An electronic file management system and publishing tool was utilized to organize, paginate, and generate the paper volumes. After the submission was created as a virtual document, the publishing tool generated all the navigational tools required for the submission, including the table of contents, cover pages, divider pages, and pagination. The submission was quality checked prior to submission.

In total, the NDA consisted of 732 paper volumes and 1200 electronic volumes, equating to approximately 2000 paper volumes. The total review time from submission to final approval was 1 day under 10 months. It was only through careful planning, and teamwork and the collaborative efforts of both the FDA and the company submission team that this became a success. The critical factors contributing to this success are summarized below.

The communication plan included direct interaction with the FDA review team to work through the issues early in the preparation of the NDA.

The communication plan mandated open communication of all issues early on, resulting in no surprises.

The applicant team worked closely with the FDA review team to provide the requested analysis and data as quickly as possible.

The appointment of an effective, empowered "submission team" that matched the key disciplines represented on the FDA review team kept the team intact and available for the entire review process.
The quality and verification of the contents was the key element for the content acceptance prior to inclusion inn the application.
Multiple review steps and independent peer review were included.
Identified experts were partnered with those with critically needed expertise early in the planning, data evaluation, and preparation process.

14 CONCLUSION

In conclusion, the contents of a regulatory submission convey the first impression the FDA will have of your product and the quality and professionalism of your organization. These "living documents" have to be kept current to reflect the profile of the product, all applicable FDA regulations, and the procedures of your organization. To ensure that necessary information is obtained within the necessary time frames, a reliable network of communication throughout your organization is paramount and essential to both preparing effective submission documents and maintaining compliance with the regulatory reporting requirements for these applications.

15 REGULATORY REALITY CHECK

With the chapter providing an instructive backdrop, several FDA observations (FDA-483s) documented during recent inspections of FDA-regulated facilities are presented below, followed by a strategy for resolution and follow-up corrective and preventative actions.

> Post approval commitments for NDA were not met in a timely manner. For example: One of the post approval commitments was to **modernize and optimize** the analytical method for the detection of **degradants**. An FDA letter reminded your firm that the firm had not yet fulfilled this commitment. Validation studies were conducted by the firm and the validation report was approved. Yet, it was not until a "Change Being Effective Supplement" was submitted by the firm to the FDA and the **improved** method implemented.

One way this company could have prevented receiving this citation would have been to establish a cross-functional team that included quality control, regulatory affairs, quality assurance, and manufacturing. This would have afforded the company a realistic assessment of the resources and the time required to fulfill this postapproval commitment. Communication between these various units, coupled with consistent communication with the FDA reviewers requesting this additional analytical work, could have easily prevented the company's loss of credibility and good compliance standing with the agency.

An appropriate remedy to regain credibility and get back on track with the FDA would be to immediately prepare a proposed plan with realistic time lines and action items to be discussed with the review team at the earliest possible date prior to implementation. This meeting should be requested by the regulatory affairs unit and the specific reviewer(s) requesting this additional work. Additionally, all regulatory affairs and quality assurance personnel should undergo training focusing on what constitutes the various categories of change currently required by the agency. For example, the training would cover the various categories of category 1, 2, and 3 changes, CBE changes, and annual reporting changes in an effort to prevent future confusion related to changes the firm wishes to implement.

16 WORDS OF WISDOM

Establish meaningful dialogue with the FDA review team early in the process during the presubmission phase.

Ensure consistency between the submission and what the investigator will find in the facilities, including nonclinical laboratory test sites, clinical sites, and manufacturing sites.

Ensure formalized change management procedures are in place during the submission process.

Appoint a "submission team" that matches the key disciplines represented on the FDA review team and keep intact for the entire review process.

Authentication, verification, and quality of the contents is a key element for the content acceptance. Include multiple quality control audits and independent peer reviews.

Identify experts and early in the planning, data evaluation, and preparation process, partner with those offering critically needed expertise.

APPENDIX A: SUBMISSION TYPES AND REQUIREMENTS

Submission type	Transmittal Form[a]	CFR reference	Purpose for submission	Mechanism for changes	Reporting requirements
Drugs (reviewed FDA's Center for Drug Evaluation and Research)					
Investigational new drug application (IND)	FDA 1571	21 CFR 312	Request for authorization to administer an investigational drug product to humans. INDs include structural formula, animal test results, and if available, prior human test results, manufacturing information, and the proposed clinical investigational plan. INDs must be submitted at least 30 days priorto the start of clinical trials; FDA does not approved INDs but will notify applicants of issues within 30 days of receipt. The effective date of an IND is 30 days from the date of receipt by FDA's central document center unless a clinical	Amendments (protocol amendments and information amendments), annual reports	Safety reports, (7-day [telephone or fax] and 15-day [written]), annual reports (within 60 days of anniversary of effective date)

			hold is placed on the study. FDA will send an acknowledgment letter with the assigned number, review division, date of receipt and corresponding effective date, and the name and telephone number of the assigned FDA regulatory project manager.		
New drug application (NDA)	FDA 356h	21 CFR 314	Marketing application submitted to FDA to demonstrate that a drug product is safe and effective prior to interstate marketing. - NDAs contain proposed labeling with sufficient information for FDA to assess the product's safety and effectiveness for the proposed use, including data from clinical trials and other required technical information	Amendments (prior to approval), supplements (post approval)—changes being effected (CBE) and Prior approval supplements (PAS)–annual reports	120-day safety update (postfiling of the NDA), safety reports (expedited–15 day–and periodic–every 3 months for 3 years after approval; annually thereafter), annual reports (within 60 days of anniversary of approval date)

APPENDIX A (continued)

Submission type	Transmittal Form[a]	CFR reference	Purpose for submission	Mechanism for changes	Reporting requirements
Abbreviated new drug application (ANDA)	FDA 356h	21 CFR 314	Marketing application submitted to FDA to demonstrate that a drug is substantially equivalent to a previously approved, eligible product. Generally omit nonclinical laboratory studies and reports of clinical trials unless that apply to the in vivo bioavailability of the new drug product.	Amendments (prior to approval), supplements (postapproval) — changes being effected (CBE) and prior approval supplements (PAS)–annual reports	Safety reports (expedited—15—day—and periodic—every 3 months for 3 years after approval; annually thereafter), annual reports (within 60 days of anniversary of approval date)
Biologics (reviewed by FDA's Center for Biologics Evaluation and Review)					
Investigational new drug application (IND)	FDA 1571	21 CFR 312	Request for authorization to administer an investigational biological product to humans. INDs contain structural formula, animal test results,	Amendments (protocol amendments and information amendments), annual reports	Safety reports (7-day and 15-day), annual reports (within 30 days of anniversary of effective date)

			structural formula, animal test results, and if available, prior human test results, manufacturing information, and the proposed clinical investigational plan. INDs must be submitted at least 30 days prior to the start of clinical trials; FDA does not approved INDs but will notify applicants of issues within 30 days of receipt.		
Biologics license application (BLA)[b]	FDA 356h or previous form 3439	21 CFR 600	Marketing application submitted to FDA to demonstrate that a new biological product is safe and effective prior to interstate marketing. BLAs contain proposed labeling and sufficient information for FDA to assess the product's safety and effectiveness for the proposed use,	Amendments (prior to approval), supplements (postapproval)—changes being effected (CBE) and prior approval supplements (PAS)—annual reports	Safety reports (Expedited—15-day—and periodic—every 3 months for 3 years after approval; annually thereafter), annual reports (within 60 days of anniversary of approval date)

APPENDIX A (continued)

Submission type	Transmittal Form[a]	CFR reference	Purpose for submission	Mechanism for changes	Reporting requirements
			including data from clinical trials and specific technical information.		
Medical Devices (reviewed by FDA's Center for Devices and Radiological Health)					
Investigational device exemption (IDE)	Cover sheet recommended	21 CFR 812	Request for authorization for shipment of devices intended solely for investigational use. IDEs include a description of the device and labeling for the investigational device, an investigational plan, manufacturing information, and	Supplements	Safety reporting (10 day), progress reports (annually to IRBs and to FDA for significant risk devices only), final report (within 6 months of completion or termination)

			investigator information. IDEs are submitted for individual clinical studies and FDA approval is required prior to the initiation of the clinical study.		
Premarket approval (PMA)	Cover sheet recommended	21 CFR 814	Marketing application for some class III medical devices. PMAs include nonclinical laboratory and clinical trial results, description and labeling of the product, and manufacturing information.	Amendments (prior to approval), supplements (postapproval)	90-day safety update (post filing of the PMA), safety reports (5-day, 30-day, baseline reports, annual certifications)

APPENDIX A (continued)

Submission type	Transmittal Form[a]	CFR reference	Purpose for submission	Mechanism for changes	Reporting requirements
Premarket notification [510(k)]	Cover sheet recommended	21 CFR 807 subpart E	Premarketing application submitted to FDA to demonstrate that class I or II or some class III medical devices are as safe and effective or substantially equivalent to a legally marketed device that was or is currently on the U.S. market and that does not require premarket approval.	Amendments	Within 3 months

[a]If the person signing the application does not live in the United States or have a U.S. business address, the name and address of an authorized agent who has a business in the United States must be included in the application.

[b]64 FR 56441, Oct. 20, 1999 FDA changed the requirements for marketing applications for biologics; in lieu of filing an establishment license application (ELA) and product license application (PLA) in order to market a biological product in interstate commerse, a manufacturer will file a single biologics license application (BLA) with the agency.

APPENDIX B: CRITICAL ELEMENTS IN SUBMISSIONS

Checklist of Required Elements for an Investigational New Drug Application (IND; 21 CFR 312.23)

Cover Letter

Cover Sheet (Form FDA 1571)

1. **Table of Contents**
2. **Introductory Statement and General Investigational Plan**
   Drug product information and the broad objectives and planned duration of clinical trials
   Brief summary of previous human experience with the drug
   Any investigational or market withdrawal
   Brief description of the overall plan for investigation
3. **Investigator's Brochure**
   Description of drug substance and formulation
   Summaries of pharmacological and toxicological effects
   Summaries of pharmacokinetics and biological disposition
   Summary of safety and efficacy data in humans
   Description of possible risks and side effects.
4. **Protocols**
   A protocol for each planned study should be provided—phase 1 protocols may be less detailed than phase 2 and 3 protocols.
5. **Chemistry, Manufacturing, and Controls Information**
   Adequate information to assure proper identification, quality, purity, and strength of the investigational drug. The amount of information required depends on the phase of the study, the study duration, the dosage form, and the scope of the proposed clinical investigation. Information should be updated throughout the development process and scale-up of drug production.

   Drug substance
   Drug product, including a list of all components and quantitative composition
   Placebo information, including a list of all components and quantitative composition and manufacturing information
   Labeling
   Environmental analysis requirements

6. **Pharmacology and Toxicology Information**
   Description of pharmacological and toxicological studies of the

drug in animals that help support a conclusion of safety for investigative use in humans.

Pharmacology and drug disposition describing pharmacological effects and mechanisms of action and any known adsorption, distribution, metabolism, and excretion (ADME) parameters.

Toxicology, including an integrated summary of toxicological effects of the drug, including results from acute, subacute, chronic, and in vitro toxicity tests, the effects on reproduction and the fetus and special toxicity tests. A detailed tabulation of toxicity data for each study should be included.

GLP statement verifying that studies conducted in accordance with GLPs or a statement of the reason for noncompliance for studies that do not comply.

7. **Previous Human Experience with the Investigational Drug**

Information on previous investigational or marketing experience in the United States or other countries relevant to the safety of the proposed investigation

Information on the individual components for combination products

Foreign marketing information, including withdrawals for safety and effectiveness reasons

8. **Additional Information**

Drug dependence and abuse potential
Radioactive drugs
Pediatric studies
Other information

9. **Relevant Information**

Checklist of Required Elements for a New Drug Application (NDA) or a Biologics License Application (BLA 21 CFR 314.50)

Traditional NDA Format
Cover Letter

Application Form (Form FDA 356h)

1. **Table of Contents**

Comprehensive table of contents indexing by volume and page numbers, the sections and supporting information provided in the application. For electronic submissions it is essential that the table of contents contain bookmarks and hypertext links and the guideline specifies that the table of contents should contain three levels of detail.

2. **Labeling**
The content and format of labeling text required under 21 CFR 201.56 and 201.57 should be provided under this item, including all text, tables, and figures proposed for use in the package insert.

3. **Application Summary**
The summary should present the most important information about the drug product and the conclusions to be drawn from this information. This should be a factual summary of safety and effectiveness data and a neutral analysis of these data. The summary should include the following items:

Proposed text of the labeling for the drug—Annotated
Pharmacologic class, scientific rationale, intended use, and potential clinical
Benefits
Foreign marketing history
Chemistry, manufacturing, and controls summary
Nonclinical pharmacology and toxicology summary
Human pharmacokinetic and bioavailability summary
Microbiology summary
Clinical data summary and results of statistical analysis
Discussion of benefit risk relationship and proposed postmarketing studies

4. **Chemistry, Manufacturing, and Controls Section**

Drug substance: Description, including physical and chemical characteristics and stability, manufacturers, methods of manufacture and packaging, specifications and analytical methods for the drug substance, solid-state drug substance forms and their relationship to bioavailability.
Drug product: Components, composition, specifications and analytical methods for inactive components, manufacturers, methods of manufacture and packaging, specifications and analytical methods for the drug product, stability.
Methods validation package
Environmental assessment
Field copy certification

5. **Nonclinical Pharmacology and Toxicology Section**

Integrated summary of data from all studies and analysis of pertinent findings for interstudy and interspecies comparisons.

Narrative summary for each study report describing the notable features and results of each study. A comprehensive study for notable findings in related studies for each species and notable species differences should be provided.

Study reports should be provided in the following order:

Pharmacology studies

Toxicological studies—acute, subacute, and chronic toxicity; carcinogenicity;

special toxicity studies

Reproduction studies

Absorption, distribution, metabolism, and excretion (ADME) studies

GLP statement verifying that studies conducted in accordance with GLPs or a statement of the reason for noncompliance for studies that do not comply.

6. **Human Pharmacokinetics and Bioavailability Section**

Integrated summary of data from all studies and analysis of pertinent findings.

Narrative summary for each study report describing the notable features and results of each study.

Bioavailability and pharmacokinetics studies

Rationale for specifications and analytical methods

Summary and analysis of pharmacokinetics and metabolism of active ingredients and bioavailability/bioequivalence of the drug product.

7. **Microbiology Section**

Biochemical basis for drug's action on microbial physiology

Antimicrobial spectrum of the drug with in vitro demonstration of effectiveness

Mechanisms of resistance and epidemiological studies demonstrating resistance factors

Clinical microbiology laboratory methods to evaluate the drug's effectiveness

8. **Clinical Data Section**

List of investigators and list of INDs and NDAs

Background/overview of clinical investigations
Clinical pharmacology: ADME, pharmacodynamic studies, including a table of all studies grouped by study type and reports of individual studies in each group.
Controlled clinical trials: Adequate and well-controlled studies, combination drug products, including a table of all studies grouped by study type and reports of individual studies in each group.
Uncontrolled clinical trials: Table of all studies grouped by study type and reports of individual studies in each group.
Other studies and information: Reports of controlled or uncontrolled study of uses not claimed in the application, reports of commercial marketing experience.
Integrated summary of efficacy (ISE)
Integrated summary of safety (ISS)
Drug abuse and overdosage information
Integrated summary of benefits and risks of the drug

9. **Safety Update Reports**
10. **Statistical Section**
    Copy of Section 8, limited to information on controlled clinical studies.
11. **Case Report Tabulations**
    Tabulations should be provided for individual patients from initial clinical pharmacology studies, effectiveness data from each adequate and well-controlled study, and safety data from all studies.
12. **Case Report Forms**
    Case report forms and narratives should be provided for all patients who died, discontinued from a study due to an adverse event, or experience a serious adverse event. Case report forms for all patients involved in pivotal well-controlled studies should be available upon request.
13. **Patent Information**
    Information on any patent(s) on the drug for which approval is sought or on a method of using the drug.
14. **Patent Certification**
    Applicants must provide a certification regarding "any patents that claim the listed drug or that claim any other drugs on which investigations relied on by the applicant for approval of the application were conducted, or that claim a use for the listed or other drug."

15. **Establishment Description**
Relevant to biological products.
16. **Debarment Certification**
Certification that the applicant did not and will not use services of any person or firm debarred under the 1992 Generic Drug Enforcement Act. In order to prepare this certification, appropriate certifications should be obtained from all contractors used uring the preparation of the submission.
17. **Field Copy Certification**
Certification that a true copy of the chemistry, manufacturing, and controls certification has been sent to the FDA field office.
18. **User Fee Cover Sheet (Form 3397)**
The FDA User Fee Office should be contacted for user fee number and payment instructions prior to filing the NDA. The complete payment amount should be mailed to the bank address at the time of or prior to submission of the NDA. This user fee number should be included on all correspondence related to the user fee and on the check.
19. **Other**
This includes all information not submitted in other sections; for example, certifications for financial disclosure of clinical investigators.
Items 1 and 13–18 should be included in the first (administrative) volume of the submission.

CTD format (Highly Recommended by FDA After July 2003)

Module 1—Administrative and Prescribing Information

All administrative documents (e.g., applications forms, claims of categorical exclusion and certifications), and labeling with all documents provided in a single volume. This section should include all the U.S. specific regional requirements.

FDA form 356h
Comprehensive table of contents (same requirement as traditional NDA format, except that documents should be identified by tab identifiers instead of page numbers)
Administrative documents
Patent information (same as item 13 in traditional NDA format)
Patent certifications (same as item 14 in traditional NDA format)
Debarment certification (same as item 16 in traditional NDA format)
Field copy certification (same as item 17 in traditional NDA format)

User fee cover sheet (same as item 18 in traditional NDA format)
Other, including financial disclosure information, waiver requests (same as item 19 in traditional NDA format)
Exclusivity information
Environmental assessment
Prescribing information (same as item 2 of traditional NDA format)
Annotated labeling text (same as item 3 of traditional NDA format)

Module 2—Common Technical Document Summaries

Summaries of the dossier with strictly defined content templates. These summaries replace the section summaries in the traditional NDA format.

2.1 Overall CTD table of contents
Comprehensive table of contents listing all the documents in modules 2 through 5
2.2 Introduction
One page summary of CTD summaries
2.3 Quality overall summary
2.4 Nonclinical overview
2.5 Clinical overview
2.6 Nonclinical summary
2.7 Clinical summary

Module 3—Quality

Includes item 4 of traditional NDA format.
3.1 Module 3 table of contents
3.2 Body of data
3.2.S Drug substance
3.2.P Drug product
3.2.A Appendices
3.2.R Regional information
3.3 Literature references

Module 4—Nonclinical Study Reports

Includes item 5 of traditional NDA format.
4.1 Module 4 table of contents
4.2 Study reports
4.2.1 Pharmacology
4.2.2 Pharmacokinetics
4.2.3 Toxicology
4.3 Literature references

Module 5—Clinical Study Reports

Includes items 6, 8, 11, 12 of traditional NDA format.

5.1 Module 5 table of contents
5.2 Tabular listing of all clinical studies
5.3 Clinical study reports
5.3.1 Biopharmaceutic studies
5.3.2 Studies pertinent to pharmacokinetics using human biomaterials
5.3.3 Pharmacokinetic studies
5.3.4 Pharmacodynamic studies
5.3.5 Efficacy and safety studies
5.3.6 Reports of postmarketing experience
5.3.7 Case report forms and individual patient listings
5.4 Literature references

Note: FDA's current recommendation concerning the ISS and ISE is to include these documents in Module 5.

Checklist of Required Elements for an Abbreviated New Drug Application (ANDA 21 CFR 314.94)

Cover Letter

Application Form (Form FDA 356h)

1. **Table of Contents**
   Comprehensive table of contents indexing by volume and page numbers, the sections and supporting information provided in the application.
2. **Basis for Abbreviated New Drug Application Submission**
   Reference listed drug: including name, dosage form, and strength
   Marketing exclusivity information for reference listed drug
   Petition information
3. **Conditions of Use**
   Statement that the conditions of use are the same as those for the reference listed drug and a reference to the proposed labeling and the currently approved labeling for the reference listed drug.
4. **Active Ingredients**
   Statement that the ingredients are the same as those for the reference listed drug and a reference to the proposed labeling and the currently approved labeling for the reference listed drug.

5. **Route of Administration, Dosage Form, and Strength**
   Statement that the route of administration, dosage form, and strength are the same as those for the reference listed drug and a reference to the proposed labeling and the currently approved labeling for the reference listed drug.
6. **Bioequivalence**
   Information that shows that the product is bioequivalent to the reference listed drug Results of any bioavailability or bioequivalence testing performed in support of a petition Methods and GLP statement for any in vivo bioequivalence studies
7. **Labeling**

   Listed drug labeling
   Copies of proposed labeling
   Statement on proposed labeling that the labeling is the same as the labeling for the reference listed drug with the exception of changes noted in the annotated proposed labeling
   Comparison of approved and proposed labeling

8. **Chemistry, Manufacturing, and Controls**

   Drug substance: Description, including physical and chemical characteristics and stability, manufacturers, methods of manufacture and packaging, specifications and analytical methods for the drug substance, and solid-state drug substance forms and their relationship to bioavailability.
   Drug product: Components, composition, specifications, and analytical methods for inactive components, manufacturers, methods of manufacture and packaging, specifications, and analytical methods for the drug product, stability.
   Methods validation package
   Environmental assessment
   Inactive ingredients: Identification, characterization, and information to show that the inactive ingredients do not affect the safety of the proposed product.

9. **Patent Certification**
   Applicants must provide a certification regarding "any patents that claim the listed drug or that claim any other drugs on which investigations relied on by the applicant for approval of the application were conducted, or that claim a use for the listed or other drug."

10. **Financial Certification or Disclosure Statement**
11. **Debarment Certification**
    Certification that the applicant did not and will not use services of any person or firm debarred under the 1992 Generic Drug Enforcement Act. In order to prepare this certification, appropriate certifications should be obtained from all contractors used uring the preparation of the submission.
12. **Field Copy Certification**
    Certification that a true copy of the chemistry, manufacturing, and controls certification has been sent to the FDA field office.

Checklist of Required Elements for an Investigational Device Exemption (IDE; 21 CFR 812.20)

Cover Letter

Cover Sheet/Checklist Recommended

1. **Table of Contents**
   Comprehensive table of contents indexing by volume and page numbers, the sections and supporting information provided in the application.
2. **Report of Prior Investigations**
   Report on all prior clinical, animal, and laboratory testing, including
   Bibliography of all publications relevant to the safety of the device
   Copies of all published and unpublished adverse information
   Summary of all other unpublished information relevant to an evaluation of safety and effectiveness of the device
3. **Investigational Plan**
   Information on the investigational plan for the product, including purpose, protocol, risk analysis, description of the device, monitoring procedures, labeling, consent materials, and IRB Information
4. **Manufacturing Information**
   Adequate information to allow a judgment about the quality control of the device, including the description of methods, facilities, and controls for manufacturing, processing, packing, storage, and installation (if appropriate)
5. **Investigator Information**
   Example of agreement entered into with investigators and list of investigators

6. **Investigator Certification**

   Certification that all investigators have signed the agreement and that new investigators will not be added until they have signed the agreement

7. **IRB Information**
8. **Identification of Contract Facilities**
9. **Sales Information**

   Information to show that if the device is to be sold, the amount to be charged and a justification that this does not constitute commercialization

10. **Environmental Assessment**
11. **Labeling**
12. **Informed Consent Materials**

Checklist of Required Elements for a Premarket Approval Application (PMA; 21 CFR 814.20)

Cover Letter

Cover Sheet/Checklist Recommended

1. **Table of Contents**

   Comprehensive table of contents indexing by volume and page numbers, the sections and supporting information provided in the application.

2. **Application Summary**

   The Summary should present the detail to provide a general understanding of the data and information in the application, including

   Indications for use
   Device description
   Alternative practices and procedures
   Marketing history
   Summary of studies: Summary of nonclinical laboratory studies and summary of clinical investigations
   Conclusion drawn from the studies

3. **Device Description**

   Device description, including pictorial representations

Description of functional components or ingredients (if applicable)

Properties of the device: relevant to the diagnosis, treatment, prevention, cure, or mitigation of a disease or condition.

Principles of operation of the device

Manufacturing information: Adequate information to allow a judgment about the quality control of the device, including the description of methods, facilities, and controls for manufacturing, processing, packing, storage, and installation (if appropriate)

4. **Reference to Performance Standards in Effect**
5. **Nonclinical Laboratory Studies**

Results of nonclinical laboratory studies, including microbiological, toxicological, immunological, biocompatibility, stress, wear, shelf life, and other laboratory or animal tests, including a GLP statement.

6. **Clinical Investigations**

Results of clinical investigations involving human subjects, including investigator and enrollment information, protocol information, study population, study period, safety and effectiveness data, adverse reactions and complications, patient discontinuations, patient complaints, device failures and - replacements, data tabulations, subject report forms for deaths and discontinuations, statistical analyses, contraindications, and precautions for use of the device. Studies conducted under an IDE should be identified.

7. **Justification for PMAs Supported by Single Investigation**
8. **Bibliography**

Bibliography of all published reports that concern the safety or effectiveness of the device Identification, discussion, and analysis of other data, information, and reports relevant to the evaluation of safety and effectiveness.

Copies of published reports or unpublished information

9. **Samples**
10. **Labeling**
11. **Environmental Assessment**
12. **Financial Certification or Disclosure Statement**

Justifications should be provided for any information omitted.

Checklist of Required Elements for a Premarket Notification 510(k) (21 CFR 807.87)

Cover Letter

1. **Device Name**

    Device name, including trade or proprietary name, common or usual name, or classification name of the device

2. **Establishment Registration Number**
3. **Device Class Classification**

    The device class classification under Section 513 of the Act and the appropriate panel (if known) should be submitted. Justification should be provided for devices not classified.

4. **Actions Taken to Comply with Performance Standards**
5. **Labeling**

    All labeling and advertisements to describe the device for its intended use, including photographs or engineering drawings

6. **Comparability Statement**

    Statement that the device is similar and/or different from other comparable products in commercial distribution accompanied by data to support the statement

7. **Identification of Significant Changes or Modifications**
8. **510(k) Summary**

    The summary should present the detail to provide a general understanding of the data and information in the application, including

    Submitter's information

    Device name and classification

    Identification of the legally marketed device to which the submitter claims equivalence

    Description of the device: including how the device functions, scientific concepts that form the basis for the device, significant physical and performance characteristics for the device.

    Statement of the intended use: Including a general description of the diseases or conditions that the device will diagnose, treat, prevent, cure, or mitigate and a description of the patient - population, if appropriate. This section should include a comparison to predicate device and a rationale for any differences.

Summary of technological characteristics in comparison to the predicate device

For submissions in which the determination of substantial equivalence is also based on an assessment of performance data, the following information should also be included.

Nonclinical tests: Brief discussion of nonclinical tests submitted, referenced, or relied on

Clinical tests: Brief description of clinical tests submitted, referenced, or relied on, including a description of the subjects and a discussion of the safety and effectiveness, including a discussion of adverse effects and complications

Conclusions from nonclinical and clinical tests

9. **Financial Certification or Disclosure Statement**
10. **Information for Eligible Class III Devices**

    Certification that a reasonable search of all information and other similar legally marketed medical devices has been conducted.

11. **Statement That the Information Submitted is Truthful, Accurate, and That No Material Facts Have Been Omitted**

APPENDIX C: SCHEMATIC OF NDA REVIEW PROCESS

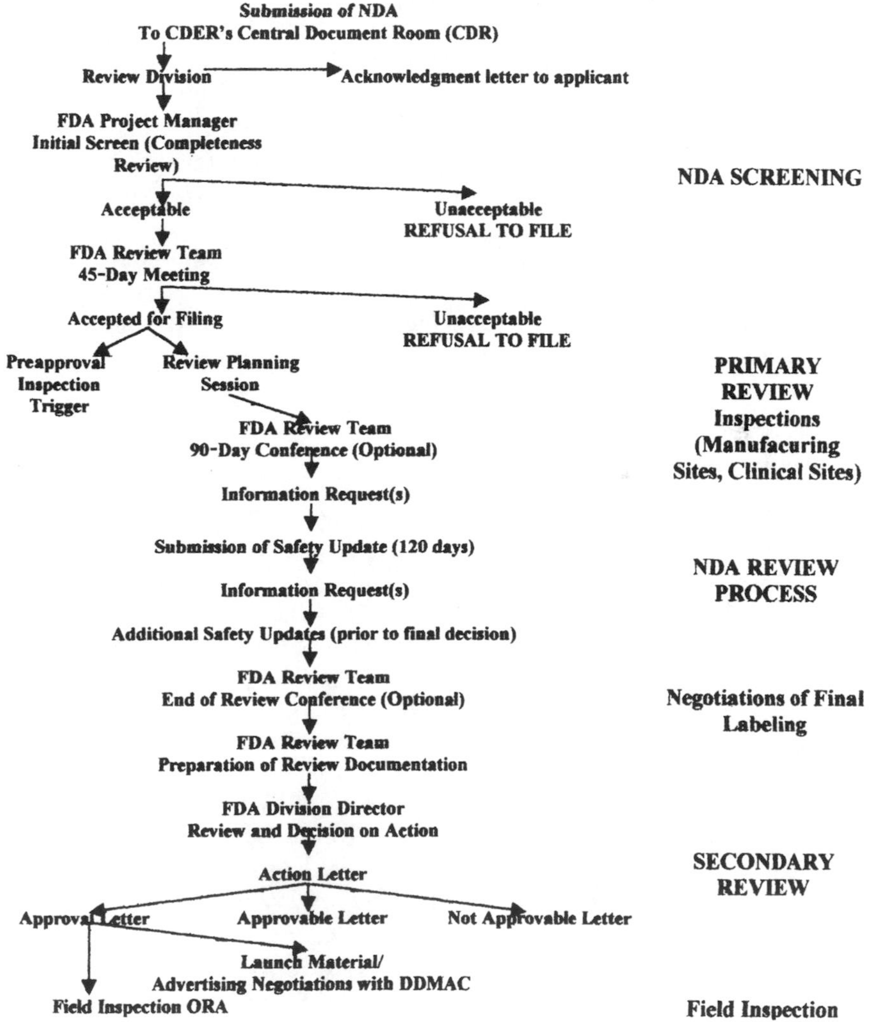

RECOMMENDED READING

Cook, J. *Plain and simple advice for improving NDAs.* Reg Aff Pharm Exec 1990.

Linberg, S.E. *Expediting Drug and Biologics Development: A Strategic Approach.* Waltham, MA: PAREXEL International, 1995.

Mathieu, M. *New Drug Approval in the United States.* Worldwide Pharmaceutical Regulation Series. Waltham, MA: PAREXEL International, 1998.

Mathieu, M. *New Drug Development: A Regulatory Overview.* rev. 3rd ed. Waltham, MA: PAREXEL International, 1994.

Quest editorial. *Optimizing interactions with FDA medical officers.* Appl. Clin. Trails, March 1996.

GMP Compliance, Productivity and Quality. SUPAC and Regulatory Consequences: The Balancing Act.

FDA PUBLICATIONS

Center for Biologics Evaluation and Research (CBER). *Guidance for Industry: Formal Meetings with Sponsors and Applicants for PDUFA Products,* March 7, 2000.

Center for Biologics Evaluation and Research (CBER). Guidance for Industry. *Providing Regulatory Submissions to the Center for Biologics Evaluation and Research (CBER) in Electronic Format—Biologics Marketing Applications [Biologics License Application (BLA), Product License Application (PLA)/Establishment License Application (ELA) and New Drug Application (NDA)],* Nov. 12, 1999.

Center for Biologics Evaluation and Research (CBER). Manual of Standard Operating Procedures and Policies, *SOPP 8001.2. Accessing the FDA Lists of Disqualified and Restricted Clinical Investigators, Debarred Individuals, and Firms Under the FDA Application Integrity Policy,* April 1, 1998.

Center for Biologics Evaluation and Research (CBER). Manual of Standard Operating Procedures and Policies, *SOPP 8401. Administrative Processing of Biologics License Applications,* Feb. 22, 1998.

Center for Biologics Evaluation and Research. *CBER Vision 2004. Strategic Plan for 2004.*

Center for Biologics Evaluation and Research. Manual of Standard Operating Procedures and Policies, *SOPP 8407. Compliance Status Checks,* Dec. 13, 1996.

Center for Biologics Evaluation and Research. Manual of Standard Operating Procedures and Policies, *SOPP 8404. Refusal to File Guidance for*

Product License Applications and Establishment License Applications, March 10, 1999.

Center for Biologics Evaluation and Research. Manual of Standard Operating Procedures and Policies, *SOPP 8411.1. Changes to an Approved Application. Administrative Handling and Review of Annual Reports*, March 10, 1999.

Center for Drug Evaluation and Research (CDER), Center for Biologics Evaluation and Research (CBER). Guidance for Industry. *Postmarketing Adverse Experience Reporting for Human Drugs and Licensed Biological Products: Clarification of What to Report*, Aug. 27, 1997.

Center for Drug Evaluation and Research (CDER), Center for Biologics Evaluation and Research (CBER). Guidance for Industry. *Regulatory Submissions in Electronic Format: General Considerations*, Jan. 28, 1999.

Center for Drug Evaluation and Research (CDER), Center for Biologics Evaluation and Research (CBER). Center for Veterinary Medicine (CVM). Guidance for Industry. *Submitting Debarment Certifications Statements*, Oct. 2, 1998.

Center for Drug Evaluation and Research (CDER), Center for Biologics Evaluation and Research (CBER). Guidance for Industry, Center for Devices and Radiological Health (CDRH). *Draft Guidance for Industry: Financial Disclosure by Clinical Investigators*, Oct. 26, 1999.

Center for Drug Evaluation and Research (CDER). Draft Guidance for Industry. *INDs for Phase 2 and 3 Studies of Drugs, Including Specified Therapeutic Biotechnology-Derived Products CMC Content and Format*, Feb. 1999.

Center for Drug Evaluation and Research (CDER). Guidance for Industry. *Environmental Assessment of Human Drugs and Biologics Applications*, July 27, 1998.

Center for Drug Evaluation and Research (CDER). Guidance for Industry. *Submitting Documentation for the Manufacturing of and Controls for Drug Products*, Feb. 1, 1987.

Center for Drug Evaluation and Research (CDER). Guidance for Industry. *Submitting Documentation for the Stability of Human Drugs and Biologics*, Feb. 1, 1987.

Center for Drug Evaluation and Research (CDER). Guidance for Industry. *Submitting Samples and Analytical Data for Methods Validation*, Feb. 1, 1987.

Center for Drug Evaluation and Research (CDER). Guidance for Industry. *SUPAC IR-Immediate-Release Solid Oral Dosage Forms: Scale-Up and Post-Approval Changes: Chemistry, Manufacturing and Controls, in Vitro*

Dissolution Testing, and in Vivo Bioequivalence Documentation, Nov. 30, 1995.

Center for Drug Evaluation and Research (CDER). Guidance for Industry. *SUPAC IR/MR: Immediate Release and Modified Release Solid Oral Dosage Forms, Manufacturing Equipment Addendum*, Feb. 26, 1999.

Center for Drug Evaluation and Research (CDER). Guidance for Industry. *SUPAC-IR Questions and Answers*, Feb. 18, 1997.

Center for Drug Evaluation and Research (CDER). Guidance for Industry. *SUPAC-IR: Immediate Release Solid Oral Dosage Forms; Manufacturing Equipment Addendum*, Oct. 21, 1997.

Center for Drug Evaluation and Research (CDER). Guidance for Industry. *Changes to an Approved NDA or ANDA*, Nov. 1999.

Center for Drug Evaluation and Research (CDER). Guidance for Industry. *IND Meetings for Human Drugs and Biologics*, Feb. 2000.

Center for Drug Evaluation and Research (CDER). Guidance for Industry. *Content and Format of Investigational New Drug Applications (INDs) for Phase 1 Studies of Drugs, Including Well-Characterized, Therapeutic, Biotechnology-Derived Products*, Nov. 20, 1995.

Center for Drug Evaluation and Research (CDER). Guidance for Industry. *Format and Content of the Clinical and Statistical Sections of an Application*, July 1, 1988.

Center for Drug Evaluation and Research (CDER). Guidance for Industry. *Format and Content of the Summary for New Drug and Antibiotic Applications*, Feb. 1, 1987.

Center for Drug Evaluation and Research (CDER). Guidance for Industry. *Formatting Assembling and Submitting New Drug and Antibiotic Applications*, Feb. 1, 1987.

Center for Drug Evaluation and Research (CDER). Guidance for Industry. *Postmarketing Reporting of Adverse Drug Experiences*, March 1, 1992.

Center for Drug Evaluation and Research (CDER). Guidance for Industry. *Conducting a Clinical Safety Review of a New Product Application and Preparing a Report on the Review*, Nov. 1, 1992.

Center for Drug Evaluation and Research (CDER). Guidance for Industry. *Submission of Abbreviated Reports and Synopses in Support of Marketing Applications*, Aug. 1999.

Center for Drug Evaluation and Research (CDER). Guidance for Industry. *Format and Content of the Human Pharmacokinetics and Bioavailability Section of an Application*, Feb. 1, 1987.

Center for Drug Evaluation and Research (CDER). Guidance for Industry. *Regulatory Submissions in Electronic Format: New Drug Applications*, Jan. 28, 1999.

Center for Drug Evaluation and Research (CDER). Guidance for Industry. *Example of an Electronic New Drug Application Submission*, Feb. 17, 1999.

Center for Drug Evaluation and Research (CDER). Guidance for Industry. *Preparing Data for Electronic Submission in ANDAs*, Sept. 1999.

Center for Drug Evaluation and Research (CDER). Guidance for Industry. *Major, Minor, FAX, and Telephone Amendments to Original Abbreviated New Drug Applications*, Aug. 1999.

Center for Drug Evaluation and Research (CDER). Guidance for Industry. *Organization of an Abbreviated New Drug Application and an Abbreviated Antibiotic Application*, March 2, 1999.

Center for Drug Evaluation and Research (CDER). Guidance for Industry. *Certification Requirements for Debarred Individuals in Drug Applications*, July 27, 1992.

Center for Drug Evaluation and Research (CDER). Guidance for Industry. *Instructions for Filing Supplements Under the Provisions of SUPAC-IR*, April 11, 1996.

Center for Drug Evaluation and Research (CDER). Guidance for Industry. *Format and Content of the Nonclinical Pharmacology/Toxicology Section of an Application*, Feb. 1, 1987.

Center for Drug Evaluation and Research (CDER). Guidance for Industry. *Fast Track Drug Development Programs: Designation, Development, and Application Review*, Nov. 18, 1998.

Center for Drug Evaluation and Research (CDER). Guidance for Industry. *Formal Meetings with Sponsors and Applicants for PDUFA Products*, Feb. 2000.

Center for Drug Evaluation and Research (CDER). Guidance for Industry. *Classifying Resubmissions in Response to Action Letters*, May 14, 1998.

Center for Drug Evaluation and Research (CDER). Guidance for Industry. *Submitting and Reviewing Complete Responses to Clinical Holds*, May 14, 1998.

Center for Drug Evaluation and Research (CDER). Manual of Policies and Procedures, *MAPP 4512.1. Formal Meetings Between CDER and CDER's External Constituents*, March 7, 1996.

Center for Drug Evaluation and Research (CDER). Manual of Policies and Procedures, *MAPP 6020.3. Priority Review Policy*, April 22, 1996.

Center for Drug Evaluation and Research (CDER). Manual of Policies and Procedures, *MAPP 6030.1. IND Process and Review Procedures (Including Clinical Holds)*, May 1, 1998.

Center for Drug Evaluation and Research (CDER). Manual of Policies and Procedures, *MAPP 7600.7. Processing an Electronic New Drug Application*, May 31, 2000.

Center for Drug Evaluation and Research (CDER). Center for Biologics Evaluation and Research (CBER). Guidance for Industry. *Submitting and Reviewing Complete Responses to Clinical Holds*, April 1998.

Title 21, Code of Federal Regulations, Food and Drugs.

Center for Devices and Radiological Health (CDRH), Guidance for Industry. *Regulation of Medical Devices: Background Information for International Officials*, April 4, 1999.

Center for Devices and Radiological Health (CDRH), *Medical Device Reporting for Manufacturers*, March 1997.

Center for Devices and Radiological Health (CDRH), *Guidance on IDE Policies and Procedures*, Jan. 20, 1998.

Center for Devices and Radiological Health (CDRH), Guidance for Industry, Review Staff, and the Clinical Community, *Guidance on Criteria and Approaches for Postmarket Surveillance*, Nov. 2, 1998.

Center for Devices and Radiological Health (CDRH), Guidance for Industry and FDA Staff, *SMDA to FDAMA: Guidance on FDA's Transition Plan for Existing Postmarket Surveillance Protocols*, Nov. 2, 1998.

U.S. Food and Drug Administration. Office of Health and Industry Programs. Division of Small Manufacturers Assistance. Project Officers James Park, Marsha Melvin, Althea Barcome. *Investigational Device Exemptions Manual.* HHS Publication FDA 96-4159. Rockville, MD: U.S. Department of Health and Human Services, 1996.

REFERENCES

1. FDAMA, FDA Plan for Statutory Compliance, Nov. 1998.
2. HHS, FDA, OMS, PDUFA II, Five Year Plan, FY 1999 Revision, July 1999.

2

Compliance Requirements During the Drug Development Process

Martin D. Hynes III
Eli Lilly and Company, Indianapolis, Indiana, U.S.A.

1 INTRODUCTION

The development of a new drug is a long, complex, and costly process. Current estimates show that it can take up to 14.7 years [1] and cost in excess of $800 million [2]. This complexity is a result of the large number of studies that need to be performed prior to the submission of a marketing application. Additionally, the complexity comes in part from the large number of regulations that govern the preclinical and clinical studies that support a New Drug Application (NDA). The U.S. compliance regulations that govern the preclinical and clinical studies that make up an NDA include Good Laboratory Practices (GLPs), Good Manufacturing Practices (GMPs), and Good Clinical Practices (GCPs). These regulations were written in the late 1970s and early 1980s to specify how preclinical safety studies, clinical trials, and development operations were to be conducted in support of an NDA.

1.1 Compliance Regulations

The understanding of these compliance regulations has evolved since their inception 20 years ago. In part, this evolution has resulted from how the

Food and Drug Administration (FDA) has interpreted these regulations through its inspection activities. During the early 1990s, FDA enforcement activities increased markedly [3]. This was evidenced by increases in warning letters, product seizures, injunctions, prosecutions, and recalls [4]. This increase in FDA enforcement activities was similarly observed across all of the regulations under which pharmaceutical companies work as they develop new drugs. For example, the FDA increased the number of clinical trial sites that it audited. In the GLPs area, there was an increase in the number of GLPs studies rejected by the FDA. This trend of evolving interpretations and heightened enforcement activity in the area of GMPs was evidenced by the initiation of pre-NDA approval inspections [3,5,6].

1.2 Drug Development

Another driver of complexity in the drug development process is the fact that most pharmaceutical companies are global; that is, drug therapies are being developed for patients on a worldwide basis. To do this means dealing with GLPs, GCPs, and GMPs that differ on a country-by-country basis. The differences in these regulations were frequently significant enough to warrant repeating many studies to meet local requirements. This had the impact of adding to the time and expense of introducing a new drug to patients in need in particular parts of the world without necessarily providing any significant new knowledge about the drugs. As a result, there has been an effort to harmonize these compliance regulations on a global basis through the International Conference on Harmonization (ICH).

The ICH was instituted as a forum for the creation of viable alternatives to the country-by-country regulatory requirements. The ICH membership consists of representatives from the Commission of the European Communities, the European Federation of Pharmaceutical Industry Associations, the Japanese Pharmaceuticals Manufacturers Association, the Japanese Ministry of Health and Welfare, the U.S. FDA, the Pharmaceutical Research and Manufacturers of America, and the International Federation of the Pharmaceutical Manufacturers Association. These groups represent both the industry and regulatory agencies for the participating countries [7]. The outcome from the various ICH working groups has impacted all of the compliance regulations (GLPs, GCPs, and GMPs). The regulatory need to conduct separate studies to meet the needs of local regulatory requirements has been dramatically reduced as a result of the harmonization activities. Whether this translates into a real decrease in the number of studies that companies have to conduct to register drugs on a global basis remains to be seen. This chapter will provide a brief overview of the GLP, GCP, and GMP requirements as they relate to the drug development process as well as present critical compliance issues related to these requirements.

2 GOOD LABORATORY PRACTICES

Good laboratory practices regulations are published in the U.S. Code of Federal Regulations (CFR), Title 21, Part 58 [8], and Title 40, Part 160 [9], for the FDA and the Environmental Protection Agency (EPA), respectively. The GLPs became an official requirement in the United States on June 20, 1979. The GLPs provide standards for the planning, performance, monitoring, recording, and reporting of preclinical safety studies conducted in support of an application to market a new drug [8]. It is only after the FDA reviews the results of the preclinical studies that the product can be judged to be safe. The FDA thus has the responsibility to review the data submitted in the NDA, while the burden for proving safety rests with the sponsor of the NDA submission. Sponsors can perform the needed studies in their own laboratory facilities, in those of a contract research organization, or in a university laboratory.

2.1 Historical Perspective

Until the mid-1970s, the assumption on the part of the FDA was that preclinical studies submitted by the sponsor were well conducted, analyzed, and reported. In the mid-1970s the FDA began to have questions about the uniformity and quality of the studies submitted to it as part of an NDA. During the process of reviewing study reports, scientists at the FDA observed data inconsistencies and evidence of unacceptable laboratory practices [10].

As a result of these FDA observations, "for-cause" inspections were performed at a number of institutions conducting preclinical safety studies. The results of these inspections were reported to the U.S. Senate in July of 1975, by the then FDA commissioner, Dr. Alexander M. Schmidt. The findings for these for-cause inspections showed problems in the design, conduct, and reporting of preclinical safety studies. These problems were deemed so serious as to question the validity of some studies. The following examples are illustrative of the magnitude and seriousness of the problems uncovered in these for-cause inspections carried out by the FDA.

One of the firms inspected by the FDA was Industrial Bio-Test Laboratories (IBT). At the time it was one of the largest contract testing facilities in the world. IBT had conducted literally thousands of preclinical safety studies that were submitted to the FDA to establish the safety of drugs awaiting approval to be marketed. The following is a partial list of the problems that were identified in the for-cause FDA inspection [10]:

Physical conditions were very poor.
Reports were not consistent with the original data.
Peer review of pathology studies resulted in different conclusions, with only the favorable one submitted to the FDA.
Food and water consumption was recorded as normal despite the fact that the animals were dead.
Drug doses could not be determined.
Histopathology reports existed for animals when tissues were not taken.

After reviewing these findings, it's easy to see why the FDA was concerned about the validity of the work conducted at IBT. As a result, the FDA declared the majority of long-term studies done by IBT to be invalid. Sponsors were then required to repeat these invalidated studies at great expense. IBT went out of business and several company officials were convicted of defrauding the government and jailed.

An audit at Searle found a number of similar problems. For example, the audit or for-cause inspection carried out by the FDA showed [10] the following:

Malignant mammary tumors were omitted from a statistical summary.
There were differences between the raw data and the final report.
Animals were dropped from the study without explanation.
Written protocols for completed studies could not always be found.

2.2 Link to Phase of Development

These findings, in conjunction with those from IBT, led the FDA to the conclusion that studies were poorly conceived, executed, documented, analyzed, and reported. Additionally, it was clear that the firm's management did not provide for adequate supervision and review of the data for accuracy prior to submission. These conclusions, which basically served to invalidate these preclinical safety studies, were deeply disturbing to the FDA, Congress, the general public, and industry. As a result, FDA Commissioner Schmidt established the Bioresearch Monitoring (BIMO) program in early 1976 to deal with the validity of data from both preclinical and clinical studies [11]. A toxicology monitoring task force set up by the BIMO dealt with the validity and reliability of all nonclinical laboratory studies conducted to support the safety of FDA-regulated products. As a result of its work, it recommended the establishment of GLP regulations to ensure the validity of preclinical safety studies. These regulations were drafted by a subcommittee composed of FDA and scientific personnel [10]. The group started its work with a rough draft that was based upon two independent sets of guidelines submitted by G.D. Searle and Company and the Pharmaceutical Manufacturers Association. After review and revision, the committee's work was published in 21 CFR, Part 58 [8].

The focus of these regulations was on the process that ensures the quality and integrity of the safety data. It should be noted that safety data were defined as "any *in vivo* or *in vitro* experiments in which test articles are studied prospectively in test systems under laboratory conditions to determine their safety"[8]. It was not an attempt to ensure good science or interfere with the judgment of scientists conducting the studies. These GLPs regulations covered the following topics [8]:

Subpart A
- General provisions
- Scope
- Definitions
- Inspection of a testing facility

Subpart B
- Organization and personnel
- Personnel
- Testing facility management
- Study director (SD)
- Quality assurance unit (QAU)

Subpart C
- Facilities
- General
- Animal care
- Animal supply
- Laboratory operation areas
- Specimen and data storage facilities
- Administrative and personnel facilities

Subpart D
- Equipment
- Equipment design
- Maintenance and calibration of equipment
- Computers

Subpart E
- Testing facilities operation
- Standard operating procedures (SOPs)

Reagent and solutions
Animal care

Subpart F
Test and control articles
Characterization
Handling
Mixture with carriers

Subpart G
Protocols and study conduct
Protocol
Conduct

Subpart H–I
Records and reports
Reporting
Storage and retrieval
Retention

Subpart K
Disqualification of testing facilities
Disqualification
Suspensions or terminations
Reinstatement

The complete regulations as well as the post conference report from the management briefing held in May 1979 can be found at http://www.fda.gov/ora/compliance_ref/ or in the postconference report [11].

The FDA has the ability to ensure compliance to these regulations through inspections conducted by its field investigator. Routine facility inspections are done every other year, while for-cause inspections can be done at any time. A refusal on the part of a firm or sponsor to allow the FDA to inspect can result in a disqualification of the studies. If during the course of an inspection the FDA finds significant deviations from the GLPs, the studies can also be disqualified. In general, at the conclusion of an FDA inspection it is more common for the FDA investigator to find deficiencies that need to be corrected but that are not significant enough to question the validity of the work. In this case, the FDA investigator documents the observations of noncompliance to the GLPs on Form 483, "Notice of Inspectional Observation." In response to these findings, firms need to respond to the FDA as to how and when these observations of noncompliance will be corrected.

2.3 Evolution/ICH Harmonization

The implementation of GLPs regulations was not limited to the United States. European GLPs were first issued in 1981 and were revised in 1997. The Japanese also set standards for preclinical safety studies in 1983. Although there is some similarity among these global GLPs, there are a sufficient number of differences, requiring additional studies to be performed, which in turn increases drug development time lines and expense with minimal return in terms of new information. These differences were the result of various legislative mandates and regulatory agendas. A great deal of work has been done within recent years to harmonize the GLPs regulations as a part of ICH [7].

These harmonized GLP's regulatory expectations are a significant advantage to pharmaceutical companies doing business on a global basis in that they allow the design and implementation of one set of standardized studies to meet all worldwide regulatory requirements, thus reducing the total number of preclinical safety studies on a drug candidate. One additional and important benefit is it helps firms in their efforts to minimize deviations from GLPs, since there is one set of standards against which to assess GLPs compliance throughout critical phase inspections and final study reports [7]. It is important to note, however, that there is a long way to go. Many differences still exist, such as the length of a chronic dog study, in vitro cardiac conduction, and Japan safety pharmacology.

3 GOOD CLINICAL PRACTICES

3.1 Overview

The federal regulation of drugs in the United States dates to the early 1900s. Although the Pure Food and Drug Act of 1906 was in large measure directed at the elimination of unclean and adulterated foods for the market, it also dealt with controlling drugs. Additional regulations were issued in 1923 that were focused on the bioassay of important drugs and their preparation.

The Pure Food and Drug Act of 1906 was amended in 1938, following the death of about 100 children. The deaths were the result of an elixir of sulfonamide, which utilized diethylene glycol, a highly toxic solvent. The elixir of sulfonamide tragedy triggered the introduction of an administrative procedure for the approval of new drugs of unknown safety prior to market introduction. Section 505 of the Food, Drug and Cosmetic Act of 1938 forbade the introduction of new drugs into interstate commerce without FDA approval. In 1962, the Kefauver–Harris amendments of the 1938 act resulted in an increase in regulatory control over many aspects of clinical research. In large measure these amendments were the result of concerns resulting from perceived deficiencies on the part of the pharmaceutical industry to adequately protect the public health in the thalidomide tragedy. Thalidomide is a hypnotic that was used in Europe in the early 1960s. Epidemiological research established that thalidomide taken early in the course of pregnancy caused a rare birth defect, phocomelia.

The 1962 amendments required sponsors to do a number of things including: demonstrate clinical efficacy in scientifically valid studies, establish informal consent guidelines, generate preclinical safety data to support clinical trials in humans, and report adverse events. The FDA was also granted the authority to regulate clinical research by requiring investigators and sponsors to maintain study records that must be made available for inspection by the FDA. A key requirement of the 1962 amendments was the need to file an Investigation New Drug application (IND) prior to the initiation of clinical trials with an experimental drug. Since that time the IND regulations have been modified to provide additional detail as to the responsibility of sponsors, investigators, and monitors. These modifications were made throughout the 1970s and 1980s.

Good clinical practices govern the approval, conduct, review, and reporting of clinical research intended for submission in an NDA. The U.S. GCPs as enforced by the FDA are delineated in the following documents:

21 CRF, Part 50—Protection of Human Subjects, Informed Consent. Effective date July 27, 1981 [12].
21 CFR, Part 56—Protection of Human Subjects; Standards for Institutional Review Boards for Clinical Investigations. Effective date July 27, 1981 [13].
21 CFR, Part 312—New Drug Product Regulations. Final rule issued 1987 [14].
21 CFR, Part 314—Applications for FDA Approval to Market a New Drug or an Antibiotic Drug [15].
Subpart C—FDA Action on Applications
314.126—Adequate and well-controlled studies

A general outline of the topics covered in the U.S. GCPs is as follows:

Part 50—Protection of Human Subjects [12].
Subpart A—General Provisions
Subpart B—Informed Consent of Human Subjects
50.20—General requirements for informed consent
Subpart C—Protections Pertaining to Clinical Investigations Involving Prisoners as Subjects
50.44—Restrictions on clinical investigations involving prisoners

50.46—Composition of institutional review boards where prisoners are involved
Part 56—Institutional Review Boards
Subpart A—General Provisions
56.103—Circumstances in which IRB review is required
Subpart B—Organization and Personnel
Subpart C—IRB Functions and Operations
56.109—IRB review of research
56.111—Criteria for IRB approval of research
Subpart D—Records and Reports
Subpart E—Administrative Actions for
56.121—Disqualification of an IRB or an institution
Part 312—Investigational New Drug Application
Subpart A—General Provisions
312.6—Labeling of an investigational new drug
Subpart B—Investigational New Drug application (IND)
312.20—Requirement for an IND
312.22—General principles of the IND submission
312.23—IND content and format
312.32—IND safety reports
312.33—Annual reports
Subpart C—Administrative Actions
312.40—General requirements for use of an investigational new drug in a clinical investigation
312.42—Clinical holds and requests for modification
Subpart D—Responsibilities of Sponsors and Investigators
312.50—General responsibilities of sponsors
312.53—Selecting investigators and monitors
312.56—Review of ongoing investigations
312.57—Record keeping and record retention
312.58—Inspection of sponsor's records and reports
312.60—General responsibilities of investigators
312.61—Control of the investigational drug
312.62—Investigator record keeping and record retention
312.64—Investigator reports
312.70—Disqualification of a clinical investigator
Subpart E—Drugs Intended to Treat Life-Threatening and Severely Debilitating Illnesses
312.83—Treatment protocols
312.84—Risk benefit analysis in review of marketing applications for drugs to treat life-threatening and severely debilitating illnesses.
Subpart F—Miscellaneous

312.110—Import and export requirements
312.120—Foreign clinical studies not conducted under an IND
Subpart G—Drugs for Investigational Use in Laboratory Research Animals or In Vitro Tests
Part 314—Applications for FDA Approval to Market a New Drug [15]
Subpart C—FDA Action on Applications
314.126—Adequate and well-controlled studies

The requirements outlined in the CFR are legally enforceable by the U.S. FDA. In addition to the above-mentioned documents, the FDA has guidance documents that are not legal requirements but do provide direction on acceptable standards for clinical research. They are as follows:

FDA, *Guidelines for Monitoring Clinical Investigators*, January 1988
FDA, *Information Sheets for IRB and Clinical Investigators*, October 1995
FDA, *ICH GCPs Consolidated Guidelines*, May 1997

In order to enforce compliance with these U.S. regulations, the FDA has a comprehensive program of on-site inspections of clinical trial investigations and data audits. These compliance activities are designed to monitor all aspects of the conduct of clinical studies intended for submission in the NDA. The goal of these inspections and audits is to ensure data quality, data integrity, and the protection of research subjects. This comprehensive program of on-site inspections and data audits is known as the BIMO program [16].

Despite the fact that the basics of GCPs have been in place for nearly 20 years, FDA inspections continue to identify compliance concerns. The most frequent GCPs compliance issues identified through inspection are the failure of investigators to follow the protocol and failure to maintain adequate and accurate case histories [17].

As part of the ICH effort, the FDA has been working on global GCPs guidelines. The draft guidelines that resulted from this harmonization effort were first published in the *Federal Register* in 1995 [18]. After a comment period, the modified guidelines, which included guidelines for investigator brochures and essential documents for the conduct of a clinical study, were submitted to the ICH steering committee in 1996. These ICH GCPs guidelines provide a unified standard for designing, conducting, recording, and reporting trials that involve the participation of human subjects. The expectation is that these guidelines will be followed when clinical trials are being conducted in support of regulatory submissions. The goal of these guidelines is the assurance that the rights, well-being, and confidentiality of trial subjects are protected. Additionally, they are designed to ensure that the data generated for an NDA submission are credible. (These guidelines can be found in the *Federal Register* [19].)

The following is a brief overview of the content of the ICH GCPs:

Introduction

1. Glossary
2. The Principles of ICH CGP
3. Institutional Review Board/Independent Ethics Committee (IRB/IEC)
4. Investigator

    Investigator's Qualifications and Agreements
    Compliance with Protocol
    Informed Consent of Trial Subjects
    Records and Reports

5. Sponsor

    Quality Assurance and Quality Control
    Trial Management, Data Handling, Record Keeping, and Independent Data Monitoring Committee
    Investigator Selection
    Manufacturing, Packaging, Labeling, and Coding Investigational Product(s)
    Safety Information
    Adverse Drug Reaction Reporting
    Monitoring
    Audit

6. Clinical Trial Protocol and Protocol Amendment(s)

    Trial Objectives and Purpose
    Trial Design
    Ethics
    Data Handling and Record Keeping

7. Investigator's Brochure

    Contents of the Investigator's Brochure

8. Essential Documents for the Conduct of a Clinical Trial

For an extensive discussion regarding compliance requirements during clinical activities, refer to Chap. 3, "Role of Quality Assurance Throughout Clinical Trials."

4 GOOD MANUFACTURING PRACTICES

The fact that GMPs apply to drug products that are in clinical trials was set out in the preamble to the GMPs (*Federal Register*, Sept. 1978). The part of the preamble that is relevant to the application of GMPs to the development of new drugs reads as follows:

> The commissioner finds that as stated in 211.1 these cGMPs regulations apply to the preparation of any drug product for administration to humans or animals, including those still in investigational stages. It is appropriate that the process by which a drug product is manufactured in the development phase be well documented and controlled in order to assure the reproducibility of the product for further testing and for ultimate commercial production. The commissioner is considering proposing additional cGMPs regulations specifically designed to cover drugs in research stages.

4.1 Historical Perspective

In addition to stating that GMPs apply to clinical trial materials, the preamble indicated that the regulations for clinical trial materials were different from those for commercial products. This fact is evidenced by the statement that the FDA was considering proposing additional GMPs to cover drugs in research. To date the FDA has not issued a separate set of GMPs for investigational drugs. Expectations for investigational new drugs have been established by the FDA through a combination of guidance documents, compliance programs, inspection guidelines, and podium policy statements [5,6].

4.2 Link to Phase of Development

Additional guidance on the application of GMPs to drug products used in clinical trials can be found in the FDAs *Guidelines on the Preparation of Investigational New Drug* (*IND*) *Products*. These guidelines clearly state that compliance to GMPs was required at the stage at which the drug was to be produced for clinical trials in humans (IND guidelines, March 1991). These IND guidelines emphasized the need for proper documentation during the drug development process. The need for control of components, product controls, process controls, equipment identification, packaging, and labeling also was covered in the IND guidelines [5,6]. The guidelines also made clear that tighter controls were expected as experience was gained with the product.

4.3 Evolution/ICH Harmonization

The fact that the FDA had jurisdiction over clinical trial material and that compliance to GMPs was expected once a drug entered human clinical trials was firmly established by the preamble to the GMPs and the IND regulations. After the generic drug scandal of 1989, however, the exceptions of compliance to GMPs in the various phases of drug development took on new importance. As a result of the generic drug scandal of 1989, the FDA issued two new documents outlining their expectations during the drug development process. They are as follows:

1. FDA Compliance Program Guidance Manual, *Pre-Approval Inspections/Investigations* (*Program 7346.832*), October 1990
2. FDA Compliance Program Guidance Manual, *Pre-Approval Inspection of New Animal Drug Applications* (*NADA*) (*Program 7346.832*), February 1991

The objectives of these compliance programs are as follows [5,6]:

1. Ensure that the facilities listed have the capabilities to fulfill the application commitments to manufacture, process, control, package, and label a drug product following GMPs.
2. Ensure adequacy and accuracy of analytical methods by proper testing.
3. Ensure correlation between manufacturing process for clinical trial material, bioavailability study material, and stability studies and filed process.
4. Ensure that scientific evidence supports full-scale production procedures and controls.
5. Have submitted factual data.
6. Ensure protocols are in place to validate the manufacturing process.

Given that these compliance programs and FDA expectations for clinical trial material are discussed in detail in Chap. 3 of this book, the topics will not be further reviewed here. The importance of GMPs compliance in development of a new drug cannot be overemphasized, as the consequences of noncompliance can be high in that the FDA can delay approval of an NDA if significant noncompliance is discovered during a preapproval inspection.

5 ASSESSING COMPLIANCE THROUGH AUDITS

Compliance to GMPs, GLPs, and GCPs is clearly required during the development of a new drug. Not only is this a regulatory requirement, it is also a good business practice. The failure to comply with these regulations can result in the delay of regulatory approval, which results in delaying the availability of important new drugs to patients in need and the generation of revenues for drug companies. It is thus in everyone's best interest to comply with these regulations during the development process. Compliance should therefore be assessed throughout the drug development process rather than waiting until an NDA submission has been made.

5.1 Internal Audits

The identification of areas of noncompliance by the FDA just prior to NDA approval can be problematic for a firm at best. All of the compliance regulations call for some type of internal quality assurance or quality control activity to ensure compliance. The purpose here is not to review these requirements in detail, but rather to stress the importance of having internal quality reviews in place to ensure a drug is being developed in compliance with the GMPs, GLPs, and GCPs. One method that is widely used in the industry is that of conducting a series of internal audits of specific areas and critical documentation.

These audits should be performed by an independent quality function that has a staff of well-trained auditors who are knowledgeable of the compliance regulations as well as the clinical or preclinical operations they will be inspecting. Firms must have documented internal audit procedures for these auditors to follow. This starts with having a well-defined process for determining what will be audited or inspected. For example, the GLPs require that all of the final study reports submitted to the IND or NDA be signed off by the quality function. Other regulations lack this degree of prescriptiveness for what needs to be audited. Firms therefore need to decide what they will audit. This is of particular importance because it is not possible to audit or inspect every drug study or clinical trial site. Most firms thus need to have a way of deciding which studies or facilities they want to audit. This can be done by developing a risk profile so that those areas that have the highest degree of risk and exposure can be audited. The areas of high risk with significant exposure should be prioritized and developed into an audit plan. Once an audit plan for the year has been developed and approved by management the auditors can execute the plan. As each audit is conducted, the auditor needs to document his or her audit findings in a formal report back to the management of the area being audited. Once the audit report has been issued it is up to the site function being audited to respond to the audit. The response needs to provide the details of management's plan for correcting the observation of noncompliance found during the audit. These responses need to be contained in a formal written document. The audit group or function needs to follow up with the site management after an appropriate period of time to ensure that all of the promised corrective actions have in fact been made.

Having a strong internal audit function throughout the drug development process will help to ensure that all of the preclinical and clinical work conducted for an NDA submission will be in compliance with the GMPs, GLPs, and GCPs. This will go a long way toward eliminating any last minute surprises during an FDA inspection just prior to NDA approval.

6 CONCLUSION

The development of a new drug is a long and costly process that is governed by myriad federal regulations. The GMPs, GLPs, and GCPs are the most notable from a compliance standpoint. These regulations grew out of a number of concerns or situations that have occurred over the past century. They include the findings at IBT and Searle in the case of GLPs, the elixir of sulfonamide and thalidomide tragedies in the case of GCPs, and the generic drug scandal of the late 1980s in the case of the GMPs for materials utilized in clinical trials. These compliance regulations were developed in the United States, as well as in a large number of countries around the world. Although these regulations were similar in many aspects, there were significant differences. These differences were large enough in many cases to warrant the conduct of additional studies to obtain approval in specific countries. In response to this the ICH was established as a forum to deal with these differences and formulate one set of global compliance regulations. Much progress has been made as a result of these efforts, which has led to the establishment of one global quality standard. This has led to the acceleration of drug regulations on a global basis and eliminated the need to conduct studies, which added little to our knowledge of the drug's efficacy or safety.

The cost of not complying with the GMPs, GLPs, and GCPs can be high for studies conducted in support of an NDA. Not only does a firm risk receiving a list of deficiencies from the FDA on a form 483, but much more important, studies can be disqualified, or worse yet, the approval of an NDA can be delayed.

Firms engaged in the development of a new drug should have strong audit functions in place to monitor compliance to the GMPs, GLPs, and GCPs throughout the development of a new drug. This will go a long way toward ensuring compliance, minimizing FDA-483 observations, and eliminating the possibility that an NDA approval will be delayed for noncompliance. More important, it will ensure the safety of patients during the development of a new drug and accelerate the ultimate approval of the drug for patients in need.

7 WORDS OF WISDOM

Drug development activities are covered by three important types of compliance regulations: GMPs, GLPs, and GCPs.

Approval of the regulatory filing is influenced by the documentation efforts put forth during development.

Pervasive quality assurance monitoring of development activities (by way of audits) will greatly impact the quality and success of the development activities.

Development activities must be captured in an official development report only for GMPs.

Noncompliance with GMPs, GCPs, and GLPs during the development phase could result in both a failed preapproval inspection and a regulatory hold of the submissions.

REFERENCES

1. DiMasi, J.A., Hanson, R.W., Grabowski, H.G., Lasagna, L. Research and development costs for new drugs by therapeutic category: A study of the U.S. pharmaceutical industry. *PharmacoEc 7*; 152–169, 1995.
2. Engel, S. R & D costs are staggering. *R&D Directions* p3, January 2002.
3. Hynes, M.D. III. The use of total quality principles to achieve regulatory compliance in research laboratories. *Qual Assur Good Prac and Law* 4(1): 34–40, March 1995.
4. Drug warning letters up, other enforcement actions down. *Washington Drug Letter*, pp 2–3, March 14, 1994.
5. Hynes, M.D. III. Developing a strategic approach to preparing for a successful pre-NDA approval inspection. Hynes, M.D. III (ed.). *Preparing for FDA Pre-Approval Inspections.* New York: Marcel Dekker, 1998; pp. 11–29.
6. Hynes, M.D. III. Introduction to food and drug administration pre-new drug applications approval inspections. Hynes, M.D. III (ed.). *Preparing for FDA Pre-Approval Inspections.* New York: Marcel Dekker, 1998; pp. 1–10.
7. Basarke, B.S., Jack, Hynes, M.D. III. Regulations, harmonization, and the industry: A complementary triangle. Bhatt, V. (ed.). *GCPs Compliance, Productivity, and Quality: Achieving Synergy in Healthcare Manufacturing.* Vol. 12. Buffalo Grove, Ill.: Interpharm Press, 1998, pp. 399–423.
8. U.S. Department of Health and Human Services, Food and Drug Administration. 21 CFR 58. *Food and Drugs Good Laboratory Practice Regulations. Final Rule.* Sept. 4, 1987 and Dec. 22, 1978.
9. U.S. Department of Health and Human Services, Environmental Protection Agency. 40 CFR, 160. *Protection of Environment, Good Laboratory Practice Standards.*

10. Taylor, J.M., Stein, G.C. Historical Perspective. Weinberg, S. (ed.) *Good Laboratory Practice Regulations.* 2nd ed., revised and expanded. New York: Marcel Dekker, 1995, pp. 1–9.
11. U.S. Department of Health and Human Services, Food and Drug Administration, Office of Regulatory Affairs, *Guidance for Industry Good Laboratory Practice Regulations Management Briefings Post Conference Report*, Aug. 1979.
12. U.S. Department of Health and Human Services, Food and Drug Administration, 21 CFR 50, *Food and Drugs, Protection of Human Subjects*; *Subpart B*: *Informed Consent of Human Subjects.*
13. U.S. Department of Health and Human Services, Food and Drug Administration, 21 CFR 56, (docket no. 77N0350) *Food and Drugs, Protection of Human Subjects*; *Standards for Institutional Review Boards for Clinical Investigations.*
14. U.S. Department of Health and Human Services, Food and Drug Administration, 21 CFR 312, (docket no. 82N0394) *Food and Drugs, New Drug Product Regulations.*
15. U.S. Department of Health and Human Services, Food and Drug Administration, 21 CFR 312, *Food and Drugs, Applications for FDA Approval to Market a New Drug*; *Subpart C*: *FDA Action on Applications.*
16. U.S. Department of Health and Human Services, Food and Drug Administration, Office of Regulatory Affairs, *Bioresearch Monitoring (BIMO), Good Laboratory Practice, Compliance Program* 7348.808.
17. Lepay, M.D., David, A. *GCP Compliance*: *FDA Expectations and Recent Findings*, slide presentation presented as part of the BIMO Conference, June 9, 1998.
18. U.S. Department of Health and Human Services, Food and Drug Administration, *Fed Reg. International Conference on Harmonization*; *Draft Guideline on Good Clinical Practice*, 60 (159):42947–42957, Aug. 17, 1995.
19. U.S. Department of Health and Human Services, Food and Drug Administration, *Fed Reg. International Conference on Harmonization*; *Good Clinical Practice*, 62 (90):25691–25709, May 9, 1997.

3

Validation: A New Perspective

James Agalloco
Agalloco & Associate, Belle Mead, New Jersey, U.S.A.

For someone who has worked in validation extensively for over 20 years, my selection as the author of this chapter might come as a surprise to some. Could someone with such extensive "history" be able to approach this subject in an objective manner? Could a graybeard such as myself view this subject with a new perspective? I sincerely hope so. Thirty years' experience in this industry and more important, my time as a consultant, has given me insights that might seem startling at times. I cannot count the number of times colleagues and clients have said to me, "Yes, but we have to do it this way because," "That's what the FDA investigator expects," "We've always done it that way," "We can't change our protocol now," "We've never done that before," or "It's corporate policy." Fill in the ending of your choice. I've heard them all, and none of them justifies doing the wrong thing. They are merely other ways of saying we are afraid to think outside the box. Often what they are rejecting is the voice of reason and common sense founded upon sound science and engineering. Well, the time has come to tell the tale the way I have always wanted to, without concession to what is politically or regulatorily correct. Here it is: validation—pure and simple, unencumbered by the trappings of pseudo-science, regulatory obfuscation, and corporate doctrine.

If there is one aspect of what we do in this industry that should be universal, it is our reliance upon science. Our products, processes, equipment, and even facilities are often the culmination of many years of rigorous scientific and engineering effort. We do these activities a severe disservice when we apply vague and irrational controls upon them in the pursuit of "validation." The immutable truths of science are used to initially define our products, and should also be used to demonstrate their validity. This principle underlies all that follows in this effort.

The reader who expects to find in this chapter a guide to the validation of every type of process, system, or product imaginable will be disappointed. The proper execution of validation belies condensation into such a brief effort. What I have endeavored to do instead is to discuss issues rather than science in an effort to address more the philosophy, compliance, and management aspects of the subject. I have provided a list of references on validation practice throughout which answers to a great number of technical questions can be found.

1 INTRODUCTION: "THE EMPEROR HAS NO CLOTHES"

Those who practice it have poorly served validation. Among the abuses this industry has witnessed are massive validation master plans without meaningful guidance on what is to be done, qualification protocols of over 80 pages for a laboratory incubator, qualification reports that are actually page after page of vendor brochures, performance qualification studies that were completed using batch record-type documentation, and myriad other useless "requirements." This is compounded by intimations by purveyors of such misinformation that if you haven't documented everything, your effort will be noncompliant. To quote one recent flyer I came across, "The volume of testing resulted in enough paper to bury the average investigator." Does any of this excess serve the firm, or even more important, the consumer? I think not, but it certainly does fatten the wallets of validation service providers, who will willingly fulfill any requirement, however unreasonable, for a fee. Is it any wonder that these providers are perhaps the worst offenders in the bloated validation efforts we are so willing to endure? Abuse of this type is unfortunately commonplace and has increased the cost and duration of validation activities without meaningful benefit to anyone except the providers of such excessive validation. Is the rote assembly of information for information's (or is it billable hours?) sake really what was intended when validation was first conceived? Ken Chapman once wrote, "Validation is little more than organized common sense"[1]. We clearly need a return to that kind of simplicity of both thought and expectation.

1.1 Definitions: "What's in a Name?"

In order to truly understand what validation is, we must briefly explore its definition. One of the clearest definitions was developed by Ted Byers and Bud Loftus in the late 1970s, and was formally adopted by FDA in 1987. "Process validation is establishing ***documented evidence*** which produces a ***high degree of assurance*** that a specific process will ***consistently*** produce a ***product*** meeting its ***predetermined specifications*** and quality characteristics" [2].

It is useful to dissect this definition to better understand its intent. The italicized words in the definition provide a clear indication of what we should expect of our validation efforts.

Documented evidence—Our efforts must be written and retained on file. This implies an organized body of information with clear conclusions.

High degree of assurance—We must be confident that the gathered information supports our conclusions. It suggests the use of "worst-case" challenges, yet recognizes that some uncertainty must exist.

Consistently—Our efforts must be reproducible. Controls must be in place to repeat the process in a consistent fashion.

Product—the focus of every validation effort. The farther we are from elements that impact critical product attributes, the less we should be concerned about the system or activity.

Predetermined specifications—Expectations must be pre-established. To be meaningful these requirements must be largely quantitative.

As interpreted within the industry, we have implemented programs based upon the classical scientific method, in which we gather information to support the premise. Where the information (read that as validation) supports the premise (that the product is of acceptable quality) we have achieved a validated state for the process. A more contemporary definition is as follows:

> Validation is a defined program, which, in combination with routine production methods and quality control techniques, provides documented assurance that a system is performing as intended and/or that a product conforms to its predetermined specifications. When practiced in a life-cycle model, it incorporates design, development, evaluation, operational and maintenance considerations to provide both operating benefits and regulatory compliance [3].

When I wrote this in 1993, I had hoped to define validation in terms of how it was to be accomplished. I also introduced the concept of a validation life cycle (see later section) as the appropriate means by which to manage its execution. I also tried to acknowledge that validation wasn't something separate and distinct from the everyday operation of the firm, nor was it something solely for use in discussions with regulators or auditors. More will be said about each of these later on in this chapter.

The last obstacle to industry understanding was a realization that the term validation itself was a source of confusion. During the early years of validation, the term had become synonymous with the activities focused on protocols development, data acquisition, and reports preparation. This narrow view did not recognize its relationship to a number of other activities already in place within the firm. As time went on, the concept came into focus of validation as being supported by a number of related activities practiced throughout the useful life of a system that provide greater confidence in the system, process, or product. To overcome the limitations of the smaller scale of the original scope of validation, many industry practitioners adopted the new term *performance qualification* for the testing phase of an overall validation program. With the introduction of this new term, the distinction between the narrower activities of validation and larger practice of validation as a program with ties to other activities has been made more evident.

1.2 Elements of Validation: "The Whole Is Greater Than the Sum of the Parts"

As introduced above, validation is dependent upon a number of activities and practices ordinarily practiced by a cGMP-compliant firm. Without these practices, it is little more than an exercise in minimal compliance and is of little value in supporting the efficacy of any process. When the proper relationship between validation and these other activities is established, there is a synergistic effect of greater compliance and some tangible operating benefits. The operational areas of the firm that link to validation are process development, process documentation and equipment qualification calibration, analytical method validation, process/product qualification, cleaning validation, and change control.

Process development—Those activities that serve to initially define the product or process. These form the basis for the product specifications and operating parameters used to achieve them. During the early 1990s the U.S. FDA mandated that firms provide a clear linkage between their small-scale development and clinical preparations and the eventual commercial-scale process. The existence of this linkage supports the efficacy of the manufacturing procedures, which must be confirmed in the validation exercise. A poorly developed process may prove unreliable (in essence unvalidatable) on a larger scale and thereby compromise patient therapy. As validation is intended to confirm the effectiveness of the defined procedural controls, it serves merely as a means to keep score. A process that cannot be validated using independent and objective means is most likely inadequately developed.

A second consideration in development is that investigational/clinical materials, while not requiring validation, can benefit from some rudimentary efforts to confirm process efficacy. Developmental materials that are intended to be sterile must be supported by validation studies that fulfill that expectation. (See Sec. 4 on sterile products for a summary of these.)

The goal of the developmental process should be to identify robust and reliable processes accommodating any expected variations in starting materials, operator technique, operational environment, and other variables. The developers must work cooperatively with operating areas to define the necessary controls to ensure commercial-scale success. The inclusion of corrective measures for common manufacturing issues (pH overshoot, temperature excursions, variations in product moisture, etc.) serves to increase reliability. Inherent in this is the establishment of proven acceptable ranges for the operating parameters, as demonstrated by success in meeting the product specifications. The gathering of information (process knowledge) must be the principal objective of the developmental effort. With that knowledge will come identification of the critical elements necessary for the process to be validated.

Process documentation—The accumulated knowledge of the firm relative to the successful manufacture of the process is maintained in a variety of documents, including raw material and component specifications, master batch records, in-process specifications, analytical methods, finished goods specifications, and standard operating procedures (SOPs). These define the product and process to ensure reproducible success in operation. Where these documents are inadequate, likely the result of insufficient developmental consideration, there are opportunities for variations that may result in process failure. A process that relies on some human knowledge not contained within the documentation is inherently unstable; a change in operator could mean a change in the product. The documents serve as guidance to the maintenance of the process, and thereby the product, in a stable state. Inherent to any fully compliant documentation system is a change control program that forces the evaluation of changes on systems to assess their impact on the regulated process [4].

The major concern with document linking to validation is with master batch records and SOPs. The defined ranges for the operating parameters (as established in development) as defined in the master batch record are confirmed to be satisfactory during the validation effort. There is no requirement, nor should there be, to establish on a commercial scale that success is possible at the extremes of these ranges. That type of confirmation is ordinarily restricted to developmental trials, in which the financial impact is less significant. The only common exception to this practice is in the validation of sterilization processes, in which the performance qualification efforts will oftentimes use worst-case conditions at or below the routine sterilization parameters.

Operating procedures for new products, processes, and equipment are prepared in draft form for the start of the qualification, and can be approved (with appropriate adjustments if required) for commercial use after successful validation and after completion of the performance qualification.

An important consideration in the preparation of any documentation is that it reflect the audience for whom it is prepared. The operators who must follow the procedure are perhaps the best individuals to write or at the very least critique it before finalization.

Equipment qualification—Definition of the equipment, system, and/or environment used for the process. These data are used to gather a baseline of the installation/operational condition of the system at the time when the performance qualification (PQ) of the system is performed. This baseline information is used to evaluate changes to the system performance over time. Intentional changes from these initial conditions must be considered and evaluated to establish that the system's performance is unaffected by the change. Unintentional changes in the form of a component or equipment malfunction or failure can be easily rectified using the available baseline data as a basis for proper performance.

Equipment qualification has been arbitrarily separated by many practitioners into installation qualification (IQ, focused on system specification, design, and installation characteristics) and operational qualification (OQ, focused on the baseline performance of the system under well-defined conditions). This separation is purely arbitrary in nature, and there is no regulatory requirement that this be the case. For smaller, simpler systems and equipment, consolidation of these activities under the single heading of equipment qualification can save time and expense with no compromise to the integrity of the effort. One of the subtle issues associated with separation of the qualification activity into subactivities is the requirement for additional documents for both protocols and reports as well as a longer execution period, as it is customary to await completion of IQ before allowing the OQ to begin. A meaningless exercise is extensive debate as to whether a particular requirement is required in IQ or OQ, or perhaps OQ or PQ. What is important is that the appropriate information is gathered without regard to the category to which it is assigned.

In general, equipment qualification is performed in the absence of the product (exceptions are made in the case of water and other utility systems) to allow the support of multiple processes, as would be the case for multiproduct equipment. It is common to utilize a checklist approach in which the information gathered is entered into a blank protocol template. The completed protocol thus becomes the qualification report without additional writing. One of the more egregious sins in qualification is to bind the protocol so closely to the equipment specification that in effect one has to prequalify the system just to prepare the protocol. In my opinion, this degree of control offers little real advantage. Provided the system as installed meets the operating requirements, minor changes in specifications, while noteworthy from a record-keeping perspective, have almost no relevance. The system is as it is, and that is all that needs to be known to establish baseline performance.

Some mention must be made of recent extensions to the jargon of validation; design qualification, vendor qualification, and construction qualification are all terms that have come into use within the last 10 years. Depending upon the scope of the project, these activities have some merit. They should all be considered as options and employed where appropriate. Their overuse can lead to the types of bloated efforts mentioned earlier; only the very largest efforts can benefit from these programs. Briefly, these activities embrace the following:

1. Design qualification—a formalized review of designs at a preliminary point in the project. Its goal is to independently confirm that the design conforms to both user requirements and regulatory, environmental, and safety regulations.
2. Vendor qualification—an evaluation of a vendor to confirm its acceptability for participation on a project. As opposed to an audit it focuses more on the technical capabilities of the vendor.
3. Construction qualification—an ongoing review of construction activities, actually more of a roving quality assurance during the construction of a facility.

One of the major pitfalls in equipment qualification (and perhaps in performance qualification as well) is the use of arbitrary criteria. Equipment and systems can only be expected to fulfill hard quantitative criteria where those requirements were clearly communicated to the supplier or fabricator beforehand. For instance, testing a compressed air system for hydrocarbons is justified only where the vendor was required by specification to supply oil-free air. Another common oversight is to accept equipment performance at the limits of current products requirements rather than to the equipment's capabilities. Consider a drying oven capable of ±5 C across the entire dryer. The qualification should measure conformance to that tolerance rather than a ±10 C requirement for the expected product. There are at least two good reasons for this. First, a premium has been paid to achieve a capability better than the process requirement. That premium should be fulfilled by the supplier. Second, a future application for the equipment may necessitate a tighter range, and checking it at the onset (for no additional expense) eliminates the need to repeat the qualification at some future time. It should be evident in all cases that qualification records should be largely numeric, as this establishes performance in more definitive fashion.

Calibration—Perhaps the simplest of all supportive activities to understand calibration ensures the accuracy of the instruments used to operate and evaluate the process. It is a fundamental cGMP requirement of all regulatory agencies. In general instruments must be shown to be traceable to proven EU/ISO/NIST standard instruments and supported by a defined program with appropriate records. This requirement is extended to include the instruments utilized in the various qualification activities, so that the generated data are of acceptable accuracy. The most prevalent error observed in calibration is a tendency to calibrate instruments in only a partial loop condition. Where this is done, the technician will use a signal generator to simulate the sensing instrument and show that the signal converter, recorder, display, and soon each has the appropriate value. This type of calibration is inadequate in that it fails to consider the effect of the sensor. Correct calibration practice should include placing the sensor at the measured condition and correcting the response at the recording or indicating location.

Analytical method validation—A prerequisite for any validation involving the analysis of the microbiological, physical, or chemical aspects of materials is the use of analytical methods that have been demonstrated to be reliable and reproducible. No meaningful assessment of product or material quality can be made without the use of validated methods. This effort should encompass raw materials, in-process testing, and finished goods, as well as any support provided to cleaning validation. The practices used for validation of analytical methods are well defined and are harmonized under ICH.

This effort must extend to the microbiological laboratory as well, in which validation of methods is essential to assure confidence in the results. This will include appropriate testing environments, laboratory sterilization/depyrogenation validation, equipment qualification and calibration, use of standards, and positive and negative controls.

Performance qualification—those activities that center on the actual product or process being considered. There are other terms, such as process qualification, process validation, product qualification, and product validation, that are sometimes used to narrow the scope of this effort. Here again, semantics have gotten in the way of more important issues. If you desire to use different terminology, go right ahead. Provided all involved understand the intent, the specific title chosen is clearly arbitrary, but the principles are the same. (See Sec. 3.1)

Under the auspices of PQ, we find much of the regulatory focus in regard to validation. Investigators worldwide are far more interested in the validation of water systems, sterilization processes, cleaning procedures, and product quality attributes than anything else in validation. While this might seem obvious, there are firms that have expended far more energy on the equipment and system qualification than they have on the far more meaningful PQ activities. While equipment qualification is important, it must play a secondary role in establishing the validation of a process. Consider the following real-life story:

After successfully operating its WFI system for more than 10 years after initial qualification/validation and ongoing sampling, a firm was inspected for the first time by a regulatory agency that had never been to the facility before. The inspector identified two or three threaded fittings on the headspace of the hot WFI storage tank, and then inquired as to the initial qualification of the system. The firm was unable to provide a sufficient response in a timely manner, and under duress agreed to replace its entire WFI system at considerable expense. All of this occurred while the chemical and microbial performance of the WFI system over its entire operating history had been nothing short of superlative. Surely the satisfactory performance of the system should have been given greater weight and the corrective action limited to a replacement of the fittings and development of an updated qualification for the system as it existed. The inspector and firm both erred in placing emphasis on deficiencies in equipment qualification that were largely unrelated to system performance as measured on a routine basis over an extended period.

There are few if any hard rules in validation practice. Much of what is cited in guidance documents, surveys, compendia, and industry presentations represents a single acceptable practice and should not preclude the utilization of other approaches to achieve the same end. The ends should largely justify the means in this regard, and thus some of the dogma associated with equipment, system, and facility design should be recognized as such. The real evidence of system acceptability is in its performance. Inspectional findings that don't relate to important product attributes should be reduced and greater weight should be placed on what is truly critical. An overview of some of the more common PQ efforts will be presented later in this chapter.

Change control—This is a simple term for what in many firms is a number of critical procedures designed to closely monitor the impact of changes of all types on the product or processes. Clearly, in any market driven company the demand for change is continuous. Moreover, there is also a drive for increased speed in all aspects of the operation. Firms must be able to evaluate changes rapidly for all aspects of their operations (analysis, equipment, environment, process, materials, procedures, software, formulations, cleaning, personnel, warehousing, shipping, components, etc.). The potential scope of changes impacts virtually every operating area and department, and as a consequence change control is considered a difficult program to manage properly. The scope of reviews that are required is such that nearly all programs are multifaceted, with separate procedures as needed to encompass the full extent of change. An aspect of change control that isn't always recognized as such is document control. As documents often serve as the primary repository of information within this industry, procedures that regulate how they are revised are effectively change control programs in another guise. The importance of change control to validated systems cannot be overemphasized. Validation for a system is not something you do, but rather something you achieve through the implementation of the programs listed above within a cGMP environment. Validation can be considered a "state" function, something akin to temperature. How a firm gets to that state of validation is open to considerable variation. The important thing, at least for validation, is that we are able to maintain that state over the operational life of the product, process, or system.

Virtually all companies have the elements described above in one form or another, yet only rarely are they integrated into a cohesive system that acts to support the validation activities within the firm. Validation is not a stand-alone activity; it is one that relies on many of the pre-existing activities within a firm. When properly linked to these other activities its execution is greatly simplified and its impact is more substantive.

2 ESSENTIAL VALIDATION DOCUMENTATION

Process development—development reports, scale-up reports, process optimization studies, stability studies, analytical method development reports, preliminary specifications

Process documentation—master batch records, production batch records, SOPs, raw material, in-process and finished goods specifications, test methods, training records

Equipment qualification—equipment drawings, specifications, FAT test plans, wiring diagrams, equipment cut sheets, purchase orders, preventive maintenance procedures, spare parts lists

Calibration—calibration records, calibration procedures, tolerances

Analytical method validation—validation protocols, validation reports, chromatography printouts, raw data

Performance qualification—validation protocols, validation raw data, validation reports, calibration results for validation instrumentation

Change control—completed change control forms

2.1 The Validation Life Cycle: "Diamonds Are Forever"

The validation life cycle focuses on initially delivering a product or process to managing a project or product from concept to obsolescence [5]. When employing the life cycle, the design, implementation, and operation of a system (or project) are recognized as interdependent parts of the whole. Operating and maintenance concerns are addressed during the design of the system and confirmed in the implementation phase to assure their acceptability. The adoption of the life-cycle concept afforded such a degree of control over the complex tasks associated with the validation of computerized systems that it came into nearly universal application within a very short period.

Applying the life-cycle concept to the validation of systems, procedures, and products is essentially an adoption of the general quality principles of Deming, Crosby, and Juran[6,7,8]. Each of these individuals recognized the inherent value in quality—that it could provide a meaningful return to the organization. Validation practiced in a comprehensive manner using the life-cycle model is wholly consistent with the quality views of the quality "gurus," and properly documented can afford compliance benefits as well. To accomplish this dual objective, validation concerns should be addressed during the design and development stage of a new process or product to afford tighter control over the entire project as it moves toward commercialization. Considering validation during design makes its later confirmation during operation substantially easier, and thus allows "quality" of performance to the greatest extent possible. The use of formal methods to control change must be an integral part of life-cycle methods, as the demand for change is constant and inflexible systems are doomed from the start.

The validation life cycle provides several advantages over prior methods for the organization of validation programs. The cohesiveness of an organization's validation efforts when the life-cycle approach is utilized as an operational model is unattainable in other operating modes.

The benefits of this concept as a means for managing validation are as follows:

Provides more rigorous control over operations
Facilitates centralized planning for all validation-related aspects
Ties existing subelements and related practices into a cohesive system
Establishes validation as a program, not a project
Offers continuity of approach over time and across sites
Affirms validation as a discipline
Results in the centralization of validation expertise
Is compatible with corporate objectives for validation

With this perspective, validation takes on an entirely different meaning. It is no longer something done to appease the regulators; instead it becomes a useful activity of lasting value to the firm.

2.2 Validation in Perspective—Keeping Score

Over the last 30 years or so, this industry has been delayed with reports of failed validation efforts. These have to be viewed from quite a different perspective than one might first adopt. A failure of a process to meet its defined validation criteria might be the result of three possible scenarios: (1) errors in which the process is deficient, (2) errors in which the validation approach itself is deficient, or (3) instances in which both the process and the validation methods are unsatisfactory. Since the desired end results of nearly all validation efforts are known—at least in general if not specifically—a failure to meet reasonable validation criteria must reflect on the process. Where the failure relates to an arbitrary or indirect requirement (as is often the case), then the validation criteria might be overstated. Barring that eventuality, which can be accomplished by adoption of only the simplest and most scientifically correct criteria, then the process itself must be at fault. The validation program thus serves as little more than a scorekeeper. It cannot by itself make a process better than the process actually is. Successful validation implies sound processes; unsuccessful validations should be attributable to underlying deficiencies in the process. Processes that are well designed, reliable, and robust and that operate in well-maintained equipment according to clear operating instructions are likely to be validatable (if there is such a term), while those processes or products that are weak in one or more of those areas will likely fail any attempt to validate them. Provided the validation effort is substantive, it only tells the score; it can't change it.

3 ORGANIZATION AND MANAGEMENT: "IF YOU DON'T KNOW WHERE YOU ARE GOING, YOU ARE LIKELY TO END UP IN SOME UNPLANNED PLACE"

The accomplishment of validation in this industry entails many different aspects of operation and quality control, and therefore raises some of the same management issues associated with the organization of any complex activity. Simply put, validation efforts must be properly managed to ensure their effectiveness. Some of the more common methods and documentation practices are outlined below. These are basic validation requirements expected by the FDA as well as foreign bodies.

Policies—At the highest level this takes the form of policy documents that broadly define an organization's values with respect to validation. These are valuable in that they establish credos by which the firm can operate as well as score as an affirmation of top management's commitment to the exercise. Policies should be written in a way that facilitates meeting global objectives, allows flexibility in implementation, and is useful over an extended period of time. As such they are generally statements of lasting value, and establish the overall tone of the validation program. Their importance is in the mandates they make for the organization to follow. Properly written, these high-level documents should endure over time.

Planning documents—Variously called master plans, validation plans, or validation master plans, these are ordinarily project-oriented to help organize the validation tasks associated with a particular project. Depending upon the scope of the project, the plan can resemble a policy document when written for a division of a firm or look much like a validation protocol when written for a small project. To accommodate these varied uses, the level of detail must vary substantially. The most useful plans are those that are quantitative in nature, since they more firmly establish the intent of the work to be performed. One of the limitations of planning documents is that they are essentially outlines of future work. In this regard, their utility once the project is complete is sometimes nonexistent, and even diminishes further with time. Despite the emphasis placed on the existence of planning documents by some investigators, their utility and importance is largely overrated once the task is completed.

Summary documents—A rather new practice is the validation master summary, in which a firm can outline the completed validation efforts that support its operations. These have the same relationship to the master plan; a validation report has to have a validation protocol. Considering that few investigators will be satisfied with reviewing a protocol when a completed validation report is available, it is surprising that there is so much reliance on planning documents rather than on summary reports to describe validation activities. Properly assembled, a validation summary report is never truly finished. As new studies are added or older ones replaced, the summary should be updated to reflect the latest information. Once fully assembled it can permit the rapid review of a large validation effort in a single document. If a validation program is to be successful it must accommodate change easily, and summary reports are vastly superior to validation plans in that regard. One can look backwards at a number of completed efforts—even those performed at various time intervals—with greater accuracy than one can look forward at future activities. We know substantially more about the past than we will ever know of the future, and therefore validation summary documents should receive far greater emphasis than they presently do. As with validation planning documents, the level of detail provided in the summary can vary with the scope of the effort being described. A concise and summarized valuation study will be audit-friendly and provide regulators with what they are looking for during an audit.

Tracking/management documents and tools—Operating a validation department is no different from other operating units, and project tracking is often required. It is commonplace in a validation department to have multiple tasks underway at various stages of completion at any given time. The department may organize its activities in a variety of ways, but operating schedules, document tracking, status reporting, resource leveling, personnel assignment, and priority designation are all necessary to properly orchestrate the activities. There are various tools that are used for these activities, including project management software (Primavera, Microsoft Project) and documentation systems (Documentum, SAP). These all form an essential part of the documents needed to effectively operate a validation organization. Essential to all of this is the recognition that the priorities within the validation unit must be the same as those of the rest of the organization. Any tool, whether it be software-based or not, that is valuable in maintaining control over the efforts will be useful in keeping the validation efforts consistent with the overall organizational goals. The best validation programs are those that can rapidly accommodate changing issues and evolving problems and that can minimize delays, maximize opportunities, and make optimal use of the organizations resources. To this end, the validation unit must maintain a close working relationship with many different parts of the firm. The use of tracking/management tools can help substantially in that effort.

Protocols—These documents, which originated as "designs of experiments" as outlined in the classic "scientific method," are the foundation for nearly all efforts. They are essential to define the requirements of the validation exercise. The first protocols in this industry were developed almost 30 years ago, and as the underlying science behind our products, processes, equipment, and systems has not changed, it would seem that the need for new protocols should diminish over time. Unfortunately, this has not always been the case, and firms have "reinvented the wheel" many times over. Firms should reuse their protocols (in actuality they constitute a valuable part of their knowledge base) as many times as possible. The validation of such common processes as sterilization and cleaning can be approached using a generic protocol and documented in project-specific reports.

In many cases it is possible to use protocols on a global basis for the same types of products, processes, and equipment. The best protocols are those that rely heavily upon concise, quantitative acceptance criteria wherever possible and avoid such terms as sufficient, appropriate, and satisfactory. The presence of subjective criteria in protocols, whether for qualification or validation, is the source of more problems than perhaps any other. One should also be wary of the excessive use of statistics in the analysis and acceptance of results. Where the underlying specification is derived from a pharmacopeial reference, the statistics may have some merit. Where they are imposed as a secondary criterion in addition to more definitive limits, they are bound to cause trouble.

If the biological indicators are inactivated and the minimum F_0 required is confirmed, there is little to be gained by requiring a tight RSD about the individual F_0 values. Adding acceptance criteria to "fatten" a protocol is perhaps the most egregious sin of all. Validation protocols should delineate a minimum of quantitative requirements linked to specific quality attributes and little else. Anything else is nothing more than useless padding of the effort, perhaps in the hope that volume will substitute for quality. Protocols are best prepared by a single individual with the appropriate education and experience. If properly written, a protocol can be used over a period of many years, because only substantive, and therefore timeless, acceptance criteria should be included. If a protocol incorporates disparate elements—microbiology and computer science perhaps—it is far preferable to prepare two separate protocols, each with its own criteria.

Reports—The validation report is certainly the most critical of all validation documents. It must provide a clear and concise discussion of the completed work that can withstand the scrutiny of reviewers over a period of many years. To that end, the report should emphasize tables and diagrams rather than written descriptions. Clarity of presentation should be the most important goal in each report prepared. The author must avoid the temptation to be creative and verbose in his or her writing. Plagiarism should be encouraged wherever possible. If a particular diagram, paragraph, or presentation model has proven effective in describing an activity, it should be reused. For instance, there should be only one way to calibrate thermocouples, and the narrative on this activity in all reports should therefore be identical. The intent of the validation report is to inform, not to entertain. Boredom on the part of the reviewer may perhaps be unavoidable, but it is preferable to the inadvertent inclusion of errors caused by original prose. Another objective in the report is brevity. (This applies to protocols as well.) There is a general tendency to write far too much and thus require the multiple reviewers to spend more time than is necessary to find the essential information.

For qualification activities, the use of fill-in-the-blank forms as both protocol (when empty) and report (when completed) is almost universal. Some firms have had success with the use of forms in the execution of PQ studies, further simplifying report preparation. Reports should have abstracts, as reviewers may be satisfied with just a perusal instead of an in-depth review. In addition, the absence of an abstract will force an in-depth review of the document. Another useful practice is to circulate only the report, keeping the raw data in a secure location. This can shorten review time substantially, provided the quality unit performs an independent audit of the data. The individual supervising the execution of the study should prepare the report. Unlike protocols, which can be utilized over long periods of time and in many general ways, reports must address a specific set of circumstances. Breaking large reports into smaller elements can be a valuable time saver in their preparation and assembly for review and approval.

Procedures—An underutilized practice in validation is the SOP, whereby repetitive activities can be defined. The use of SOPs increases reproducibility of execution and allows for further brevity in both protocols and reports. Procedures make everyone who is involved with the project substantially more efficient, and should be employed wherever possible. Practices such as calibration of instrumentation, biological indicator placement, sampling of validation batches, and microbial testing are clear candidates for inclusion in SOPs. Among the more innovative uses is the inclusion of standardized validation acceptance criteria for similar products.

Approvals—Each of the documents described in this section is subject to formal control and approval. The best practices minimize the number of approvers, with an ideal maximum of no more than four to six individuals who have the appropriate technical understanding. Of course this must include the quality control unit, which of necessity invests in sufficient training to be able to review and approve a broad range of documents extending over all of a firm's products, equipment, processes, and systems.

Approval by an excessive number of personnel does not mean the quality of the documents is any higher. When a large number of individuals approves a document, there is often a sense that one need not read the document too closely, as if there were any errors someone is bound to catch it. Three or four critical reviews are far more meaningful than cursory signatures from a larger number of reviewers.

3.1 Performance Qualification: Make It Meaningful

The essence of validation is PQ, those studies designed to establish confidence in the product, process, or system. As stated earlier, it is this core activity that all of the other elements support. A "qualified" piece of equipment has no value until it is evaluated in a structured study to "validate" its performance under a specified set of conditions to effect the desired result on a specific product. Without a meaningful PQ, all of the other efforts establish little more than a capability, as opposed to an effective reality.

Qualified technical personnel should oversee PQ studies; however, the execution can be left to specially trained hourly personnel. For larger projects it may be necessary to use a team approach to ensure that the requisite technical skills are available. Evaluation of dose uniformity may require the skills of a quality control analyst, formulation expert, and statistician. In these circumstances, report preparation may have to be split among the team members.

In this era of restricted headcount (and for any major project) it is common to bring in outside assistance to do some or all of the work. This is certainly acceptable, but firms should maintain some degree of internal expertise to oversee any external support. Once the validation project is completed, the firm is required to maintain the validated state. This is only possible when the core capabilities exist within the firm. The use of prepared forms for data entry during execution can be useful as a means to ensure that the correct information is gathered and promptly recorded.

The execution of PQ studies (and even some EQ studies) will often entail reliance upon analytical and/or microbiological testing. As stated earlier, this mandates validated methods in the laboratory to ensure the acceptability of the results. Not to be overlooked here is the laboratory workload itself. As validation testing generally includes an expanded sampling of the product and materials both as in-process and finished goods, each validation batch may represent as much as 10 or more times the testing required in a routine batch. Analyst time and laboratory capacity must be available to accommodate the testing requirements. This is even more critical in microbiological testing, as samples should be tested with minimal delay to avoid perturbing the results upward or downward. For new facilities this suggests that laboratory construction be considered a first priority to ensure their readiness for the testing of samples from the operating facility. This can be offset by the use of outside laboratories to offset peak demand, but this entails other complications, including transfer of methods between laboratories, sample shipment, and so forth. The author has seen extreme cases, in which firms' validation progress have been restricted by analytical/microbiological testing limitations, an unfortunate circumstance which should be avoided if possible.

3.2 Process or Product Validation: "Which Came First, the Chicken or the Egg?"

A common misunderstanding in validation practice is the true relationship between the process and the product. This appears to have its origin at the very beginnings of validation in this industry. The first studies performed focused heavily on the system and methods used in the preparation of parenterals. As such there was considerable attention paid to WFI systems, steam and dry heat sterilizers, aseptic processing, and other processes that assure the sterility of the finished products. As a consequence, little if any attention was paid to the physical and chemical aspects of the filled formulation. The term used for all of this effort was process validation, and thus was borne the impression that confirming (validating) the nonsterility-related aspects was to be accomplished in some other manner. This perception is clearly erroneous; ensuring product quality must embrace all of its key attributes. These attributes are established through conformance to all of the required process controls. The processes used to ensure proper potency, pH, moisture, dissolution, and every other product attribute must also be subject to validation.

The validation of these processes can be established in a manner comparable to the methods used for sterilization procedures. Independent verification of process parameters is used on the commercial-scale process to confirm that the operating parameters (mixing speed, compression force, blending time, etc.) are consistent with the batch record requirements. The sole difference in the validation of these processes is that sampling of the production materials can provide a direct indication of process acceptability. As a key element of process validation, the limitations of product sampling and testing that are a consequence of "sterility" concerns are not present when these other quality attributes are confirmed. Sampling and testing of materials and finished products as used for routine release are inadequate for the validation of these processes.

Product quality is assured through the collection and analysis of samples taken from the process (see Sec. 6 on sampling) to establish the acceptability of the process. The optimal approach to validation considers process parameters and product attributes, as well as their relationship. The link between the defined independently established parameters (either variable or fixed process equipment set points) must be established during the developmental process. The PQ of a pharmaceutical process should demonstrate how conformance to the required process assures product quality. The supportive data should include process data (temperature measurement, addition rate, etc.) and product data (content uniformity, moisture content, impurities, etc.). In many ways this effort resembles the validation of utility systems, in which samples taken from the system are used to establish its acceptability. For sterile products this information must be augmented by the validation studies that support sterility assurance. The nature of sterility makes this support inferential rather than direct, but all of these activities are examples of process validation, and all of them support key product quality attributes.

3.3 Utility and Environmental Systems: "Don't Drink the Water! Look What It Does to the Pipes!"

Among the simpler validation efforts that are performed in this industry are those used for process utilities (water systems, steam systems, compressed gases) and controlled environmental (particle, microbially classed, and/or temperature-regulated environments) systems. In these, the equipment qualification effort documents the baseline operating condition for the system, and in some programs this is supplemented by samples taken from the system or environment under idealized conditions (also called static testing). Some firms include worst-case studies with increased intervals between system regeneration; however, this is by no means universal. Following the EQ, the system passes into a PQ phase in which a defined sampling regime is followed to assess the performance of the system under "normal" use. This phase lasts from 10 to 90 days, depending upon the size of the system and its intended use. Upon completion, a PQ report is issued and the system is accepted for routine usage. In the best firms, periodic trend reports are issued supporting the continued suitability of the system over time. The periodic reports can be issued monthly (environmental systems used in aseptic processing), quarterly (water systems), or annually. It must be recognized after several of these review periods have passed that the system is defended more by its ongoing monitoring than by the initial PQ studies. An older system with well-established controls is thus less likely to experience excursions outside the expected range than a new one with only limited operating history. Newer is only new; it may not in fact be better.

Change control must of course be present to evaluate intentional changes to the system over its operational life. A large change could result in the execution of a new EQ and PQ, while smaller changes can be managed with less intensive efforts.

Some practical suggestions for water system design and validation that will also make it easier to comply with the regulations include the following:

Bring water systems online early in the project; many of the subsequent efforts will rely on the assured reliability of the water system in processing and cleaning as well as in microbial control.

Sample pretreatment locations during the PQ phase to develop a baseline of normal system performance. This sampling is informational rather than directed toward attaining a specific limit.

Focus on microbial attributes rather than chemical sampling in the PQ. After initial flushing, the chemical results at various locations are unlikely to vary significantly, while microbial variances (perhaps due to ease of sampling) are more common.

If at all possible design the system to keep the water hot ($>50^{\circ}C$) and in motion at all times. Biofilms are more likely where water is allowed to stagnate and cool.

Don't bother with sterilization or sanitization of hot systems. These systems are largely self-protective as a result of their temperature, and the added complication of sterilization is not warranted.

4 STERILIZATION PROCEDURES: "THE BUGS DON'T LIE!"

It is widely acknowledged that the first validation efforts in this industry were those directed at sterilization processes, given the pre-eminence of sterility as the most essential of all product attributes. As a consequence, after nearly 30 years of validation activity, sterilization and depyrogenation procedures are perhaps the most thoroughly documented processes within our industry. Within the United States this largely led to a perspective that evolved from the teachings of Dr. Irving Pflug, who has served as the predominant source of sterilization validation "know-how"[9]. Dr. Pflug has schooled a large portion of the world's industry in the principles of sterilization, and a recurring theme in his many lectures and papers is the principle of "The bugs don't lie." The principle, so ingrained to many of us, is that the micro-organism is the best arbiter of the conditions to which it is exposed. This tenet leads directly to the use of appropriately selected biological indicators positioned within the items to be sterilized as a means to directly assess the lethality of the process. Were we to properly use resistant BIs (biological indicators) alone (and virtually no one does), we should be able to establish process effectiveness for sterilization procedures with little or no ambiguity.

4.1 Biological Indicators and Physical Measurements: "He's Not Dead, He's Just Mostly Dead"

Unfortunately, things are rarely so simple. With sterilization processes the opportunities for physical measurements of lethal parameters (temperatures, relative humidity, gas concentration, belt speed, etc.) abound. The ability to collect these data has increased greater, over the years, and with this increase has come a loss of perspective. Since we can so easily collect large amounts of data, we have become increasingly reliant upon them, resulting in sterilization validation protocols that are often cluttered with arbitrary numerical criteria for the collected data. Unfortunately, these criteria are generally given the same weight as the microbiological challenge studies mentioned previously. This should never be the case. Physical data can never be more than circumstantial evidence that a process is effective. Consider the following simplistic example of a rather unique sterilization process:

> A hunter shoots at a standing turkey 100 meters away using a shotgun. The spread of the shot is known to be 2 meters at 100 meters from the gun. An excellent shot, the hunter centers his shot on the turkey and pulls the trigger, and the shotgun fires. Can we conclude from this alone that the turkey will in fact be killed (sterilized), or would we be better served to look at the turkey to see if in fact it has actually been hit by one or more of the pellets and died from its wounds? Tempted though we might be to rely on the technical data, direct evidence of process can only be established by examination of the target.

Knowledge of the physical conditions near an object we desire to sterilize suggests that conditions are appropriate for the intended result, but cannot truly establish that fact. Biological indicators positioned in or on the surfaces we intend to sterilize provide a clarity of result that is hard to dispute. Properly sited the BI must experience the lethal effect of the sterilizing agent in order to succumb. Predicting its death, or worse yet explaining away its survival on the basis of some physical measurement, is wholly inappropriate.

So what value, if any, do physical measurements have with respect to the validation of sterilization processes? Their primary utility is in the comparison of one process to another. This can be done in myriad ways (process to process, load to load, item to item, etc.), and forms the basis for claims of uniformity and reproducibility for the process.

Sterilization validation procedures thus should rely primarily on the results of appropriately designed microbial challenge studies, with physical measurements serving solely as corroborative, but certainly not definitive, evidence of process effectiveness.

Some of the more salient points in the validation of sterile products include the following:

The bugs don't lie; the results of BT studies must be considered more indicative of the process effectiveness than any physical data.

Information on the bioburden present in or on the product is essential to truly understanding the level of sterility assurance provided.

Microbiological tests are substantially less reproducible than chemical tests, thus the quantifiable results of any microbial test are less reliable.

Sampling of materials, surfaces, and so on for micro-organisms can perturb the results; not all contamination originates with the process.

The use of proper aseptic technique is essential in the maintenance of sterility in aseptic processing and is the largest contributor to success.

Much of what has been proposed as finite standards for successful sterile product manufacture is little more than documented prior success. Alternative conditions might be equal to or even superior to the prior success in their performance.

The sterility test is notoriously imprecise and might be more aptly termed the "test for gross microbial contamination."

Isolators need not be perfect for them to supplant manned cleanrooms; they only have to perform at a higher level. Some isolators are approaching that now.

5 PRODUCT QUALITY ATTRIBUTES: "99 AND 44/100% PURE"

This industry makes its profit from the sale of products, the quality of which should be the real focus of the validation efforts. If anything is to be taken to excess in the practice of validation it should be the support we provide to the quality of our products.

> Some years ago when I was head of validation for a major manufacturer, we had a single product that made up 40% of our corporate sales and perhaps 60% of the profits. It was a very simple product formulation; each strength was a different size tablet made from the same blended granulation. Nevertheless, each strength was validated by multiple lots, with at least three of both the largest and

smallest tablets. In addition, retrospective validation studies were completed annually, and even the smallest changes were fully evaluated in additional studies. We even joked that the passage of a large truck nearby might be cause for a new performance qualification study.

That type of fixation on product quality for a single product family might seem excessive, but is certainly preferable to the more prevalent benign neglect that passes for validation of product quality in many firms. Each of the key quality attributes of the product should be established in a validation effort that establishes the consistent conformance to the specifications. As mentioned previously the PQ effort should consist of independent confirmation of process parameters coupled with in-process and finished goods sampling of production materials. The combination of parameter verification with product sampling ties the process to the product.

Figure 1 outlines priorities relative to the validation of products within a firm. The "jewels" are those products of higher quality and profitability, whose value to the organization should be protected with validation studies

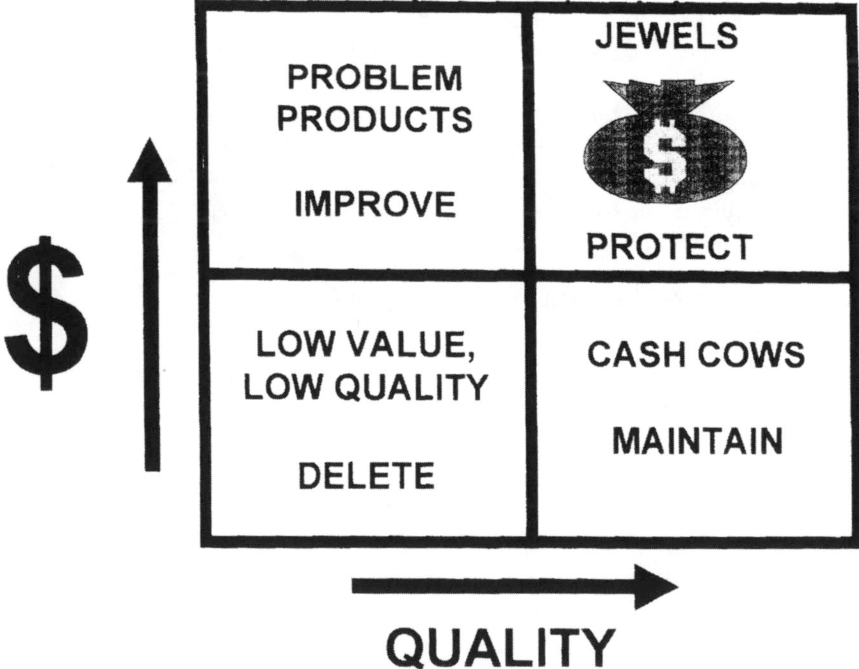

FIGURE 1 Validation—cost and quality implications.

in an effort to maintain their viability. Typically, management gives these products the highest priority. Profitable products with quality issues—"problem products"—should be supported by developmental efforts to resolve their quality problems and convert them to jewels. This effort is usually the second priority. Quality products with limited profitability—"cash cows"—have their status quo maintained to ensure continued compliance. Products that are neither profitable nor of high quality are candidates for deletion from the product line. These clearly have the lowest priority of all and perhaps should not even be on the market.

The conduct of product PQ must be done using the routine process controls, materials, environment, and personnel. The use of worst-case-type conditions, as is common in sterilization, is not warranted. The developmental effort should be focused on establishing "proven acceptable ranges" (PARs) for both operating parameters, which can be selected independently, and product attributes, which are dependent upon those process parameters [10]. For instance, a vacuum drying process might be specified as 12 to 20 h, with the expectation that it will result in a final loss on drying value for the dried material of 1–5%. The developmental effort must establish that material with 1–5% LOD at that stage of the process is acceptable for further processing. Once this is confirmed, the independent process parameter PAR of 12 to 20 h has now been correlated to the dependent moisture content of 1–5% in the material. The experiment can be conducted in either direction. The drying targets may be established first and then the drying times that support it identified, or vice versa. With this knowledge in hand, the firm can now choose to dry every batch for 16 h with the confidence that the final moisture value will meet the required specification limit. Figure 2 depicts the relationship between the independent and dependent variables.

The selection of the independent process parameters (which include the choice of specifications for the raw materials and intermediates) is made during development in an effort to ensure the appropriate response of the dependent parameters. The relationship between the independent and dependent variables need not be linear, and may be inverted. The key is to recognize that the selection of the independent variable influences the dependent variable. While this description is simplistic and ignores the possible influence of other variables, it accurately describes the symbiotic relationship between process and product. Without a process (as defined by the selection of the independent variables), there is no product (with its dependent product attributes). Without a product, there is no reason for the process. The PAR approach describes how one is to develop the relationship between the process and its resultant product. There is no reason to choose one over the other; consideration and confirmation of both is necessary to validate a product.

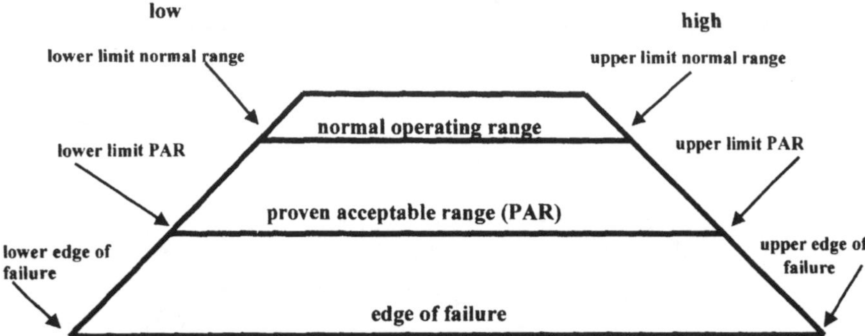

FIGURE 2 PAR approach to process validation.

6 SAMPLING/TESTING: "BLIND MAN'S BLUFF"

To this point the discussion of process and product validation has focused on parameters and attributes. The situation is somewhat more complex with regard to product attributes, as the source of the sample used to assess the attribute can have a profound effect on the results. To understand this, one must explore the genesis of sampling practice as performed in this industry.

The pharmacopiea, such as *United States Pharmacopica* and *Pharmacopica Europe*, are generally considered the most complete guidance on sampling. After all, samples taken from production materials, whether they are raw materials or finished goods, are required to conform to the specifications provided. The focus of this sampling is release-oriented and appears to be driven from the perspective of the retail pharmacist. At that point in the product supply chain there is virtually no distinction among the dosage form units. Each tablet is expected to contain the required potency of active and fulfill all its other quality attributes. On this basis, sampling and testing is clearly random. The origin of the samples tested is unknown, and as a consequence the results can merely confirm or deny the acceptability of the material. The same situation prevails in the plant when samples drawn from the process are composited before analysis. Once the sample location has been obscured, its utility in validation is substantially reduced.

To utilize a blind approach to sampling or random sampling in the validation of a process or product is essentially worthless. Having established the concept or approach of worst case during the design of the validation plan, the corresponding worst-case sampling approach seems fundamentally sound and more appropriate than random sampling. Samples should be taken from the production process materials in an effort to examine those locations within the process equipment or process execution in which the greatest variability might exist. This can be termed "targeted" sampling, with the intent of sampling those locations and events that are more likely to vary from the expected norm. If these samples meet the requirements, one gains substantially more confidence in the robustness of the process. Consider the following real-life process:

> A tablet compression process operates over multiple days on a single shift. A single operator is responsible for the operation of the press over the course of the entire period. During the compression, the operator resupplies the feed hopper and checks tablet weights, thickness, and hardness periodically. The operator also has two short breaks and a lunch period daily. Random sampling of this process might be performed on a timed (every hour) or container (every fifth bulk tablet container basis). Using either time- or container-based sampling (both are "random" methods), substantial variation in the process can be easily overlooked. A targeted sampling might be quite different. Samples would be taken from the press just before and immediately after each new drum is added to the feed hopper, just after each restart (the press being stopped during break, lunch period, and overnight), and every time the feed frame is depleted (as it might be at the end of each working shift). Some of these events might have coincided with either time- or container-based samples, but forcing the sampling toward expected worst-case events and clearly documenting them can substantially enhance the utility of the samples taken.

Targeted sampling can be applied to many different process and equipment situations. A simple analysis of potential process- and equipment-related process variation is used to identify locations to be used in a targeted sampling. Identification of each sample is retained through testing. Sampling from worst-case locations in this manner can significantly increase confidence in the process's robustness and ruggedness over random sampling. Some additional examples of its application are as follows:

Sampling all four corners and the center of every shelf in a freeze dryer

Sampling a suspension-filling process after every interruption in the filling process longer than 1 min

Sampling the bottom of the first drum and top of the last drum in a bulk subdivision

Perhaps the greatest utility of targeted sampling occurs when the expected results are not achieved. In a random sampling mode, the absence of sample source information means that troubleshooting must begin without any insight into potential causes for the aberrant result. In contrast, if targeted sampling had been performed, process correction is already simplified, as knowledge of which location, step, or event led to an unwanted outcome can lead to more rapid resolution of the problem.

In any case, it should be clear that the validation exercise should encompass significantly more samples than are ordinarily used for product release. According to USP, typical production batches can be released based upon the results of as few as 20 dosage form units. Few if any practitioners would accept such a small sample size in a validation exercise, in which the number of samples is typically at least tripled (3×) over the normal release size. Other firms have adopted even larger multiples, with 5×, 10×, and even 20× having been reported. From a statistical perspective these larger sample sizes provide substantially more confidence than is available in a typical release decision.

7 CONCURRENT VALIDATION: "ONE AND ONE AND ONE IS THREE"

In a perfect world—one with unlimited resources—all validation is performed prospectively; three trials are performed and the results are reviewed and approved before commercial use of the process or system. In actuality, there are numerous instances in which concurrent approaches must be adopted, including preparation of clinical supplies, manufacture of orphan and expensive drugs, manufacture of low-volume products, and minor process changes to established products. For reliable processes, there is actually little difference between prospective and concurrent approaches. The results of the validation exercise, whether available from three batches produced over a longer period of time or closely spaced, should be the same if the underlying process is in a state of control (recalling that validation is merely a means of keeping score).

For most products, a properly structured PQ protocol will require rigorous acceptance criteria, whether employed concurrently or prospectively. A protocol with vague requirements will provide very little information about a process or system. The application of targeted sampling approach methods in conjunction with 3× or higher sample sizes will result in the strongest support to process reliability. Concurrent validation can be used almost interchangeably with prospective approaches. There are, however, some constraints with regard to its application.

It is not appropriate for sterilization or depyrogenation processes.
It requires the use of use rigorous quantitative acceptance criteria.
Sampling should be performed using a worst-case approach.
No intentional change to the process should be permitted without considering recall of prior materials.
Failures require documented investigation and possible recall.
Consider larger sample sizes than prospective efforts to achieve greater confidence in the temporary absence of corroborative data from subsequent production runs.
FDA prefers prospective approaches because process reliability is established before release to the marketplace.

Properly structured concurrent validation is nothing more than prospective validation in slow motion. The use of rigorous criteria and a worst-case sampling approach can assure the suitability of the process and acceptability of the product. The fact that it should be performed three times is more a product of regulatory safeguards and FDA guidance than a scientific necessity.

8 CLEANING VALIDATION: "THE BABY OR THE BATHWATER?"

Cleaning validation emerged in the early 1990s as a *cause celebre* across the industry. When first discussed by regulators, industry often adopted a "deer in the headlights"-type response. The task was perceived to be so overwhelming that there was little or no chance of avoiding an adverse comment on the part of the investigator. Over the course of the intervening decade, the issues have largely been resolved. Sampling techniques, analytical methods, and limit selection has been the subject of enough discussion that little mystery remains.

Perhaps the simplest advice that can be given in with regard to cleaning and its eventual validation is to recognize that cleaning is an essential component to the process that must result in effectively cleaned equipment. The equipment, material controls, procedures, documentation practices, and personnel aspects thus should be equivalent to those used for the execution of the production processes. If addressed in that fashion, the cleaning processes can be validated far more easily because the controls applied to them are identical to those used for the production process. An investment in this level of detailed control and knowledge of cleaning procedures is justifiable since it ensures reproducibility of the process or end result and the concomitant ease of validation that results. (Refer to Chap. 8 for an extensive discussion on cleaning validation.)

9 CHANGE CONTROL: "I JUST VALIDATED MY PROCESS; I CAN'T DO A THING WITH IT!"

As a dynamic and interrelated function, validation is subject to disruption if there is a change to any of the elements that define that dynamic state. Changes to equipment, materials, procedures, or personnel could all result in disturbance of the validated state. According to the cGMP regulations, firms must have in place procedures that assess and document the impact of recognized changes to their operating systems and controls. The results of these assessments can be relatively benign (where the change has been deemed to have no impact on the process or product) or extensive (where the extent of the change is so widespread as to force an entirely new qualification or validation effort). Change is so pervasive in our industry and within any given operation that the establishment of an effective mechanism to evaluate all of its nuances is essential to maintaining a validated status. The best change control programs are all-inclusive and capture all types of changes, while at the same time ensuring the rapid evaluation of each change by appropriately qualified personnel. Considering that changes to component and material specifications, physical equipment, and computer software could all impact the validated state suggests that firms assess change in the broadest manner possible. (Refer to Chap. 10 for a comprehensive discussion on change management.)

10 REVALIDATION: "WHEN IS IT REALLY NECESSARY?"

As described in an earlier section, the validation status of a product, process, or system can be significantly altered by a change impacting any of the surrounding or core conditions that contribute to a state of validation. This suggests that a firm should be able to validate its operations once and rely on its change control program(s) to ensure their ongoing sustainability and acceptability. While technically correct, this approach is largely frowned upon by regulators who require that firms establish a revalidation program in which validated systems and processes undergo periodic reassessment of their validated conditions.

Current industry practice is to revisit sterilization and depyrogenation processes on an annual basis. This should be done using a preselected worst-case load in each sterilizer. Reliance on empty chamber studies is considered by most to be inadequate, as those studies typically lack the sensitivity to assess subtle changes in the sterilization process efficacy.

Utility and HVAC systems are among the easiest to address with respect to revalidation. Their performance is evaluated on a near continuous basis through the collection and testing of samples taken from the system. This affords a direct and ongoing assessment of the system's acceptability for use. Coupled with effective change control, calibration, and preventive maintenance programs, the collected data from the system should support its continued use. The preparation of summary reports on results from the system on a monthly or quarterly basis precludes the need for further evaluation.

Production processes and cleaning represent the last major concern with regard to revalidation. There is no widely accepted period to be used for revalidation for these types of processes, with periods ranging from 1 to 5 years mentioned at industry gatherings. A shorter period between revalidation is warranted for the highest volume/profitability products and processes, with successively longer periods for lesser products, processes, and systems. These studies usually take the form of a single-lot study using the same acceptance criteria as the initial study.

Beyond the periodic evaluation of the product, process, or system it is important to assess that the other elements that contribute to the validated state are still in place. This would include an assessment of change controls, calibration, product annual reports (if applicable), process deviations (waivers, alerts, etc.), physical inspection of the equipment, and an evaluation of relevant regulatory guidance. All of these help to augment the data generated in the revalidation study and significantly support the dossier proving successful validation maintenance.

Demonstration of satisfactory validation maintenance is an exercise of ongoing monitoring and documentation. Comprehensive change control documents with linkages to the relevant qualification or validation summary report that supports the acceptability of the change will adequately serve as proof of validation maintenance. In the absence of change (which can only be discerned through ongoing monitoring of the process or system), a periodic and well-documented audit of the validated system or procedure may be sufficient. The audit should include as a minimum a review of system, product and process performance, production variance, preventive maintenance, calibration, equipment downtime, relevance of existing qualification and validation documents to current standards, and physical inspection of the production environment.

11 COMPUTERIZED SYSTEMS VALIDATION: "IS THE INDUSTRY INTERPRETING FDA'S CURRENT EXPECTATIONS CORRECTLY?"

A major area of activity in the mid-1980s—computerized systems validation—evolved to a relatively calm area of validation by the early 1990s. With the double whammy of Y2K and 21 CFR 11, its cause for concern re-emerged. With the change of the century, Y2K mania largely subsided; however, the challenges associated with 21 CFR 11 compliance have not. A seemingly simple issue, establishing the validity of electronic records and signatures, has mushroomed out of control. A lack of perspective, first on the part of the FDA and later on behalf of the industry, has resulted in huge issues (and commensurate expenses) for users of computerized systems across FDA-regulated industries.

Narrowly focused, the tenets of 21 CFR 11 are indisputable: that firms must establish and maintain the integrity of their electronic information. What has been lost is the clear indication of which electronic information the requirements apply to. In my opinion, data generated outside a computerized system, that are manipulated by that system and are ultimately available in hard copy should not be subject to this ruling. Validation of the computerized system should be more than sufficient to establish that the final documentation accurately reflects the input information. In that instance, the computerized system is little more than a tool whose functionality can be readily established, yet firms are endeavoring to assure 21 CFR 11 compliance for numerous systems in which the computer is little more than an adjunct to the cGMP activity and corresponding hard copy. Batch record preparation, SOP and test method archives, and many process equipment control systems are examples of systems in which requirements for 21 CFR 11 compliance appear excessive. A system that processes or communicates data or records and subsequently retains and stores hard copies should not be subject to the 21 CFR 11 requirements. Far too many systems are being unnecessarily held to the very restrictive portions of 21 CFR 11. Electronic record retention as defined in 21 CFR 11 has its place, but not necessarily in every computerized system used within the industry.

One of the other difficulties with computerized systems validation is the range of system validation requirements for different types of systems. A simple approach is to separate the computerized systems according to a defined hierarchy. This serves to reduce the validation requirements for systems with minimal cGMP impact. A possible approach is outlined below.

All computerized systems are reviewed from a validation perspective. Depending upon the extent of the cGMP functionality performed by the computerized system and the criticality of those functions, the extent of the validation performed would vary. The following three categories of validation performed on computerized systems seem to fit most situations:

Validation review—Systems considered for validation review will likely have minimal cGMP involvement, and failure of the system is expected to have no significant effect on cGMP compliance. Typical systems in this category include conveyors, refrigerators, accumulation tables, bail banders, and palletizers. Annotated source code and operating code (where customary for this project), version numbers, and complete vendor documentation are required for all systems. It is expected that the proper functionality of these systems can be confirmed during EQ activities. Given the proliferation of computerized controls for even simple equipment, this category may make up the majority of systems in a facility.

Targeted validation—Systems suitable for targeted validation are those performing some cGMP functions and whose failure during operation could affect cGMP compliance. Systems falling into this category include filter integrity apparatus, electronic scales, vial capper, labeling, and machines. Systems in this category will be subjected to all of the requirements for validation review delineated earlier. In addition, each computerized system will be subjected to focused factory acceptance testing to confirm its conformance to the design criteria. In most cases, specific software program steps or modules will be reviewed to ascertain the identified cGMP functions. Acceptance testing of these systems from a control standpoint shall focus on those functions of the equipment that impact cGMP. Some of the systems included under this category are "packaged systems," in which the software and hardware are duplicated in other systems sold by the same vendor.

Comprehensive validation—Systems in this category may perform a number of critical cGMP functions, (e.g., formulation, washing, sterilizing, and filling). Systems within this category are PLC-assisted manufacturing, washers, sterilizers, and filling machines. These systems will be subject to validation of their complete functionality in accord with established industry practices. As these systems may have numerous refinements and modifications specific to a particular facility, they require extensive documentation and testing during system design, development, and integration. Validation packages for these systems will be representative of industry norms for systems designed for a single customer. In addition, validation of these systems will include an audit of the vendor's software quality assurance program, to include at a minimum change control, personnel qualifications, and documentation standards.

Once the systems from the three levels are validated, the same levels of security, change control, and disaster contingency will be applied to all. (See Chap. 7 for a thorough examination of computer systems validation and additional discussion about some of its more challenging compliance aspects. Also see Chap. 7 for a comprehensive examination of FDA's expectations as they relate to computer system validation).

12 THE QUALITY AND COMPLIANCE BENEFITS OF VALIDATION

Those of us who worked in this industry when validation was first introduced in the mid-1970s, were at first dismayed by the imposition of this new requirement. Once we came to grips with what had to be done, there was a growing sense that validation was perhaps much more than a regulatory requirement. Awareness of the teachings of Juran, Deming, and Crosby led many in this industry to believe that validation could become an inherently beneficial activity related to true quality and forward thinking.

The following are some positive quality and compliance outputs often resulting from validation efforts:

Substantial reductions in batch rejections, reworks, reblending, refiltration, resampling, and resting. This is a clear benefit resulting from a validated process or system, coupled with the establishment of proper process controls.

Reductions in process cycles and consequently utility costs is certainly possible with optimized processes using appropriate nonarbitrary controls.

Increased throughput resulting from the elimination of excessive controls that non-validated processes are frequently subjected to.

A streamlined process, offering enhanced troubleshooting when needed.

Product complaints are often reduced, since a validated process offers more consistency and opportunity for streamlining.

Validation of a process often results in the reduction of in-process and finished-product testing requirements. Process controls established during validation may prove sufficient to assure product quality without excessive quality markers throughout the process.

Awareness of how process parameters and product quality attributes are related often results in more rapid investigations into process failures, glitches, and upsets.

Formalized qualification efforts can provide more rapid start-up of equipment and processes.

The same equipment and utilities qualification efforts can serve to facilitate the maintenance of said equipment, because of the availability of well-structured, concise, and meaningful qualification documentation coupled with monitoring data.

Validation activities throughout a facility force greater personnel awareness of established and formalized procedures and controls. The fact that the equipment is qualified and processes and systems are validated often fosters enhanced performance by everyone involved. The burden of reliability and reproducibility has been shifted from the employee onto the validated system or process.

Validated processes allow for future automation of that same process. Automating a nonvalidated process or system is largely meaningless and potentially disastrous.

Across the industry, there is a clear impression that since process validation is a regulatory requirement the possibility of any financial return has been eliminated. As a consequence, it has taken the industry a long time to recognize that there is a commercial advantage and tangible financial benefit to validation activities.

If we accept that Juran, Deming, and Crosby are correct in their views on quality it should be clear that *validation must be an inherently valuable activity.* It must be viewed as more than a regulatory requirement a mechanism for significantly enhanced process control. The pressures to increase profitability without compromise to product quality require enhanced methods for product preparation and production. This can be achieved today through the employment of a sound validation program.

13 BENEFITS OF VALIDATION

Reduction in rejection, reworks, resamples, retests, reblends, refiltration, etc.

Reduction in utility costs

Increased throughput

Fewer complaints

Reduction of in-process and finished-product testing

Expeditious investigations

Speedier start-up and consistent performance of new equipment

Easier scale-up from development

Enhanced employee awareness

Reduced failure rates

Rapid troubleshooting
Automation potential

14 PERSPECTIVES ON IMPLEMENTATION AND EXECUTION

The remainder of this chapter will address the management of validation in a contemporary setting. The challenges associated with executing validation are similar to those faced by the industry in other areas. In no particular order, these challenges are ensuring compliance for all operations, minimizing time to market for new products and managing to accomplish these in the face of resource limitations, and maintaining the systems and processes validated in order to avoid revalidation.

15 MAINTAINING COMPLIANCE

Compliance is so central to industry's perspective on validation it is sometimes difficult to remember that there are other motivations for it. (see preceding section). Compliance is nevertheless essential and can be aided by following some basic tenets.

15.1 Maintain a cGMP-Compliant Facility

Some of the measures used to keep a facility in compliance require a firm to

Execute frequent internal and/or third party audits.
Provide adequate personnel training in applicable regulations, SOPs, and guidance.
Establish and enforce meaningful quality standards throughout the operation.
Actively partner with and involve the firm's vendors, suppliers, and contractors.
Ensure that all subcontracted production activities meet the same standards as internal operations and that their operations are treated as an extension of the contracting company.

15.2 Honor All Commitments

When working with regulatory agencies make every effort to:

Fulfill the company's obligations by meeting all promised dates for any corrective compliance actions.

Respond to regulators in a timely manner (generally within 14 to 21 days of any communication from them).

Be nothing but honest in all dealings. This does not mean it is necessary to disclose everything.

15.3 Work Proactively

Staying ahead of the curve is always a good idea.

Be vigilant and stay fully apprised of new developments by actively following regulatory actions imposed upon other firms. This allows a firm to anticipate potential new requirements and specific areas of interest during inspections.

Build meaningful relationships with regulatory officials. In working with any regulatory body, establishing a constructive working relationship will prove extremely beneficial. Learn the prevailing perspectives of the investigator, compliance officer, and district director.

Build a cooperative partnership between the firm and the district and, when applicable, headquarters.

Seek guidance and clarification from the FDA early on in the process. Do not attempt to guess at its expectations.

15.4 Minimizing Time to Market

The importance of being first to market with a new product in the pharmaceutical, device, or biologics industries almost invariably leads to a larger market share and higher profits. The premium is such that being able to launch products rapidly can translate to enormous profits. Some areas that significantly support rapid launch of new products are as follows:

Project planning and management—Develop master plans for larger validation projects to provide formalized structure to the effort. Assign clearly defined roles and responsibilities to personnel or consultants for all qualification and validation tasks. Recognize when it's a small enough task that can be managed without a formalized planning effort. Integrate tasks wherever possible. (Combine installation and operational qualification into equipment qualification and include PQ for simple equipment.)

Demand more from R&D—There is no replacement for good science, yet at the same time some critical R&D objectives must be consistent with those of operations where the development of a commercially viable and robust process is concerned. This can be aided by continued interaction between R&D and operations during the early stages of development. Wherever possible, use standard technology transfer templates for all new products and develop processes to fit preexisting production capabilities rather than aggressively adopting new technologies.

Standardize production and quality control practices—The use of consistent methods, materials, and practices for formulations, and packaging keeps costs down and decreases the time required to bring new products on-stream. A logical extension of this is the use of identical process and test equipment in all plants, consistent test methods, and SOPs (including cleaning). A single set of documentation practices across the entire firm can go a long way toward making everything progress more rapidly.

Enhance Communication—The key players of any project should be in continual communication with one another. Production, quality control, R&D, engineering, and regulatory affairs must be able to work effectively as a team. Holding periodic technology transfer meetings for each product led by launch facilitators can help ease communication and accelerate the launch.

Accept some risk—Adopting the most conservative approach in all decisions is certainly safe, but it is rarely ever quick. Firms must be willing to accept some degree of uncertainty in their planning and execution. More data can always be gathered, but the ability (or perhaps the conviction) to make sound decisions based upon less information should be encouraged. This may mean re-education of internal and external personnel when necessary. If the decisions are based on sound science, then the firm should be prepared to stand behind and defend what it believes.

Track and report your progress frequently—The priorities in the validation area must match those of the operating departments. This can be achieved through the distribution of updated, periodic issuance of validation status and activity reports.

Shorten approval processes—The time required to secure approvals of validation documents should be kept to a minimum. One way to accomplish this is to minimize the number of review and approval signatures required. Typically no more than five or six should be necessary. All reviewers must have sufficient technical background to understand the documents they are reviewing. In every instance, the quality assurance unit must be a part of the formal approval process.

15.5 Resource Limitations

Most firms are concerned with the appropriate management of limited resources; therefore it is essential that firms effectively explore the various ways of saving wherever they can without cutting quality. The following are a few ways in which this can be accomplished:

Avoid excessive documentation—While documentation is one of the keys to cGMP compliance (at least according to FDA), there is nothing that requires that paper be used in excess. Incorporation of previously documented work should be encouraged. A well-articulated presentation of a process or procedure should be reused wherever possible. Pictures, diagrams, and tables are worth substantially more than 10,000 words in the presentation of technical information. Use procedures, templates, and forms extensively to reduce both execution and documentation time periods. Use a minimum of clear and concise prose at all times. "War and peace"-type efforts should be avoided at all times. Quality and clarity are far more important than quantity.

Leverage Resources—Make maximum utilization of all available resources. Several firms have been successful with part-time workers who have been specially trained to conduct validation studies (e.g., place and recover BIs, calibrate thermocouples, take samples). The validation workload can be shared with other departments by having them complete documentation and collect samples. Use external help such as graduate students and retirees during peak workload periods, provided they are adequately trained prior to setting foot in the plant. Obtain vendor and supplier assistance in the I/O and even PQ of equipment and processes.

Accept only value-added activities—The execution of design qualification is generally necessary only in the largest-scale projects. Specification qualification as a separate activity is rarely necessary. Both of these activities have only limited utility in the majority of validation projects. The repetition of FAT/SAT tests requirements in I/OQ efforts is extremely wasteful. The equipment is highly unlikely to change dimensions in transit from the vendor's site to yours. Avoid superfluous requirements; the color of the insulation is useful to the electrician repairing the system, but is largely meaningless in a qualification effort. The repetition of I/OQ tests is rarely if ever beneficial. Qualification and validation execution is not batch manufacturing; double signatures are not required on all entries.

Make effective use of contractors—All contractors should be provided with a well-defined scope of work. Their performance against requirements should be monitored frequently. Hire real expertise wherever possible; the low bidder may not be fully capable, and true talent is rarely inexpensive. Be on the alert for add-on charges such as administrative fees. Be wary of bait and switch tactics where the A team bids on the job, but only the C team is ever on site. Use a completion-based payment schedule to keep contractors performing fully.

Out-source activities—Don't try to do everything internally; consultants, retirees, college professors, and graduate students can all be used to supplement your permanent workforce for critical tasks. Contract services are more available than ever and can provide calibration, maintenance, and analytical testing support. Contract manufacturing firms frequently can assist a firm with the validation of activities.

Validation maintenance—A comprehensive program supporting validation maintenance can be of substantial benefit in reducing resource requirements. It seems logical that the revalidation of a process, product, or system should be accomplished with less effort than the initial effort. Additionally, maintaining a system, utility, or process validated will be the most cost-effective approach. By maintaining systems, equipment, and products in a validated state (with the support of a monitoring program, change management, calibration, and preventative maintenance), only periodic revalidation will be required. This has the added benefit of substantially improving the firm's compliance posture, along with its bottom line.

16 REGULATORY REALITY CHECK

The following FD 483 observations were drawn from real-life inspections. Included with each is a brief analysis and recommendation for correcting it and perhaps preventing its occurrence in the first place.

Observation—"The revalidation of the ... tablet manufacturing process was inadequate in that only one batch was subjected to limited validation testing for a change in the order of charging raw materials and the addition of a screening process for the pre-blending step. This process change was initiated after two batches had failed blend uniformity specifications."

Analysis—The firm had a substantial production problem and took corrective action; however, in its review of the process it considered the changes it had made as being of minimal impact and subjected only a single batch to a formalized validation study. The investigator was clearly of the opinion that additional batches should have been produced to support the changes made. In this instance I concur, as a change in order of addition can have a profound effect on the process. I suppose the firm believed that since the problem was resolved by the changes they made that only a single batch would suffice. By failing to support the reproducibility of the process in repeat studies they erred. Changes to processes, equipment, and products can be major or minor, and it should be easy to distinguish between them when we encounter extreme circumstances such as this. In general one should err on the side of safety and do more than the minimum required.

Observation—"The firm failed to complete an equipment qualification (IQ/OQ) for their new encapsulation machine prior to use. The firm lacked data to assure that this encapsulation machine was installed and working correctly per manufacturer's specifications."

Analysis—Requirements for equipment qualification have been in existence for many years. The FDA outlined the basic requirements in the *Guideline on General Principles of Validation* [11]. There is really no excuse for the firm to have omitted this for a new piece of equipment. Note that FDA used the term equipment qualification, suggesting a combined approach would be acceptable. The FDA guidance mentioned above speaks only to equipment qualification, not IQ or OQ.

17 CONCLUSION

What can be done to ensure validation is properly managed? The application of sound science and proven engineering principles with a healthy dose of

Validation myths	Validation realities
Invented by regulators.	Closely related to TQM concepts.
Driven by an absolutist mentality.	No absolutes—only with validation of sterile processes.
Not considered a value-added activity.	
Costs too much and offers no financial reward.	Provides tangible benefits.
Performed to appease the regulators.	The long-term benefits out weight the initial costs.

common sense can go a long way to keeping a firm's validation program on track. Like any other important activity in a complex organization, validation must be properly managed. This is especially true given the multi-disciplined and cross-functional nature of many qualification and validation activities. Validation is substantially easier to manage and support when it is an integral part of the day-to-day operation of the facility.

18 VALIDATION MISCONCEPTIONS AND REALITIES

Some typical misconceptions exist relative to validation that by this day and age should be clarified. Among them are the following:

Myth—Validation is one-time activity, which once completed can be largely ignored.

Truth—Validation is a journey, part of a dynamic, long-term process, not a final destination or stopping point. It's actually a beginning or starting point for continuous improvement and streamlining efforts. It should be recognized as a required and integral part of every procedure, process, or system used for a cGMP purpose. Validation is best performed when it is practiced in a life-cycle model, using a cradle-to-grave approach. This provides the maximum benefit in compliance and finance. Remember that validation maintenance is essential even when the validation process has been completed.

Myth—A small cadre of individuals can accomplish validation with minimal intrusion on the rest of the organization.

Truth—Validation impacts everyone's job to some degree. Each portion of the organization must contribute to the overall validation effort to ensure success. No single organization unit can hope to satisfy all the requirements (and reap the benefits) alone.

Myth—Too much is made of validation in this industry. If we are GMP-compliant, following procedures and making quality products, validation can be kept to a minimum.

Truth—Validation does not replace the need to do any of those things, it merely helps us do them in a more consistent manner. Validation is an essential part of forward-thinking quality and compliance.

Implementing and maintaining a firm in a validated state can at times seem like an impossible task, but the steady application of sound science and proven engineering principles throughout the program, coupled with a healthy dose of common sense, should ensure success. Validation should not be performed to appease the regulators. We must recognize that validation can be an inherently useful activity that can provide meaningful benefits to a firm when applied in a rational manner. Application of sound science with due consideration of cGMP should result in validation efforts focused on the important aspects of equipment, processes, and products. Acceptance criteria should target the collection of meaningful data rather than requiring the accumulation of every conceivable piece of information. Acceptance criteria should also be largely quantitative and not couched in such terms as suitable, appropriate, and reasonable. The focus of the effort should be on those aspects that directly impact on the identity, strength, quality, and purity of the product. Practiced in this fashion, validation is an activity that is beneficial to any organization: "The difficulty that the industry is experiencing with validation is not in the basic precepts of validation, which are laudatory and difficult to dispute. The difficulties arise when those principles are implemented in excess by individuals who do not understand them" [12].

19 WORDS OF WISDOM

Validation is a journey, not a destination. A firm is best served when it recognizes that validation is a required activity necessary throughout the life of a process, piece of equipment, or system.

Validation is an inherently beneficial activity if we can overcome its regulatory heritage. It has a lot in common with the total quality management concepts prevalent in other industries.

The focus of validation should be on the product and its defined quality attributes. The greatest abuses have come where validation has been applied to aspects remote from any impact on quality, identity, purity, and strength.

Some risk is unavoidable in every validation study. The only way to be totally certain that every dose is safe is to destroy it during analysis.

Microbial tests are substantially less certain than chemical ones, especially when they entail quantification. Microbial limits must be looked at with some skepticism; the real value might be something quite different—either higher or lower—than the reported one. Consider any "absolute" number with some caution; the answers are never as clear as they might seem with microbiology. When it comes to enumeration, microbiology can be less than a pure science.

Ongoing monitoring, change management, calibration, and preventive maintenance are essential components of validation maintenance.

RECOMMENDED READING

Validation

Agalloco, J., Carleton, F., eds. *Validation of Aseptic Pharmaceutical Processes.* 2nd ed. New York: Marcel Dekker, 1998.

FDA. *Guideline on General Principles of Process Validation.* 1987.

Sterilization—General

Pflug, I.J. *Syllabus for an Introductory Course in the Microbiology and Engineering of Sterilization Processes.* 4th ed. St. Paul, MN: Environmental Sterilization Services, 1980.

Perkins, J.J. *Principles and Methods of Sterilization in Health Sciences.* Springfield, IL: Charles Thomas, 1973.

Russell, A.D. *The Destruction of Bacterial Spores.* New York: Academic Press, 1982.

Steam Sterilization

Ball, C.O., Olson, F.C.W. *Sterilization in Food Technology.* New York: McGraw Hill, 1957.

Parenteral Drug Association. *Validation of Steam Sterilization Cycles.* technical monograph no. 1. Philadelphia, 1978.

Parenteral Drug Association. *Moist Heat Sterilization in Autoclaves: Cycle Development, Validation and Routine Operation.* PDA technical report 1, revision, draft 11. Bethesda, MD, May 2001.

FDA. Current Good Manufacturing Practices—Large Volume Parenterals (proposed). Title 21, Part 212. June 1, 1976.

Owens, J. Sterilization of LVPs and SVPs. In: *Sterilization Technology: A Practical Guide for the Manufacturer's and Users of Health Care Products.* Morrissey, R.F., Philips, G.B. (eds.) New York: Van Nostrand Rheinhold, 1993.

Agalloco, J. Sterilization in place technology and validation. In: Agalloco, J., Carleton, F.J. (eds.). *Validation of Pharmaceutical Processes: Sterile Products.* New York: Marcel Dekker, 1998.

Dry Heat Sterilization

Wood, R. Sterilization with dry heat. In: Morrissey, R., Phillips, G.B. (eds.). *Sterilization Technology: A Practical Guide for Manufacturers and Users of Health Care Products.* New York: Van Nostrand Reinhold, 1993.

Parenteral Drug Association. *Validation of Dry Heat Sterilization & Depyrogenation Cycles.* technical monograph no. 2, Philadelphia, 1980.

Colman, L., Heffernan, G. Dry heat sterilization and depyrogenation. In: Agalloco, J., Carleton, F. (eds.). *Validation of Aseptic Pharmaceutical Processes. 2nd ed.* New York: Marcel Dekker, 1998.

Gas Sterilization

Bekus, D. Validation of gas sterilization. In: Agalloco, J., Carleton, F. (eds.). *Validation of Aseptic Pharmaceutical Processes. 2nd ed.* New York: Marcel Dekker, 1998.

Burgess, D., Reich, R. Industrial ethylene oxide sterilization. In: Morrissey, R., Phillips, G.B. (eds.). *Sterilization Technology: A Practical Guide for Manufacturers and Users of Health Care Products.* New York: Van Nostrand Reinhold, 1993.

Sintin-Damao, K. Other gaseous sterilization methods. In: Morrissey, R., Phillips, G.B. (eds.). *Sterilization Technology: A Practical Guide for Manufacturers and Users of Health Care Products.* New York: Van Nostrand Reinhold, 1993.

Radiation Sterilization

Herring, C., Saylor, M. Sterilization with radioidotopes. In: Morrissey, R., Phillips, G.B. (eds.). *Sterilization Technology: A Practical Guide for Manufacturers and Users of Health Care Products.* New York: Van Nostrand Reinhold, 1993.

Cleland, M., O'Neill, M., Thompson, C. Sterilization with accelerated electrons. In: Morrissey, R., Phillips, G.B. (eds.) *Sterilization Technology: A Practical Guide for Manufacturers and Users of Health Care Products.* New York: Van Nostrand Reinhold, 1993.

Reger, J. Validation of radiation sterilization processes. In: Agalloco, J., Carleton, F. (eds.). *Validation of Aseptic Pharmaceutical Processes, 2nd ed.* New York: Marcel Dekker, 1998.

Gordon, B., Agalloco, J., et al. Sterilization of pharmaceuticals by gamma irradiation. *J. Parenteral Sci Tech* technical report 11. 42 (suppl.): 1988.

REFERENCES

1. Chapman, K.G. A history of validation in the United States: part I, 15 (10), 82–96, Oct. 1991, and part II. *Pharm Tech* 15(11): 54–70, Nov. 1991.
2. FDA. *Guideline on General Principles of Validation.* 1987.
3. Agalloco, J. Validation—yesterday, today and tomorrow. *Proceedings of Parenteral Drug Association International Symposium*, Basel, Switzerland, Parenteral Drug Association, 1993.
4. Agalloco, J. Computer systems validation—Staying current: Change control. *Pharm Tech*, 14(1), 1990.

5. Agalloco, J. The validation life cycle. *J Parenteral Sci Tech*, 47(3), 1993.
6. Deming, W.E. *Out of Crisis*, Cambridge, MA, MIT Press, 2000.
7. Crosby, P.B. *Quality Is Free: The Art of Making Quality Certain.* New York, McGraw Hill, 1979.
8. Juran, J.M. et al. *Juran's Quality Handbook*. New York: McGraw Hill. 1999.
9. Pflug, I.J. *Microbiology and Engineering of Sterilization Processes, Tenth Edition*. Environmental Sterilization Laboratory, Minneapolis, MN, 1999.
10. Chapman, H.G. *The PAR Approach to Process Validation*. 1984.
11. FDA. *Guideline on General Principles of Validation*. 1987.
12. Agalloco, J. Letter to the editor. *PDA Journal of Pharmaceutical Science & Technology* 50(5): 276, 1996.

4

Validating Analytical Methods for Pharmaceutical Applications: A Comprehensive Approach

Paul A. Winslow and Richard F. Meyer
Quantitative Technologies, Inc., Whitehouse, New Jersey, U.S.A.

1 INTRODUCTION

The mission of any ethical pharmaceutical company and the charter of every regulatory agency worldwide is to provide safe and effective drugs to the marketplace. Analytical methods play a vital role in supporting every facet of the drug development and approval process, from discovery through formulation, process development, manufacturing, packaging, and ultimately the release of both active pharmaceutical ingredients (APIs) and finished drug products. This importance has been recognized in the United States by the Food and Drug Administration (FDA), which has published guidances concerning analytical methods [1,2], and internationally, with the International Conference on Harmonisation (ICH) devoting an entire "quality topic" to analytical methods [3,4].

While regulatory agencies have been criticized for their lack of consistency from topic to topic, region to region, and in some cases department to department, globally they agree that analytical methods, which are used in support of product registration, must be formally proven accurate and reliable. The process by which this occurs is validation. In fact, FDA, ICH, and the U. S. Pharmacopeia and National Formulary (USP/NF) consistently define validation of an analytical method as "the process by which a method is tested by the user or developer for reliability, accuracy and preciseness of its intended purpose" [2].

Regulatory professionals too often assume that any analytical method can undergo the steps necessary to validate its use in marketing a pharmaceutical product. In many cases the need for validation of a particular analytical method is often revealed late in the drug development process by corporate regulatory and quality assurance (QA) professionals who are responsible for compliance with the regulatory requirements associated with product registration. Commonly these individuals view the requirements and parameters of the validation processes as independent of the actual analytical chemistry and technical objective of the method itself.

In fact, method validation is merely the final step in the dynamic process of method development. The emphasis must be placed on the development stage, since any well-developed method can be successfully validated. Initial method development must therefore be undertaken with both the regulatory and technical requirements of validation in mind. While the emphasis is placed on method selection, sufficient development time is provided to ensure that the method meets both its technical and regulatory requirements. Only after this development stage is the testing procedure and validation protocol documentation finalized.

Emphasis is placed on preparing a validation protocol in which the specific validation experiments and associated acceptance criteria substantiate that the method meets its technical and regulatory objectives. Only after these steps does the process conclude with the performance of the formal validation and generation of the validation report. Finalization of method development (by way of the method validation) is never truly complete, as validation is a living process that encompasses the ongoing use of the method in various laboratory settings.

Method validation must be performed in a regulatory-compliant environment. In particular, the organization must have a QA unit (QAU), adequate laboratory equipment and facilities, written procedures, and qualified personnel. Since a successful validation requires the cooperative efforts of each of these organizational elements, successful fulfillment of the regulatory and technical objectives of validation requires senior management support. Additionally, it is essential that the organization have a well-defined validation master plan (VMP) for analytical methods, which defines the steps necessary to effectively validate methods.

Presented within this chapter are the organizational requirements necessary to validate methods and a multistep VMP that has been designed and proven effective in the validation of hundreds of analytical methods from a wide range of scientific disciplines. The development and validation of technically sound analytical methods in the early stages of drug development may ultimately prove invaluable in the approval process. Lack of validation can most certainly result in approval delays and possible rejection of data dependent upon the analytical results.

2 ORGANIZATIONAL REQUIREMENTS

Similar to the production of a finished pharmaceutical, validation of analytical methods must be performed in a suitable facility. Any organization responsible for developing and validating analytical methods must have the "quality elements" shown in Table 1 to demonstrate suitability and control.

TABLE 1 Quality Elements That Define a Suitable Environment for Method Development and Validation

Quality element	Description
Quality assurance unit (QAU)	Proactive functional group responsible for the duties related to QA
Adequate laboratory facilities	Facility with adequate space, appropriate environmental conditions, security, and control systems, including analytical instrumentation and equipment that is qualified and suitable for performing method development and validation
Qualified personnel	Sufficient employees with suitable education, training, and experience to perform laboratory experimentation
Training program	Comprehensive program that provides effective regulatory, safety, procedural, and proficiency training
Written procedures	Documents such as Standard Operating Procedures (SOPs), testing procedures and policies which provide written instruction to ensure compliant and consistent performance
Document control	Systems that define data handling and management, report generation, record retention and retrieval, and security
Change control	Chronicles changes made to all of the quality elements
Internal audits	Demonstrate compliance and effectiveness to all of the quality elements

2.1 Quality Assurance Unit (Quality Control Unit)

The term quality control unit is listed in 21 Code of Federal Regulations (CFR) Part 210, "Current Good Manufacturing Practices for Finished Pharmaceuticals," as a department of full-time employees who are dedicated to developing corporate policies and procedures that ensure quality throughout the organization and guarantee consistency and control. Responsibilities include regulatory compliance, training, review and release of analytical data, approval of methods and protocols, and auditing. The reporting line for the quality control unit must be independent of the laboratory personnel.

In general, the pharmaceutical industry has used the term quality control in reference to the laboratory function in which finished pharmaceuticals or APIs are tested for stability and release. The term quality assurance has replaced quality control as defined above. The terms (QAU) will be used throughout this article in reference to a proactive quality control unit.

2.2 Adequate Laboratory Facilities

All laboratory facilities must be of adequate space and design to provide a suitable work environment for experimentation and testing. The facility must provide an appropriately controlled environment (temperature, humidity, venting, etc.) to allow for a consistent laboratory function. A secure environment with limited and controlled access is required to assure result integrity. Suitable instrumentation and equipment must be installed and qualified as per defined procedures. Scheduled periodic calibration must be performed to demonstrate proper instrumental suitability. Such procedures must be appropriately documented. Reagents and standards must be stored and handled in accordance with good laboratory procedures.

2.3 Qualified Personnel

The company must have sufficient employees with suitable education, training, and experience to perform method development, validation, and performance. Also, the laboratory staff will require such supporting staff as administrative, supervisory, maintenance, safety, and shipping and receiving. The company should therefore have an organizational chart that clearly defines the role and responsibilities of each employee. Additionally, each employee must have a well-defined job description and a curriculum Vitae, that delineates the employee's work history and educational experience.

2.4 Training Program

Employee training should be part of the corporate culture. A comprehensive training program should include safety, regulatory compliance, procedures, and analyst proficiency. The program should be a continuous process that begins on the day of employment and continues throughout the employee's tenure. The program must be well documented to demonstrate effectiveness.

2.5 Written Procedures

Aside from the regulatory requirements, good science dictates that employees work from written procedures. These procedures should cover every repetitive function as well as specific testing methods. They must define responsibilities and be of sufficient detail to guarantee consistency. The organization must have sufficient controls in place for the generation, approval, and distribution of procedures. The company must have a clearly defined policy for reviewing the procedures to ensure that they accurately reflect current industry standards and regulatory requirements. Table 2 provides a list of essential or core SOPs.

2.6 Document Control

The current good manufacturing practices (GMP) state that "Laboratory records shall include complete data derived from all tests necessary to assure compliance with established specifications and standards" [5]. This includes all electronic and hard copies of raw data, laboratory notebooks, and/or worksheets and reports. Additionally, the regulations require that records "shall be readily available for authorized inspection" [6]. Consequently, the organization must have a well-defined audit trail for the generation, storage, and retrieval of reports and raw data. It must also be shown that there is adequate document management and security of all raw data and reports in terms of both disaster recovery and prevention of falsification of results.

In an effort to prevent reporting false or inaccurate data, organizations must make a routine practice of periodically authenticating reported results against raw data.

2.7 Change Control

All changes to approved procedures must be documented to provide a complete history of the method. Likewise, changes to equipment, software, and computer-controlled equipment that has been installed,

TABLE 2 Essential Standard Operating Procedures

Topic	Description
Procedures	Define the generation, approval, distribution, revision, and review of SOPs and analytical testing procedures
Procedural deviation	Defines how to document deviations from written procedures
Investigation of out of specification results	Clearly defines responsibilities and investigation requirements
Employee training	Includes the training requirements, frequency documentation, and storage of training records
Facilities/security	Includes general housekeeping, dress requirements, and security-related issues
Data handling	Defines the necessary steps for the generation, storage, archival, and retrieval of raw data
Review and release of analytical reports	Defines the steps necessary to ensure accuracy of calculated data and reported results
Use and storage of laboratory notebooks	Defines the items that must be included in laboratory notebooks
Chemical calculations	Provides the analyst with the pertinent chemical calculations and instructions for consist use
Storage and handling of reference standards	Defines the proper storage handling and expiration date requirements for reference standards
Storage and handling of reagents and chemicals	Defines the proper storage handling and expiration date requirements for chemicals and reagents
Installation and qualification of laboratory equipment	Defines the requirements for the selection, installation and qualification of laboratory equipment
Calibration of laboratory equipment	Defines the frequency and specific requirements for the calibration of laboratory equipment
Instrument logbooks	Defines the calibration and maintenance items that must be included in instrument logbooks
Validation of analytical methods	Instruction for performing method vadidation

TABLE 2 *(Continued)*

Topic	Description
Transfer of analytical methods	Provides, instruction for performing method transfers
Installation and qualification of computer systems for laboratory applications	Defines the steps necessary to select, audit, install, qualify, and validate computer-controlled equipment and software
Validation of spreadsheets and databases	Defines the steps necessary to validate and document spreadsheet and database applications
Change control	Defines documentation requirements for changes to methods, equipment, and computer systems and software
Conducting general compliance audits	Defines the responsibilities and roles of QA personnel during a general compliance audit
Conducting internal quality assurance audits	Defines the frequency and requirements for conducting internal QA audits

qualified, and validated must be documented as per a company SOP. Any changes in production or to individual product formulations must include an assessment of the effect the change has on associated validated methods.

Change control is a tool that can be used both prospectively and retrospectively. When used retrospectively, it can reveal the complete history of the analytical method and any changes that have been made to it. When used prospectively, it provides the appropriate parties the opportunity to evaluate the impact the proposed changes would have on the method.

2.8 Internal Audits

Internal QA audits must be conducted and documented at a defined frequency to ensure overall compliance, control, and effectiveness of the quality elements. Such audits should be conducted by members of the QAU or third-party compliance specialists and the results reported directly to the senior management of the corporation. The senior management should prepare an action plan to address any deficiencies and follow up to confirm adequate implementation.

3. VALIDATION MASTER PLAN (VMP) FOR ANALYTICAL METHODS

3.1 Defining the Validation Master Plan

Validation, as previously defined, is a dynamic process intended to satisfy different but complementary objectives (technical and regulatory). It is not an isolated event. The validation process is designed to provide confirmation that a particular analytical method is applicable for its intended use. To accomplish this, the planning and implementation of the validation experiments require the input of different individuals throughout the drug's development and manufacture. For example, the production plant may require the validation of a cleaning procedure to support the use of certain process equipment. A validation of this nature will require input from the process engineer to define the scope of the project, the formulator to provide the formulation, the toxicology department to set the limits of detection and quantitation based on the toxicity of the actives and excipients, chemists to develop and validate the method, a regulatory professional to assist in the development and approval of the validation protocol, and a QA professional to review and release the laboratory data and final reports.

Due to the diverse nature of analytical methods and the potential complexity of the validation process, it is essential that any firm that is engaged in the development and validation of analytical methods have a well-defined VMP. The VMP should be a corporate-level document and have the full support and endorsement of the senior management of the company. As an example, Fig. 1 outlines the individual action items of a multistep VMP. At the start of this sample VMP, a validation team is selected that includes the necessary representation from the technical and regulatory sectors of the organization. Once in place, the team defines the technical and regulatory objectives as well as the roles of each team member. Typically, analytical methods are developed to support a critical phase of development or to address a specific problem. Often timing is crucial, and at this stage a time line listing projected milestones should be included.

From this point the analytical department assumes leadership and begins by selecting the analytical technique. The selection process is dependent upon several related factors, including the analyte itself and the level of measurement precision required. Once the technique has been chosen, a suitable method is identified either from modification of an existing method or the development of a new method. Upon successful demonstration of method feasibility, the technical team members, in collaboration with the regulatory professionals, draft a test procedure and a validation protocol.

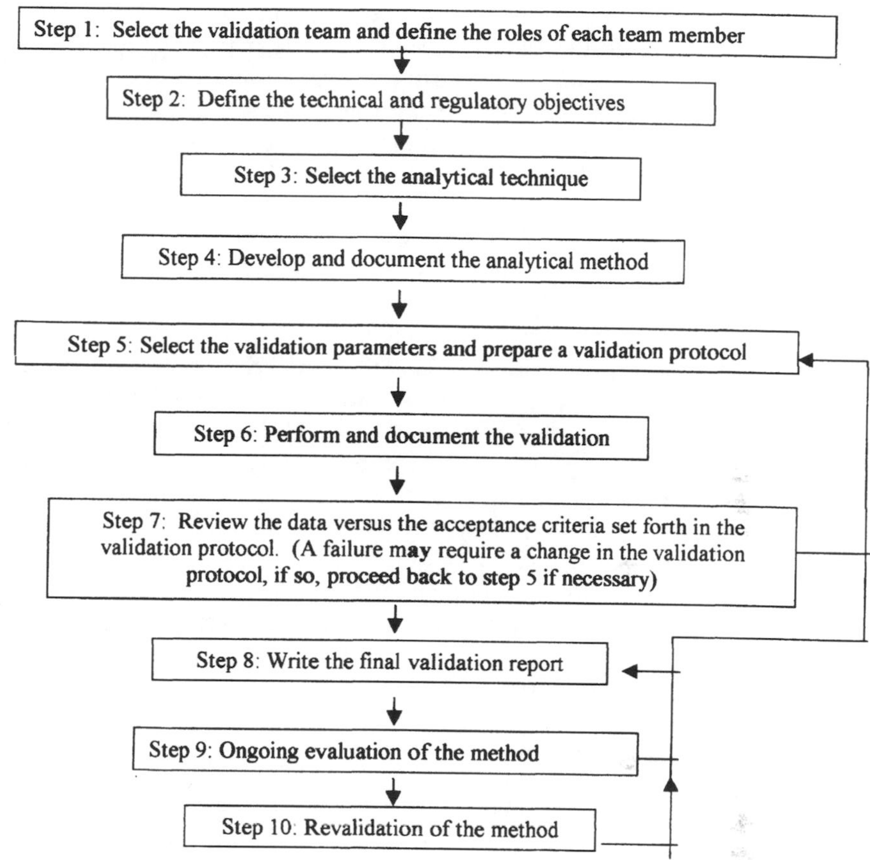

FIGURE 1 A multistep validation master plan (VMP).

The analytical chemist must perform the experiments outlined in the validation protocol and write the final validation report, which is presented to the QAU for review and release.

The dynamic nature of the validation process requires a constant re-evaluation of the method. An appropriately developed and validated method is continuously substantiated by the successful performance of method system suitability on the day of use and up to the point of method transfer. Changes to the manufacturing process or analytical instrumentation may require revalidation, depending upon the degree of change. Depending upon the impact of the change, the validation may necessitate additional development and complete revalidation of the method or simply repeating those validation parameters that may be affected. For example, a change in drug product excipients requires revalidation of such items as analyte specificity and recovery, while a change in the high performance liquid chromatography (HPLC) column for a dissolution end analysis may require demonstration of system suitability parameters if alternate columns were investigated during method development and feasibility.

3.2 The Role of Regulatory Professionals

3.2.1 Regulatory Responsibilities

A well-developed VMP must clearly define the roles and responsibilities of each department involved in the validation of analytical methods. Preferably, this is outlined in the corporate QA manual. The regulatory department must define the regulatory objective of the method. To do so, it must define the type and overall purpose of the method required. The departments requiring the analytical results, however, (e.g., production, toxicology, formulation, and manufacturing) must provide input for regulatory to define the objective. For example, methods can range from those used to identify a raw material received from a supplier to those used to support a cleaning validation to those that quantify a drug substance in a multicomponent solid dosage formulation.

The regulatory department also must determine the method's scope. Is the regulatory requirement a simple limits test, such as a pass/fail or a present/absent test for an impurity, or is it intended as a stability-indicating purity method, that would be concerned with actual concentrations of impurities? Both methods deal with quantitation but in vastly different ways. Once the objective and scope of the method have been determined, the regulatory professional may consult the published FDA, ICH, and/or a Pharmacopoeia (i.e., USP, BP, EP, JP, or HPB) guideline defining the specific parameters, such as linearity, accuracy, precision, specificity, limit of detection, limit of quantitation, ruggedness, and robustness, that make up a validation. While the literature defines the performance parameters, the regulatory department must set all applicable criteria for each parameter, such as the actual limit of detection based on toxicological data and purity specifications based on potency and end-use requirements.

3.2.2 Regulatory Review

Currently, the CFR Title 21, Current Good Manufacturing Practice (cGMP), part 211, subpart I, entitled "Laboratory Controls," section 211.165(e) states that "the accuracy, sensitivity, specificity and reproducibility of test methods employed by a firm shall be established and documented" [5]. Establishment (validation) and documentation may be accomplished in accordance with section 211.194(a)(2), "Laboratory Records." Additionally, proposed amendments to the cGMP regulations include section 210.3(b)(25), which defines methods validation "as establishing, through documented evidence, a high degree of assurance that an analytical method will consistently yield results that accurately reflect the quality characteristics with the product tested" [6].

The FDA also has proposed adding a new subpart L to part 211 entitled "Validation," which would consist of two regulations: section 211.220 for "process validation," and section 211.222 for "methods validation." Proposed section 211.222 (methods validation) would require the manufacturer to establish and document the accuracy, sensitivity, specificity, reproducibility, and any other attribute necessary to validate test methods. Validation would be necessary to meet existing requirements for laboratory records provided in section 211.194(a)(2).

Moreover, FDA has prepared two guidance documents concerning the validation of analytical methods. The first guideline, entitled "Guideline for Submitting Samples and Analytical Data for Method Validation," was published in February 1987. This document was intended to "assist applicants in submitting samples and data to the Food and Drug Administration (FDA) for methods evaluation" [1]. Section 5, "Information Supporting the Suitability of the Methodology for the New Drug Substance," however, and section 6, "Information Supporting the Suitability of the Methodology for the Dosage Form," detail the validation characteristics required for each type of method.

The second guidance, published in November 1994, is a "reviewers guidance" entitled "Validation of Chromatographic Methods." The purpose of this document was "to present the issues to consider when evaluating chromatographic test methods from a regulatory perspective" [2]. This guidance deals specifically with the validation of chromatographic methods. It defines the types of chromatographic methods, as well as providing details for each of the following validation parameters:

Accuracy
Detection limit and quantitation limit
Linearity
Precision, Repeatability, Intermediate, Reproducibility
Range
Recovery
Robustness
Sample solution stability

Specificity/selectivity
System suitability specifications and tests, Capacity factor (k′), Precision/injector repeatability (RSD), Relative retention (α), Resolution (R_s), Tailing factor (T), Theoretical plate number (N)

Historically, pharmaceutical companies wishing to market products internationally had to meet the stringent requirements of FDA and a number of different regulatory agencies worldwide. This arduous process not only increased the cost of introduction but also delayed and in some cases prevented the release of life saving medicines to the global community. In an effort to streamline the process, the first International Conference on Harmonization was held in Brussels in November 1991. Harmonization under ICH involves the European Union, Japan, and the United States. Additional assistance is provided from the World Heath Organization (WHO), the European Free Trade Association (EFTA), and the Therapeutic Products Program (formerly Canadian Health Protection Branch). The overall objective of ICH is to make recommendations concerning the harmonization of technical and regulatory requirements for the development and approval of new drug substances and products. The intent is to improve the efficiency of the drug development and approval process by effectively using resources to eliminate unnecessary duplication of efforts without compromising on product quality, safety, and efficacy.

The ICH is organized by a steering committee (SC) consisting of 14 members. The SC is advised by expert working groups (EWGs), which are made up of experts from industry and regulatory agencies on technical issues concerning the harmonization topics. The original topics included safety, quality, efficacy, and regulatory communications. Quality is dedicated to pharmaceutical development and specifications. The quality topic is further subdivided into six sections: Q1, "Stability," Q2, "Validation of Analytical Methods," Q3, "Impurities Testing," Q4, "Pharmacopoeial Harmonization," Q5, "Quality of Biological Products," and Q6, "Specifications."

"Validation of Analytical Methods" (Q2) consists of two guidelines: Q2A, "Validation of Analytical Procedures: Definitions and Terminology" was adopted in October 1994, and published in the Federal Registry (FR) on March 1, 1995 (60 FR 11260). This guideline lists analytical methods by type and "validation characteristics" which must be investigated during method validation and included as part of a product registration. Table 3 summarizes the various types of analytical methods and their relevant validation characteristics.

The Second guideline, Q2B, "Validation of Analytical Procedures: Methodology," was adopted in November 1996 and published in the FR on May 19, 1997 (62 FR 27464). This guideline is an extension of Q2A and

TABLE 3 ICH Analytical Methods and Validation Characteristics

Validation characteristics	Type of analytical procedure: Identification	Testing for impurities: Quantitation	Testing for impurities: Limit	Assay (content/potency) dissolution (measurement only)
Accuracy	−	+	−	+
Precision				
Repeatability	−	+	−	+
Intermediate Precision	−	+[a]	−	+[a]
Specificity[b]	+	+	+	+
Detection limit	−	−[c]	+	−
Quantitation Limit	−	+	−	−
Linearity	−	+	−	+
Range	−	+	−	+

Note: − signifies that this characteristic is not normally evalutated; + signifies that this characteristic is normally evaluated.

[a] Cases in which reproducibility has been performed, and intermediate precision is not needed.

[b] Lack of specificity of one analytical procedure could be compensated by other supporting analytical procedures.

[c] May be needed in some cases.

includes a discussion of the actual experimental details and statistical interpretations for each of the validation characteristics.

3.3 The Role of Analytical Chemists

Undoubtedly the major responsibility for generating a validated analytical method falls on the shoulders of the analytical chemist. The chemist must select an analytical technique that will fulfill the regulatory and technical requirements set forth by the regulatory professionals. This includes not only the analyte to identify or quantitate, but also the purity specifications for assay methods and the impurity specifications and the limit of detection and quantitation for related compound methods. The analytical technique chosen will depend upon the degree of precision, linearity, range, and accuracy necessary to meet the regulatory requirements. Once a new method is developed or the feasibility of an existing method is established, the test procedure is documented. The analytical chemist must then propose experiments that meet the regulatory requirements for validating the precision, accuracy, and other technical criteria of the method. Following regulatory acceptance of the validation experiments, the analytical chemist, in collaboration with the regulatory professionals, is responsible for writing the final test procedure and the validation protocol. Upon approval of these documents, the chemist performs the validation experiments and prepares a final report which details the experimental results. This complex process is outlined in Fig. 2.

3.3.1 Defining the Technical Objective

Ultimately the technical objective of an analytical procedure, which is included as part of a registration application for pharmaceuticals, is the responsibility of the analytical department. This objective is directly related to the regulatory objective as defined by the regulatory professionals. For example, the regulatory professionals require a means to quantitate an API in a finished pharmaceutical for release between 95–105% of label claim. This translates into a technical objective for the analytical chemist that involves the selection of an analytical technique and the development of a method that has the required accuracy and precision to meet the requirements for release.

From the viewpoint of the analytical department, meeting the technical objective can only be accomplished by understanding the regulatory objective's ultimate application. For this, cooperative communication between the various departments involved (e.g., production, toxicology, formulation, and manufacturing) is essential. Additionally, the end users of the method (e.g., quality control [QC], contract manufacturers, contract analytical laboratories, and foreign subsidiaries) must be considered since they will be responsible for performing the validated method and generating accurate and reliable results.

The analytical chemist will choose the appropriate analytical technique (e.g., chromatography, spectroscopy, or titration) to satisfy the technical objective based upon his or her expertise and past experiences with similar analytical problems. Often, however, the analyte itself dictates the kind of analysis method to be used. For example, a residual volatile solvent would most probably be analyzed by gas chromatography (GC), while a residual catalyst, such as palladium, would best be analyzed by atomic absorption or emission spectroscopy.

Considering the Required Precision. The level of precision necessary to meet the material specifications must be included in a determination of the

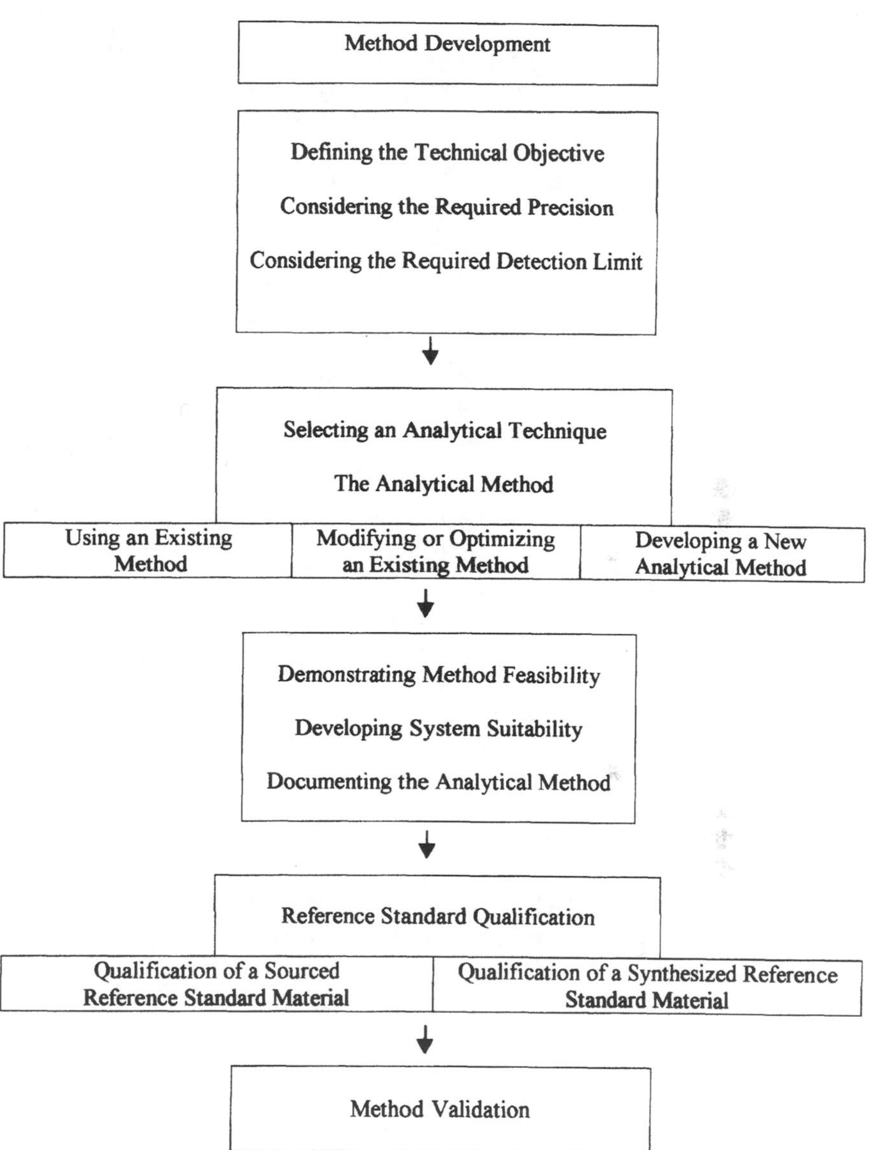

FIGURE 2 The validation process from method development to method validation.

analytical technique. For example, the manufacturing plan is to produce a physical blend of two salt forms (potassium and sodium) of an API to be used in a new marketed formulation. The production specifications are set at ± 2.0% for the ratio of the salt forms. Although atomic absorption (AA) spectroscopy is an obvious analytical method that can be used to quantitate these cation salts, it typically has a precision of ± 5–10%.

While this spectroscopic technique may accurately and reliably analyze the salt forms, it lacks adequate precision to meet the release specifications. A method developed from this technique would not therefore meet the regulatory requirement and could not be validated for this application. Consequently, an ion chromatographic (IC) method should be developed to separate and quantitate both potassium and sodium, which theoretically can meet the precision requirement of ± 2.0%. This chromatographic technique would meet the regulatory requirement and could be successfully validated.

Considering the Required Detection Limit. The toxicity of a potential impurity will impact the product specification for the impurity and therefore the required method's limit of detection. In some instances, this may result in allowing very relaxed precision requirements. Typically, when the potential for toxicity from an impurity is documented to be low, the specification may be set significantly higher with much tighter method precision requirements. For example, the impurity specification for a residual solvent such as isopropanol in a drug substance will be vastly different from the specification for residual aluminum in a parenteral drug product. While methods developed for both components can be categorized as impurity determinations, the specification for isopropanol might be set at 2000 ppm, while because of its potential toxicity, the specification for aluminum might be set as low as 10 ppb. The individual validation requirements for such parameters as accuracy, precision, and limit of detection may be dramatically different for the methods used to quantify these two impurities.

A chromatographic technique such as GC, which might be used to quantitate the residual isopropanol, may be accurate to ± 5% at the required specification level of 2000 ppm. Conversely, a technique such as graphite furnace atomic spectroscopy, which would be used to quantitate residual aluminum, may be accurate to ± 25% at the required specification level of 10 ppb. Although the accuracy of both techniques differs greatly, methods derived from each of these technologies may be acceptable based on the overall regulatory objective.

3.3.2 Selecting an Analytical Technique

Once the technical objective has been clearly defined, the analytical chemist will begin the selection process for an appropriate analytical technique. As stated previously, the technique is often decided by the analyte and specification level. In other instances, there is a great degree of freedom. For example, residual trifluoroacetic acid (TFA) in a neutral drug substance in which fluorine is not present in the empirical formula, may be detected and quantitated by many completely different analytical techniques. Table 4 lists and describes six techniques that can be used to analyze residual TFA.

TABLE 4 Possible Analytical Techniques for the Determination of Residual TFA in a Drug Substance (DS)

Item number	Technique	Description
1	Ion chromatography	Sample is dissolved and TFA is separated from the DS using an anion exchange column with conductivity detection.
2	Reversed phase chromatography	Sample is dissolved and TFA is separated from the DS using a reversed phase column, a highly aqueous eluent, and low-wavelength UV detection.
3	Gas chromatography	Sample is dissolved and TFA is derivatized and analyzed by capillary GC with flame ionization detection.
4	Titration	Sample is dissolved and TFA is potentiometrically titrated using a weak base.
5	Ion specific electrode analysis of fluorinea[a]	Sample is fully combusted in an oxygen-rich environment that generates fluoride from the TFA. Total fluoride is determined by ion-specific electrode and back-calculated to TFA in the DS.
6	Ion chromatography of fluorinea[a]	Sample is fully combusted in an oxygen-rich environment that generates fluoride from the TFA. Total fluoride is determined by ion chromatography using an anion exchange column and conductivity detection. TFA is back-calculated from the fluoride results.

[a]Assuming flourine is not listed in the DS empirical formula and no other source of fluorine in the DS is possible.

In a review of Table 4, the ion chromatographic technique appears to be the obvious method of choice because of lower detection limits and reduced run times. It may not be the best method, however, if significant counterions such as acetate, benzoate, nitrate, or phosphate are present. In situations in which the DS has limited solubility with the potential for TFA to be present in inclusion complexes, combustion techniques may prove superior.

When a choice in analytical methods is possible, it is important that the analytical chemist reviews the various techniques. This may be accomplished in a variety of ways. In many instances it is prudent to perform a literature search that includes such sources as the USP/NF and the Association of Analytical Communities (AOAC) International's *Book of Methods*. Additionally, technical journals, such as the *Journal of Analytical Chemistry, Journal of Chromatography, and Journal of Pharmaceutical Sciences*, may be searched through *Chemical Abstracts* or *Analytical Abstracts*. Most of these references are available through online services such as STN International. Frequently, this approach saves time and safeguards against "reinventing the wheel." In combination with the literature search, the analytical chemist may draw upon previous experience as well as the expertise of colleagues.

The availability of instrumentation in the QC laboratory or at the production facility will often influence the choice of the analytical technique. For example, the trace analysis of a DS for three different metal elements (iron, copper, and nickel) can be simultaneously performed by an inductively coupled plasma (ICP). The cost of this instrument, however, is $100,000 or more. For this example, the same analysis can be performed to the level of precision and detection defined in the technical objective by an AA spectrometer. Unlike the ICP, the AA analysis is sequential, and therefore is significantly more time-consuming. The choice of the AA method may be desirable, however, since the instrumentation cost is a fraction of the cost of an ICP, and often is an instrument already available in a QC or production laboratory.

3.3.3 The Analytical Method

The purpose of the analytical method is to generate reliable and accurate data that are evaluated in making decisions regarding the acceptability of a drug substance or drug product for use in humans. The analytical method determines and defines a testing procedure suitable for the specific analytical application. The actual analytical method itself can be obtained in one of three basic ways: (1) use an existing method, (2) modify or optimize an existing method, or (3) develop a new method. The commonality between each of these options is the need to confirm applicability through method feasibility experimentation. While method feasibility is an integral part of method development, it is usually a necessary stand-alone study when an existing method is used, and will be covered in a later section.

Using an Existing Method. This is the easiest of the three method-generation options. An example of this would be the use of a Karl Fischer method developed for moisture determination in DS hydrochloride being used in DS sulfate. Regardless of the apparent suitability, the decided use of an existing method must be confirmed with method-feasibility experiments.

Modifying or Optimizing an Existing Method. Method modification or optimization is a broad category and can cover a range of changes made to an existing method. (Note: For discussion purposes, optimization is addressed here as the improvement to an existing validated method. Optimization is also a process of changing conditions to generate a usable or validatable method. Since this is a development process, it is included as an integral part of method development.) In its simplest form, method modification or optimization can be a change in a sample preparation step to accommodate the newly formulated half-strength tablet drug product. In this scenario, the actual method for content uniformity would require a change in which the half-strength tablet is placed in a volumetric flask that is one-half the size of that used for the full-strength product. Since no change has occurred to the concentration of the API in solution, the method is expected to perform as well with the half-strength product as it did with the full-strength product. This should be confirmed by performing a sample reproducibility experiment.

At the other end of the method-modification spectrum, significant changes to the existing method may be necessary in both sample preparation and the analytical method conditions. For example, an HPLC method that was originally developed for the assay of a DS might be the starting point for a method that will eventually assay an extended-release combination drug product that contains the DS. In this example, the chromatographic conditions may be significantly altered to accommodate the elution of the second API. Additionally, sample preparation will become more complicated due to the addition of formulation excipients and a sample matrix designed to provide extended release characteristics.

Developing a New Analytical Method. Analytical method development is necessary when no existing method proves suitable for generating the required analytical result. This is often the case when a new drug entity is discovered, or for an existing API, when a new formulation renders the current analytical method unsuitable. Significant method development resources have been utilized recently for many pharmaceutical companies in developing procedures for existing DS and drug products. This has been made necessary because the current regulatory requirements for such items as specificity, precision, and detectability are more stringent compared to those in place when the method was originally developed. As an example, many assay methods for older pharmaceuticals were not developed as stability indicating methods. A stability-indicating method must differentiate the analyte from degradants. Either the methods were titrative procedures, or if chromatographic in nature, were not validated using current standards. Many older HPLC methods had not been tested using purposefully degraded samples or with known degradants, and once tested in this manner are shown not to be stability-indicating. Additionally, FDA's desire to continually reduce the acceptable levels of related compounds has necessitated modernization to methods that are unable to meet the new requirements.

True method development calls upon the technical expertise of the analytical department and more specifically the project analytical chemist to assess the nature of the chemical entity under investigation and the matrix in which it is present. Success or failure in method development is directly attributable to the analyst's overall technical competence and breadth of knowledge in various analytical techniques. Technically challenging analysis problems are usually handled by more senior members of the laboratory staff or those possessing advanced education and training.

The most successful method-development chemists and analytical development laboratories include method feasibility and various aspects of method validation in their method-development experiments. Early stages of development focus on the analyte neat or in well-defined and well-behaved sample matrices. It is important to evolve the development program to include evaluation of the analyte with synthesis precursors, process contaminants, and degradants. Once the general scheme of a successful development program for a particular analyte has begun to emerge, method feasibility at the detection limit required or in the actual sample matrix must be performed. These prevalidation experiments will confirm the "validatability" of the method and signal the start of formal documentation of the method.

In a perfect world, this process encompasses a time frame that parallels drug development. There is ample time to develop the method and several opportunities to modify and improve the developing method based on its continuing use in the research laboratory. The results of successful and failed feasibility studies and prevalidation studies performed during method development will allow fine-tuning of all the sample preparation and method parameters. The use of the method over time by a varied group of analysts, however, provides a significant contribution to information on the accuracy and reliability (ruggedness and robustness) of the method in actual laboratory environments.

Unfortunately, few analytical methods are developed in a perfect world. A more likely situation has method development and validation scheduled on a tight time line. Using an HPLC assay and related compounds method for a novel DS as an example, Table 5 outlines a realistic development scenario that would provide for a method that could be successfully validated and utilized.

Generally scientists involved in method development are not the same individuals who will ultimately use the method on a day-to-day basis. As listed in steps 5, 6, and 8 of Table 5, it is quite necessary to consider the analysts who will ultimately be called upon to use the method as well as the equipment and instrumentation in the laboratory. Methods that are destined for the QC laboratory should, when possible, utilize standard laboratory equipment, be independent of the technician performing it, and be easily transferable. The following is an actual example of a method that was developed without consideration of the aforementioned concerns.

An HPLC assay, content uniformity and related compounds method for a blockbuster new drug product, was developed that utilized concave gradient elution, a flow rate of 1.25 ml/min and no temperature control of the column. Samples were placed in 1000-ml volumetric flasks, sample diluent was added, and the flasks were sonicated for 10 min followed by 30 min of mechanical shaking. This method was to be run in a QC laboratory at the contract manufacturing facility in Puerto Rico. While the developed method worked flawlessly in the development laboratory, the QC laboratory had many problems in performing the procedure.

The HPLC instrumentation available to the QC laboratory could only run linear gradients and could only control the flow rate to ± 0.1 ml/min (not 0.05 ml/min as the method would require). Additionally, the contract QC laboratory did not have adequate control of the room temperature so in the summer the temperature could rise as high as 32°C. Simply providing the development laboratory with the instrument specifications prior to method development would have allowed the developed method to be directly transferable to this QC laboratory.

A second aspect of this method, which specifications alone would not have addressed, is the issue of sample preparation. The method-development chemist purposefully developed a sample preparation procedure that did not require a sample dilution step. Although this resulted in a 1000-ml sample flask, the development chemist thought the absence of a dilution step would be appreciated by the QC chemist who would be using the method. In the QC environment, however, the use of such a large sample flask was seen as a burdensome inconvenience. A 1000-ml volumetric sample flask made for slow sample preparation. Over 12 liters of sample diluent were needed to test

Table 5 Steps in the Successful Development of an HPLC Assay and Related Compounds Method

Step	Item	Description
1	Obtain a UV scan of analyte	Scan from 200 nm to 400 nm, looking for max or a unique wavelength appropriate for analysis. Generally, selectivity increases with wavelength.
2	Develop separation conditions	Working only with the DS, develop conditions that provide acceptable retention and peak shape.
3	Determine feasibility for related compounds	Analyze all related compounds available (process impurities, degradation products, etc.) and perform purposeful degradation on the API. Determine interferences.
4	Redevelop separation (if necessary)	If there is interference, redevelop analysis conditions and repeat feasibility.
5	Determine final chromatographic conditions	Evaluate alternate column lots. Based on this information and considering the instrumentation and personnel who will be using the method, set final chromatographic conditions, including mobile phase preparation.
6	Perform prevalidation linearity study	By analyzing sufficient data points to determine the linear range, sample preparation concentrations and related compound detection and quantitation limits can be estimated. Consideration must be given to the capability of the HPLC instrumentation that will eventually run the method.
7	Develop basic sample preparation scheme	Determine preparation solvents and conditions that would dissolve both sample and known related compounds
8	Finalize sample preparation scheme	Based on linearity and considering the equipment and personnel who will be preparing samples, set the final standard and sample preparation scheme.

TABLE 5 *(Continued)*

9	If possible, work with method over time to detect problems and modify accordingly (robustness evaluation)	This important aspect of method development is often left out because of deadlines and a need to quickly move to validation.
10	Write the final analytical method	Prepare the final version of the method to be validated.

a single lot of drug product. The flasks were so large that only a few could be sonicated at one time. Only one-half of the positions available on the wrist action shaker could be accessed, and the weight of the sample flasks was such that the shaker needed to be tied to the benchtop to prevent it from "walking." Because of their size and weight, the flasks could only be moved around the laboratory two at time.

Had the QC laboratory been consulted during the development stage, the chemist would have suggested that sample preparation be done in a 100-ml volumetric flask followed by a 5-ml to 50-ml dilution. With this procedure, the QC chemist would be more productive since more flasks could be sonicated and shaken at once, 10 to 12 flasks could be easily carried using a sample tray, and 85% less sample diluent needed to be prepared. This more than made up for the dilution step.

Demonstrating Method Feasibility. Regardless of the source of an analytical method, method feasibility is recommended prior to any attempt at method validation. While method feasibility is an essential part of the method development process, it is generally a separate experiment that must be applied when considering the use of an existing method. All methods that fail feasibility will require some level of further development or optimization. Even a seemingly insignificant change to a formulation requires investigation of such validation parameters as specificity and purposeful degradation of the placebo, which may necessitate changes in system suitability requirements. For example, a manufacturer changes the supply source of the cherry flavor for a multicomponent cough syrup. There is seemingly no need to confirm feasibility of the HPLC procedure used for both assay and related compounds since no changes have been made to the active ingredients or the preservative, and the flavor remains cherry. Upon investigation, however, it is found that a component in the new cherry flavor exhibits near coelution with one of the actives by the existing HPLC assay method. This observation indicates that at a minimum, method optimization is required. A slight weakening of the mobile-phase to increase the resolution of these components may require a re-evaluation of the system suitability requirements. Plate count and peak tailing may be effected by the mobile phase change. Additionally, a new resolution requirement may be necessary if the active and new flavor component constitute the "critical resolution pair."

In many ways, a method-feasibility study can be considered a prevalidation study to determine if the method can be validated as written or if modifications are necessary before the method can be formally validated for its intended purpose. For example, Table 6 lists some method-feasibility experiments that should be performed before a DS assay method is used for a corresponding tablet drug product.

The filter compatibility and accuracy experiments should provide analysis values within the precision of the DS method. The specificity experiment should also show no interference from placebo components or degradants. If these criteria are met, there is a high level of confidence that method validation will prove successful. Any values outside the expected range would necessitate method modification with a subsequent feasibility study.

An additional phase of method feasibility that is often overlooked is an assessment of method items that may not have an equivalent replacement. In HPLC, an obvious item that falls under this category is the HPLC column. Other items, however, such as the actual instruments, equipment, reagents, and supplies used, must be evaluated during feasibility. It is important that

Table 6 Suggested Experiments for Determining Feasibility of the Drug Substance Assay Method for the Tablet Drug Product

Feasibility experiment	Description
Filter compatibility	Analyze filtered vs. nonfiltered standard solution
Specificity	Prepare placebo sample and determine if interfering placebo peaks are present. Purposefully degrade placebo and determine if interfering degradant peaks are present.
Accuracy	Set sample preparation conditions such that a 100% label claim (LC) sample solution has the same concentration as the 100% drug substance sample solution. Spike placebo with drug substance at 80% and 120% LC. Determine recovery.
Solution stability	Retain 80% LC accuracy sample and evaluate over time.

even items such as the type (plastic or glass) and make of autosampler vials be investigated.

Developing System Suitability. An area that requires special attention during method development and feasibility is system suitability. System suitability should define those method parameters that can demonstrate that the analysis results are accurate and reliable. The criteria for system suitability should provide confirmation that the results generated throughout the analytical run are unquestionable. This requirement has been included in the most recent revisions of the USP [7]. Additionally, proper system suitability should demonstrate the suitability of the instrumentation at the time the analysis was run. For example, standard industry practice is to calibrate laboratory instrumentation at a set frequency of every 3 to 6 months. If the instrument fails calibration testing during the performance of a scheduled calibration, then the integrity of the results generated by that instrument since the last passing calibration are in question. If the instrument was an HPLC detector, then a limit of quantitation (LOQ) requirement in the system suitability system could demonstrate the detector's performance at the time of analysis. If the instrument was an HPLC pump, then a system suitability requirement for analyte retention time could confirm the pump's flow rate performance at the time of the analysis.

The method parameters to include in system suitability are based on the analytical technique itself and on the particular requirements of the individual method. Most quantitative analytical techniques will have a reproducibility of measurement requirement for system suitability. Quantitative impurity analyses should have a limit of detection (LOD) or LOQ measurement. This is necessary because response may be instrument-dependent or may change from day to day. Table 7 lists suggested system suitability requirements for chromatographic methods from November 1994 Center for Drug Evaluation and Research (CDER) reviewer guidance [2].

It should be noted that the tests and specifications listed in Table 7 are provided for guidance and should be set for each individual method based on its application and performance. For example, there may be no resolution requirement for an HPLC dissolution assay. For the same method, it may also be impractical to set the injector repeatability requirement at an RSD of $\leq 1\%$ if the peak response for the API is very small. Again, in the same example it is also impractical to require a check standard to be within $\pm$ 1% of the working standard if the RSD requirement for replicate analyses of the working standard is ≤ 2 %. Some ion exchange columns may not provide plate counts of > 2000 or tailing factors of ≤ 2.

TABLE 7 Recommended System Suitability Tests and Specifications for Chromatographic Methods (11/94 CDER Reviewer Guidance)

System suitability test	Recommended specification
Capacity factor (k′)	>2
Precision/injector repeatability (RSD)	$\leq 1\%$ for $n \geq 5$
Relative retention (α)	Method-specific
Resolution (R_s)	>2
Tailing factor (T)	≤ 2
Theoretical plate number (N)	>2000

Considering the importance of system suitability in determining the accuracy and reliability of method results, it is improper to set an RSD requirement of $\leq 2\%$ for a method that consistently yields values of $< 1\%$. If peak tailing is always ≤ 1.5, then a requirement of ≤ 2.0 does not provide feedback to notify the chemist of a poor or failing column. Resolution requirements must be required for the "critical pair" of components where a loss of resolution would impact method performance. By addressing these parameters with a more critical eye, the suitability of the analysis system can be accurately determined.

Documenting the Analytical Method. Once a method has been developed, it must be formally documented into a final analytical testing procedure. While there is no specific set of rules that dictate how a method must be written, there is general agreement that the document must be detailed enough to ensure that it can be reproduced by a qualified technician with comparable equipment. This concept takes on added significance when a method is developed by one company and is intended to be run at the analytical laboratories of several other companies. What may be the corporate culture at the originator's company is not necessarily an industry standard and must be clearly communicated through the test method. For example, it may be the originator's policy to test all samples in duplicate and to analyze the standard as the final sample run. Therefore, they do not include these statements in any of their methods. Another company may never test in duplicate unless otherwise instructed, however, and a third company may only calibrate with standards and never analyze a standard as a sample. Because of this, the documented method should include all information necessary to reproducibly perform the analysis. This would include items such as the preparation and use of sample blanks and check standards, the type and manner in which calibration is performed, and an assay sequence. Example 1 is a general outline for an analytical method that incorporates these important documentation items.

EXAMPLE 1: AN OUTLINE FOR AN ANALYTICAL TESTING PROCEDURE

Analytical Testing Procedure

Title:

Effective date:

Purpose or scope—A brief description of the method objective.

Summary of methodology—A general description of the methodology used.

1. Instrumentation and Equipment

   List of Instrumentation: A listing of all instrumentation required. Standard instrumentation should be listed as "or equivalent" only if it has been scientifically determined that other instruments can be used. Critical instrumentation must be clearly defined with make, model, and specifications, if applicable. Instrumentation must be qualified and calibrated (if applicable) prior to use.

   List of Equipment: A listing of all additional equipment needed to perform the procedure. Standard equipment should be listed as "or equivalent" only if it has been scientifically determined that alternate equipment can be used. Critical equipment must be clearly defined with make, model, and specifications, if applicable. Equipment must be qualified and calibrated (if applicable) before use.

   Description of configuration

   For certain procedures, a picture or illustration may be helpful.

   Accessories and Supplies: A listing of all nonroutine items, such as the special filters, low-actinic autosampler vials, or solid-phase extraction cartridges required.

2. Reagents

   List of reagents: All chemicals (including water) are required, as well as their purity or grade, and the source, allowing for grade or source equivalency where applicable.

   Preparation of reagents: Complete description of quantities and procedures used. The expiration date and storage conditions should be listed.

3. Reference standards

   A listing of reference standards, system suitability standards, impurities, and degradants, where applicable. List the purity and source, allowing for equivalency where applicable.

4. Preparation of standards and samples

Comprehensive description including all necessary weights (including weight range) and glassware needed to make dilutions (if applicable). Following validation, the stability of the standard and sample solutions will be documented here as well. Include confirming or check standards. This section will also include resolution solutions, sensitivity, or LOQ solutions, where applicable.

Preparation of standards

Preparation of samples

5. Operating conditions

   Reference operation, calibration, and maintenance SOP, or include unique conditions here.

6. Procedures

   List procedures to perform assay.

System suitability: Including a requirement for the confirming standard that is matched to the RSD of replicate analysis of the working standard. For example, the RSD of five replicate measurements must be $\leq 1\%$, and the confirming standard must agree within $\pm 1\%$.

Identification: Include, if applicable

Assay sequence: A detailed list of the analysis sequence to follow, including system suitability, calibration, and frequency of standard reanalysis.

7. Calculations

   All calculations necessary to obtain the assay result. The reporting format and the number of significant figures or decimal places are included here. Also, include the convention used for reporting between the LOD and LOQ, and for undetected components.

8. Approval

   This section should include the names, titles, date, and signatures of those responsible for the review and approval of the analytical test procedure.

9. Reason for revision

   If this is a later version of the method, a listing of the changes made to the method should be included in this section.

3.4 Reference Standard Qualification

Before an analytical method can be validated, it is necessary to qualify those materials that will serve as reference standards. To allow for expeditious validation, it is important to source and/or synthesize and characterize reference standard materials (RSM) for APIs, impurities, and known degradation products concurrently with method development. Reference standard materials are typically obtained in one of two ways. They can be synthesized by the user or from a contract manufacturer, or they can be sourced from a supplier who manufactures standards or chemicals. While all synthesized RSMs must be fully characterized and qualified, the degree of characterization or qualification for sourced materials depends on the supplier. For example, regulatory agencies typically accept National Institute for Standards and Testing (NIST) and compendial source materials, such as those from the USP, as primary source standards without further qualification. Caution should be applied in taking this approach, however, since the USP does not provide certificates of analysis (COA) with standards, and certain lots of reference materials have shown differing numbers and levels of trace impurities.

All RSMs obtained from secondary sources must be qualified. The Sigma–Aldrich family of companies is an excellent example of a secondary standard source, offering over 200,000 chemical compounds in their standard catalogues and an additional 70,000 in various specialty catalogues.

3.4.1 Qualification of a Sourced Reference Standard Material

Qualification of a sourced RSM begins with obtaining a COA, the synthesis pathway, if available, and a list of methods used in product manufacturing. This information will help in determining the essential parameters for qualification. The identity of the material should be confirmed with a "fingerprinting" technique such as FT-IR to a library source, or by elemental analysis to confirm the molecular formula. Once identity has been established, the quality of the material must be ascertained. Usually, this will simply be a confirmation of the purity of the material prior to use. Techniques such as elemental analysis, GC, or liquid chromatography might be used for purity determination. Based on the purity results, the material may require further purification by distillation or recrystallization. Additional testing may also be required to identify and quantify known or potential impurities that may have been overlooked during the manufacturer's assessment of the material. This may depend on the specification application of the RSM.

3.4.2 Qualification of a Synthesized Reference Standard Material

In those instances in which an RSM is not available from a commercial source, the material must be synthesized. For APIs, this material may start out as simply a lot of DS. This lot of material may be of sufficient purity to be designated as the RSM, or it may require further purification. Known impurities or degradants will require custom synthesis; however, their purity requirements are generally not as stringent. Qualification of these synthesized materials requires a "full" characterization. While there is no set guideline to the characterization of an RSM, Fig. 3 depicts a decision tree approach that uses broad-range analytical techniques that are useful in the characterization of organic materials.

3.5 Method Validation: Developing the Validation Protocol

Whether the method for the analytical application was an existing method, or a modified existing method, or a newly developed method, the requirement under the FR remains the same. Chapter 21, part 211.194, concerning "Laboratory Records," requires that the "suitability of all testing methods shall be verified under the actual conditions of use" [5]. Therefore, the analytical chemist must experimentally demonstrate the method's ability to achieve the regulatory and technical objectives that were originally set forth.

Developing the validation protocol is a crucial step in the validation process. It is the culmination of the regulatory and technical accomplishments to this point in the development of the method. The protocol must define which validation parameters are needed and the specific experiments necessary to demonstrate the validity of the analytical method. The protocol must contain all of the acceptance criteria for each of the relevant validation parameters. Additionally, the protocol must define the number of replicates, the reporting format, and the number of significant figures. In short, the validation protocol instructs the analyst on how to validate the analytical method.

Because of their importance in pharmaceutical analyses, much attention has been focused on harmonizing the parameters necessary for the validation of chromatographic methods. While some of these parameters are applicable to other analytical techniques, it is the responsibility of the analytical chemist to select and tailor the appropriate parameters and acceptance criteria for the particular method to be validated. Since most analytical chemists are not experts on regulatory matters, it is essential for the regulatory affairs professional to understand the requirements of method validation and work closely with the analytical chemist to select appropriate validation parameters and meaningful acceptance criteria.

The ICH has attempted to harmonize the parameters allowing for global uniformity. Table 8 lists individual validation parameters as they appear in the various guidances.

Since FDA and the HPB have adopted the ICH Q_2A and Q_2B guidelines, the discussion will be limited to the ICH parameters. To this end, there has been a clarification to the definition of ruggedness, precision,

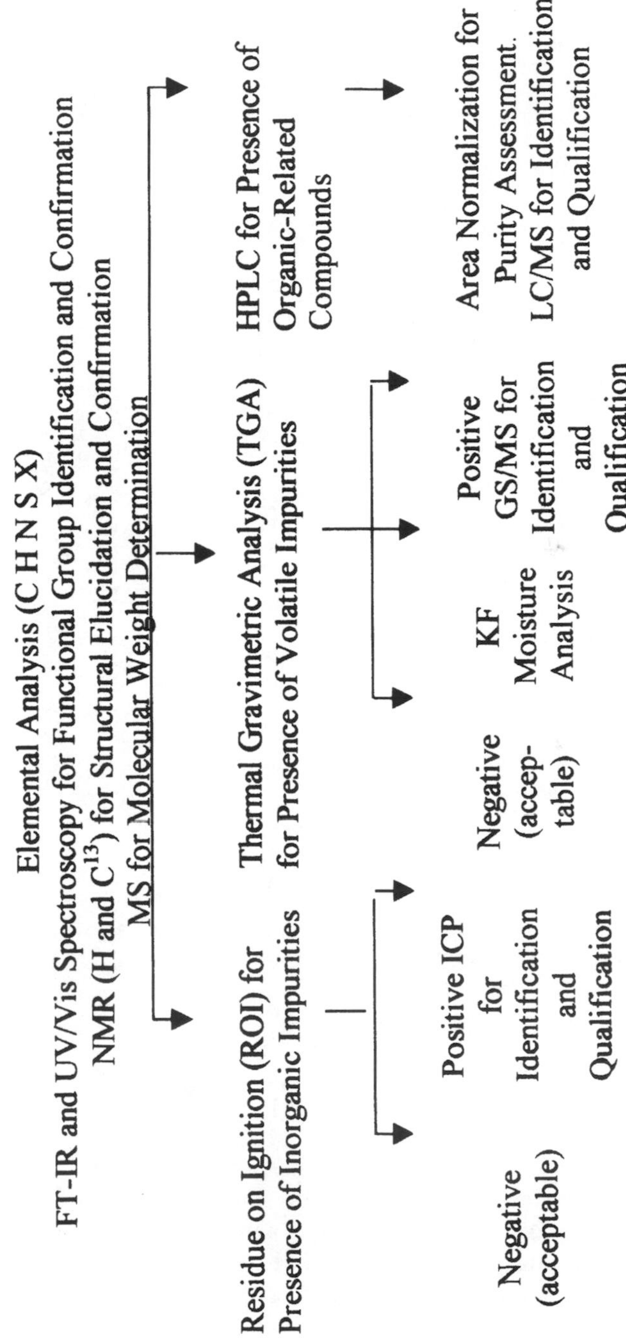

FIGURE 3 Decision tree for reference standard qualification.

Table 8 Validation Parameters Which Appear in Regulatory Guidance Documents

Validation parameter	ICH [3,4]	FDA [2]	USP [7]	HPB [8]
Accuracy	✓	✓	✓	✓
Range	✓	✓	✓	✓
Precision				
Repeatability	✓	✓	✓	
Intermediate Precision	✓	✓		
Reproducibility	✓	✓	✓	
Precision System				✓
Precision				
Method				✓
Specificity	✓	✓	✓	✓
Selectivity		✓		
Linearity	✓	✓	✓	✓
Limit of defection (LOD)	✓	✓	✓	✓
Limit of quantification (LOQ)	✓	✓	✓	✓
Robustness	✓		✓	
Ruggedness	✓	✓	✓	✓
System Suitability[a]				
System precision		✓	✓	✓
Resolution		✓	✓	✓
Tailing factor		✓	✓	✓
Number of theoretical plates		✓	✓	✓
Capacity factor		✓	✓	✓
Relative retention		✓	✓	
Solution stability				
Sample		✓	✓	
Standard		✓	✓	

[a]ICH states a requirement but does not give any specific parameters or refer to the pharmacopoeia for additional information.

limit of detection, and quantitation, and a stipulation for the requirement of robustness.

Earlier guidelines defined precision in terms of system precision and method precision. System precision was the measure of reproducibility based on multiple measurements of a single sample preparation. Method precision was the measure of reproducibility based on analysis of multiple sample preparations. Ruggedness was a measure of day-to-day, analyst-to-analyst, and instrument-to-instrument variation.

The ICH has broadened and redefined these terms to more accurately describe the method's ability to reproducibly generate analytical results. Precision is defined as a combination of repeatability, intermediate precision, and reproducibility. Repeatability is system precision, as defined previously. Intermediate precision includes multiple analyses by multiple analysts on different days using different equipment within a given laboratory. This is only the first step in demonstrating the ruggedness of the method.

Reproducibility encompasses the variation in analytical results between laboratories and provides a second level of method ruggedness. This is becoming an increasingly important part of method validation as the pharmaceutical industry becomes more specialized and diversified. Major manufacturers may develop and validate a method in a corporate research center for use in a foreign manufacturing site or at a contract testing laboratory. It is therefore critical that the validation demonstrates that the method is free of analyst or instrument bias.

A new term, robustness, describes the ability of the method to withstand minor but deliberate changes to operating parameters and sample preparation while delivering accurate and reliable results. The validation protocol must consider all of the pertinent method operating parameters and sample preparation procedures that could impact the final result. Peters and Paino [9] detail a modified fractional factorial testing of seven such factors in only eight experiments. Such matrix studies allow extensive robustness testing in a controlled and limited set of experiments. The degree of method robustness, which is both method- and sample-dependent, must be defined by the chemist through such testing.

If analytical measurements are susceptible to variations in the analysis parameters or sample preparation conditions, the method must be suitably controlled or a precautionary statement must be included in the written procedure that alerts the chemist to the susceptibility. The method's system suitability parameters should be defined in such a way that meeting all system suitability criteria would ensure that the method is currently being performed within the acceptance window provided by validation robustness testing.

Detection limit is defined in the ICH Q_2A guideline as the "lowest amount of analyte in a sample which can be detected but not necessarily quantitated as an exact value" [2]. ICH guideline Q_2B expanded upon this statement to identify three approaches to the actual testing methodology [3]. The detection limit determination can be based on (1) a visual evaluation, (2) a signal-to-noise measurement, or (3) a standard deviation measurement of the response and slope of the calibration curve. The standard deviation may be of a blank sample or of the residual of a regression line from a calibration curve in the range of the detection limit. The ICH Q_2B guideline also allows approaches other than those listed. In all cases, the method for the determination must be listed in the validation protocol and supported by the validation experiments.

Quantitation limit is defined in the ICH Q_2A guideline as the "lowest amount of analyte in a sample which can be quantitatively determined with suitable precision and accuracy" [2]. As with the detection limit, ICH guideline Q_2B expanded upon this statement and listed the same three approaches to the testing methodology [3]. Again the method for the determination must be listed in the validation protocol and supported by the validation experiments.

3.5.1 The Validation Protocol for Chromatographic Methods

High-performance liquid chromatographic methods are the most common form of analytical technique used to support drug product registration. An example of a validation protocol for an HPLC assay and related compounds method is provided in Example 2.

EXAMPLE 2: AN EXAMPLE OF A VALIDATION PROTOCOL FOR A CHROMATOGRAPHIC METHOD

Validation Protocol Determination of Compound X and its Related Compounds in Product X Tablets

Introduction

The determination of compound X and its related compounds in product X tablets will be validated according to the tests described in this protocol. The chromatographic parameters for these experiments are as stated in the method, "HPLC Assay and determination of related compounds in product X tablets."

1. Specificity

   Prepare individual solutions of the diluent blank, placebo blank, compound X, impurity Y, and degradant Z. Inject the blank and then each of the individual solutions on columns from two separate column lots. Prepare a solution containing all of the above compounds and inject on both columns to determine resolution between each peak.

   Acceptance criteria

   Each compound must have a unique retention time and must have a minimum resolution of 1.5 from each other.

2. Linearity

Compound X

This study will cover the range of 50–150% of the nominal assay concentration for compound X. Prepare five standard solutions at 50%, 75%, 100%, 125%, and 150% of the nominal assay concentration, and make one injection per preparation.

Determine the linear regression of the peak area versus concentration. Calculate the slope, y intercept, bias, and correlation coefficient.

Related compounds

This study will cover the range of 0.05–5% of compound X, impurity Y, and degradant Z in the drug product. Prepare five standard solutions at 0.05%, 0.2%, 0.5%, 2%, and 5% of the nominal concentration of compound X in the drug product and make one injection per preparation.

For all the compounds listed above, determine the linear regression of the peak area versus concentration. Calculate the slope, y intercept, and correlation coefficient.

Acceptance criteria

Assay

The correlation coefficient for compound X must be not less than 0.999.

The bias of the y intercept must be bias less than or equal to 2.0% of the theoretical 100% assay concentration value.

Related compounds

The correlation coefficient for compound X, impurity Y, and degradant Z must be not less than 0.99

3. Range

The range is derived from the linearity and accuracy of the method and will be demonstrated for compound X over the range of 80–120% of the nominal assay concentration, and for related compounds over the range of 0.1–2% of the related compound in the drug product.

4. Accuracy

Assay

Accuracy will be evaluated by performing the recovery study for compound X over the range of 80–120% of the nominal tablet levels and analyzing as per the method. This will be accomplished by spiking product X placebo with compound X at 80%, 100%, and 120% of the nominal tablet levels. The samples will be prepared in triplicate at 80% and 120% levels and six replicates at the 100% level.

Related compounds

Accuracy will be evaluated by performing the recovery study for compound X, impurity Y, and degradant Z over the range of 0.1–2% of the related compound in the drug product and analyzing as per the method. This will be accomplished by spiking product X placebo with compound X, impurity Y, and degradant Z at 0.1%, 0.5%, and 2% of each related compound in the drug product. The samples will be prepared in triplicate at 0.1% and 2% levels and six replicates at the 0.5% level.

Acceptance criteria

Assay

The recovery of compound X must be between 98–102% of the theoretical value at each level.

Related compounds

The recovery of each component must be between 80–120% of the theoretical value at the 0.5–2% levels.

The recovery of each component must be between 75–125% of the theoretical value at the 0.1% level.

5. Precision

Analysis repeatability

Assay

The assay precision data is obtained from the 100% level of assay accuracy. (See Sec. 4.1).

Related compounds

The related compounds precision data are obtained from the 0.5% level of related compound accuracy. (See item 4.)

Acceptance criteria

Assay

The RSD of the 100%-level assay accuracy values must not be more than 2.0%.

Related compounds

The RSD of the 0.5%-level related compound accuracy values must not be more than 15% for each component.

Intermediate precision

Assay

Six samples will be prepared by spiking product X placebo with compound X at 100% of the nominal tablet level and analyzing as per the method. These samples will be prepared and analyzed by a different analyst using a different column from a different column lot on a different day with a different instrument from those used for performing item 4.

Related compounds

Six samples will be prepared by spiking product X placebo with placebo with compound X, impurity Y, and degradant Z at 0.5% of each related compound in the drug product and analyzing as per the method. These samples will be prepared and analyzed by a different analyst using a different column from a different column lot on a different day with a different instrument than those used for performing item 4.

Acceptance criteria

Assay

The RSD of the assay percentage recovery must not be more than 2.0%.

Related compounds

The RSD of the related compound percentage recovery must not be more than 15% for each component.

Reproducibility

The reproducibility of this method will be determined as part of the method transfer exercise to the QC laboratory.

6. Detection limit

Estimate the detection limit for compound X, impurity Y, and degradant Z by the following equation:

$$DL = \frac{3.3\sigma}{S}$$

Where DL = the estimated detection limit

σ = the standard error in the y intercept of the regression line for the component using the data provided in accuracy item 4.

S = the slope of the regression line

Confirm this value by preparing a solution of each component at the estimated detection limit concentration and chromatograph in triplicate as per the method.

Acceptance criteria

The peaks for each component must each yield a response at least three times the baseline noise.

Quantitation limit

Estimate the quantitation limit for compound X, impurity Y, and degradant Z by the following equation:

$$QL = \frac{10\sigma}{S}$$

Where QL = the estimated detection limit

σ = the standard error in the y intercept of the regression line for the component using the data provided in accuracy item 4.
S = the slope of the regression line

Confirm this value by preparing a solution of each component at the estimated quantitation limit concentration and chromatograph in triplicate as per the method.

Acceptance criteria

The peaks for each component must each yield a response at least 10 times the baseline noise.

Robustness

Duplicate chromatograms of the system suitability resolution solution and the sensitivity solution will be collected using the following method changes:

Vary flow rates, column temperature, and acetonitrile concentration of the mobile phase by 5% (relative to the method values).
Vary the pH of the aqueous buffer portion of the mobile phased by 0.1 pH unit (relative to the method values).

Acceptance criteria

The system suitability criteria for resolution, peak tailing, and plate count must be met for both injections of the resolution solution.
The system suitability criteria for peak response must be met for both injections of the sensitivity solution.

Solution stability study

Assay the standard solution and a sample solution stored at ambient temperature and light, and refrigerated at 0, 24, 48, and 72 h against a freshly prepared standard.

Acceptance criteria

The standard solution is stable if the assay results are within 2.0% of the initial value.
The sample solution is stable for assay if the assay results are within 2.0% of the initial value.
The sample solution is stable for related compound determination if all related compounds at or above the LOQ level are within 15% of the initial value, and no new impurity peaks above the LOQ are observed.

Forced degradation

Light

Placebo and drug product will be photolytically stressed by placing it in a thermostated light chamber at 25°C and subjecting the samples to a UVA light source of not less than 200 watt h/meter and a fluorescent light source of not less than 1.2 million lux h.

Heat

Placebo and drug product will subjected to 75°C until between 10–30% degradation of compound X has occurred (up to a maximum of 128 hours).

Acid

Placebo and drug product will be subject to 1 N hydrochloric acid until between 10–30% degradation of compound X has occurred (up to a maximum of 48 hours). If a minimum 10% degradation of compound X has not occurred, the procedure will be repeated at a temperature of 60°C.

Base

Placebo and drug product will be subject to 0.5 N sodium hydroxide until between 10–30% degradation of compound X has occurred (up to a maximum of 48 hours). If a minimum 10% degradation of compound X has not occurred, the procedure will be repeated at a temperature of 60°C.

Oxidation

Placebo and drug product will be subject to 10% hydrogen peroxide until between 10–30% degradation of compound X has occurred (up to a maximum of 48 hours). If a minimum 10% degradation of compound X has not occurred, the procedure will be repeated at a temperature of 60°C.

Acceptance criteria

No observable interference of any degradation products with compound X, impurity Y, and degradant Z. This will be determined by checking peak purity of compound X, impurity Y, and degradant Z using a photodiode array detector.

System suitability

Perform the System Suitability portion of the method by two different analysts on two different days using two different columns from two different column lots on two different instruments.

Acceptance criteria

All system suitability criteria are met.

The appropriate signatory to meet compliance requirements would be the director of quality control. Additional signatures from quality assurance are not necessarily required but would provide an opportunity for regulatory oversight.

Approved by:

____________________ ____________________

Signature Date

____

Title

3.5.2 The Validation Protocol for Nonchromatographic Methods

There are many instances in which a nonchromatographic method of analysis is the preferable technique to generate the data necessary for making a decision regarding the acceptability of a drug substance or drug product for use in humans. It is therefore necessary that nonchromatographic methods also be validated using the same performance parameters required for chromatographic methods. An example of a validation protocol for an ICP spectroscopic method is provided in Example 3.

EXAMPLE 3: AN EXAMPLE OF VALIDATION PROTOCOL FOR AN ICP SPECTROSCOPIC METHOD

Validation Protocol Determination of Trace Levels of Element A in Product B Tablets

Introduction

The determination of element A in product B tablets will be validated according to the tests described in this protocol. The ICP spectroscopic parameters for these experiments are as stated in the method "ICP determination of trace levels of element A in product B tablets."

1. Specificity

   Prepare individual solutions of the diluent blank, method blank, placebo blank, and element A, and analyze as per the method.

   Acceptance criteria

   None of the blanks shows a response for element A greater than 2 ppb.

2. Linearity

   This study will cover the range of 10 ppb to 500 ppb of element A in solution (equivalent to 0.1 ppm to 5 ppm of element A in the drug product). Prepare five standard solutions, including a blank (0 ppb), 10 ppb, 50 ppb, 200 ppb, and 500 ppb, and analyze twice each as per the method.

   Determine the linear regression of the emission response versus concentration. Calculate the slope and correlation coefficient.

   Acceptance criteria

   The correlation coefficient for element A must be not less than 0.995.

3. Range

The range is derived from the linearity and accuracy of the method and will be demonstrated for element A over the range of 0.2 ppm to 2 ppm of element A in the drug product.

4. Accuracy

Accuracy will be evaluated by performing the recovery study for element A over the range 0.2 ppm to 2 ppm in the drug product and analyzing as per the method. This will be accomplished by spiking product B (known to contain < 0.05 ppm of element A) with element A to 0.2 ppm, 0.5 ppm, and 2 ppm of element A in the drug product. The samples will be prepared in triplicate at 0.2 ppm and 2 ppm levels and six replicates at the 0.5-ppm level.

Acceptance criteria

The recovery of element A must be between 80–120% of the theoretical value at the 0.5-ppm and 2-ppm levels.

The recovery of element A must be between 75–125% of the theoretical value at the 0.2 ppm level.

5. Precision

Analysis repeatability

The assay precision data are obtained from the 0.5-ppm level of accuracy. (See item 4.)

Acceptance criteria

The RSD of the 0.5-ppm level assay accuracy values must not be more than 15%.

Intermediate Precision

Six samples will be prepared by spiking product B (known to contain < 0.05 ppm of element A) with element A to 0.5 ppm of element A in the drug product and analyzing as per the method. These samples will be prepared and analyzed by a different analyst on a different day from those used for performing item 4.

Acceptance criteria

The RSD of the assay percentage recovery must not be more than 15%.

Reproducibility

The reproducibility of this method will be determined as part of the method transfer exercise to the QC laboratory.

6. Detection limit

Note: Since all measurements are made relative to a blank, the baseline noise is very small and leads to an artificially low detection limit based on a signal-to-noise ratio measurements. The detection limit therefore must be determined by visual inspection of diluted standards.

Perform serial dilutions of a standard and analyze to determine the lowest concentration that can be visually distinguished from the background. Confirm this value by preparing a placebo blank spiked with element A at the estimated detection limit concentration and analyze six times as per the method.

Acceptance criteria

The RSD of the response for element A must be less than or equal to 25%.

Quantitation limit

The quantitation limit is estimated at three times the detection limit. Confirm this value by preparing a solution of element A at the estimated quantitation limit concentration and analyze six times as per the method.

Acceptance criteria

The RSD of the response for element A must be less than or equal to 10%.

7. Robustness

Robustness of the sample preparation scheme

Three samples will be prepared by spiking product B (known to contain < 0.05 ppb of element A) with element A to 0.5 ppm of element A in the drug product.

Diluent volumes:

Change the initial dilution volume to 15 ml and then the final volume to 45 ml after acid digestion.

Change the initial dilution volume to 25 ml and then the final volume to 45 ml after acid digestion.

Microwave digestion time

Microwave the samples for 5 min

Microwave the samples for 7 min

Acceptance criteria

The recovery of element A must be between 80–120% of the theoretical value.

Robustness of the sample analysis scheme

Three samples will be prepared by spiking product B (known to contain < 0.05 ppm of element A) with element A to 0.5 ppm of element A in the drug product.

Lower the forward power of the ICP plasma to 1400 watts.

Evaluate the samples at a pump speed of 0.9 and 1.1 ml per min.

Acceptance criteria

The recovery of element A must be between 80–120% of the theoretical value.

8. Solution stability study
   Assay the standard solution and a sample solution stored at ambient temperature at 0, 24, 48, and 72 hours against a freshly prepared standard.
   Acceptance criteria
   The standard and sample solutions are stable if the assay results are within 10% of the initial value.
9. System suitability
   Perform the system suitability portion of the method by two different analysts on two different days.
   Note: Due to the specialized nature of the instrumentation, both analysts may use the same instrument.
   Acceptance criteria
   All system suitability criteria are met.
   Final approval, should be provided by the director of quality control, with a supporting quality assurance counter-signature.

3.6 Method Validation: Validation Performance

Once the validation protocol has been documented and accepted, the actual validation becomes a matter of method performance. The results of the validation hinge on the completeness of method development and the appropriateness of the validation protocol experiments. As we have mentioned throughout this chapter, the validation process involves many different parts of the organization and can span a time frame from months to years. Also, since the validation is usually the focus of regulatory agencies, it is recommended that the QAU perform a prevalidation audit prior to the performance of the validation. The QAU should consider developing a checklist that includes such critical items as method confirmation, protocol approval, equipment suitability, and expiration dates of reference standards and reagents. This prevalidation review is important as a means to uncover potential oversights that might otherwise go undetected until the validation is complete and the final report is under review by the QAU or possibly a representative of a regulatory agency. A well-designed checklist will root out commonly occurring problems that could result in time delays or even a recommendation to withhold an application. Example 4 contains a checklist of the critical components that should be confirmed prior to the performance of a validation.

EXAMPLE 4: AN EXAMPLE OF A PREVALIDATION CHECKLIST

Analytical method—Confirm that the method is documented and approved as per the organization's policies and procedures.

Reviewed by: ________ Date: ________

Validation protocol—Verify that the validation protocol has been written, reviewed, and approved by all pertinent members of the validation team.

Reviewed by: ________ Date: ________

Personnel qualifications—Review the training records of all analysts that will be involved in the validation experiments.

Reviewed by: ________ Date: ________

Equipment suitability—Review the calibration records of all equipment and instruments to be used to confirm their suitability.

Reviewed by: ________ Date: ________

Reagents and standards—inventory all reagents, standards, and supplies to determine acceptability with method requirements and specifications.

Reviewed by: ________ Date: ________

Quality assurance approval

Approved by: ________ Date: ________

Upon successful completion of the prevalidation audit, the analytical chemist simply performs the experiments listed in the validation protocol using the analytical method as written. Based on the results of the validation, it may be necessary to revise the method to include details such as solution stability, relative retention times, relative response factors, or cautionary statements resulting from the robustness experiments.

3.6.1 Substantiation of the Validation

The FDA's guideline, "Reviewer Guidance, Validation of Chromatographic Methods," which was issued November 1994, states that "method validation should not be a one-time situation to fulfill Agency filing requirements, but the methods should be validated and also designed by the user to ensure ruggedness or robustness." Additionally, the guideline goes on to state that "methods should be reproducible when used by other analysts, on other equivalent equipment, on other days or locations, and throughout the life of the drug product."

In practice, the sheer nature of the dynamic validation process results in a substantiation of the original validation throughout the useful life of the method. As previously mentioned, a well-developed and-written method includes system suitability that must be met each time the method is performed. Additionally, the pharmaceutical industry has standardized on the need of formal method transfer exercises whenever an analytical method is to be performed by a different laboratory or in another facility. Also, change control procedures may require the revalidation of part or all of the method in the event of changes to the method, the process, or the formula of a drug product. The QAU should therefore have a system to formally monitor the data obtained during these activities and include them as addenda to the original validation report.

3.6.2 Method Transfer Exercises

The term method transfer does not formally appear in the current FDA regulations or guidance documents. The ICH requirement of "reproducibility", however, is intended to demonstrate the precision of analyses between laboratories. As a successful part of the total method validation, this analyst-to-analyst comparison at different laboratory sites serves to prove the method validity. Also, this portion of validation can occur during the original validation experiments or at a future date. As an example, a method is developed in an analytical R&D group to be eventually transferred to QC labs, production facilities, or contract laboratories worldwide. These reproducibility experiments would be performed as method-transfer exercises.

A method transfer consists of the following three objectives:

Analyst proficiency
Equipment suitability
Method validity

Sometimes the original validation is performed by the same analyst or group of analysts who were involved in the original method development activities. Consequently, by the time the validation occurs the analysts have become very experienced using the method and any idiosyncrasies have become second nature, and thus may be inadvertently omitted from the written procedure. For example, those chemists directly involved with method development and validation understand the sample preparation statement "mix until finely dispersed" to mean vigorous vortexing and sonication until a hazy solution is obtained with no visible particles. A group of analysts in the QC laboratory, however, having no personal experience with the method, may interpret the same statement to mean simple wrist-action shaking until all large clumps of material have disappeared. In this scenario, the QC analysts are using a less vigorous means of sample extraction and consequently obtain a low recovery of the API.

A common industry practice is to list the actual equipment used during the original validation in the validation report and analytical method. Quite frequently "or equivalent" will follow the listing. Since certain instrumental parameters can differ between manufacturers of basically the same equipment (e.g., UV detectors for HPLC), it is essential to include experiments in the method-transfer exercise that are specifically designed to qualify the equipment as equivalent or suitable.

The reproducibility portion of the validation is the truest test of a method's validity. Since the ICH states that reproducibility data "are not part of the marketing authorization dossier" [4] the parameter is often overlooked or omitted from the original validation. As a result, the method validity is not confirmed until a formal method transfer to another facility is required. Additionally, since FDA will normally perform the method validation in one or more of their regional laboratories following the new drug application (NDA) submission, it would be prudent to have transferred the method or proven interlaboratory precision prior to the submission.

The process of method transfer must follow a "method-transfer protocol" which defines the experiments and acceptance criteria necessary to demonstrate the analysts proficiency, equipment's suitability, and true ruggedness of the analytical method. If we assume that any quality analytical laboratory has proficient analysts who operate suitable equipment, then the method transfer stands as an ongoing means to substantiate the suitability of the original method validation. Example 5 contains an example of a method-transfer protocol for a chromatographic procedure.

EXAMPLE 5: AN EXAMPLE OF A METHOD-TRANSFER PROTOCOL

Method-Transfer Protocol Determination of Substance J and Related Compounds in Tablets

Introduction

The determination of substance J and related compounds in tablets will be transferred according to the tests described in this protocol. The chromatographic parameters for these experiments are as stated in the method, "assay of substance J and the determination of related compounds in tablets by high-performance liquid chromatography."

1. System suitability

   Perform the system suitability portion of the method on two different days by two different analysts using two different columns on two different instruments.

   Acceptance criteria

   All system suitability criteria for injection reproducibility, standard confirmation, resolution, sensitivity determination, and peak tailing must be met on each day.

2. Linearity

   Substance J

   This study will cover the range of 50–150% of the nominal assay concentration for substance J. Prepare five standard solutions at 50%, 75%, 100%, 125%, and 150% of the nominal assay concentration, and make one injection per preparation.

Determine the linear regression of the peak area versus concentration. Calculate the slope, y-intercept, bias, and correlation coefficient.

Related compounds

The linearity of related compounds Q, R, and S will be determined over the range of 0.1% to 2% of each related compound relative to the amount of substance J in tablets. Prepare five standard solutions of related compounds Q, R, and S at 0.1%, 0.2%, 0.5%, 1%, and 2% of the nominal substance J concentration in the sample solution. Make one injection per preparation.

For each of the related compounds, determine the linear regression of the peak area versus concentration. Calculate the slope, y–intercept, and correlation coefficient.

Acceptance criteria

Assay

The correlation coefficient for substance J must be not less than 0.999.

The bias of the y intercept must be less than or equal to 3.0% of the theoretical 100% assay concentration value.

Related compounds

The correlation coefficient for each related compound must not be less than 0.99.

The peak height response of the 0.1%-level solution for each related compound must be at least ten times the noise.

3. Accuracy

Assay

Accuracy will be evaluated by performing a sample analysis on three lots of substance J tablets. Each lot of substance J tablets will be prepared in triplicate and analyzed as per the method. The assay results will be compared to those obtained from the corporate QC laboratory.

Related compounds

Accuracy will be evaluated by performing a recovery study for related compounds Q, R, and S in substance J tablets. This will be accomplished by spiking triplicate sample solutions of a single lot of substance J tablets with each of the related compounds. The sample solutions will be spiked at a level equivalent to 0.5% of the related compound relative to the substance J concentration in the sample.

Acceptance criteria

Assay

The average assay values for substance J tablets must be within ±

2.0% of the value obtained by the corporate QC laboratory for each lot.

The RSD of the triplicate assay values for each lot must be $\leq$ 2.0%.

Related compounds

The average recovery of each related compound must be between 80–120% of the theoretical spike value.

The RSD of the triplicate recovery values for each related compound must be $\leq$ 10%.

Approved by:

Signature	Date
Title	

Similar to the methods validation protocol, it is recommended to obtain approval signatures from both the director of quality control and quality assurance because of the regulatory criticality of a method transfer exercise.

Formal documentation of the method-transfer results as addenda to the original validation report would further substantiate the overall validation process. Thus, each new laboratory setting would either confirm the original method validation, or indicate a possible need for method modification with revalidation.

3.6.3 Revalidation

At some point in the life cycle of an analytical method it may become necessary to revalidate. Revalidation is simply the process by which a method that was previously validated is validated again. This may be necessary because of a change in an incoming raw material, a manufacturing batch change, a formulation change, or any change to the method itself. Based on the degree of change, the revalidation may involve a reperformance of a single validation parameter or a repeat of the entire validation protocol.

Revalidation of Minor Changes. A minor change to the drug product may require revalidation of only a single validation parameter. For example, a change in the supplier of a cherry flavor for a multicomponent cough syrup would require a revalidation of specificity in the HPLC assay procedure to confirm the lack of interference from the new flavor source. Only if there were an interference would method modification and further revalidation be required.

Revalidation After Method Development. A major change to a method following development may require a complete revalidation. For example, a change in the excipient for a tablet drug product might result in incomplete recovery by the existing HPLC assay method. If this necessitated a change in the sample preparation, including sample diluent, all parameters of the method would require revalidation.

3.7 Method Validation Report

From a regulatory perspective, the most important document is the validation report. This is because the report is typically the first document that a regulatory agency will review. If the validation report is error-free and complete, the regulatory body has no need to investigate further. A well-written validation report should include the following three essential items.

A summary of all data and findings for the experiments listed in the validation protocol. This should include a complete listing of specific instrumentation, actual reagent lots, standards, equipment and supplies used in the performance of the validation. All results should be provided with references to the original notebook entries. Example chromatograms, spectra, or other instrument outputs should be provided.

A copy of the approved validation protocol.

A copy of the method as it existed at the time of validation.

The organization must have a well-documented audit trail to raw data that will facilitate the retrieval of any supporting data that may be required during a regulatory inspection. In case of any off-site inspection, it may be prudent to include a copy of all notebook pages and all raw data.

The validation report should be a living document that reflects the dynamic validation process. Therefore, it should be updated using addenda to report method transfer results, ongoing system suitability and any revalidation efforts.

4 AN UNCONVENTIONAL APPROACH TO VALIDATION

In today's pharmaceutical marketplace the goal of corporations is to get new products to market faster and reduce the overall cost of the drug-development process. The VMP presented in this chapter can be considered a classic or conventional approach to generating validated methods and fits into the typical corporate structure and philosophy of many of the ethical pharmaceutical manufacturers. Often a pharmaceutical firm splits method development and validation into two distinct and mutually exclusive events. Method development is commonly performed in analytical R&D groups that support the research efforts, preclinical studies, and other early-stage development processes. These groups stress the fulfillment of the method's technical objectives. Validation is generally performed later in the drug-development process, and supports the release and stability of finished pharmaceuticals.

As mentioned earlier, validations must be conducted in a regulatory-compliant environment. Many research departments lack this level of compliance in their development laboratories to allow analysts the freedom to explore various options without having to deal with the strict change control requirements of a regulatory-compliant laboratory. Once a method has been selected, developed, and optimized, the developing chemist will perform several of the validation parameter experiments, such as specificity, accuracy, and precision to demonstrate method validation feasibility. This exercise is a prevalidation study that, because of the noncompliant nature of the commingled method development data, must be repeated during the formal validation. While this approach has been proven effective in generating a compliant validation study, it is time-consuming and very costly.

An innovative approach that allows the development data to be used in the final validation report would save a considerable amount of time and significantly reduce the overall cost of validation. Figure 4 represents a schematic representation of an unconventional approach.

In order to utilize this approach the firm must develop systems that result in a compliant organization without restricting the required creativity or flexibility needed for method development, therefore, all of the quality elements listed in Table 1 must be implemented in the analytical research laboratory. Table 9 lists the modifications required to utilize developmental data.

4.1 Quality Assurance Unit and Internal Audits

The QAU must perform routine internal audits to assure that the laboratory facilities are in compliance. Internal audits should include calibration procedures and records, chemical and reagent labeling and expiration dating, and laboratory record keeping and data handling.

4.2 Laboratory Equipment and Instrumentation

Research laboratories conducting experiments to be included in the final validation report must use instrumentation and equipment that has been

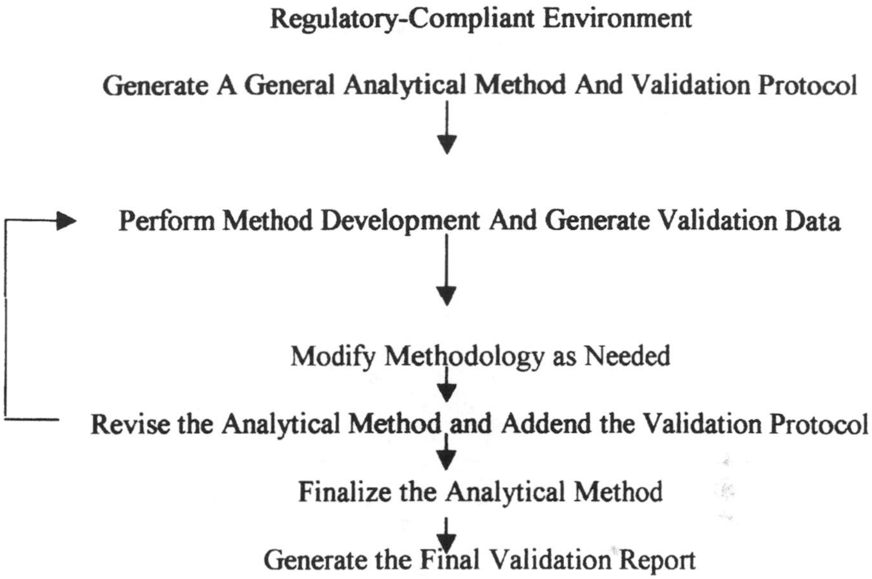

Regulatory-Compliant Environment

Generate A General Analytical Method And Validation Protocol

↓

Perform Method Development And Generate Validation Data

↓

Modify Methodology as Needed

↓

Revise the Analytical Method and Addend the Validation Protocol

↓

Finalize the Analytical Method

↓

Generate the Final Validation Report

FIGURE 4 A nonconventional approach to method validation.

installed and qualified per defined procedures. All instrumentation and equipment must be in calibration and suitable at the time of use.

4.3 Training Program

In many cases method development chemists hold advanced degrees and possess years of laboratory experience. Their cGMP training is often limited, however, and thus inadequate for performing validation. The training program, therefore, must emphasize regulatory compliance, especially in notebook documentation skills.

4.4 Written Procedures

The corporation must have an SOP that combines the method development and validation activities. Also, SOPs for the preparation of analytical method and validation protocols must stress change control and procedural deviations. The SOPs must allow the analytical chemist to revise methods and protocols to reflect changes encountered during the drug development process. Additionally, an accurate audit trail of changes must exist to track all changes and modifications.

Table 9 Modifications to Quality Elements That Define a Suitable Environment for Utilizing Developmental Data in Validation

Quality element	Description
Quality assurance unit (QAU) and internal audits	Must perform scheduled audits to ensure regulatory compliance.
Laboratory equipment and instrumentation	All laboratory equipment (including analytical instrumentation) must be qualified and calibrated at all times.
Training program	Comprehensive program that provides effective regulatory, safety, procedural, and proficiency training is required.
Written procedures	Written procedures stressing documentation, validation protocols, change control, and procedural deviations must be in place.
Document control	Systems that define data handling and management, report generation, record retention and retrieval, and security must be suitable for a compliant environment. Systems must be in place to insure data accuracy and integrity with sufficient staff for data review.
Change control	A comprehensive change control program must be in place to chronicle changes made to methods and protocols.

4.5 Document Control

Conventional method development records are typically not the subjects of regulatory review. Failures, which are commonplace in early method development, are rarely investigated or fully explained. Laboratories that plan to use development data in a validation must document their data in accordance with regulatory standards. Therefore, all entries must be recorded in laboratory notebooks that "shall be readily available for authorized inspection" [6].

4.6 Change Control

A stringent change control system must be developed and implemented to provide a complete history of the analytical method and validation protocol. This system must fully document each of the changes made to a method and whether or not the change affects previous validation results. Often the changes will impact these validation experiments, requiring their repetition. The change control system must permit lengthy explanations to allow auditors to easily reconstruct the events surrounding the change.

An effective change control system must allow a firm to proactively evaluate proposed changes to a method before they are implemented.

4.7 Time and Cost Savings

The benefit to this unconventional approach to method validation is that many of the standard method-development experiments lend themselves to the validation. For example, the method changes that normally surround the optimization of a method provide important robustness data for the validation.

If properly executed, the validation is essentially completed at the conclusion of method development. The experiments that provide confidence to the analytical chemist that the method is suitable are generally the same experiments that a method validation requires.

5 CONCLUSION

The drug-development process, from discovery of the chemical entity through development to a finished pharmaceutical drug product, often takes 7 to 15 years at a cost of $100 million to $500 million. The reward for this expenditure of time and money is the hope of a product that will earn $500 million a year or more in sales. It is, thus, crucial that the approval of a regulatory submission proceeds smoothly without avoidable delays. Each week that approval is delayed costs the company millions of dollars in lost sales and gives other manufacturers more time to get their competing pharmaceuticals to market. It is, therefore, of utmost importance that drug approval is not denied or withheld because of a validation issue with an analytical method.

The comprehensive approach presented within this chapter is a guidance, that if properly followed, results in validated analytical methods that will substantially improve the chances of approval by regulatory agencies worldwide. Also, with the proper controls in place an organization may implement this unconventional approach that combines method-development and -validation activities and further expedites the validation process without assuming additional regulatory liability.

6 REGULATORY REALITY CHECK

With the chapter providing an instructive backdrop, several FDA observations (FDA-483s) documented during recent inspections of FDA-regulated facilities are presented below, followed by a strategy for resolution and follow-up corrective and preventative actions.

FDA-483 issued for API. In the HPLC gradient procedure, used also as the firm's stability-indicating method, the firm lacks data that demonstrate the suitability of a resolution requirement of 3.0 between X and Y to establish the suitability of the chromatographic system.

Appropriate response. From a review of all retained lots and stability samples of API, no impurities were observed that eluted near the API with a resolution that was less than the resolution of known impurities X and Y. No other known impurity pairs have a resolution that is less than that for X and Y, thus these two components form the critical resolution pair. If these components are resolved to 3.0, then all known impurities will be resolved from the API and from each other.

Preventive Strategy. The firm should have an SOP for the validation of analytical methods that contains specificity requirements. If so, the firm would have shown in the specificity portion of the validation that X and Y indeed form the critical resolution pair. If this were proven in the specificity section with confirmation of the resolution of unknown impurities during a purposeful degradation study, this FDA-483 citation would have been avoided.

FDA-483 issued for drug product. An impurity tested for and limited to X% in Y tablets is quantified by comparing its response directly to that of a standard Z. There is no response factor included in the method to correct for the differences in absorptivities between these two compounds or data to demonstrate that the response factors are similar or that the amount of this impurity will be overestimated if no response factor is included.

Appropriate Response. It was assumed that the structure of the impurity is similar to that of the standard, since the impurity is a simple hydrolysis product of the standard. The portion of the molecule affected is distant from the chromophore, therefore the absorptivities of these compounds can be considered similar. (Note: If necessary, an experiment can be run in which a single-point comparison of absorptivity can be made for the impurity and standard.)

Preventive Strategy. The firm should have shown in the related compound linearity portion of the validation that the slope of the line for both the impurity and standard are similar (within 10% of each other). The response of the impurity can, therefore, be considered equivalent to that of the standard.

FDA-483 issued for drug product. The HPLC assay of X, has not assured that the test results are consistent and reliable. Since the approval of the product on Y, there have been two major changes to the testing procedure following out of specification (OOS) results:

Following the OOS result for lot X, a pipette rinsing technique using 50 ml of reagent alcohol was clarified to assure standardization of the sample preparation.

Following the OOS result for Lot Y, the test method was updated to allow for a gravimetric sample measurement (weighing) in place of the volumetric method (pipetting).

Appropriate response

Item a. The method was not changed. The procedure used was clarified to assure that each analyst associated with the method was performing this step as it was written and validated. Once the method change was made, a supplemental training session was held to instruct on the method wording change.

Item b. The method was changed to allow for gravimetric sample preparation. A revalidation experiment will be performed as per a QAU-approved protocol. This revalidation experiment will repeat the accuracy and precision sections of the original validation. The acceptance criteria will be the same as for the original validation.

Preventive Strategy. The firm should have performed a proficiency training exercise for new analysts in which the "unclear" pipetting step was explained. If each successive analyst found the explanation inadequate, then a method revision should have been made with a statement explaining that the clarification did not affect the original validation.

The firm should have a well-defined change control system that clearly defines change criteria for methods that require revalidation. In this instance, the change control system would have required a revalidation and an update to the original validation report.

7 WORDS OF WISDOM

Method Validation is an essential component of the global harmonization of analytical methods.

Method Validation must be performed in a regulatory compliant environment.

Any organization involved in the validation of analytical methods should have a well defined Validation Master Plan.

Comprehensive method development is critical to insuring a successful validation.

Ongoing system suitability and method-transfer exercises should be monitored to substantiate the original validation.

Innovative approaches can be used to reduce the burden of validation.

REFERENCES

1. Food and Drug Administration. Guideline for Submitting Samples and Analytical Data for Method Validation Feb 1987.
2. Food and Drug Administration. Reviewer Guidance, Validation of Chromatographic Methods. Nov 1994.
3. Department of Health and Human Services, Food and Drug Administration.International Conference on Harmonisation: Guideline on Validation of Analytical Procedures: Definitions and Terminology; Availability. docket no. 94D-0016. March 1, 1995.
4. Department of Health and Human Services, Food and Drug Administration. International Conference on Harmonisation: Guideline on Validation of Analytical Procedures: Methodology; Availability. docket no. 96D-0030. May 19, 1997.
5. Department of Health and Human Services. Food and Drug Administration. 21 CFR Part 210. Current Good Manufacturing Practices for Finished Pharmaceuticals revised April 1, 1999.
6. Department of Health and Human Services, Food and Drug Administration. 21 CFR Parts 210 and 211. Manufacturing Practices: Proposed Amendment of Certain Requirements Finished Pharmaceuticals. docket no. 95N-03621 RIN 0910-AA45. Current Good March 29, 1996.
7. U.S. Pharmacopeia 24/National Formulary 19, General Information <1225> Validation of Compendial Methods, <621> Chromatography, National Publishing, 1999.
8. Health Protection Branch, Health Canada. Drugs Directorate Guidelines: Acceptable Methods. 1994.
9. Determination of the LOD and LOQ of an HPLC method using four different techniques. Paino, T.C., Moore, A.D., Pharm Tech, Oct 1999.

5

The Stability Testing Program

Maria A. Geigel
MAG Associates, LLC, Washington Crossing, Pennsylvania, U.S.A.

1 INTRODUCTION

Stability testing and ancillary analytical work are undoubtedly the most resource-intensive activities in the chemistry, manufacturing, and controls (CMC) area of drug development. Stability studies are required in the preclinical phase, throughout the clinical trials, in support of regulatory registrations, in support of changes to the approved product, and for as long as the product is marketed. Stability data are a key element in assuring the quality of both the active pharmaceutical ingredient (API) and the formulated drug product during all phases of development and marketing.

Stability testing is primarily associated with the establishment of a retest or expiration dating period; however, it goes much farther. Stability testing helps to provide the fundamental understanding of the chemical and physical properties of the API and the drug product that are essential to the design and development of a stable drug substance and drug product formulation.

Early stability work is used to understand degradation pathways and chemical and physical interactions that continually occur during the processing, storage, and use of the product. Knowledge about these interactions and mechanisms of degradation provides guidance in formulation, manufacturing, process, and container-closure development that helps prevent or reduce undesirable transformations in the drug product.

The International Conference on Harmonisation (ICH) Q1A process established principles and conditions that are acceptable in the United States, the European Community, Japan, and many other countries. Nonetheless, it is the sponsor, who, having the most knowledge, information, and data on the product, must ultimately define the most appropriate program to evaluate the stability of its products using the defined regulatory principles.

One should remember that stability testing is but a surrogate process based on the premise that a product with the same composition, made in the same way, using the same raw materials, and stored in the same container-closure system under the same conditions will exhibit the same stability profile. As such, it is only valid in the absence of change. Consistency and proper evaluation of all changes, as well as process validation, appropriate process controls, and sampling plans, are key elements that provide confidence about the stability, and thus the safety and efficacy, of the product over time. Nonetheless, it is an inescapable fact that as a surrogate process, the dosage unit actually used by the consumer is never tested.

This chapter discusses the various stability activities undertaken during drug development, through regulatory approval, to postapproval changes. It provides reference to and sometimes a synopsis of the key regulatory guidance and discusses both the areas to be addressed and issues that may be encountered during development. It also addresses common compliance pitfalls and ways to prevent them.

The goal of this chapter is to provide an understanding of the overall stability study process and its corresponding compliance challenges. It is hoped that such an understanding will result in stability studies that not only meet all regulatory and compliance requirements, but also improve and increase the assurance of the quality of the marketed product.

The key stability regulatory guidances are given in Example 1.

EXAMPLE 1: KEY REGULATORY STABILITY GUIDANCES

FDA Guidance for Industry—Content and Format of Investigational New Drug Applications (INDs) for Phase 1 Studies of Drugs, Including Well-Characterized, Therapeutic, Biotechnology-Derived Products, November 1995 (FDA IND Phase 1 Guidance)

FDA Draft Guidance for Industry—INDs for Phase 2 and 3 Studies of Drugs, Including Specified Therapeutic Biotechnology-Derived Products, Chemistry, Manufacturing, and Controls Contents and Format, February 1999 (FDA IND Phase 2/3 Guidance)

ICH Guideline for Industry—Q1A Stability Testing of New Drug Substances and Products, September 1994 (ICH Stability Guidance)
ICH Draft Revised Guidance for Industry—Q1A(R) Stability Testing of New Drug Substances and Products, April 2000 (ICH Draft Revised Stability Guidance)
ICH Guidance for Industry—Q1B Photostability Testing of New Drug Substances and Products, November 1996 (ICH Photostability Guidance)
ICH Guidance for Industry—Q1C Stability Testing of New Dosage Forms, November 1996
FDA Draft Guidance for Industry—Stability Testing of Drug Substances and Drug Products, June 1998 (FDA Draft Stability Guidance)
FDA Guidance for Industry—Changes to an Approved NDA or ANDA, November 1999 (FDA Guidance for Changes)
Current Good Manufacturing Practices (cGMP), 21 CFR 211

2 DEVELOPMENT STABILITY STUDIES

Stability studies parallel the clinical drug development program, thus as the clinical development program moves from phase I, in which the safety of the drug is evaluated in humans, to phase II dose-ranging studies, to phase III safety and efficacy studies, stability testing also progresses. Stability studies go from early testing, which supports the use of the clinical trial materials for the length of the clinical studies, to focused testing, which determines the retest date of the API and the expiration dating period of the drug product.

Stability studies on both the API and drug product are carried out from the very start of the development program and provide increasingly valuable information. It should be remembered that development is an iterative process during which many changes occur that are mirrored in additional stability studies.

2.1. Regulatory Requirements

The FDA's stability requirements for INDs are provided in FDA IND Phase 1 Guidance and FDA IND Phase 2/3 guidance. These guidances recognize the progressive nature of manufacturing and controls information and prescribe a limited amount of stability information initially, augmented by additional data as the clinical trials proceed.

The FDA IND Phase 1 Guidance specifies that during phase 1 only a brief description of the stability study and test methods and preliminary tabular data based on representative material in the proposed container-closure system be submitted; neither detailed stability studies nor stability protocols are required at this phase. This applies to both the API and drug product. For phase 2, FDA IND phase 2/3 guidance indicates that a stability protocol, which includes the tests, analytical procedures, time points for each of the tests, duration of the study, and preliminary stability data on representative material, should be submitted for both the API and the drug product. All stability data on clinical trial material used in phase 1 should also be submitted.

For phase 3, FDA phase 2/3 guidance indicates that the following should be submitted for both the API and the drug product:

- Stability protocol for accelerated and long-term studies, including all the information in the phase 2 stability protocol (tests, analytical procedures, time points for each of the tests, and duration of the study) plus temperature and humidity conditions and a detailed description of the material under investigation, including packaging
- Short description of each parameter being investigated, demonstrating that appropriate controls and storage conditions are in place to ensure the quality of the material used in clinical trials.
- Well-defined description of any tests unique to the stability program
- Proposed bracketing and matrixing protocol, if applicable
- Tabulated data that include the lot number, manufacturing site, and date of manufacture for the material used in clinical trails, and for drug products, the lot number of the drug substance used
- Representative chromatograms and spectra
- Dissolution profiles in physiologically relevant media with reasonable speeds of agitation, where appropriate

The FDA phase 2/3 Guidance also points out that stress studies for both the API and drug product should be carried out if not performed previously. In addition, a container-closure challenge test that demonstrates that the container-closure maintains an effective microbial barrier during the product's expiration dating period should be developed for sterile drug products. This challenge test on the container-closure system, in conjunction with drug product sterility testing, provides assurance of the sterility of the drug product.

2.2 API Stability Studies

Stability studies on the API include the evaluation of various forms of the active moiety, (e.g., different salts, hydrates, solvates, and polymorphs). After the API has been defined, stability studies are carried out as the manufacturing process is optimized and/or scaled up. The API stability information can also be important in guiding the design of the formulation (e.g., if the API is not stable in the presence of certain excipients or at certain pHs).

2.3 Drug Product Stability Studies

2.3.1 During Preclinical Testing

During preclinical studies, the drug is tested in animals to identify any toxicity issues. Chemical analysis includes identification of the number and amount of impurities by area percentage. The first stability studies on a drug product, even before initiation of the IND, are those carried out to demonstrate that the drug used in the preclinical studies was stable throughout the test period. These very short-term stability studies may be no more than an assay of the drug (and microbiological testing if required) carried out before initiation and after completion of the toxicity study.

2.3.2 During Phase 1 Clinical Testing

In order to assure the quality of the drug product during phase 1 studies, special precautions may be necessary for drugs that are unstable in their early formulations (e.g., use of the most protective packaging and refrigerated conditions). For such products, in addition to full real-time stability studies, it may be desirable to test the drug product just prior to use in the clinical trial and after completion of the dosing. As with preclinical studies, this strategy removes any concerns about the drug's stability during the clinical trial.

2.3.3 During Phase 2 and Phase 3 Clinical Testing

Stability studies during the IND phase provide not only assurance of the quality of clinical trial material but also the basis for the drug product development program. Stability data are used to evaluate different formulations, methods of manufacture, and container-closure systems, as well as to determine storage requirements, expiration dating periods, and specifications.

To Support Clinical Studies. Many different formulations may be used during phase 2 and phase 3 clinical studies. Stability studies must be carried out to assure that the strength, purity, and quality of all clinical trial materials are within certain established ranges and specifications. Clearly, dose-related and efficacy conclusions drawn from clinical trials can only be as good as the data on the strength of the drugs under investigation, and safety conclusions can only be as good as the information about their purity, therefore accurate, precise, and stability-indicating analytical methods that provide good estimates of appropriate expiration dating periods are critical to "protect the clinical trial" (i.e., assure that the results of the clinical trials are based on valid data and sound assumptions).

Every batch of clinical trial material is not necessarily included in stability studies; however, because so many factors that can affect stability (e.g., formulation, manufacturing, container-closure) change during product development, many if not most clinical trial materials are included in some form of stability study. The results of the stability studies and the rationale for not conducting stability studies should be documented and organized appropriately in anticipation of regulatory review during pre-approval inspections (PAIs).

Several steps are often taken by firms to turn an inordinate amount of stability testing into a manageable program (e.g., bracketing, matrixing, application of data from one material to another, evaluation of the worstcase scenario, and extrapolation of accelerated data). Bracketing is usually used for different container-closure sizes (i.e., testing the same formulation or strength in the smallest and largest container), but may also be used for different formulation strengths (i.e., testing the lowest and highest strength of the same formulation in the same container-closure systems). Data from one formulation may be applied to a similar formulation in the same container-closure, or data for a worst case may be applied to a lesser case. For example, the expiration-dating period of a formulation in a certain container-closure, such as a polyvinyl chloride (PVC) blister, may be applied to the same formulation in a more protective container-closure of the same type, such as a PVC/Aclar blister, which provides a better moisture barrier.

Extrapolation of data from one material to another must always be based on sound scientific principles and all available data. Even so, extrapolation always bears risk. To demonstrate that every batch of clinical trial material was within applicable specifications for the duration of its use in clinical studies, some firms test every batch of clinical trial material at expiry. This is especially valuable when an expiration-dating period was not based on real-time data for the precise clinical trial material in the precise container-closure used in the clinical trials.

To Support Product Development. In addition to supporting the use of clinical trial materials, stability studies are carried out on both the API and different formulations in different container-closure systems to guide the development of the final formulation and container-closures. Stability studies may be carried out on either the formulation to be marketed or a representative formulation to evaluate different API suppliers, drug product manufacturing processes and sites, and drug product container-closures.

Container-closure studies may evaluate

Type of container-closures (e.g., blisters and bottles for solid dosage forms; vials, ampules, and prefilled syringes for injectables)
Container-closures of the same type (e.g., poly vinyl chloride [PVC], PVC/Aclar, and aluminum blisters; high density polypropylene [HDPE] containers with different caps and/or seals)
Suppliers of the same container-closure made using different resins

More protective container-closure systems generally allow a longer expiration-dating period. Since more protective systems are usually more expensive than less protective systems, these studies provide information for choosing the optimum balance between packaging costs and expiration dating period.

2.4 Compliance Issues

All stability studies on clinical trial materials must be carried out in full accordance with cGMPs, even if a research department carries out the studies. All studies must be carried out by adequately trained personnel under adequate work conditions. The personnel must use properly qualified and calibrated stability chambers, instruments, reagents, and standards. They must follow validated analytical methods and approved written procedures, and they must properly document all work. There must be proper sample and data traceability, change control, and go on.

Failure to comply with cGMP requirements is just one of the areas during clinical trials that may result in compliance issues. Improperly assigned expiration dates, inadequate procedures related to the extension of expiration dates, and a breakdown in any of the steps from manufacture of a clinical trial material to its use in a clinical trial could result in compliance issues. Starting with the manufacture of clinical trial material, there must be well-designed and documented procedures that effectively involve all appropriate persons and departments to assure that only material within specifications is used in clinical trials. There must be procedures that ensure the timely availability of clinical supplies, timely and appropriate extension of expiration dates, timely investigation of out-of-specification (OOS) results and rapid notification of appropriate persons, timely recall of all OOS and expired materials, and timely resupply with fresh materials. Depending on the severity of the compliance issue, the clinical trial could be considered compromised.

To enable a clinical trial to start as soon as possible, stability testing is carried out in parallel with the clinical trial, but is initiated 3 to 6 months earlier. Three-to-six-month accelerated and real-time data, which are used to project or support the expiration dating period for the batch(es) to be used in the trial, then become available before the trial begins. In addition, based on other available data, an expiration date beyond the available realtime data may be justified.

Some countries require that clinical supplies be labeled with their expiration date. When that date is extended based on additional data, appropriate persons are notified and stickers with the new expiration date are distributed to those who hold the drug so that it may be overlabeled with the new expiration date. In this scenario, drug that is near its expiration date should be obvious and would not be used in a clinical trial.

Other countries, including the United States, do not require that clinical supplies be labeled with an expiration date. Instead, systems are created to assure that any drug that fails to meet specifications will not be used in a clinical trial. This requires good, documented communication among all groups involved in clinical supplies. These groups are usually those that generate the stability data and extend expiration dates, those that manufacture, package, label, and manage clinical supplies, and those that monitor clinical trials. If the drug does not degrade and the expiration date is extended, there are few problems. If the expiration date cannot be extended, however, clinical study monitors must be notified immediately so that all drug that is about the expire can be recalled from clinical trial sites. Clearly, during clinical trials there are many potential compliance pitfalls related to stability studies and the role of the responsible analytical person. These must be continually kept in mind and not underestimated.

3 FORCED DEGRADATION/STRESS STUDIES

Forced degradation/stress studies elucidate the inherent stability characteristics of the molecule under study and determine its degradative pathways. Forced degradation studies are needed to support method validation, determine product protection requirements, and guide formulation development.

The sponsor should consider evaluating forced degradation products in animal studies to determine if they present a toxicity issue, and, if so, an appropriate limit to ensure safety. Failure to identify toxic degradation products early in the development program can be costly mistake if the drug has to be abandoned later because of potential toxicity concerns.

A comprehensive literature search should be carried out before initiating any laboratory work related to forced degradation studies. Such a search may uncover the needed information if the API is a well-known entity or guide the choice of conditions to be used if information is available for related compounds.

To support method validation. Forced degradation studies provide the degradation products that are the basis for the development and validation of both API and drug product methods. Forced degradation products are a prerequisite for assessing the stability-indicating properties of a method (i.e., that the proposed method is capable of separating and accurately quantifying the analyte of interest in the presence of potential degradation products).

Do all of the forced degradation products need to be evaluated during method validation? How long do you continue to validate a method using degradants that have never been observed? (This is especially relevant when new methods are developed for old products.) These questions are significant because of the impact of forced degradation products on the complexity of the method validation work, and because forced degradation products can be difficult to obtain.

The fact that some degradation pathways are complex and under forced conditions may result in products that are unlikely to be formed under accelerated or long-term conditions is recognized in ICH Draft Revised Stability Guidance and FDA Draft Stability Guidance. These guidances point out that it may not always be necessary to examine specifically for all degradation products if it has been demonstrated that they are not formed in practice. Remember, however, that the onus remains on the sponsor to demonstrate that they are not formed.

To determine product protection requirements. Forced degradation studies provide valuable information related to the type of protection (i.e., container-closure system and storage conditions) that will be necessary, not just for thc API, but also for the drug product. These studies also help to determine whether or not special precautions are required during shipping.

To guide formulation development. Forced degradation studies guide formulation development when evaluation of degradation products shows that a degradation product may interact with certain excipients. In such a case, boundaries can be placed on the choice of excipients rather than later uncovering inexpicable instability if the degrandants had not been generated and evaluated.

Without forced degradation studies, instability that is observed later in the development program and is due to excipient-degradant interactions might be attributed to heat and/or moisture. The proposed solution might then be to use more costly packaging, shortening the expiration-dating period, or restricting the storage conditions. Instead, early recognition of excipient–degradant interactions could be more easily resolved by modifying the formulation.

3.1 Forced Degradation Studies on the API

Forced degradation studies on the API are generally done early in the development program on a single batch. The FDA IND Phase 2/3 Guidance recommends that these studies be carried out in phase 2.

During forced degradation studies, the API is subjected to various extreme conditions. Forced degradation studies should address the effect of extreme temperature (e.g., in 10°C increments above the accelerated temperature test), high humidity, where appropriate (e.g., 75% or greater), acidic and basic conditions across a wide range of pH, oxidative conditions using both hydrogen peroxide solution and/or oxygen gas, and photolysis conditions, which are discussed later.

The desired outcome of forced degradation studies is the generation of not only degradation products that may be observed during normal storage of the API and the dosage form, but also degradation products that are not normally observed but could be formed under certain strenuous conditions.

Some firms are satisfied when no degradation is observed under the chosen forced degradation conditions and consider this evidence that the API is inherently stable. It should be emphasized, however, that the main purpose of forced degradation studies is to generate degradation products, not to test the stability of the product, thus if the conditions initially chosen do not result in degradation, the severity of the conditions (i.e., time, temperature, and/or concentration) should be increased. If the substance does not react because it is insoluble, organic cosolvents should be used to solubilize the material.

As determined by decrease of assay, about 20% degradation is desirable. Greater than 20% degradation may result in isolating degradants of the primary degradation products, while less than 20% degradation may not result in production of all degradation products. If no degradation is observed after reasonable attempts and fairly severe conditions, the attempts and resulting data should be presented in the regulatory submissions.

3.2 Forced Degradation Studies on the Drug Product

Forced degradation studies on the drug product are carried out on a single batch of the formulation to be marketed to determine if there are any additional degradation pathways and degradants formed due to interactions between or among the API and the excipients.

The FDA IND Phase 2/3 Guidance recommends that these studies be carried out in phase 3. The timing of forced degradation studies on the drug product is a trade-off. On the one hand, the sponsor would prefer to carry out these studies only once, which would be after the formulation to be marketed has been identified. On the other hand, forced degradation products are needed for method validation, which may take place before the final formulation is chosen.

If forced degradation studies on the drug product are done too early (i.e., before the final formulation is chosen), they may have to be repeated if the formulation changes. If forced degradation studies on the drug product are done too late (i.e., after method validation) and new degradation products are observed, method validation may have to be repeated. From a compliance perspective, if forced degradation studies are carried out after method validation and it is determined that a method is not specific for a new degradation product, the analytical results carried out using that method could be called into question.

3.3 Photolysis Studies

Photolysis studies are described in ICH Photostability Guidance and FDA Draft Stability Guidance. There are two types of photolysis studies on the API: forced degradation testing and confirmatory testing. The former is used to evaluate the overall photosensitivity of the substance and to generate degradants. The latter is carried out on API that is stored under practical conditions to determine appropriate light conditions during handling, packaging, and use of the API, and any necessary light precautions that may be required on API labels.

Photolysis studies on the drug product are carried out in a sequential manner, starting with testing the fully exposed product then progressing as necessary to the product in the immediate container-closure and then as provided for market. Testing is continued until results demonstrate that the product is adequately protected from light. As with photolysis confirmatory testing on the API, these studies determine the appropriate light conditions during handling, packaging, and use of the drug product, and any necessary light precaution that may be required on the labeling.

If the literature states that a material is not photosensitive, confirmatory photolysis studies may suffice and preliminary studies may not be necessary.

3.4 Compliance Issues

Failure to carry out appropriate forced degradation studies can result in compliance issues. If forced degradation products are not generated and included in method validation studies, those methods may be deemed unvalidated and all results generated using those methods could be questioned. If an unknown peak is observed later in the development program as the result of a stability study or after approval of the product, investigations related to the unknown peak might be deemed inadequate if they do not include a review of forced degradation products.

4 REGISTRATION STABILITY STUDIES

The requirements for registration stability studies for API and drug products (i.e., studies that support regulatory submission) are outlined in ICH Draft Revised stability guidance and FDA Draft Stability Guidance. Stability studies are also required by cGMP and are cited in 21 CFR 211.137 and 211.166, which discuss the requirements for expiration dating and stability testing, respectively.

The FDA Draft Stability Guidance includes all the ICH requirements as well as further elaboration and discussions on other stability-related topics, such as stability protocols, content and format of stability reports, bracketing and matrixing, mean kinetic temperature, container-closures, microbiological control, sampling, manufacturing sites, site-specific stability studies, degradation products, thermal cycling, considerations for specific dosage forms, and stability required for postapproval changes.

Only some of the topics in these guidances are summarized here. The reader is encouraged to become familiar with the full details in the original documents.

4.1 Selection of Batches

Perhaps the most important criterion for registration stability studies (i.e., the studies used to determine the retest date for an API and the expiration-dating period for a drug product) is the selection of batches, which must be representative of the material to be marketed.

4.1.1 API

A minimum of three API registration stability batches should be made at a minimum of pilot scale by the same synthetic route using a process that simulates the process to be used at the manufacturing scale. They should be packaged in a container-closure system that is the same as, or a simulation (small scale) of, the packaging for marketing storage and distribution. The overall quality of the API should be representative of the quality of material to be made on a manufacturing scale.

The main criteria for selecting batches for API registration stability studies are given in Table 1.

TABLE 1 Selection of Batches for API Registration Stability Studies

Number of batches	Minimum three
Batch size	Minimum pilot scale
Synthetic route	Same as manufacturing scale
Manufacturing process	Same as, or simulation of, manufacturing scale
Container-closure	Same as, or simulation (small scale) of, container-closure for material at manufacturing scale

Source: See ICH Draft Revised Guidance for Industry-Q1A(R) Stability Testing of New Drug Substances and Products, April 2000, and FDA Draft Guidance for Industry-Stability Testing of Drug Substances and Drug Products, June 1998.

4.1.2 Drug Product

Drug product registration stability batches should be the same formulation and dosage form as the marketed product. A minimum of three batches should be manufactured by a process that meaningfully simulates the process for the marketed product and provides product of the same quality as is intended for the marketed product. Where possible, batches of the drug product should be manufactured using identifiably different batches of the API. The batches should be packaged in the container-closure systems intended for the marketed product. Two of the three batches should be a minimum of pilot scale. The third batch may be smaller (e.g., 25,000 to 50,000 tablets or capsules for solid oral dosage units).

If the drug product in the registration stability study is not the same as the product used in the clinical trials, the differences must be assessed. Likewise, any differences between the drug product in the registration stability study and the product to be marketed must be assessed. All differences should be explained and justified. Release data comparisons and/or a bioequivalence study may be required, depending on the significance of the differences.

Sometimes container-closure development is not complete when the registration stability studies are initiated. Firms may then include multiple container-closure systems in the registration stability study and choose the one(s) to be used for the marketed product based on the results of the stability study. Although this means additional work, it is usually preferable to a delay in the start of the stability study.

Criteria for selecting batches for drug product registration stability studies are given in Table 2.

TABLE 2 Selection of Batches for Drug Product Registration Stability Studies

Number of batches	Minimum three, using, where possible, identifiably different batches of API
Batch size	Minimum two pilot scale batches plus one additional batch, which may be smaller (e.g., 25,000 to 50,000 solid oral dosage units), but not laboratory scale
Formulation/dosage form	Same as the marketed product
Manufacturing process	Meaningful simulation of process for the marketed product
Container-closure	Same as the marketed product

Source: See ICH Draft Revised Guidance for Industry-Q1A(R) Stability Testing of New Drug Substances and Products, April 2000, and FDA Draft Guidance for Industry-Stability Testing of Drug Substances and Drug Products, June 1998.

4.2 Test Procedures

Test procedures should include all parameters that are susceptible to change during storage and that are likely to influence quality, safety, and/or efficacy. These include, as appropriate, physical, chemical, biological, and microbiological characteristics, and for drug products, loss of preservative and functionality. Validated, stability-indicating methods should be used.

4.3 Acceptance Criteria

For APIs, the registration stability study acceptance criteria should be derived from the quality profile of the material used in preclinical and clinical studies. The acceptance criteria should be numerical limits, ranges, and other criteria for specific tests, and should include limits for individual and total impurities and degradation products.

For drug products, the following considerations should be taken into account in setting acceptance criteria for registration stability studies:

They should be based on all available stability information.

They may differ from release specifications, if justified.

They should include specific upper limits for degradation products based on the levels observed in the material used in the preclinical and clinical studies.

Acceptance criteria for other tests, such as particle size or dissolution rate, should be based on results observed for material used in bioavailability and/or clinical studies.

Any differences between the release and shelf-life acceptance criteria for antimicrobial preservatives should be supported by the results of preservative efficacy testing.

4.4 Testing Frequency

The testing frequency for stability studies was clarified in ICH Draft Revised Stability Guidance. The testing frequency is given in Table 3.

The FDA Draft Stability Guidance, which predates the ICH Draft Revised Guidance, sdoes not provide testing frequency for intermediate studies or accelerated studies for API. It indicated a minimum of four test stations (e.g., 0, 2, 4, and 6 months) for accelerated studies for drug products.

4.5 Testing Dates

Regulatory guidances specify appropriate testing frequency for stability studies but do not discuss how to define zero-time, analysis date, analysis time, or how much deviation from scheduled test dates is acceptable.

4.5.1 Zero-Time

Different companies define zero-time in different ways. It may be the date corresponding to the completion of the manufacture of the bulk product, the start or end of packaging, the release by quality assurance (QA), the placement of samples in the stability chambers, or other measures. These dates can differ significantly for a variety of reasons, such as scheduling conflicts in the manufacturing and packaging departments and the time required to initiate a stability program.

An standard operating procedures (SOP) should define zero-time as well as when release data may be considered zero-time data and when testing

Table 3 Testing Frequency for Registration Stability Studies

Long-term studies	Every 3 months for the first year, every 6 months for the second year, and annually thereafter (e.g., 0, 3, 6, 9, 12, 18, 24, and 36 months)
Intermediate-condition studies	Minimum of four points, including the initial and end points (e.g., 0, 6, 9, and 12 months)
Accelerated studies	Minimum of three points, including the initial and end points (e.g., 0, 3, and 6 months)

Source: See ICH Draft Revised Guidance for Industry-Q1A(R) Stability Testing of New Drug Substances and Products, April 2000.

must be repeated to generate zero-time data. This is usually dependent on the amount of time between release testing and zero time. Zero-time should be assigned consistently in accordance with the SOP and clearly described in the stability report. To this end, FDA's draft stability guidance recommends entry of the manufacturing, packaging, and stability start dates on each stability data table.

Another consideration is whether zero-time data for a batch in one container-closure may be used for the same batch in a different container-closure system. This would depend on the amount of time between packaging of the product and zero-time as well as the stability profile of the product. The data generated for zero-time for a batch in one type of container-closure are usually also used to evaluate the same batch in the same container-closure under different storage conditions.

However zero-time is defined, it is important to consistently honor that time point, as this has been a recurring compliance concern for the FDA. The notion that there is a "grace period" for a zero-time point or any other test-time point can have serious regulatory and compliance implications.

4.5.2 Analysis Dates

Different analytical tests for a given time point may be carried out on different days. What then is considered the analysis date for a sample? Again, there is no universal rule, and the important thing is a justifiable SOP, consistency, and adherence to the SOP. The start date for the most critical test, usually impurities and/or assay but sometimes dissolution or release, is a logical point.

4.5.3 Analysis Time

An SOP should also govern the allowable time from the start of the first test to the completion of all testing. This interval should be justified and not prolonged, which could skew interpretation of the results. Proper oversight to assure timely completion of all tests is especially important when different tests are carried out by different analysts or departments with different workloads, schedules, and priorities.

4.5.4 Testing Date Window

Once a stability study is started, testing dates are defined by the testing frequency. Occasionally, however, most laboratories will experience difficulty testing precisely on the scheduled test date. An SOP therefore should describe the allowable time between the scheduled test date and the actual analysis date. A tiered approach is common, with a narrow testing window for short-term testing points and a wider window for long-term testing points. Testing date windows should take into account how much data are likely to differ, based on the analysis date.

Significant recurring deviations from scheduled test dates should be investigated. A periodic review of these investigations may assist in identifying recurrent problems and implementing corrective actions. The investigation should include the impact of the delay in testing on the results generated and the impact on the interpretation of the results. Deviations between scheduled test dates and analysis dates are often covered during laboratory inspections.

Significant discrepancies between the scheduled test date and the actual analysis date should be explained in the stability report, including the impact of actual analysis dates on the results generated. To ameliorate the consequences of late analysis dates some companies generate stability curves using equations that incorporate the actual analysis dates rather than the scheduled test dates. This should be explained in the stability report.

4.6 Storage Conditions

Registration stability studies should be carried out under storage conditions and for a length of time that covers the conditions for storage, shipment, and subsequent use. The effect of temperature and moisture (either high or low humidity) should be evaluated as appropriate. Registration stability studies include long-term and accelerated storage conditions, and if necessary, intermediate storage conditions. Long-term conditions provide data for determining the retest date for an API and the expiration-dating period for a drug product. Data at accelerated conditions assess the impact of short-term excursions outside what will be the label storage conditions, and provide the technical confidence needed to project expiration dating periods when long-term data are not available.

The conditions for both API and drug products registration stability studies are summarized in Table 4. In addition, cycling studies to determine the effect of temperature variations on certain drug products should be considered.

The FDA Draft Stability Guidance, which predates the ICH Draft Revised Guidance, does not provide specific storage conditions for temperature-sensitive APIs. The conditions for drug products to be stored in a refrigerator are the same as in the ICH Draft Revised Guidance. The conditions for drug products to be stored in a freezer are $-15^\circ \pm 5^\circ$C for long-term studies and $5^\circ \pm 3^\circ$C for accelerated studies.

TABLE 4 Storage Conditions for API and Drug Product Registration Stability Studies

General case
Long-Term: 25° ± 2°C/60% ±5%RH[a]
Intermediate: 30° ± 2°C/60% ± 5%RH[a]
Accelerated: 40° ± 2°C/75% ± 5%RH[a]
Material to be stored in a refrigerator
Long-Term: 5° ± 3°C
Accelerated: 25° ± 2°C/60% ± 5%RH[a]
Material to be stored in a freezer
Long-Term: -20° ± 5°C plus data for one batch at 5° ± 3°C or 25° ± 2°C to support use outside the freezer

[a]Relative humidity (RH) depends on the material and the container-closure. See humidity discussion.
Source: See ICH Draft Revised Guidance for Industry-Q1A(R) Stability Testing of New Drug Substances and Products, April 2000.

4.6.1 Humidity

Specific relative humidity conditions are not necessary for products stored in impermeable container-closure systems that provide a permanent barrier to passage of moisture or solvent (e.g., liquids in sealed glass ampules and semisolids in sealed aluminum tubes).

The high relative humidity conditions cited in Table 4 apply to APIs and solid dosage forms. Low relative humidity conditions should be used for liquid drug products in semipermeable container-closure systems. The ICH Draft Revised Stability Guidance suggests low RH conditions of 40% + 5% relative humidity for long-term studies, 60% + 5% relative humidity for intermediate studies, and not more than 25% RH accelerated studies.

4.6.2 Intermediate Storage Conditions

Intermediate storage conditions are used when a significant change occurs under the accelerated conditions for products that use 25°C as the long-term condition.

For an API, significant change is defined as failure to meet specifications. For a drug product, significant change is defined as

Five percent potency change from the initial assay value
Any specified degradant exceeding its specification limit
pH exceeding its specification limits
Dissolution exceeding the specifications for 12 dosage units (i.e., USP stage 2)

Failure to meet specifications for appearance and physical properties (e.g., color, phase separation, resuspendablity, delivery per actuation, caking, hardness, and as appropriate to the product type)

Stability studies at intermediate storage conditions should include testing of all parameters included in the accelerated stability studies, not just the parameter(s) that failed the significant change criteria.

4.6.3 Duration of Studies

Accelerated studies are carried out for 6 months, intermediate studies for 12 months and long-term studies for the duration of the proposed retest period for APIs and expiration dating period for drug products.

Registration applications should include 6 months accelerated data, 6 months intermediate storage conditions data if applicable, and 12 months, long-term data.

For ANDA's, accelerated studies are carried out for 3 months. Available long-term data are included in the original submission. If a significant change is observed under the accelerated conditions, 12 month data at the intermediate condition or long-term data through the expiration-dating period are required.

4.7 Evaluation

Stability data (not only assay but also degradation products and other attributes as appropriate) should be evaluated using generally accepted statistical methods. The time at which the 95% one-sided confidence limit intersects the acceptable specification limit is usually determined. If statistical tests on the slopes of the regression lines and the zero-time intercepts for the individual batches show that batch-to-batch variability is small (e.g., p values for the level of significance of rejection are more than 0.25), data may be combined into one overall estimate. If the data show very little degradation and variability and it is apparent from visual inspection that the proposed expiration dating eriod will be met, formal statistical analysis may not be necessary.

Extrapolation of an expiration-dating period beyond real-time data may be justified if supported by data from accelerated and supportive stability studies. The extrapolated expiration-dating period must be confirmed by real-time stability studies on the marketed product. Although a long expiration-dating period is most desirable, caution should be exercised in extrapolating beyond real-time data, as this could put marketed product at risk. Batches that fail to maintain the required quality parameters throughout the extrapolated period could be subject to a recall.

4.8 Bracketing and Matrixing

Bracketing is the testing of the outside values of a range of a parameter to support the complete range. Testing only the smallest and largest container-closure size or testing only the lowest and highest strength of a dosage from are examples of bracketing.

Matrixing is a statistical design of a stability schedule so that only a fraction of the total number of samples are tested at any specified sampling point. At subsequent sampling points, different sets of samples from the total number are tested.

Bracketing and matrixing reduce the amount of testing that is carried out. These approaches are appropriate and highly advantageous in large programs that include multiple strengths of the same formulation in multiple container-closures. These techniques are not always appropriate, however and a basic understanding of the product's stability is a prerequisite. Samples in a bracketed study should be the same in all respects except for the one being abbreviated. For example, if the testing of different strengths of a formulation will be abbreviated, the container-closure for all of the strengths should be identical. Care must be taken to assure that seemingly identical container-closures do not have small but significant differences that could impact stability (e.g., different bottle resins or the use or lack of desiccants). For liquids, a difference in the surface to volume ratio may be important for a permeable container.

A bracketing plan, in which only the smallest and the largest or the lowest and the highest value of a range are tested, is fairly easy to develop. However, matrixing must be based on a rational, scientifical statistical plan. However, care must be taken to assure that the data resulting from the plan are sufficient to support the proposed expiration-dating period. Various matrixing plans are discussed at length in FDA Draft Stability Guidance.

It should also be remembered that extrapolation of the expiration-dating period beyond real-time data may not be possible with an abbreviated testing program. The five basic rules presented in Example 2 should be considered in the development of any abbreviated testing program.

EXAMPLE 2: BASIC RULES WHEN CONSIDERING BRACKETING OR MATRIXING

Basic understanding of the product's stability
Rational statistical plan
Documentation of rationale
Sufficient data to support proposed expiration dating period
Discussion with FDA

Because of the many different possible issues and the many ways in which a stability program can be bracketed or matrixed, the sponsor should not embark on such a testing plan without discussing it with FDA.

4.9 Special Studies on the Drug Product

In addition to stability studies on the product as it will be marketed, the product's conditions-of-use should be examined to determine if special studies are necessary. Such studies only need to be as long as the duration of expected conditions-of-use, or to be conservative, some multiple of the duration of the expected conditions-of-use.

Products requiring condition-of-use studies include reconstituted products, infusion mixtures listed in the labeling for parenterals, and products supplied in primary packaging enclosed in additional packaging that protects the product during long-term storage. The last category includes moisture-sensitive solid dosage forms packaged in blisters enclosed in pouches impervious to moisture to assure adequate protection during long-term storage. When more than one blister unit is contained in the secondary packaging, the stability of the blistered product after it is removed from the secondary packaging must be determined.

4.10 Abbreviated New Drug Application

Much of the information discussed is applicable to ANDAs. Depending on the complexity of the dosage form and the availability of information, however the amount of information required may be different.

For APIs, stability data should be generated on a minimum of one pilot-scale batch made using equipment of the same design and operating principle as the manufacturing-scale equipment (with the exception of scale).

For simple dosage forms, the following data package is recommended:

Minimum of one pilot-scale batch
Accelerated data at 0, 1, 2, and 3 months
Long-term data available at the time of filing
If significant change is observed, 12-month data at intermediate conditions or long-term data through the proposed expiration dating period

A minimum of three pilot-scale batches (two at least pilot scale and one smaller batch, e.g., 25,000 to 50,000 units for solid oral dosage forms) are required for the following dosage forms, and other exceptions, which should be discussed with FDA:

Complex dosage forms, such as modified-release products, transdermal patches, metered-dose inhalers

Products without a significant body of information
New dosage forms submitted through the ANDA suitability petition process

4.11 International Stability Programs

The identification by ICH of conditions for stability studies that are acceptable to three major regions of the world and also accepted by many other regions is invaluable to an international drug development program. Previously, individual stability studies were conducted for each region, often under different conditions, making comparison of data difficult. It is now possible to design a stability program that can support registrations in all the major areas of the world.

It must be remembered, however, that ICH guidelines stipulate that several parameters of the material under study must be the same as, or a meaningful simulation of, the parameters of the product to be marketed.

For products that will be manufactured in different parts of the world, a careful comparison must be made of the API and the drug product to be included in the stability program and those manufactured in different facilities. For the API, the comparison should include the route of synthesis, manufacturing process, and container-closure; for the drug product, the formulation, manufacturing process, and container-closure. This comparison is especially critical when raw materials and container-closure components are obtained from different sources.

Example 3 lists important questions that should be examined for a stability program designed to support marketed product that will be manufactured in different facilities.

EXAMPLE 3: QUESTIONS FOR INTERNATIONAL STABILITY PROGRAMS

Are the manufacturing process in different facilities, which may use different equipment, the same? If not, are there any differences that could result in different stability profiles?
Are the container-closure systems the same?
Are the resins and additives used in the container-closure components the same?
Are the resins and additives obtained from different manufacturing facilities of the same supplier the same?
Do excipients meet both USP and Ph.Eur. criteria?
Are there any differences in the excipients that could affect the stability of the drug product?

4.12 Stability Section of Registration Applications

By the time preparation of a registration application has begun, voluminous amounts of stability information and data are available. What should be included in the registration application and what is the best way to present the information and data?

The FDA Draft Stability Guidance includes a discussion on the content and format of stability reports. The API stability section is generally straightforward; however, the drug product stability section can have many parts and should be especially well organized. In addition to the elements discussed in FDA Draft Stability Guidance, the items included in Example 4 should be considered in order to facilitate review.

EXAMPLE 4: ITEMS THAT FACILITATE REVIEW OF THE STABILITY PROGRAM

Introduction that explains the overall organization of the stability section

Separate sections for the thermal studies, photolytic studies, special studies, supportive studies, and forced degradation studies/stress studies

If thermal studies are extensive, a table that provides an overview, including batch number, batch size, manufacturing site and date, drug substance batch number used, container-closure, storage conditions, and amount of data available, with reference to table number where data are summarized

Discussion of the bracketing or matrixing plan used, if appropriate, including justification

In addition to tabulations of all data for a given batch by storage conditions as specified in FDA's draft stability guidance, tables and/or graphs that include data across batches for assay, impurities, and dissolution

Clearly specified storage statements and expiration dating periods, especially when these are different for different product strengths and/or container-closures

Clearly specified stability commitments

Clearly specified stability protocol for the first three production batches and annual batches thereafter

Clearly marked methods used in stability studies when these are different form the regulatory methods, with an explanation of their relationship to the regulatory methods

4.13 Compliance Issues

All registrations stability studies, both accelerated and long-term, must of course be carried out in full compliance with cGMPs. Example 5 presents stability areas often cited as deficient during regulatory inspections.

EXAMPLE 5: COMMON COMPLIANCE DEFICIENCIES RELATED TO REGISTRATION STABILITY STUDIES

Improper sampling of registration batches
Inadequate tracking and accounting of samples
Improper storage of samples
Inadequate labeling
Improperly-qualified and/or calibrated stability chambers
Improper definition of zero time
Deviations from scheduled test dates
Use of unvalidated methods
Inadequate laboratory procedures
Inadequate investigations of OOS results
Lack of procedure for out-of-trent (OOT) results
Expiration dating period not supported by available data

5 POSTAPPROVAL STABILITY STUDIES

It appears that stability studies never really end. Indeed, development and registration stability studies that culminate in approval to market a drug may be considered just the start of stability studies for that product. Example 6 shows the various types of postapproval stability studies.

EXAMPLE 6: TYPES OF POSTAPPROVAL STABILITY STUDIES

Completion of registration stability studies
Stability commitments
Marketed product stability studies
Studies requested by QA
Studies in support of proposed changes

5.1 Stability Commitment

As allowed by ICH guidelines, to avoid incurring the considerable cost of production-scale material that may expire before it can be marketed, registration stability studies are often carried out on pilot-scale batches. The retest date and the expiration dating periods approved in the application based on these data are then considered tentative until they are confirmed by data on three production-scale batches. Under these circumstances, the application should include a commitment to carry out stability testing on the first three production-scale batches using a protocol that is specified in the application and becomes known as the "approved stability protocol."

The approved stability protocol, which may be used later to extend the retest or expiration dating period, usually specifies only the long-term storage conditions specified in the product's labeling and no accelerated conditions. It usually specifies only testing parameters that are included in the regulatory product specifications. This is in contrast to the registration stability protocol, which may have included additional parameters not included in the regulatory specifications.

5.2 Marketed Product Stability Studies

Marketed product stability studies are required to provide assurance that the drug product continues to exhibit reproducible quality over its shelf life and that accumulated minor changes over time have not adversely affected the product. These studies are initiated annually on one batch of each marketed product in each marketed container-closure system using the approved stability protocol.

For products that are available in multiple strengths, each in multiple container-closure systems, the marketed product stability program can be dauntingly large. Many companies therefore wish to abridge the stability program in any of a number of ways (e.g., bracketing or matrixing of drug product strengths, container-closure sizes, and/or fill size; matrixing of time points; reduction of the testing frequency for all or certain parameters; or elimination of testing of certain parameters.

Abridged marketed product stability programs should be submitted to FDA in a prior-approval supplement, but usually not until after the approved expiration dating period has been confirmed through satisfactory stability data on the first three production batches.

5.3 Quality Assurance Studies

Quality assurance may request postapproval stability studies on specific batches of product to confirm that those batches have the same stability profile usually associated with that product. Special stability studies may be requested for batches that experienced manufacturing deviations, exceeded internal alert limits even if they passed regulatory specifications, exhibited an usually large variation in a critical parameter such as dissolution, exhibited an atypical yield, or were stored or shipped under new conditions.

Additional stability studies may also be undertaken if there is an increase in the number of complaints for a product or if there is a change in the grade and/or supplier of a raw material, manufacturing process, and/or equipment, even if stability is not required for a regulatory filing.

5.4 Stability Studies to Support Changes

Many changes may be proposed after approval for a variety of reasons, such as quality, marketing financial or other business considerations. Stability studies of one sort or another are required for most changes.

5.4.1 Reporting Categories for Postapproval Changes

The FDA Guidance for Changes discusses the filing requirements for changes. The guidance divides changes into reporting categories according to the likelihood that the change will affect the quality or performance of a product. The three categories for postapproval changes are outlined in Table 5.

A major change has substantial potential to adversely affect the identity, strength, quality, purity, or potency of the product. Major changes requires prior-approval supplement (PAS; i.e., FDA approval for the change must be obtained before distributing product made using the change).

A moderate change has a moderate potential to adversely affect the identity, strength, quality, purity, or potency of the product. Moderate changes are divided into the following two categories:

Changes that require submission of a supplement at the same time product using the change is distributed (supplement-changes being effected-CBE)

TABLE 5 Reporting Categories for Postapproval Changes

Category	Assessment	Filing requirement
Major change	Substantial potential to adversely affect the product	PAS
Moderate change	Moderate change with moderate potential to adversely affect the product	CBE or 30-day CBE
Minor change	Minimal potential to adversely affect the product	Annual report

Source: FDA. Guidance for Industry. Changes to an Approved NDA or ANDA, Nov. 1999.

Changes that require submission of a supplement at least 30 days before distributing product made using the change (supplement-changes being effected in 30 days–30-day CBE)

A minor change has minimal potential to adversely affect the identity, strength, quality, purity, or potency of the product. Level 1 changes may be reported to FDA in an annual report.

5.4.2 Stability Requirements for Various Categories of Change

The amount of stability data required to support various categories of change is described in FDA Draft Stability Guidance, which presents five types of data packages. These range from type 0, which includes no stability data at the time of submission, to type 4, which includes 3 months of comparative accelerated data and available long-term data on three batches of product made using the proposed change. The type of data package required is dependent on the proposed change.

The five types of stability data packages are outlined in Table 6.

5.4.3 Changes Requiring Stability

Detailed discussions, including the stability data requirements, are included in FDA Draft Stability Guidance for the changes listed in Example 7.

EXAMPLE 7: EXAMPLES OF POSTAPPROVAL CHANGES REQUIRING STABILITY

Manufacturing process of the API
Manufacturing site for the API or drug product
Formulation of the drug product
Addition of new strength of the drug product
Manufacturing process and/or equipment for the drug product
Batch size of the drug product
Reprocessing of the drug product
Container-closure of the drug product
Approved stability protocol

5.5 Compliance Issues

All stability studies on a marketed product must of course be carried out in full compliance with cGMPs. Example 8 presents stability areas often cited as deficient during regulatory inspections.

TABLE 6 Stability Data Packages to Support Postapproval Changes

Type	Stability data at time of submission	Stability commitments
0	None	None beyond regular annual batches
1	None	First production batch[a] and annual batches thereafter on long-term stability studies[b]
2	Three months' comparative accelerated data and available long-term data on one batch[c] of product made with the proposed change	First production batch[a] and annual batches thereafter on long-term stability studies[b]
3	Three months' comparative accelerated data and available long-term data on one batch[c] of product made with the proposed change	First three production batches[a] and annual batches thereafter on long-term stability studies[b]
4	Three months' comparative accelerated data and available long-term data on three batches[c] of product made with the proposed change	First three production batches[a] and annual batches thereafter on long-term stability studies[b]

[a]If not submitted in the supplement.
[b]Using the approved stability protocol and reporting data in annual reports.
[c]May be pilot scale.
Source: FDA. Draft Guidance for Industry—Stability Testing of Drug Substances and Drug Products, June 1998.

EXAMPLE 8: COMMON COMPLIANCE DEFICIENCIES RELATED TO POSTAPPROVAL STABILITY STUDIES

Failure to complete registration stability studies
Failure to include each marketed product in the stability program
Stability data not evaluated as part of the annual product review
Stability results do not support expiration dating period
Batches evaluated under QA requested studies released before sufficient data were available
Stability data do not support proposed changes

6 STABILITY PROGRAM

A well-designed stability program meets all regulatory requirements and attains its objectives with minimal expenditure of resources. It provides all necessary data in a form that can be easily interpreted and evaluated, and distinguishes between analytical variability and instability. It specifies a testing frequency, which will provide early detection of instability and support the desired expiration-dating period.

A well-designed stability program has many aspects. Some aspects are general and best addressed in a stability program SOP. Other aspects are specific and best described in product-specific stability protocols.

6.1 Standard Operating Procedure

A stability program SOP should define all general aspects of stability studies and serve as the basis for preparation of specific protocols. The SOP should include sections on (or reference other SOPs for) the procedures for the stability protocol, samples, testing, chambers or rooms, and final report. Items to be included in each section of a stability program SOP are outlined in Examples 9–15.

EXAMPLE 9: STABILITY PROGRAM SOP—PROTOCOL

Content and format
  Introduction/objective
  Description of batches
  Test parameters
  Testing frequency/parameter

- Rationale for bracketing and matrixing
- Storage conditions
- Container orientations
- Test methods
- Acceptance criteria
- Retest/expiration dating period

Storage conditions for different types of protocols
- Clinical trial material
- Registration stability
- Annual batches
- Postapproval changes
- Special studies

Test Parameters
- Discussion for different dosage forms

Testing frequency for different types of protocols
- Clinical trial material
- Registration stability
- Annual batches
- Postapproval changes
- Special studies

Microbiological concerns
- Testing
- Preservative efficacy testing

Preparation, review, approval, and revision
- Stability coordinator, laboratory head, QA, and others as appropriate
- Revisions to be approved by all original signatories

Cancellation of stability studies
- Procedure
- Request/authorization forms
- Disposition of samples

Protocol deviations
- Procedure or forms for investigating, reporting, corrective action

Forced degradation/stress studies
- Material to be tested
- Conditions
- Extent of degradation
- Identification of degradants

Switch to ICH conditions
- Allowable circumstances
- Procedure
- Request/authorization forms

EXAMPLE 10: STABILITY PROGRAM SOP—SAMPLES

- Batches to be placed on stability
  - Registration stability commitment—first three production batches
  - Annual batches—one batch of each product in each container-closure system or as approved in the application
  - Multiple manufacturing sites
  - Registration stability studies for new chemical entities
  - Registration stability studies for changes to approved products
  - Special studies
- Orientation
  - Liquids: with and without cap contact
  - Semisolids: vertical
- Sampling plans
  - Sampling batches from production for stability evaluation
  - Sampling units from batches for stability studies
  - Sampling units from stability study for testing
  - Sampling analytical samples from units selected for testing
- Number of units
  - Requirement for full testing
  - Additional units
- Labeling
  - Minimum information required
- Tracking
  - Procedure or forms for logging in and pulling and tracking samples
  - Validated laboratory information management system (LIMS)
- Handling
  - Storage after removal from stability chamber
  - Time between receipt of samples and start of testing
  - Holding of samples and sample solutions for investigations of OOS
- Disposition of unused samples
  - Documentation

EXAMPLE 11: STABILITY PROGRAM SOP—TESTING

- Methods
  - Validation or method transfers commensurate with use of method
- Equipment
  - Qualified and calibrated equipment, including robotic methods
  - Validated LIMS, computer programs for calculation, and other data-handling systems

- Testing dates
  - Definition of zero-time
  - Definition of analysis date
  - Definition of analysis time
  - Testing date windows
- Data handling
  - Rounding, averaging, reporting
  - Security
  - Review and approval
- Out-of-specification (OOS) results
  - Procedure or forms for investigating, reporting, corrective action
- Out-of-trend (OOT) results
  - Definition of OOT
  - Procedure or forms for investigating, reporting, corrective action
- Deviations from methods
  - Procedure or forms for investigating, reporting, corrective action
- New or revised methods
  - Allowable circumstances
- Contract labs
  - Method transfers
  - Audits
  - Coordination of results
  - Oversight

EXAMPLE 12: STABILITY PROGRAM SOP—STABILITY CHAMBERS OR ROOM

- Temperatures and humidities
  - Definition
  - Tolerances
  - Documentation
  - Alarm systems
  - Definition of temperature or humidity excursion
  - Investigation and reporting of excursion
- Light chambers
  - Light source
  - Output from lamps
  - Age of lamps
  - Positioning of samples
- Calibrations and maintenance
  - Procedure

Location of probes
Timing
Documentation

EXAMPLE 13: STABILITY PROGRAM SOP—STABILITY REPORT

- Content and format
  - Objective
  - Description of batches
  - Relevant protocol
  - Data presentation
  - Tables and graphs
  - Individual results versus averages or high and low values
  - Data evaluation
  - Conclusion
- Data evaluation
  - Statistical programs
  - Criteria
- Application of stability results
  - Labeling of storage conditions
  - Retest or expiration dating period
  - Shipping and warehousing conditions
- Preparation, review, approval, and revision
  - Stability coordinator, laboratory head, QA, and others as appropriate
  - Revisions to be approved by all original signatories

6.2 Product-Specific Stability Protocols

A product-specific stability protocol supplements the information provided in the stability program SOP. In some instances, the requirements in a product-specific stability protocol may differ from those specified in the stability program SOP for scientific, regulatory, or business reasons. In all cases, a product-specific stability protocol takes precedence over the general stability program SOP. The product-specific stability protocol should clearly define and justify both the conditions to be used and any deviations from the stability program SOP, including reason and justification.

A product-specific stability protocol should include the items outlined in Example 14.

EXAMPLE 14: CONTENTS OF A PRODUCT-SPECIFIC STABILITY PROTOCOL

Protocol number
Introduction objective
Batches to be tested, including product name, strength, formulation number, batch numbers
Batch information
  Drug product: batch size, manufacturing date, manufacturing process, manufacturing site
  Drug substance: batch number, batch size, manufacturing date, synthetic route and process, manufacturing site
Container-closure
  Description, size, fill/count, supplier, resins, packaging site and date
  Container orientation (e.g., inverted or upright)
Storage conditions, including temperature, humidity (if applicable).
Testing frequency, including time points for each storage condition
Tests to be performed at each test station
Test method, which must be validated and stability-indicating
Acceptance criteria, which may be "for information only"
Proposed or approved expiration dating period
Sample requirements, based on amount needed for each test
Preparer or reviewers' approval signatures, including stability coordinator, laboratory head, QA, and others as appropriate

6.3 Approved Stability Protocols

The FDA Draft Stability Guidance defines an approved stability protocol as a detailed plan described in an approved application that is used to generate and analyze stability data to support the retest period for a drug substance or the expiration dating period for a drug product.

Approved stability protocols are important for the sponsor because they may be used in developing data to support an extension of an approved retest or expiration dating period via annual reports in accordance with 21CFR314.70(d)(5). To change an approved stability protocol or to define one in an application requires a prior approval supplement.

6.4 Compliance Issues

A review of the stability program and stability data is a key part of inspections by regulatory agencies. Common compliance deficiencies related to stability programs are outlined in Example 15.

EXAMPLE 15: COMMON COMPLIANCE DEFICIENCIES RELATED TO STABILITY PROGRAMS

- General
  - Compliance with cGMP, ICH, and other regulatory guidelines
  - Approved stability program SOP
  - Adherence to internal company policies, SOPs, protocols
- Stability Protocol
  - Scientific basis
  - Relationship of shipping and storage conditions to stability studies
  - Compliance with internal company and regulatory requirements
  - Review, approval, and revision process, especially QA involvement
  - Reason and justification for protocol amendments, especially changes in methods used
  - Investigation and reporting of protocol deviations and timelines, i.e., the fact that it is timely of corrective actions
  - Reasonable test parameters, testing frequency, and acceptance criteria
  - Microbiological tests and testing frequency
  - Statistical sampling of plans and testing of representative samples
  - Scientifically sound bracketing and matrixing designs
  - Transition plan to ICH conditions for old products
- Samples
  - Written sampling plans
  - Representative samples
  - Authenticity of batch information provided
  - Storage, labeling, transferring, and tracking of samples
- Testing
  - Closeness of pull dates to analysis dates; adherence to relevant SOP
  - Missed time points
  - Authenticity of data, especially impurity profiles
  - Data handling (e.g., rounding, averaging, data security, reporting, review, and approval)
  - Use of validated, stability-indicating methods
  - Evaluation of forced degradation products
  - Properly executed method transfer protocols
  - Use of qualified and calibrated equipment by trained analyst
  - Use of qualified and calibrated stability chambers or rooms
  - Handling of OOS and OOT results
  - Timely implementation of corrective actions
- Stability chambers/rooms

- Qualification and maintenance
- Temperature and humidity records
- Investigation and reporting of excursions

Stability reports
- Impurities profiles
- Test points used for data evaluation
- Selective reporting of data
- Data trends
- Expiration dating period based on real-time results
- Justification for amount of preservative in the formulation
- Lack of QA review

Other
- Relationship between stability studies and proposed labeling, shipping, and warehousing conditions
- Change control
- Stability information provided in annual product reviews

7 REGULATORY REALITY CHECK

With the chapter providing an instructive backdrop, a FDA warning letter citation and two observations (FD-483s) documented during recent inspections of FDA-regulated facilities are presented below, followed by a strategy for resolution and follow-up corrective and preventative actions.

Warning Letter Citation. The firm failed to adequately assess the stability characteristics of drug products in that samples representing all container-closure systems used in packaging in a given year are not included in the annual stability program.

We (FDA) acknowledge that the cGMP regulations are not explicit about annual stability testing; however, it should be noted that the cGMP regulations are not all inclusive and that what determines a manufacturing practice to be "current" and "good" is if it can be considered feasible and valuable. In the case of annual stability testing, the agency has determined that such a practice is feasible and valuable and, thus, enforceable under section 501(a)(2)(B) of the FD&C act.

Preventative Action. This citation could have been avoided by a stability testing program SOP, which specifies batches that are to be entered into the annual stability program (i.e., one batch per year of each product in each container-closure made that year). If this is an inordinate amount of work, a PAS could be submitted for a reduced program. The reduced program could be reduced testing of all required batches or elimination of certain batches justified by the history of the product and the similarity of the container-closure system. In this case, every product in each container-closure system would be tested on a periodic, but not annual, basis.

FD-483 Observation. "The firm failed to conduct a stability testing program using the current marketed container-closure system. Additionally the firm failed to establish a written stability test program that analyzes for impurities and degradants in their multiple products."

Preventative Action. This observation could have been avoided by establishing a stability testing program SOP that specifies batches that are to be entered into the annual stability program, including all container-closure systems, and that requires that all commercial container-closure systems be tested. The stability program must include an evaluation of degradants and impurities if it is intended to assess stability. Additionally, a product-specific stability protocol should have been developed that specified the methods to be used.

FD-483 Observation. "The firm failed to maintain complete records of all stability testing performed in accordance with 21CFR211.166 as required by 21CFR211.194(e). For example, there was no assay data recorded in the analyst's notebook or automated stability database for the 18 and 24-month stability test points."

Preventative Action. This observation could have been avoided by developing a laboratory SOP that describes proper documentation for all analyses conducted, including stability. Additionally, QA within the lab should have provided adequate oversight and monitoring of stability data generated by individual analysts through periodic reviews of laboratory notebooks.

8 WORDS OF WISDOM

Protect the clinical study by ensuring there are no outstanding compliance issues related to stability studies performed on clinical trial materials.

If degradants are not formed under initial forced degrations study conditions, repeat the studies using more stringent conditions.

Stability data must support the entire expiration dating period.

Ensure that stability studies continue postcommercialization.

Ensure that any changes to ongoing stability studies are preceded by the appropriate change-management mechanisms.

Conversely, ensure that all changes during development and post-approval are accompanied by the necessary stability studies.

6

Computer Validation: A Compliance Focus

Timothy Horgan and Timothy Carey
Wyeth BioPharma, Andover, Massachusetts, U.S.A.

1 INTRODUCTION

Today it seems that everything has automation hiding in it somewhere! The focus of this chapter on computer validation compliance, however is what we will term the application-based computer-related system (CRS), as opposed to computer-controlled process equipment. Examples of a CRS may include laboratory information management systems (LIMS), supervisory control and data acquisition (SCADA) systems, document control systems, calibration data management systems, and any other of the myriad systems that mate a portable software application to commercially available computer hardware. This is different, for example, from a modern water-for-injection still that employs computer software and hardware but is inextricably linked to specialized mechanical process equipment. The validation testing for such equipment is typically focused on the mechanical performance of the equipment, rather than on the performance of the software itself. Many of the concepts discussed in this chapter (validation plans, specifications, etc.), however have analogous counterparts related to equipment validation, and the goal is the same: verification that a system consistently performs its intended function throughout its usable life.

2 GAMP GUIDE

The most prominent and widely recognized guide to CRS validation practices is the Good Automated Manufacturing Practices (GAMP) guide [1]. While the concepts surrounding software testing and quality assurance have been discussed for years (in fact, a seminal book in the field was published in 1979) [2], the GAMP guide is a crucial reference because it is written from the specific perspective of the regulated pharmaceutical industry. The GAMP guide was first published in Europe in 1994, and was written by a small consortium of European professionals in response to European regulatory agency concerns. It did not take long for industry professionals worldwide to appreciate this publication and recognize the industry's need for it. The GAMP guide has since become an international collaboration between pharmaceutical industry validation and compliance professionals. The guide has gained acceptance worldwide as *the* pharmaceutical industry guideline on the validation of software and automated systems. While of course one size never fits all, the GAMP guide is an indispensable resource for discussion of validation strategies. While this chapter will not extensively review information already available in the GAMP guide, the following are some highlights of what can be found in the GAMP guide:

- Categorization of types of systems and guidelines on the extent of validation required for each
- Supplier and vendor guidelines on software development expectations within the pharmaceutical industry
- A helpful glossary
- Numerous valuable appendices with concrete recommendations on system development, implementation phase management, and ongoing system operation

Copies of the GAMP guide can be obtained through industry professional associations or directly from the GAMP forum organization (www.gamp.org).

3 SOFTWARE DEVELOPER AUDITS

The International Standards Organization (ISO) definition of audit is "Systematic, independent, and documented process for obtaining audit evidence and evaluating it objectively to determine the extent to which agreed criteria are fulfilled." The bottom-line goal of the software developer audit process is to allow you to assess the developer's quality assurance (QA) system.

3.1 Why Audit?

The FDA regulations clearly require evaluation of suppliers as evidenced by the following current regulations:

From the device regulations: 21 Code of Federal Register (CFR) 820.50 Quality System Regulation Subpart E—Purchasing Controls: "Each manufacturer shall establish and maintain procedures to ensure that all purchased or otherwise received product and services conform to specified requirements. (a) Evaluation of suppliers, contractors, and consultants. Each manufacturer shall establish and maintain the requirements, including quality requirements, that must be met by suppliers, contractors, and consultants."

21 CFR Part 11, Sec. 11.10 (i). "Determination that persons who *develop*, maintain, or use electronic record/electronic signature systems have the education, training, and experience to perform their assigned tasks."

In business terms, auditing of software developers will allow you to assess the vendor's technical competence, vendor reaction to your company's user requirements specification (URS), vendor QA system adequacy, supplier experience with GXP systems, and quality level of vendor-prepared validation and qualification protocols. In short, vendor auditing is a regulatory expectation and auditing provides a means of assessing the supplier's ability to deliver a validatable system that will achieve the requirements of your company's URS.

3.2 The Preaudit

Assess the need for an audit. What is the criticality of the software product in business terms? Evaluate the risk to the pharmaceutical product, production process, and quality data associated with the software.

Prior to a site visit, assess the vendor remotely through product literature and submission of a preaudit questionnaire. When a vendor site audit becomes necessary, plan the audit's scope and focus and identify the audit team. Audit team members should include user group lead, information services groups (IT/IS/MIS), compliance, purchasing, and validation personnel.

Schedule the audit with the vendor to ensure that key development, quality, and management personnel will be onsite and available to the audit team. Verify how much on-site time the vendor will allow. Make sure that the goals of your audit can be completed in the allotted time. Schedule a closing meeting with key developer personnel for the end of the audit. Your company should develop a vendor audit checklist. The checklist should be your plan for proceeding through a thorough audit.

Elements should include

General company information
Standard operating procedures (SOPs)
Customer support
Quality management/standards
Software/system development methodology
Testing methods/verification and validation
Technical personnel
Change control, configuration, and distribution management
Security features
Documentation
21 CFR Part 11 compliance assessment

3.3 The Audit

Don't allow the opening meeting to turn into an extended sales show. Inspection and interview should constitute the bulk of your work. Ask open-ended questions; don't set up for simple yes or no answers. These open-ended responses will often lead you to unforeseen concerns.

Where documentation of a process is found to be substandard, describe to the vendor how to comply with your standards. This is a free GXP consulting service for the developer, which is usually eager to receive some feedback.

Plan an interim meeting with your team to check focus, issues, and progress. Remember that you, the pharmaceutical manufacturer, bear the ultimate responsibility for regulatory compliance, including the compliance level of the software you implement.

3.4 Postaudit Activities

Produce an audit report. Typical report sections will include

Purpose: State the company and division that was audited—when, where, and by whom. List audit team members by name, title, and department. List key representatives of the developer company.

Overview: Describe focus issues. Summarize findings. Refer to the audit checklist for more detailed notes. Note that audit observations require a response/action plan from the software summary.

Review summary.

History of company and product. Note that level of development staffing.

Software development (SW Dev) and quality Assurance manual: Describe purpose. For example, is it intended to facilitate validation by including protocols? What programming standards and development life-cycle models are cited? The software quality assurance (SQA) program manual should be version/revision controlled and address an overview of the vendor's quality program, the vendor's programming standards, version control (SDLC), and maintenance procedures. It should also describe development documentation, such as the data sheet used by developers when executing tests, system requirements that detail all features and functionality of the software and final system specifications.

SW Dev/SW unit testing note: Version, operational software (VB 6.0, MS Access 8.0, Crystal Reports 7.0) and operating environment/platform (NT or Linux Server). Plans to version product how soon and in what manner. What is the future development direction of this product? Is an SQL server version in development? Is an Oracle-based product coming? How closely linked are the content of the functional specs and the unit testing? What steps are defined in their development cycle (design, develop, test, implement)? Is the documentation consistent with their SOPs and purported development models? Describe testing review policy and procedures. Who is responsible for what? What does review failure or success mean? What happens in each case? Is there evidence of review failures? Ensure that the review is meaningful. Describe the testing approach and routine. Is documentation signed? Are comments dated and initialed? Are deviations and failures pursued to conclusion per SOP? How is the product protected by an adequate backup, recovery, and disaster recovery plan? Where and how is the product secured?

Change control: What policies are in place? Are they adequate? Are they observed? Does the vendor rely on the development tools to control/ revisions, new functionality, and new functionality changes?

Employee training: Are the SOPs clearly written? How are they reviewed and approved? Are they maintained in the work area or available to employees? Are they numbered and versioned? Is training documented and assessed? How are training records stored and filed? Have they been audited? Is there evidence of GXP training? There should be resumes on file at a minimum.

Customer support: How many persons available/per week? How many calls are handled per day or per week? Number of minutes per call what is the average? Is there a formal problem/bug log? How is it followed up? Is a previous product version supported? What is the cost and coverage of the support service package? Are other corporate services provided, such as database conversion or migration? Is a statistical method or standard used to assess successful conversion or migration? [See ANSI/ASQC Z1.4–1993 "Sampling by Attributes"; military standard 105E has been canceled (obsolete).]

Security: Is there server backup policy or routine? Is the backup log being used accurate and reviewed? Is there offsite storage? Are there source code security and storage, source code escrow arrangements, and facility access control? Is there physical and logical control of the computing environment? How is the development server networked? Is there an open or closed system? Is it firewalled? What is the password policy and control? Has the backup and recovery or disaster recovery plan ever been tested? Was it documented?

CFR 21 Part 11: Do key personnel understand the rule clearly? Run through a detailed list of the requirements of Part 11 and attempt to determine

What Part 11 requirements does the vendor concede that the product does not meet?

In your own judgment, what Part 11 requirements does the software fail to meet? Beyond the specific requirements of part 11 is your business and operational context. How configurable is the record review and approval signature functionality? Assess the risk of the noncompliance level. Risk is best assessed in terms of risk to the patient, proximity to the drug product production process, quality data, and dispositioning process. Risk of implementing a less than fully compliant system is also relative to the risk of continuing to use the even less compliant system being replaced.

3.5 Possible Audit Observations

Communicate and document your findings to the vendor. Findings such as the following may be identified:

Internal quality activities, including personnel training files, need to be up to date and documented on a regular basis.

Internal audits should be performed and documented to ensure SOPs are observed.

Software is not fully 21 CFR Part 11-compliant. Document your plan for bringing the product into compliance.

Documentation for software testing (release to production testing) does not clearly indicate the version being tested.

There is no formal revision change control tracking method.

Formally communicate your summary report to the vendor. Ensure that a corrective action plan from the vendor will be provided. Update the report in accordance with the vendor response. Use this audit summary report as an integral part of the validation plan, test protocols, and Part 11 remediation and assessment plan.

4 THE VALIDATION MASTER PLAN

The validation master plan (VMP) is the roadmap and gatekeeper of the CRS validation process. It needs to answer several questions.

What specifically will be validated?
How are we going to validate?
How will we know when the system is ready?

It should identify what validation protocols are required and everything else that is needed before the system can be considered validated and ready for use.

While a VMP is typically drafted under the auspices of the validation/quality group, the end users and engineering groups involved with project implementation should be involved in review and approval of this document. Everyone needs to understand and agree to the objectives that must be satisfied before the CRS is put into GMP-related use. The following are points and topics to consider when drafting a VMP.

4.1 Scope of Computer-Related System

The scope of the CRS validation must be defined. For a stand-alone application on a stand-alone computer system, this may be straightforward. If there are any interfaces with other systems, however, the scope becomes more challenging. A clear definition of the VMP scope will help prevent misunderstandings and "scope creep" that can cause significant schedule delays. Consider such questions as

Will this CRS include a data backup system that needs to be validated or will it utilize an existing validated backup system?
Is this networked CRS tying into an existing network infrastructure or does it include its own network?
Does this CRS provide data to an existing system, and will the existing system require any revalidation?
What group or groups does this CRS serve and who is the ultimate "owner" of the system?

As with most project plans, it is important for the VMP to include a definition of personnel roles and responsibilities. Numerous documents may need to come together to validate a system (e.g., specifications, SOPs, protocols), and team members must agree upon who will be responsible for specific deliverables, as well as approve the final documentation.

4.2 Documentation Needs

All documentation that must be produced to support the validation process should be listed in the VMP. At the time of drafting a VMP, it may not be known, for example, exactly how many SOPs will be written for the system, but the types and categories of SOPs needed should be delineated. Early on, the need for operation, administration, and maintenance SOPs should be apparent. Additionally, be sure to consider the following as they apply to your project:

Specifications
SOPs
Vendor documentation
Vendor audit reports
Engineering peer reviews
Module testing
Protocols
Protocol reports

In addition to listing the required testing documentation, describe the philosophy behind the testing documents. Is the CRS a simple stand-alone system requiring only a single installation qualification (IQ) and operational qualification (OQ)? If so, briefly state the objective of these documents. Is the CRS a complex, multifunctional system requiring several layers of testing, from discrete software unit/module tests up to a fully integrated system performance qualification (PQ)? If so, describe these layers of testing and the requirements and objectives of each. In particular, note who will be performing each layer of testing (the design engineers may be required to perform some unit module testing before the validation/quality engineers continue with higher-level testing) and the type of documentation required at each level (possibly vendor-designed engineering review forms at the earliest stage, leading to validation protocols at later stages). Consider creating diagrams to help explain the testing methodology. These can be invaluable to quickly convey the basics of the testing methodology to new or peripheral project team members. For example, Fig. 1 illustrates a complex, multifaceted system development and validation process, but communicates the basic methodology very efficiently.

4.3 Training

Ensure training requirements are addressed within the VMP. Define the training that must be performed and documented before the system can be put officially into use. Determine how training will be documented. Refer to organizationwide training policies that are applicable. Be sure to require

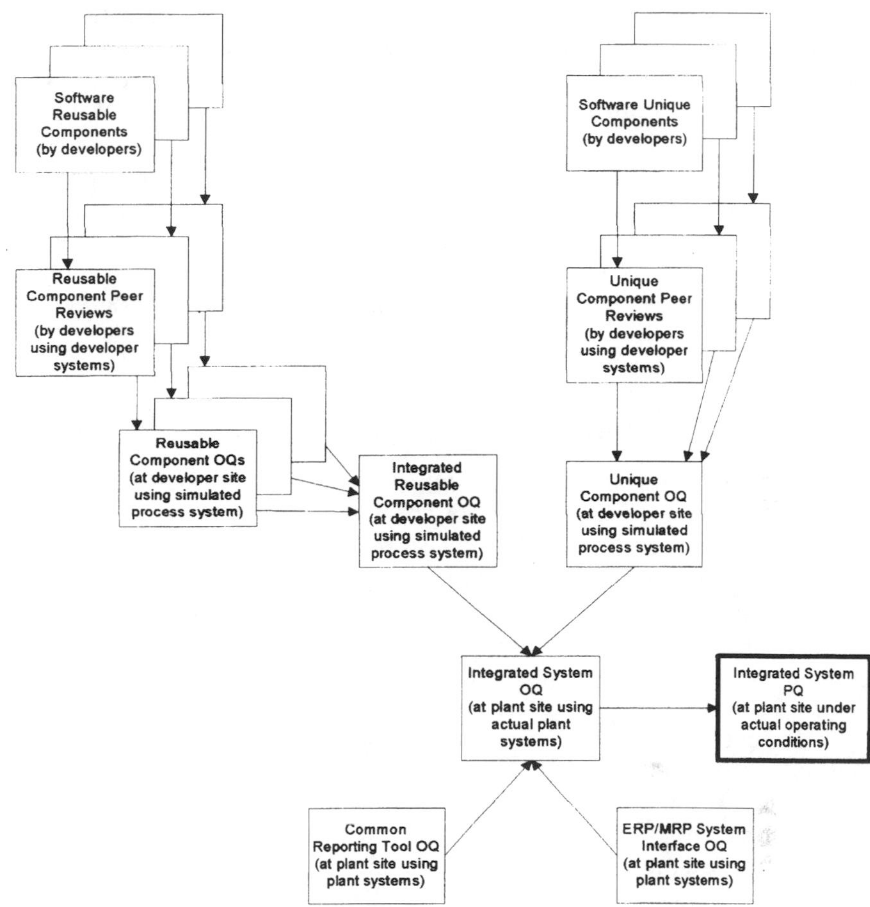

FIGURE 1 Complex multifaceted system development and validation process.

verification that CRS user security administration (if any) accurately reflects the training documentation. For example, untrained users may have been allowed access to the CRS for testing purposes. These users must be inactivated before the system is initiated into GMP uses.

4.4 System Acceptance

In order to determine whether all VMP requirements have been met, the CRS VMP should require what will be referred to as a master validation summary report (MVS). The MVS should discuss each deliverable required by the

VMP for its successful execution. Reference the location of each deliverable and provide the detail necessary for retrieval at a later date (e.g., SOP and protocol numbers). Identify any conditions surrounding the use of the system. Were some features of the system found to be unsatisfactory for use? Clearly state what aspects of the CRS are not approved for use until they are re-engineered and tested and what formal controls are in place to enforce this (SOPs, security programming, etc.). Design the VMP so that approval of the MVS document is the end point that releases the CRS for GMP use by appropriately trained users. This end point is called system acceptance and signals the transition to the validation maintenance phase of the system life cycle.

5 SYSTEM SPECIFICATIONS

System specifications are essential to the CRS validation process. This does not mean, however, that the process of creating specifications need be an onerous one. In fact, the specification creation and approval process can be one of the most valuable aspects of the system life cycle, as it spurs everyone involved to understand and agree upon what the system is ultimately expected to do. Regardless of how well-thought-out a system may be in the preliminary design stages, differing perceptions are always brought to light once specifications are put in writing.

In contrast to the VMP, system specifications are typically created by the end user(s) and system engineering organizations. Since the system specifications will be the basis for much of the validation testing, validation and quality representatives should be brought into the specification development process whenever possible. If these representatives are left out of the specification process, the result will often be specifications that are only intelligible to the design engineers and principal end users because much of the wording in the specifications will be based on perceptions that have not been written into the documents. When validation and quality systems' representatives are excluded from developing specifications, extensive revisions are likely to be needed in order to make it suitable for the validation.

Specifications typically required for a custom-designed application are known as the URS, functional requirements specification (FRS), and the detailed design specification (DDS). A description of each specification and its function follows.

User requirements specification—The URS is an overview of the system functionally from the perspective of the end users' needs. This document is often the first document generated to kick off a new CRS design and implementation project. It can be drafted early on by a key individual or group and used to help clarify the business needs to toplevel management. Once the system concept has the necessary management support it can be used to introduce the project to a wider audience and solicit additional user input as to the desired features of the system. If an outside firm will do system engineering, the URS can be used as the defining document for the engineering services bid process. If the engineering will be performed internally, the engineering group can use the URS to project estimated resources needed to support the system development and implementation process. The requirements in the URS will typically be used as the basis for PQ testing. A typical statement in the URS might be "The system must allow simultaneous use of different system features by multiple material handlers."

Functional requirements specification—The FRS is a detailed description of required system functionality from the end users' perspective. This document should discuss all required functionality by the end user since it will be used by the engineering team to establish the system design. While the URS may be written entirely with end user input, the FRS is typically a collaborative effort involving both end user representatives and engineering team representatives. The FRS must address the users' desired functionality, but must also take into account what functionality can be feasibly designed into the system given the constraints of the schedule, budget, and development platform. The requirements in the FRS will typically be used as the basis for OQ testing. A typical statement in the FRS might be: "The user must be required to enter a control code for the raw material lot that is being dispensed; the system must verify that this control code is valid and that the lot disposition status allows dispensing of this material."

Detailed design specification—The DDS is a detailed description of how the requirements identified in the FRS will be met from the perspective of the system design engineer. This document must be descriptive enough to facilitate a consistent programming effort across the entire (potentially large) team of engineers. This document is typically drafted by engineers who understand the programming platform that will be used to create the system. While this document may be very technical, end user review is still necessary to help ensure that the requirements of the FRS have been properly interpreted by the engineering team. Design information contained in the DDS may include such things as

User interface screens

Database names and field definitions

Security scheme implementation

Specific features of the development platform to be used for specific functions of the application

The specifications in the DDS will be used as the basis for IQ and some OQ testing. As mentioned above, it is important that these specifications are reviewed by validation and compliance team members throughout the process in addition to the applicable end users and design engineers.

6 SPECIFICATION NOMENCLATURE

The specification names used here are one example, but there are many ways to refer to the same documents. The URS might instead be called the business requirements specification (BRS). The FRS might be functional specification (FS) or the system requirements specification (SRS). The DDS might be a design specification (DS) or an engineering specification (ES). The intended purpose of the specification must be delineated in the document to avoid confusion.

7 COMBINING SPECIFICATIONS

What if the project is small and the specifications are not very complex? Can these documents be combined? Absolutely. On simpler projects, it is not unusual for the goals of the URS and FRS to be incorporated into one user-focused document with a separate engineering-focused design document. It is also not uncommon for the user to generate the high-level URS and submit it to design engineers familiar with the process for creation of a combined FRS/DDS document.

8 OFF-THE-SHELF/COTS APPLICATION

What if you are implementing a commercially available application (commercial off-the-shelf, or COTS application) rather than designing a custom application? Do you still need these documents? Yes, but probably not all of them. The FDA has made it clear [3] that URS are expected, even for COTS applications. Without a URS, there would be no definition or required system functionality on which to base initial validation activities. The COTS URS will typically end up as a blend of the URS and FRS, as described previously. This blend of requirement types is appropriate for the COTS situation, in which you are specifying what is necessary for your business operations but do not need to write specifications sufficient to actually design the application from scratch. After selection of the application, additional specification documents should be written focused on the choices that will be made in your specific implementation of this application. A configuration specification is recommended to define the configuration choices made in implementation of the software. A security specification is recommended to define the security implementation scheme. An installation specification is also recommended to define the associated hardware and software requirements necessary to run the application in the chosen environment.

9 TEST EXECUTION HARDWARE

One interesting variation of CRS validation when compared to equipment validation is that software is not inherently linked to a unique piece of computer equipment, yet it does require some form of suitable computer equipment for operation. One can streamline the CRS validation process by testing the software using computer hardware other than the final hardware on which the software will eventually be installed for use. This means that validation testing can start using a pilot hardware system while the final hardware is, for instance, still being procured and installed.

10 INSTALLATION, OPERATION, AND PERFORMANCE QUALIFICATION

When a pilot system is used for validation, it is preferable to perform an IQ on the system to verify that it actually satisfies the hardware requirements of the software. If an IQ is not performed on this system, at a minimum the details of the hardware (manufacturer, model, serial numbers configuration) should be recorded as part of the OQ. There absolutely needs to be a record of the test bed hardware in case the validity of testing on the pilot hardware is called into question at a later date. Full OQ testing can then be performed on the pilot system. Execution of the OQ on a pilot system does not mean, however, that use of the final hardware system becomes a "plug-and-play" affair. There still needs to be full confidence that the system performs in a satisfactory manner on the final hardware.

The final integrated hardware/software system must achieve a full IQ to verify that the predetermined hardware and software configuration specifications are satisfied. Be sure to verify that the software version installed on the final system is the same as the one used for OQ testing on the pilot system. If not, there must be formal change control documentation in place that verifies that the validated status of the software has not been affected. There must also be some operational testing done on the final system. If the validation plan has identified the need for IQ, OQ, and PQ, it would be reasonable to perform the OQ on the pilot system and PQ on the final system. If only IQ and OQ have been planned for the system, a second OQ should be prepared that repeats some portions of the OQ that were performed on the pilot system.

If the system being implemented is replacing or upgrading an existing critical system, use of a pilot system for validation testing can pay off with greatly reduced downtime for the critical system, as less testing needs to be performed on the critical system before it is put back into GMP use. If the system being implemented is entirely new, validation testing on a pilot system can pay off by shortening the validation schedule as testing and final hardware procurement and installation activities are allowed to run in parallel rather than consecutively.

10.1 21 CFR Part 11

The FDA release and enactment of the Part 11 final rule in 1997 was met with overwhelming confusion and reticence on the part of the pharmaceutical industry. There was almost a sense that, given enough reluctance and procrastination, the rule would be recalled. There was good reason for procrastination. The expense of complying with the regulation was and continues to be great. In some cases the technology still does not exist to comply with the "letter" of Part 11 without turning your pharmaceutical manufacturing organization into a software development company, but procrastination is no longer an option. Part 11 must be taken into account when specifying and implementing all new electronic systems. In addition, strides toward compliance must be made for legacy computer systems.

The first place to turn when coming to terms with Part 11 compliance is to the regulation itself. While the regulation covered only about two pages in the *Federal Register*, it was accompanied by a 30-page preamble upon its issue. While it does not bind the FDA, the reasoning and clarification offered in the preamble is the best available written insight into the goals and intent behind the creation of Part 11. It is a good starting point.

The most important step in appropriately implementing Part 11 is understanding that the FDA requires the electronic records used in GMP activities to be as reliable and trustworthy as traditional paper records. This concept has been expressed at numerous public speaking engagements by FDA officials.

The Part 11 rule can be broken down into three major concepts.

Audit trails and related controls necessary to maintain record integrity (11.10)
Requirements for valid electronic signatures (11.50–11.200)
System security (11.30 and 11.300)

While there is some overlap among these, the regulatory requirements are easier to evaluate when taken in these separate parts. Break up your evaluation of any new or existing system into these three areas. Create a generic list of questions or preferred features based on these three areas, keyed to specific sections in the regulation, then apply these questions or research these features for the systems in question. Remember that not all electronic records require signatures, so the electronic signature controls may not even be relevant to the system under review.

When it comes to assessing the compliance level of existing ("legacy") electronic systems, don't shy away from a critical review. If you operate in an established manufacturing environment, you will need to upgrade and/or replace some systems. A critical review is necessary to appropriately prioritize the systems that need attention.

Plan and undertake a sitewide or companywide review of all legacy electronic systems. Evaluate them based on your list of compliance questions or features keyed to the Part 11 regulation. If you work at a large site or company, it will be more efficient in the long run to execute this large-scale evaluation rather than enacting it in smaller portions. You will likely find some commonality among many noncompliant systems across the organization. That commonality can lead to common solutions. For example, it is far more efficient to implement a compliance solution for 100 spreadsheets used in GMP operations than to implement separate solutions for 20 spreadsheets in each of five workgroups. As another example, you may find 10 systems that require customized solutions. You may be able to gain resource efficiency by building those 10 solutions on common core software functionality, however (e.g., the same database platform).

Once you have identified the legacy systems that need to be brought into compliance, documented plans must be put in place to remediate them. Thoughtful prioritization is critical to the remediation process. Most firms do not have the resources to undertake all Part 11 remediation project at once. Analyze and rank the records managed by these systems according to their risk to quality systems and product integrity. The highest risk systems should be improved first. Similarly, do not apply resources to prioritized projects that offer little benefit. For example, if a particular noncompliant system will be phased out over the next few years, do not implement a 1 to 2-year effort to develop and validate a computerized replacement for that system. Instead, investigate the possibility of accelerating the phase-out.

When evaluating system upgrades or replacements, do not necessarily expect complete Part 11 compliance on the part of the new system. You may not find it. Compare solutions to each other and choose from among the ones that offer the highest levels of compliance. Ensure that each of the three major concepts noted above has been appropriately addressed to some degree. Do not use the lack of a fully compliant solution as the reason for not implementing any solution. The near-term goal for legacy systems is improvement, not perfection.

10.2 Change Control

There should be written procedures to establish systems to manage and control changes that may impact the development, validation, or implementation or affect the maintenance of a validated state for computer systems. Such procedures and controls should apply to all GXP operations and all the systems that support GXP operations.

Change control procedures should be

Observable
Adequate to maintain a CRS's validated state
Capable of maintaining the validity of CRS documentation
Able to ensure product quality and patient safety

The purpose of change management and control procedures is to document, evaluate, and manage proposed changes in such a way as to maintain the validated system.

Change management is typically a cross-functional activity involving the vendor, QA, and development and engineering personnel. The developer's (or your own) QA unit should approve and occasionally audit conformance to the change management process.

Configuration control may be defined as an element of configuration management, consisting of the evaluation, coordination, approval or disapproval, and implementation of changes to configuration items after formal establishment of their configuration identification (IEEE 610.12–1990).

Changes are proposed, documented, requested, evaluated, approved, and tested prior to implementation. The quality unit should be responsible for managing both the entire process and all corresponding documentation associated with the change. Change control typically begins at an SDLC milestone defined in a project validation plan. The CRS should at least be fully designed, documented, validatable, and implementable.

The CRS's VMP (or corporate policy) should state during the validation process that a system will be managed under the firm's change control policy or system. For example, the subject system will be considered to be under change control upon the acceptance of the OQ summary report by the quality unit. Consider when the system in question will actually begin to be used to support GXP production.

Multiple change management processes may exist within a company—documentation, equipment, CRS, and project-or system-specific processes.

In compliance terms effective change management and change control are a clear regulatory expectation.

21CFR211.68, Subpart D—*Automatic, mechanical, and electronic equipment. "(b) Appropriate controls shall be exercised over computer or related systems to assure that changes in master production and control records or other records are instituted only by authorized personnel."*

The ICH's Q7A Good Manufacturing Practice Guidance for Active Pharmaceutical Ingredients, V. Process Equipment (5), D. Computerized Systems (5.4)—*"Changes to computerized systems should be made according to a change procedure and should be formally authorized, documented, and tested. Records should be kept of all changes, including modifications and enhancements made to the hardware, software, and any other critical component of the system. These records should demonstrate that the system is maintained in a validated state."*

Change control—"A formal change control system should be established to evaluate all changes that could affect the production and control of the intermediate or API. Written procedures should provide for the identification, documentation, appropriate review, and approval of changes in raw materials, specifications, analytical methods, facilities, support systems, equipment (including computer hardware), processing steps, labeling and packaging materials, and computer software."

Beyond the compliance imperatives, effective change management is also good business practice. Benefits include maintaining knowledge of complex CRS within your department or the organization. Concise and detailed system documentation protects against system failures. Change management supports traceability of the system's evolution. It also helps track the costs associated with a system's life cycle.

The essential compliance characteristics associated with change control include evaluation of the change, qualification testing, documentation update, and approval and implementation.

10.3 The What, Why, When, Who, and How of Change Control

What? Identify all of the CRSs that are covered under your change control and management SOP. Classify and document what systems are GXP and non-GXP. Make certain to include a process for implementing emergency changes. Document the change, the evaluation and assignment of a change level, the testing evidence, and a regulatory audit trail. An example of a typical change would be when an individual leaves the company. The first step should be to disable the account. Ensure that systems' users understand that user rights are disabled. Additionally, all system history related to that former user should be maintained. One might consider if there are training implications associated with changes. Are all users aware of the change? Will SOPs need updating? Pay particular attention to changes of systems used in product manufacture or that generate, analyze, archive, and report data to be used in submissions to the FDA.

Why? Change management in development, pilot, and production environments. The development environment is where system changes, such as software upgrades, are first examined. The pilot system is essentially a qualification and validation test environment. Essentially, identical pilot and production environments allow you to protect your production data, environment, and business process from the risks associated with the manipulation of the system that occurs in OQ/validation testing. It is essential to protect against any risk to production environments. Make certain the test environment is adequately segregated from the other production environments. Identify what data will be used and exposed to testing. Distinct yet identical development, pilot, and production environments will allow for OQ in the pilot rather than in the production environment.

When? A formalized change management process is required throughout the entire software and system development life cycle. The VMP should state when the corporate change management policy will initially apply to a new system, usually at the end of validation, when the final summary report is signed off or a validation certification is issued. Some systems are covered by the corporate change control policy when PQ begins. Change management assures the maintenance of the application system in a validated state. Change control steps should be logical, flexible, and applicable to contingencies. During posttesting determine what documents should be routed and approved as a package. Remember that there may be a need to update specifications and validation summary reports as changes occur during posttesting. Install a periodic review and evaluation—perhaps annually. It is also necessary to plan for system retirements. Develop a formal plan when system retirement is needed. Do not forget to be prepared for unplanned and emergency changes.

Who? The system owner or user initiates changes and is responsible for adherence to change control policy and procedures; however, evaluation and review should be cross-functional. The quality unit should always be responsible for managing the change management system, even if there are several processes for different systems (documentation change, facilities change control, etc.). Ensure the system's user is in the position to take full responsibility of the system's functions, and ensure it is used according to its corresponding documentation and specification. Users should define the system's operational features in such a way that it can only be used as described in the systems' SOPs and can request changes to specifications or SOPs as needed.

The information technology (IT) group should provide such network services as backup, recovery, continuity planning, and disaster recovery.

How? Change control documentation should be readily traceable for every step in the process; for example, during initiation, evaluation, testing, and implementation. It is particularly critical when the review process for batch release is involved. Require that regression testing be part of all change management. When change is of a magnitude that requires testing, specifically test the new or changed functionality as well as associated and related systems functionality. Consider how changes may impact data integrity and critical user interfaces. Ensure that a risk assessment also accompanies change management. Assess the impact on the overall system, the business, and the consumer. Integrate your help desk application and system error logs into the change control process. These systems are often change initiators. Systematically characterize changes as major or minor (or multiple levels), based on whether the change is a low-level one, such as a database maintenance activity, or one in which one or more GXP systems will be affected.

Validation is a process of verifying documented system features and conducting qualification testing. Interestingly, after a CRS is validated, its validation status can be effectively maintained through comprehensive change control.

11 REGULATORY REALITY CHECK

FD-483 observation 1: SOP *X*, "GMP Computer Systems." The procedure describes establishing a written security policy, maintain an access control roster, and virus protection will be installed. There is no written security policy, however, and there is no virus protection installed for the Y system.

FD-483 Observation 2: The Z workstation is considered GMP equipment and as such generates electronic records that are not backed up or stored for retrieval. The OQ document states that "since reports are printed after each run and attached to the original laboratory data document, no data is stored long term and data security is not an issue .... Data will not be stored on the system long term since analysts will printout and attach copies of reports to their original laboratory data documents. Therefore backup and archiving of data is not necessary."

In the first observation noted here, the firm was clearly cited for not adhering to its own computer systems SOP. In such cases, it is helpful to evaluate whether the firm wrote an unnecessarily stringent SOP and developed a habit of not following it or whether the SOP is reasonable and was ignored. This citation may offer some of both. All computer systems that handle GMP electronic records should have a written security policy and written security procedures, as made clear by 21 CFR Part 11, Section 11.10—"Such procedures and controls shall include the following: (d) Limiting system access to authorized individuals." It is quite possible, however that virus protection is not necessary on every GMP computer system at the firm, and that this was an excessive requirement that the firm put on itself and subsequently did not follow. The firm needs to write a procedure for security controls on this system and consider whether or not virus protection is an appropriate SOP requirement for *all* of the computer systems in the scope of this policy.

The second observation involves either lack of attention to or misinterpretation of the electronic records requirements. The FDA defines that 21 CFR Part 11 applies to "records in electronic form that are created, modified, maintained, archived, retrieved, or transmitted, under any records requirements set forth in agency regulations" (Section 11.1 b). Since the firm feels the need to print out and retain these records, the records are presumably "created under requirements set forth in agency regulations," therefore the electronic records on the computer system are bound by 21 CFR Part 11 whether or not the firm feels the need to retain them. Ignoring the state of the electronic records and relying on the paper printout is in violation of at least one other clearly stated requirement of 21 CFR Part 11, Section 11.10: "Such procedures and controls shall include the following: (c) Protection of records to enable their accurate and ready retrieval throughout the records retention period." This firm needs to implement a secure backup system for these electronic records.

FDA warning letter citation: "Your firm failed to establish and maintain procedures . . . in order to ensure that specified design requirements are met. For example, the software designed by your firm was developed without design controls."

This citation talks to a core concept of computer systems validation: development procedures and system specifications. Note that the firm was not cited for lack of validation testing of the software but for lack of design controls. Effective design controls would have included written design procedures. Adherence to these procedures would have produced well-written system specifications. The firm may have had validation testing documentation for the software. In the absence of robust system specifications, however, there is little evidence that a quality process was followed in the development of the software.

12 WORDS OF WISDOM

Involve the user in development and validation.

Many organization take a "change control committee" approach; however, a comprehensive change control system based on a key change control form or cover sheet and related documentation may prove more efficient.

As with pharmaceuticals themselves, remember that *you can't test quality into the finished software product*; it must be produced utilizing quality processes throughout. Require evidence of quality assurance and development processes within your firm and during vendor audits.

Involve technical staff in software vendor quality audits.

A COTS application may have a large number of exciting features that management and users will be eager to implement. Beware. Plan a phased implementation of the system modules and features. Temper COTS enthusiasm with the degree of customization required. It is rarely the case that COTS can be used without some customization.

Expect that the functionality of "off-the-shelf" software will not be a perfect fit for your business process. Chances are a firm will have to negotiate between the ability to configure and modify the functionality of the software and changing the business process.

When it comes to Part 11 compliance, don't let the perfect be the enemy of the good! Have a long-term view and focus on product quality and product risk.

APPENDIX I: COMPLIANCE GUIDEPOSTS FOR A VALIDATION TEMPLATE

The following is a brief outline of the critical compliance parameters in a software validation template.

Purpose

To plan and describe the validation activities for the deployment of the very useful system (VUS) at a particular manufacturing site.

Scope

Describe the scope of the VUS validation effort. What are the boundaries? Are there interfaces to other systems included in this validation effort? (This is very important to identify and define.)

System Description

Provide a clearly worded description of the VUS and its purpose in the organization. Also describe the makeup of the VUS (multiple servers, networks, software packages, etc.). This does not need to be technically in depth, but should provide a basic framework for understanding the system concepts. Consider including an explanatory diagram.

Responsibilities

List all parties who have responsibilities for actions or deliverables described in the VUS validation plan. Describe their responsibilities in general terms (e.g., "will review and approve all system specifications").

Developer Acceptance

If an outside developer is being utilized (whether developing custom software or providing a packaged COTS product), describe the steps taken to qualify the software developer. If an audit was performed, briefly state the results and reference the audit report.

Standard Operating Procedures

List the SOPs that must be developed in order to use the VUS in an appropriately compliant fashion. Include SOPs on operation, maintenance, administration, security, backup, and disaster recovery. Some of these may be combined into one document. If the exact document names and number are known, list them; otherwise use generic names.

Specifications

List the VUS specifications that must be developed in order to properly control system development and implementation. See Sec. 5 for a discussion of specification types.

Qualification Protocols and Reports

List the qualification protocols that must be developed in order to qualify the system for use. Note whether each protocol will have a corresponding report or whether multiple protocol results will be discussed in one report. If the exact document names and number are known, list them; otherwise use generic names.

Training

Describe the training requirements for the VUS. Reference any appropriate SOPs.

Change Control

Describe or reference the change control system that will be used to maintain the validated state of the VUS after the initial qualification.

Validation Document Management

Describe or reference the documentation system that will be used to control validation documentation.

Validation Methodology

Describe in clear, concise terms the steps that will be taken to qualify the VUS for use in the manufacturing environment.

System Acceptance

Describe the process by which VUS acceptance for use will be achieved. If the VUS will be phased in and accepted on a modular basis, describe the acceptance procedure for each phase.

Traceability Matrix

Here is an example of a traceability matrix as defined by the IEEE. By definition, a matrix records the relationship between two or more products of the development process; for example, a matrix that records the relationship between the requirements and the design of a given software component (Std. 610.12–1990).

As commonly used in CRS validation, a traceability matrix verifies the relationship between system specifications and testing protocols. The goal of matrixing is to establish the adequacy of protocols. More specifically, all specified system characteristics and functionality should correspond to verification and qualification testing in protocols.

This example uses a small portion of a traceability matrix for validation of a database application, here named "your CRS."

Good practice would require citing exactly what specifications are being matrixed to the test protocols. In our example, four specification documents were matrixed to the contents of IQ and OQ validation protocols.

Reference specification requirement title	Document number
Functional requirement specification for your CRS	Record document number
Installation specification or detailed design specification for your CRS	Record document number
Security specification for your CRS	Record document number

Specifications			Protocols		
Functional requirements specification section number	Installation specification section number (for COTS) or detailed design specification section number (for non-COTS)	Security specification section number	IQ section number	OQ section number	Comments/ Conclusion
Cite specification section number(s) and title or not applicable (N/Ap).	Cite specification section number(s) and title or not applicable (N/Ap).	Cite specification section number(s) and title or not applicable (N/Ap). May be multiple specific citations.	Cite protocol section number and title. Consider stating test objective. The objective of this column is to verify that the "specific system feature" (such as system hardware) will be verified as installed per the installation specification or design specification	Cite protocol section number and title. Consider stating test objective. The objective of this column is to verify that the specific system requirement (report printing) will be verified as functional per the functional requirements	Does the protocol contain verification or qualification testing to provide documented evidence that each specified system feature will be adequately tested?

8.1—Computer Hardware Requirements (8.1 1)	7.1.2—Computer **Server**—Minimum Hardware Requirements 7.1.3—Application **Clients**—Minimum Hardware Requirements	N/Ap	9.0—Hardware Components Installation Verification The objective of this test is to verify that the "your CRS" database system hardware has been installed as per the installation specification.	N/Ap	OK

BIBLIOGRAPHY

Food and Drug Administration. Guidance for Industry. 21 CFR Part 11. *Electronic Records; Electronic Signatures. Glossary of Terms—Draft Guidance.* Washington, D. C., Aug. 2001. (Later withdrawn by the FDA.)

Myres, G. J., *The Art of Software Testing*, New York: Wiley, 1979

ICH. Harmonized tripartite guideline. *Good Manufacturing Practice Guide for Active Pharmaceutical Ingredients.* ICH Steering Committee. Nov. 10, 2000.

IEEE. *Standard Glossary of Software Engineering Terminology.* IEEE Std 610.12-1990. IEEE, Sep. 28, 1990.

ISPE/GAMP Consortium. GAMP 4 Guide, *Validation of Automated Systems.* Tampa, FL: ISPE, 2001.

Validation of computer-related systems technical report no. 18. PDA Committee on Validation of Computer-Related Systems. PDA J Pharm Sci Tech 49 (1; supplement): Jan.–Feb. 1995.

INTERNET REFERENCES

Freedom of Information Act warning letters are available on the FDA Website (currently linked at http://www.fda.gov/foi/warning.htm).

Informative FDA documents and communications related to 21 CFR Part 11 compliance are posted to public dockets on the FDA Website (currently linked at http://www.fda.gov/ohrms/dockets/dockets/dockets.htm). The relevant dockets are 00D-1538 through 00D-1543.

REFERENCES

1. ISPE/GAMP Consortium. GAMP 4 guide. *Validation of automated systems.* Tampa, FL: ISPE, 2001.
2. G. J. Myers. *The Art of Software Testing.* New York: Wiley, 1979.
3. Food and Drug Administration. Guidance for Industry. 21 CFR, Part 11. *Electronic Records; Electronic Signatures validation. draft issues.* Sept. 2001. (Later withdrawn by the FDA.)

7

Compliance Issues Associated with Cleaning in the Pharmaceutical Industry

William E. Hall
Hall and Pharmaceutical Associates, Inc., Kure Beach, North Carolina, U.S.A.

1 INTRODUCTION

Cleaning pharmaceutical equipment and facilities has a direct impact on the quality of the products, and thus is closely associated with compliance. Most of the manufacturing of pharmaceutical products in the United States and other developed countries is accomplished using multiuse equipment; that is, the same equipment is used to manufacture several different products. Because of this, there is ample opportunity for the cross-contamination between one product and the subsequently manufactured product. Even for products manufactured in dedicated equipment, contamination is possible by environmental and microbiological contaminants. The current good manufacturing practices regulations [1,2] require that products be made using clean equipment and facilities. Since cleaning is generally regarded as a critical step in the manufacturing process, the potential compliance issues associated with cleaning are multifold and very important to the manufacture of high-quality and safe pharmaceutical products. This chapter will be devoted to the identification and discussion of some of the specific compliance issues associated with cleaning in pharmaceutical facilities.

There are common elements for cleaning programs of all pharmaceutical facilities. Examples are adequate cleaning procedures and good training programs. It would appear fundamental that cleaning in any facility cannot be done in a suitable and reproducible fashion without having these essential elements. However, just as there are common elements in all programs, there are also unique features in each facility that make it different from other facilities. For example, it is very difficult to expect the types of cleaning that are very effective for dosage form facilities to be equally effective for facilities that manufacture active ingredients. This is because active ingredient facilities often involve processes with very different manufacturing parameters (e.g., high temperature, pressure), and typically utilize organic solvents as reaction media, whereas dosage form facilities typically involve mixing of powders, dissolving of solids in liquids, and room temperature. It is sometimes difficult for auditors or regulatory investigators to relate to the specific type of facility, and they typically look for templates and expect all facilities to have exactly the same approach or same type of cleaning program. It is all too easy to expect that just because dosage form facilities utilize aqueous-based cleaning procedures that active pharmaceutical ingredient (API) facilities should also utilize them. On further examination, one usually finds that there are very good reasons for the different approaches. For example, if a residue is insoluble in water, then it makes very little sense to "hit it" with a stream of water for 10 hr. Another way of looking at this is that water may be great for removing water-soluble residues but other residues may require different solvents. The caution and the challenge for auditors and investigators thus is not to prejudge a situation until all the facts are known and understood.

2 DEDICATED VERSUS NONDEDICATED FACILITIES

Impacts approach and strategy
Required for penicillins and cephalosporins; expected for cytotoxics
Stability of residues a consideration even for dedicated equipment

The degree to which equipment and facilities are dedicated has a major impact on potential contamination. With regard to dedication, there are three possible manufacturing situations.

1. Equipment dedicated to a single product
2. Equipment dedicated to a class of products (e.g., cytotoxics)
3. Multiuse equipment (i.e., equipment not dedicated at all)

When equipment is dedicated to a single product, the potential for cross-contamination is eliminated except when there would be an intentional contamination (e.g., industrial sabotage by a disgruntled employee). The dedication of equipment to production of a single product would appear to be the most desirable situation except for the fact that resources are often not available except for the most successful of products. Certain products, however, such as penicillins and cephalosporins are required to be manufactured and packaged in dedicated facilities because of their highly allergenic nature. In addition, FDA has indicated that certain other toxic classes of drugs, such as cytotoxic products, should also be considered as appropriate for dedicated facilities.

To dedicate equipment to products that are only infrequently manufactured or packaged would tend to make products extremely expensive and noncompetitive in the marketplace. The only type of contamination that is of concern for dedicated equipment is contamination by environmental agents (dust, debris, microbial) and contamination that results from instability of the product itself. The latter type of contamination deserves a further explanation. In many manufacturing situations it is a common practice to manufacture many batches of the same product one right after another. This practice is usually referred to as a campaign mode and is common for manufacturing such nonsterile products as APIs and nonsterile oral dosage forms. The argument is that even if cross-contamination occurs, the contaminant is the same product, so it actually does represent a mixing of two different products. The possibility exists, however, that material from the first batch might remain on the equipment for a very long period of time and that this residual material might be repeatedly subjected to the harsh conditions of manufacturing (temperature, pressure, humidity, etc.), thus causing it to chemically break down into a different material. A major compliance issue for dedicated manufacturing equipment is thus establishing that the materials are stable during the length of the entire manufacturing campaign. The proof of such stability falls to the company, its cleaning program, and the inherent stability of the product, in this case. The company would need documentation demonstrating that the material was stable under manufacturing conditions during the entire length of the manufacturing campaign.

3 TYPES OF CONTAMINANTS

Since the history of use for each piece of equipment is well established, cleaning should be targeted to remove specific residues. In simple terms, in order to establish meaningful testing to prove that equipment is clean you must know the nature of the residues being removed during the cleaning process. A simple definition of a contaminant is any material that should not be in the product. The author's experience is that anything that can get into a product, will. A list of potential types of contaminants includes but is not restricted to the following:

Active ingredients of other products made in the same equipment
Precursors
Intermediates
Degradation products
Cleaning agents
Microbiological
Environmental agents
Excipients
Foreign solvents
Manufacturing materials

The major compliance issue regarding contamination is the selection of the types of contaminants that offer a significant threat to the specific product being manufactured and focusing the cleaning program toward those particular contaminants. It goes without saying that not all of the list need be considered for all equipment and facilities, and it is therefore incumbent on the individual company to evaluate the various possible contaminants and determine which are possible given their specific environments and situations. A contaminant may be significant in one type of manufacturing situation but not in another. For example, precursors and intermediates may not be significant for the manufacture of oral dosage forms but could be a potentially significant contaminant for an API manufacturer. In a similar vein, cleaning agents would be a potential contaminant only in companies that use them, and not all cleaning involves cleaning agents. Foreign solvents would almost always present a greater contamination threat to the manufacture of APIs than to the manufacture of tablets.

Microbiological, endotoxin, and environmental agents (particulates) would typically be a more serious contaminant for sterile products than for topical products (e.g., ointments, creams). That does not mean that microbial issues are not important for nonsterile dosage forms, since ointments and creams are often applied to abraded or compromised skin. It just means that in the risk spectrum, contamination of sterile (typically injectable) products by microbiological contaminations tends to result in more serious injury to the patient.

Manufacturing materials can and often are a source of contamination. Various materials, such as filtering aids and charcoal, are used especially in chemical synthesis of active ingredients. These materials can carry over to the finished product and can represent a significant source of contamination. For many years, companies would occasionally see tiny "black spots" on white tablets. Often those black spots could be traced using very sensitive analytical techniques back to the active ingredient and positively identified as charcoal. Other examples of contamination by manufacturing materials are diatomaceous earth (a filter aid), fibers from brushes, small slivers of stainless steel, glass particles from glass-lined tanks, plastic particulates from gaskets, paper particles from filters, lubricants from motors and bearings, and even fibers from personnel uniforms. Any material involved in manufacturing a product has the potential to become a contaminant.

Excipients are also listed as potential contaminants. This is rather surprising, since most of us think that excipients are inactive materials and play no role in the product. Excipients can represent a potential type of contamination, however. Imagine, for example, a facility that manufactures sterile solutions. Possibly an initial product contains an excipient that is poorly soluble in water. Possibly some other solvent such as alcohol may be present to keep the material in solution. If a subsequently manufactured product is completely aqueous, then any remaining residue of the excipient might become a suspended particulate in the next product, an unacceptable situation for sterile injectable products. Before you dismiss this possibility as remote, let me say that this author has faced this exact situation. It was even more alarming that the production manager tried to dismiss the contamination as "not a failure since the acceptance criteria only required no active to remain on the manufacturing surfaces." In his mind, it was not a legitimate failure since the excipient had not been selected as a potential contaminant.

An auditor or investigator should investigate the rationale for the selection of potential contaminants addressed in the cleaning program. Did the company select the most appropriate ones (i.e., the ones presenting the greatest threat and the ones most likely to occur)?

4 ELEMENTS OF A COMPREHENSIVE CLEANING PROGRAM

When we think of cleaning, most of us think of cleaning validation. A good cleaning program is much more than validating the cleaning process, however. Cleaning should be a long-term program and neither begins nor ends with validation. Some of the elements of a cleaning program are

Master plan
Product matrix
Product grouping and worst-case approach
Approach/strategy
Documentation
Training
Standard operating procedures (SOPs)
Change management
Cleaning procedures

Validation of cleaning
Monitoring of cleaning
Validation maintenance
Revalidation (when appropriate)

A comprehensive cleaning program usually begins with a master plan. A master plan should function as a tool to ensure that activities occur in the most efficient and timely manner and should delineate the individuals responsible for the execution of the various activities; otherwise the entire program would eventually begin to fade and invariably fail without the formalization that a master plan brings into play.

The sequence of activities is important. An ideal sequence could be represented as follows:

Formulation of a master plan

↓

Development of good cleaning procedures

↓

Selecting on of contaminants to focus on and
determination of suitable carryover limits

↓

Development of appropriately sensitive analytical methods

↓

Preparation and implementation of cleaning validation protocols

↓

Placement of cleaning process into change management program

Monitoring of cleaning

↓

Validation maintenance of all cleaning systems and procedures

↓

Revalidation of cleaning process when necessary

The author has encountered several facilities that want to proceed directly to cleaning validation. Upon close examination, however, the cleaning procedures are found to be poorly written, resulting in a procedure that cannot be validated. In a similar fashion, what sense does it make to begin developing analytical methods to verify the effectiveness of a cleaning procedure if the limits still need to be established? The assay developed may not be sensitive enough to pick up the levels of contaminant at the required levels. As a result, the analytical development would need to be repeated to develop a suitably sensitive assay.

It is also interesting to note that many professionals consider cleaning validation as a destination instead of a journey ("Once completed, you're done"). As difficult as it is to reach a validated condition of cleaning processes, however, it is equally difficult to maintain the validated condition. There must be virtually no change to cleaning agents, scrub times, temperature of cleaning solutions, rinse times, or drying conditions once validation is complete without first subjecting those changes to a change control review mechanism. Once reviewed, it may be discovered that the proposed changes will necessitate a revalidation effort.

5 POTENTIAL SHORTCOMINGS OF TYPICAL CLEANING PROGRAMS

Inadequately written cleaning procedures
Inefficient cleaning processes
Lack of scientific basis for establishing limits
Not identifying the most difficult-to-clean (worst-case) locations in equipment
Ignoring the role of heating, ventilation, and air conditioning (HVAC) and utilities in cleaning
Poor training programs
Inadequate documentation of actual cleaning and validation of cleaning
Ineffective monitoring
Lack of support from development group
Inadequate management of incremental and significant changes to the cleaning program

This section will discuss some of the compliance difficulties typically encountered in today's cleaning programs. These are areas that continue to cause the industry major regulatory challenges.

5.1 Inadequately Written Cleaning Procedures

As mentioned in the previous section, some companies have cleaning procedures that are written so generally that they cannot be carried out in a repeatable, consistent manner. When a cleaning procedure can be interpreted in more than one way, it is not surprising that the implementation will be different, depending upon the interpretation of the individual operator. Cleaning procedures should make every effort not to be ambiguous. One very important aspect of a cleaning procedure is an accurate description of equipment disassembly and the extent to which it should be disassembled. If this is not specified, the equipment may be fully disassembled by one operator but incompletely disassembled by another. A useful technique to specify the extent of disassembly and thus not leave it open to interpretation is to use a photograph or a diagram of the disassembled equipment.

5.2 Inefficient Cleaning Procedures

Many companies use the same cleaning procedure for all products (i.e., a generic cleaning procedure). This "one size fits all" concept may work well for men's socks, but it does not work well for the cleaning of pharmaceutical products, some of which may require considerable effort to effectively and adequately clean. The development of cleaning procedures should be product-specific and ought to receive the same consideration during the development phase as the formulation for the new product.

5.3 No Scientific Justification for Establishing Limits

Another very important aspect of cleaning from the compliance viewpoint is the evaluation of the basis or justification of the limits for the cleaning process. The limits may be known by various terms, such as "maximum allowable carryover," "acceptable daily intake," and so on. Basically, it refers to how much of the residual product may safely remain on the equipment after cleaning and not cause an adulteration problem in the next product to be manufactured in the same equipment; such adulteration may or may not cause a medical or toxic response in the user. Again, many companies may use what they consider to be a gold standard as the limit, such as 10 ppm, or 0.1%. There is no such thing as a gold standard; establishing the appropriate limit depends entirely on the drug, active ingredient, allowable carryover, and even the drug's dosing schedule.

5.4 Failure to Identify Most Difficult-to-Clean (Worst-Case) Locations in Equipment

Although the selection of the worst-case locations for demonstrating that equipment is clean is somewhat subjective, it does deserve serious consideration. Selection of easy-to-reach and easily cleaned surfaces without also sampling difficult-to-clean areas is a sure mechanism to undermine a cleaning program. Even if certain areas are impossible to access, they can still be sampled by rinse samples using a solvent in which the residues have appreciable solubility. The use of rinse samples for hoses, transfer pipes, and intricate equipment is much more preferable than trying to use a swab sampling technique that simply is not reproducible or representative.

5.5 Ignoring the Role of HVAC and Utilities in Cleaning

Ventilation and air conditioning systems have the capacity to spread contamination not only from one piece of equipment to another but even from one manufacturing area to another. It is important to be able to evaluate complex air distribution and balancing systems, especially with regard to their cleanability. The air systems have been rigidly evaluated and controlled in sterile facilities to prevent microbial contamination, but often little thought is given to the possible movement of micronized powders from one manufacturing area to another. The seriousness of this is that the air system has the potential to not only contaminate equipment but also to contaminate the breathing air for the entire facility, including administrative areas in which personnel do not wear face masks or respirators. Surprisingly, many HVAC systems are still not filtered, or else the filters are inappropriate or located in the wrong locations, proving ineffective at preventing contamination of the supply air.

Water systems also represent a source of potential contamination. We simply cannot assume that water is good enough to rinse with or wash with. It is absolutely necessary to establish the condition and purity of water used for cleaning through seasonal testing and by thoroughly validating any corresponding water systems used for cleaning and manufacturing.

5.6 Poor Training Programs

Two major problems typically associated with training programs and captured in recently issued FD-483s are

1. Facilities that conduct adequate training but poorly document it
2. Facilities that conduct inadequate training but manage to document it

Change resulting from altering any aspect of a cleaning program should be accompanied by retraining and complete documentation of the training. An auditor or investigator will be suspicious of any procedure that is more than 5 years old. There may not be changes in the written procedures, but more than likely there have been changes in the way the procedures are implemented in practice (e.g., "shortcuts" or changes in cleaning agents, equipment, and personnel).

5.7 Inadequate Documentation of Actual Cleaning Activities and Validation of Cleaning

Critical activities in pharmaceutical manufacturing and packaging processes are described by procedures (e.g., SOPs, batch records, validation documents). Documentation is critical in substantiating what occurred, yet much of our documentation demonstrates a lack of attention to detail. Empty blanks, cross-outs, inadequate explanation, and missing pages are still common occurrences. In the United States, cleaning steps are considered a critical step in the manufacturing process, and there is an FDA expectation that there will be two signatures, one for the operator who performed the cleaning and a second signature of a colleague who verified the cleaning. In other countries where only one signature is required, an even greater compliance emphasis is placed on that single signature.

In addition, there are certain closed systems cleaned by clean in place (CIP) procedures in which the equipment is not opened and visually inspected after each cleaning event. One might question what the signature actually represents in such cases. The signature means that every phase of cleaning occurred as intended, and that is enough, providing the CIP procedure was appropriately validated prior to utilization on a commercial batch.

In the event that the CIP procedure does not proceed in accordance with the validated procedure, a deviation report and investigation would need to occur. Unfortunately, documentation associated with deviations, pharmaceutical exceptions, and failures is often not handled with the extent of detail and urgency required by regulatory agencies.

5.8 Ineffective Monitoring

Once a cleaning procedure is validated, we may be complacent that no further testing is required. Experience demonstrates, however, that the efficiency of the cleaning process may drift in and out of a State of control due to variables that might not be initially identified. For example, the equipment may age over the years, and scratched equipment is more difficult to clean than shiny new equipment. A batch-to-batch variation might occur in the cleaning agents themselves, thus altering the effectiveness of the cleaning process. The potable water might be drawn from different sources by the local utility company (switch from surface water to artesian wells). Sometimes the effect is subtle and might give a different pH or a slightly different mineral content. For these and many other subtle reasons, it is good to occasionally monitor the effectiveness of the cleaning process. For toxic or potent drug situations, it may be necessary to monitor cleaning constantly (i.e., after every cleaning).

5.9 Lack of Good Support from the Development Group

When a new product is handed over from development to production, the handover is usually accompanied by a development transfer report. Unfortunately, many times the cleaning procedure is not included in the report and production must develop the procedure concurrently with producing the first few batches. This is a major oversight and the author suggests that production not accept such a report until the recommended cleaning procedures are included in the report.

5.10 The Dangers of Not Including Cleaning Under the Change Management Program

Any procedure that is validated should be placed under the auspices of a change management program. This includes the cleaning procedure, because it is simply too important that any change be reviewed prior to implementation. The author has seen many instances in which the written cleaning procedure bears no resemblance to the way equipment is actually cleaned. If the procedure is not given a sacred status, small changes will creep into the actual practice and eventually undermine the validation of the cleaning procedure. As emphasized earlier, it is very important that the cleaning procedures be written with sufficient detail. The author has encountered several companies that allow the cleaning operator to select any available cleaning agent according to availability or personal preference. How can such a situation be considered validated?

There are many subtle and insidious changes that may inadvertently creep into a cleaning procedure and essentially change it. Some of the changes that can undermine a cleaning validation program are expressed in Table 1.

6 VALIDATION OF CLEANING PROCEDURES

How does one determine when to validate cleaning procedures?
What are the critical elements of the cleaning validation protocol and report?
When is validation of cleaning not appropriate?
How does one select the best approach to validation of cleaning procedures?

TABLE 1 Typical Changes That Will Assault the Status of a Validated Cleaning Procedure

Procedural changes
Small changes in the cleaning procedure
Changes in the manufacturing procedure
Changes in the equipment storage procedure
Changes in the maintenance schedule
Material changes
Changes in the equipment configuration
Changes in the composition of the product contact materials
Changes in raw materials supplier
Changes in potency of products
System changes
Changes in the water system
Changes in the HVAC system (especially ventilation and air filtering)
Personnel changes
New personnel leads to new interpretations of procedures

6.1 Knowing When to Validate Cleaning Procedures Can be Tricky

It is important to know the proper time in the development sequence to validate the cleaning procedure. There are still many variables during the development process that have not been finalized (e.g., the dosage, formulation, specific manufacturing equipment, and scale of the manufacturing equipment). Surprisingly, many companies do not perform formal validation of the cleaning process until after the Pre-Approval Inspection (PAI). This is because the variables mentioned previously have not been finalized and often three full-scale batches have not yet been manufactured. There thus has not been the opportunity to clean after three full-size batches have been manufactured. Often the cleaning process is validated concurrently with the validation of the manufacturing process.

6.2 Critical Elements of the Protocol and Report

The most important parts of the cleaning validation protocol and final report are

Justification of the limits. This should be a logical and scientific explanation of how limits were established.

The raw data themselves. The calculations and assumptions must be absolutely correct and within the acceptable limits.

Explanation of any deviations and/or failures. Again, any failures should be accompanied by extensive investigation, assignment of cause of the failure, and statement of the corrective actions.

7 CLEANING REQUIREMENTS IN THE DEVELOPMENT ENVIRONMENT

Cleaning "verification" is more likely to occur, as opposed to full-blown validation.
The need to determine limits when the next product is unknown.
Cleaning associated with the manufacture of Clinical Trial Materials (CTMs).

7.1 The Concept of Cleaning "Verification"

For research and development facilities, the concept of standard validation often does not make sense for the following reasons. First, the product formulation, dosage, equipment to be used, and batch size have not been finalized. The manufacturing parameters therefore have not been fixed, but are still being "tweaked." It is unlikely that even two batches of the product would be manufactured in the same manner, and extremely unlikely that three batches would be made in the same manner. For this reason, the whole development process does not lend itself to validation of either the manufacturing process or the cleaning procedure. Many companies, however, subject the cleaning process to a concept referred to as cleaning verification [3]. Cleaning verification treats each manufacturing and cleaning event as a unique entity. Individual protocols are written, limits are established, samples are taken of the cleaning equipment, and samples are analyzed. Each piece of equipment is thus cleaned and tested against predetermined standards. The main difference is that there is only a single set of data to be evaluated versus three sets in the case of validation.

7.2 Determining Limits When the Next Product is Unknown

Unlike production facilities, in research and development facilities the next product to be manufactured in the equipment is often unknown. One of the significant factors in determining the limits for a typical cleaning are the parameters of batch size and dosage of the following product. It is important for the research or development facility to develop a strategy of how to handle this situation not knowing the following product. The strategy may be to review the equipment history log, use the worst-case combination of parameters from previous manufacturing situations, and choose the smallest batch of product previously produced in order to use the largest daily dose of products made in this equipment for computing limits. Another approach has been to go ahead and clean the equipment and test the cleaned surfaces but not make a judgment about the pass or fail of the cleaning process until immediately prior to the next use of the equipment. At that point, the calculations could be done and it could be determined whether the equipment was suitably clean or should be recleaned prior to use. Regardless of the approach, the strategy and cleaning procedures must be documented.

7.3 Cleaning Associated with Manufacture of Clinical Trial Materials (CTMs)

The manufacture of CTMs often falls within the domain of the research and development department and is often manufactured in development facilities. This is a typical example of development batches. These batches will be used in human patients, however. It is very important that the equipment be proven clean prior to the manufacture of the CTM. The previously manufactured material may have been developmental in nature, but could seriously alter the effect of the CTM if contamination were to occur, thus undermining essential pivotal studies. Likewise, cleaning after the manufacture of the CTM should be subjected to cleaning verification and thoroughly documented.

8 CLEANING IN ACTIVE PHARMACEUTICAL INGREDIENT AND BIOTECHNOLOGY FACILITIES

There are several unique aspects to these facilities, along with the following distinguishable features:

- Cleaning with organic solvents
- Early steps versus terminal steps
- Use of boilouts as a cleaning mechanism
  - Hard-to-clean areas
  - Transfer pipes and hoses
  - Centrifuges
  - Patch panels
  - Chromatography columns

8.1 Unique Aspects of These Facilities

Facilities that manufacture APIs, either by chemical synthesis or biological processes, are quite different from dosage form manufacturing facilities. The processes usually occur in reactors or fermentation tanks in which materials may be introduced and physical parameters such as temperature and pressure may be carefully monitored. These processes also end with materials that are quite different physically from the starting materials [4].

8.2 Cleaning with Organic Solvents

Because of the chemical and biochemical processes associated with the formation of the active ingredient, very often the residues are complex and difficult to dissolve with aqueous solutions, even with the aid of cleaning agents. Traditionally these companies have often resorted to organic solvents for cleaning. Many of the facilities were originally designed to be very similar to chemical manufacturing facilities because of the nature of the processes. The equipment thus is often very difficult to disassemble and therefore it is difficult to access all product contact surfaces.

8.3 Early Steps Versus Terminal Steps

These processes almost always involve a series of steps, progressing from initial starting materials through a set of one or several intermediates and finally resulting in the finished active ingredient as represented by the following diagram.

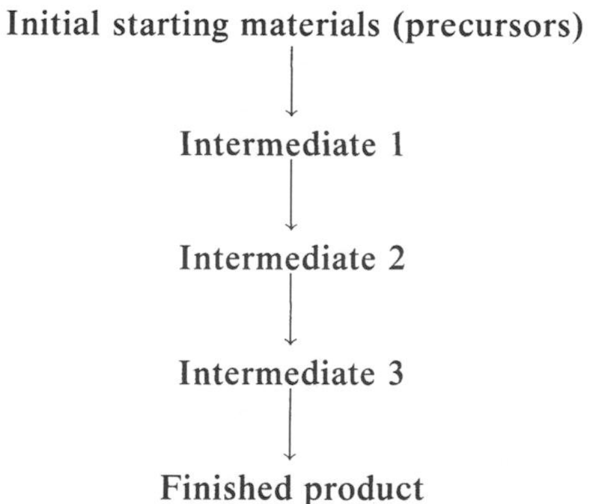

Because these steps are essentially chemical or biochemical reactions, they are often incomplete, meaning that there may be part of the material from the preceding step in each step of the overall synthesis. It is therefore necessary to have various purification steps at critical locations throughout the overall synthesis. These purification steps mean that potential contaminants are being removed throughout the synthesis, therefore a contaminant may be at an unacceptably high level in an early step only to be subsequently removed by a purification step and not represent a significant risk to the finished product if the process is under a state of control. Typically, then, the level of equipment contamination may not be as significant in the early steps of the process as in the latter or terminal steps of manufacture. The process should be constantly monitored, however, to determine that the purification steps are in a state of control and that the capacity of the purification steps is not exceeded or swamped by a high level of impurities.

8.4 Use of Boilouts as a Cleaning Mechanism

One of the results of the nature of the API facilities is that they often have areas that cannot be accessed for direct and thorough cleaning either by hand or by other mechanical means. Examples of these areas are the extensive piping that cannot be disassembled and may or may not drain freely. Another example are the so-called patch panels, in which various pipes and hoses are switched to divert liquids to various pieces of manufacturing and holding tanks. One mechanism that has been used for many years to clean these difficult-to-access regions is the use of boilouts. The equipment is actually cleaned by placing organic solvents (typically acetone or alcohol) into a reactor and increasing the temperature to bring the solvent to a boil. At that point the solvent becomes a vapor that permeates the piping and related equipment and subsequently condenses on the cooler surfaces. The condensed liquid then runs back down the pipes and returns to the reactor. This situation functions essentially as a very large distillation column, and the theory is that the repeated distillation is also a cleaning mechanism that will clean areas that cannot be accessed. The point is that sometimes this is an effective means of cleaning and sometimes it is not; it needs to be evaluated on a case-by-case basis.

8.5 Hard-to-Clean Areas

There are areas in API equipment that are known to be difficult to clean. Some were mentioned in the previous section. Other difficult-to-clean areas are hoses, centrifuges, and chromatography columns. Along with transfer pipes, hoses, and patch panels, these constitute difficult-to-clean areas and should be the focus of any investigation as to the adequacy of cleaning in these types of facilities.

9 MANUAL VERSUS AUTOMATED CLEANING

Consider the following:

Pros and cons of manual cleaning
Clean in place (CIP)
Clean out of place (COP)

9.1 Manual Cleaning

It is amazing how so many speakers at conferences expound on the virtues of automated cleaning, yet in reality the overwhelming majority of companies use manual cleaning procedures either completely or in part. In fact, in the author's opinion it is often much wiser in certain circumstances to use manual cleaning procedures. Manual cleaning offers certain advantages in that the equipment is usually fully disassembled and it is less likely that inaccessible areas will escape good cleaning. In contrast, automated procedures, sometimes do not guarantee good cleaning of equipment that is not fully disassembled but cleaned in place. (See Sec. 10.) In particular, areas such as gaskets are very difficult to clean even by automated procedures if not fully disassembled. There is virtually no substitute for actually being able to visually examine a piece of equipment "up close and personal" to determine if it has been cleaned.

The main problem with manual cleaning is that often the cleaning procedures and training program are not rigorous enough to ensure that the cleaning is carried out in a consistent fashion. Cleaning procedures should be just as detailed as manufacturing procedures, and there should be no room for "interpretation." Most companies now regard cleaning as a critical step in the manufacture of high-quality products. It is thus a regulatory expectation that equipment be cleaned by one operator and that the cleaning be verified by a second operator, resulting in two signatures on critical cleaning documentation.

9.2 Clean in Place

Having discussed some advantages offered by manual cleaning in certain situations, it should be noted that automated CIP cleaning also offers some advantages in certain situations. Particularly in sterile manufacturing areas, CIP has the major advantage of less invasive and less disruptive cleaning, thus if systems could become contaminated by bacterial contamination by opening and disassembly during manual cleaning automated CIP cleaning offers a major advantage. It should be mentioned that CIP was originally developed by the dairy industry and was specifically developed to clean closed systems and prevent microbial contamination from batch to batch. As such, most of the early systems were designed to removed milk and other colloidal solutions. There has been tremendous refinement of CIP to adapt it to pharmaceutical cleaning, in which the residues to be removed may be powders, ointments, and other physical "challenges."

CIP cleaning may be either manually controlled or controlled by a computer. For the manually controlled CIP, compliance or control of the cleaning process is as much an issue as it is for manual cleaning. For CIP controlled by a computer, there is an added requirement that the automation of the process must also be validated, so in a manner of speaking, automation of the cleaning process is both a "pro" and a "con." Automation offers the advantage of consistency, but is an additional validation burden. Each cleaning situation should be evaluated individually. It does make sense to have automated CIP cleaning for a development laboratory in which the manufacturing system may consist of a 500-ml round-bottom flask. On the other hand, trying to clean a very complex sterile manufacturing system manually may be virtually impossible.

9.3 Clean Out of Place

Somewhere between manual and CIP cleaning, there is an intermediate type of cleaning referred to as "clean out of place." COP cleaning is most easily described as that cleaning that occurs when equipment is disassembled and then placed in a washing machine that automatically cleans it, much like a home washing machine. Many companies use COP cleaning because it appears to offer the combined advantages of both manual and automated cleaning. COP cleaning also offers the advantages of disassembly of equipment as well as the consistency of cleaning by a reproducible, automated cleaning cycle.

There are certain subtleties for COP, however, that should also be mentioned. Very often companies have neglected to fully appreciate how important it is to define the loading patterns of the washing machines, with the results that some areas may be "shadowed" by other parts in the washing machine and may not be fully cleaned. Loading patterns should therefore be validated. There is also a new generation of washing machines available that are specifically designed to ensure the cleaning of specific types of manufacturing equipment and laboratory glassware. Internal parts of the washing machine may be changed according to the type of equipment being cleaned, thus ensuring good cleaning for the particular equipment.

10 HOLDING TIMES ASSOCIATED WITH CLEANING

Issues to consider include

Time between manufacturing and cleaning
Time between cleaning and next manufacturing

10.1 Time Between Manufacturing and Cleaning

One issue that continues to emerge during inspections is whether or not time has been factored into the validation plan for the cleaning process. The time between the end of the manufacturing event and the initiation of the cleaning process may greatly affect the efficiency of the cleaning. Certain types of residues will be relatively easy to remove if cleaned immediately, but will become relatively resistant to the same cleaning process if allowed to remain on the equipment for an extended period of time. For example, proteins fall into this category. If a company manufactures a protein and because of various business constraints does not clean the equipment until days later, the question becomes "How do we know that the same cleaning process that is effective if equipment is cleaned immediately is also effective after the material has remained on the equipment for some time?" The simple answer is that we do not know the answer without factoring that lag time into the validation process for the cleaning procedure. Each company should establish the longest period of time in which the equipment can be expected to remain in the "soiled" condition. If that is, for example, a period of 2 days, the equipment should be held in the soiled condition for 48 hr and then cleaned and this cleaning should be included as a part of the validation study. This represents the worst-case situation for holding or allowing the equipment to sit in the uncleaned condition.

10.2 Time Between Cleaning and Next Manufacturing

Just because equipment is cleaned does not imply that the equipment will stay clean indefinitely. Depending on the length of the storage period and the condition of the storage environment, the equipment may or may not stay clean, especially for extended periods of time. It is not uncommon in pharmaceutical facilities for equipment to remain idle for weeks or even months at a time. Also, storage facilities may or may not be able to maintain the clean condition, depending on such environmental factors as air systems, facility design, and traffic in area. The time between cleaning and the next use of the equipment therefore should also be included in the design of the protocol for cleaning validation. Some companies have dealt with this issue by simply recleaning the equipment immediately prior to the next manufacturing event, thereby eliminating the need to establish a holding time during validation.

11 ESTABLISHING LIMITS FOR ACTIVE INGREDIENTS AND CLEANING AGENTS

Address the following aspects:

The need for scientific versus arbitrary limits
An approach based on therapeutic dose
An approach based on toxicity
Selection of the best approach
Microbial limits

11.1 The Need for Scientific-Based Versus Arbitrary Limits

The author continues to encounter companies that have used limits for their cleaning processes without any scientific justification. Limits should be practical, achievable, and scientifically justified. The scientific rationale for the limits is probably the most important aspect of the cleaning program [5–8]. All pharmaceutical companies can expect regulatory agencies to evaluate the rationale for selected limits in great detail. This is extremely important because a large number of warning letters have been issued and will continue to be issued until companies can leverage adequate justification. In the author's opinion, any company not having an adequate justification for the limits of its particular cleaning process can almost expect an automatic Warning Letter.

11.2 An Approach Based on Therapeutic Dose

Setting limits based on allowing some fraction of the therapeutic dose to carry over to the following product is probably the most common approach to setting limits. This method was developed several years ago by scientists at Eli Lilly and Co. and is often referred to as the Lilly approach. It is based on the widely accepted view of physicians, toxicologists, and other medical professionals that giving 1% or less of the lowest therapeutic dose of a product will not produce any significant medical effect on the patient. This reduction factor is often referred to as a "safety factor" or "risk assessment factor." Commonly used safety factors are 1/100, 1/1000, and 1/10,000. The specific safety factor chosen may be selected based on the inherent risk in the cleaning situation. For topical products and nonproduct contact surfaces, the safety factor of 1/100 is often used. For research compounds and newly developed products still in the clinical stages of development, a safety factor of 1/10,000 or even lower may be appropriate. The reason for the lower safety factor is that there may be unfavorable properties (e.g., allergenicity) that may not be known at the early stages of product development.

In simple terms, the limit or maximum allowable carryover of a product into the next product can be quantitatively established by the following relationship:

$$\begin{aligned}\text{Total limit} &= (\text{therapeutic dose of product}) \times \text{safety factor (SF)} \\ &\quad \times \text{number of doses in the next product batch}\end{aligned}$$

As noted, this calculation of the limit takes into account not only the medical dose of the current product (i.e., the potential contaminant) but also certain information about the next product to be manufactured in the same equipment. This relationship may be represented mathematically as

$$\begin{aligned}\text{Total limit} &= (\text{dose of product A}) \times (\text{SF}) \\ &\quad \times \frac{(\text{batch size of product B})}{(\text{daily dose of product B})}\end{aligned}$$

In many cases, because of the uncertainty about the subsequent product, companies often choose to use the worst-case following product, which would be the product having the smallest batch size and the largest daily dose for calculation purposes. In that case, the limit would be valid no matter what product is subsequently introduced in the manufacturing sequence and the company could go ahead and finish its cleaning validation study.

11.3 An Approach Based on Toxicity

There are cases in which the limit cannot be based on the therapeutic dose of the potential contaminant; for example, precursors and intermediates used in chemical or biochemical synthesis of active ingredients or for establishing limits for residues of cleaning or sanitization agents. In these cases, a limit may be established based on the relative toxicity of the material. One approach that has been popular has been based on the "no observed effect level" (NOEL) of the compound or drug. This approach is based on the work of scientists at the Environmental Protection Agency (EPA) [9] and uses animal models and toxicity data in the form of LD_{50} (dose required to kill $\frac{1}{2}$ of the animal population in the study). Other approaches have also based the limit on the toxic dose with a large safety factor applied.

11.4 Selecting the Best Approach

Selecting the most appropriate approach is largely driven by the cleaning situation. The most common method is to base the limit on the therapeutic dose of the product where possible and to base the limit on the toxicity of the material when medical or therapeutic dose information is not available.

11.5 Microbial Limits

Setting limits for microbial and endotoxin residuals is another individual situation. It is also more difficult to establish meaningful limits for these materials, since unlike chemical residues, the levels of bacteria and endotoxin are continually changing and are also affected by the presence of the product. Some products will actually inhibit bacterial growth, while others will promote bacterial growth. Evaluating microbial levels at a certain time (e.g., immediately following cleaning) yields only one single data point and is of dubious value in this author's opinion.

Bacterial levels are critically important for sterile products. The multiple data points obtained in a comprehensive environmental monitoring program may give a much greater level of information on the adequacy of the cleaning and sanitization program, however. For nonsterile products, the microbial and endotoxin requirements are even more diffuse. There have been very few attempts to define what is considered to be a "safe" level (i.e., limit) for microorganisms following cleaning and/or sanitization. Future endeavors, however, will no doubt more fully define acceptable levels of micro-organisms. As recently as 1999, the United States Pharmacopeia (USP) proposed the following as suitable levels for micro-organisms for nonsterile pharmacopeial articles:

Solid oral: <1000 CFU/g
Liquid oral: <100 CFU/g
Topicals: <100 CFU/g

It should be noted, however, that this proposal was not adopted for USP 24.

12 SAMPLING AND ANALYTICAL ISSUES ASSOCIATED WITH CLEANING

The following issues must be considered:

Choosing the appropriate sampling method
The sampling protocol
Sensitivity versus limits
Validation of the analytical method as a precursor to cleaning validation

12.1 Choosing the Appropriate Sampling Method

The sampling method selected should be based on the consideration of the nature of the equipment and the product being manufactured. It is foolish to select a single sampling method for all products and equipment and just stick to that one method no matter what. In a single cleaning validation study, it may be necessary to use two or more different types of samplings. Regardless of the sampling method selected, it is necessary to document the rationale for the choice. This is often so obvious that companies forget to document the reason for the choice.

A few years ago, the Parenteral Drug Association (PDA) Task Force on Cleaning issued a "Points to Consider" document [10] that enumerated various types of samplings currently in use in the industry. They were (at that time)

Swabs
Wipes
Rinses
Coupon
Solvent
Product
Placebo
Direct surface monitoring

Of these eight types of samplings, only two (swab and rinse sampling) have been accepted and included in regulatory guidelines. A third type, direct surface monitoring, offers promise of future acceptance once fully mature methods and instruments are available.

It is beyond the scope of this chapter to describe in detail each method and its subtleties. Suffice it to say that there should be a logical choice of the sampling method for each particular situation. In many cases, the method of sampling will in fact be dictated by the equipment. Inaccessible equipment parts such as hoses and transfer pipes must be sampled by rinse sampling since representative surfaces cannot be reached by swabbing. In other cases, the residual material may not be removed by simple rinsing and will need to be swabbed in order to remove the residue from the equipment and onto the swab.

Some simple prevalidation studies may be necessary to establish which sampling method is appropriate and also to develop a technique that will give adequate recovery of the residues from the equipment. If experimentation demonstrates a very low percentage of recovery of spiked residues from equipment by rinse sampling, then there may be two possible alternatives: (1) develop a better, more complete rinse by recirculating the rinse back over the equipment using multiple passes to increase contact time, or (2) change to a swab sampling technique. Typical swabbing may yield poor results during the initial empirical studies. It may be necessary to swab the same site multiple times from multiple directions or even to use two or three new swabs to remove a suitable percentage of the material.

Now you may be wondering what a suitable percentage of recovery is. Regardless of rumors to the contrary, there is no guideline as to an acceptable percentage of recovery. While we would all like to have recoveries of 99%, that is not usually feasible or even possible. In most sampling situations, there will be a maximum percentage of recovery that will be very difficult to significantly improve upon.

12.2 The Sampling Protocol

Concurrent with determining how to sample (e.g., swab, rinse) a decision must be made regarding where to sample. The sampling location should be determined by evaluating what would fairly represent all the product contact surfaces,–easy-to-clean areas as well as difficult-to-clean locations. Most companies use diagrams or photographs to identify sampling locations. The sampling locations should definitely include the most-difficult-to-clean locations. The easy way to determine the most -difficult-to-clean locations is to ask the operators who have cleaned the equipment for years.

There are two types of most-difficult-to-clean locations; namely "hot spots" and "critical sites," and they are defined in the following table.

Term	Definition	Example
Hot spot	A difficult-to-clean location which, if contaminated, would result in the contaminant being uniformly distributed throughout the next product	Residue on the mixing blade of a mixer in a liquid formulation tank
Critical site	A difficult-to-clean location which, if contaminated, would be oncentrated in only a few doses of the next product	Residue in filling needles of a liquid filling machine

12.3 Sensitivity Versus Limits

In terms of the assay for cleaning residues, one of the most important parameters is the sensitivity of the analytical method, often expressed as LOD (limit of detection) or LOQ (limit of quantitation). Simply put, the assay must be demonstrated to detect residues below the level of the acceptable limits, otherwise the method invalidates itself and is not appropriate for the evaluation of the cleaning samples. For that reason, the limits ideally should be developed at the same time as the analytical method or prior to the development of the assay. Many times analytical methods have been developed without taking the LOD and LOQ into account only to have to be repeated because of a lack of sensitivity.

Assays should also be chosen based on the actual situation. Assays for cleaning samples for new equipment that has just arrived at the facility should be nonspecific, so that any potential foreign material could be detected. A specific assay should not be attempted when the potential contaminant(s) are unknown. Ideally, assays for product cleaning validation either should be specific or an assumption should made that any residue detected is due to the most potent or toxic material potentially present (worst-case assumption). Consider an example to demonstrate how the assumption approach might work. Assume that we have manufactured a product containing one active ingredient and several excipients, and we have subsequently cleaned the equipment using a cleaning agent. We have taken samples and are trying to decide on an analytical method. We don't have a very sensitive High Performance Liquid Chromatography (HPLC) method available, but we do know that the active ingredient is a carbon-containing compound with a good response to total organic carbon (TOC) analysis [11–13]. The question is "Could we use TOC as our analytical technique given that the method is non-specific and will detect any residue containing carbon, including active, excipient, or cleaning agent:" The author feels that TOC is an appropriate technique because of its sensitivity if the following assumption is made. The assumption is that all carbon detected by TOC is attributed to the most potent material in the group (i.e., the active ingredient). If the sample's assay result is less than the previously established limit for the active, then the requirements of the protocol have been met using the worst-case assumption, namely that all the carbon detected was attributed to the active. If the limit were exceeded using this worst-case assumption, then it would be necessary to switch to a product-specific assay to factor the total residue into the various possible components.

12.4 Validation of the Analytical Method as a Precursor to Cleaning Validation

Once the stage has been reached at which the product has been developed and the formulation and batch size determined, it is time for validation of the manufacturing process and the cleaning process. By that point in the development process, the analytical method should be fully validated, otherwise(the impending cleaning validation studies will be for naught and will have to be repeated after the analytical method has been validated. The sequence in which these activities occur therefore is very important from a compliance perspective. The assay or method to be used must be developed prior to implementation of the cleaning validation.

13 CLEANING IN THE LABORATORY

Essential elements to consider include the following:

Cleaning of instruments
Cleaning of glassware
Cleaning of non-product contact surfaces

13.1 Cleaning of Laboratory Instrumentation

An often overlooked area of cleaning validation is cleaning in the laboratory. The activities that occur in the laboratory have the potential to seriously assault many quality control (QC) programs, including the cleaning program. It doesn't take a great imagination to visualize the impact improperly cleaned instruments would have when attempting to evaluate cleaning validation samples. Even with the use of positive and negative-controls, the presence of residues on cells (e.g., in spectrophotometers) and columns (HPLC) can wreak havoc with results. Although there has been minimum regulatory comment thus far concerning cleaning validation as applied to lab instruments, the potential is certainly present for greater regulatory scrutiny. It should be emphasized that the main weaknesses will probably not be the poor cleaning of instruments but rather inadequate documentation corroborating that the instruments were properly cleaned after use.

13.2 Adequate Cleaning of Glassware

Similar to laboratory instrumentation, laboratory glassware offers the potential to seriously impact results, depending on the cleanliness of the glassware. Many companies have switched to disposable glassware and plasticware to avoid the potential for contamination of analytical materials. Years ago, plastic disposable pipette tips hit the market, and a major marketing claim was the fact that there was no need to clean them since they are disposable. This cleaning aspect has turned out to be a blessing since these sample-contact surfaces do not require cleaning validation.

13.3 The Cleaning of Nonproduct Contact Surfaces

Nonproduct contact surfaces include

Lyophilizers
Laminar flow hoods
Ductwork, HVAC, and air filters
Walls, floors, and ceilings

The subject of nonproduct contact surfaces has been an area of great uncertainty for quite some time. Regulators routinely ask for the rationale related to the cleaning or noncleaning of these areas, for the justification of not having established limits, for not having a program or policy regarding these areas, and for the derivation of limits. Let us examine some nonproduct contact surfaces individually.

13.3.1 Lyophilizers

During "normal" processing in a lyophilizer, the ceiling, walls, and shelves would be considered not to have contact with the product. There is a remote possibility, however, that residual material, including microbial material, could become dislodged and enter the product through the extremely small opening between the neck of the vials and the readied stopper sitting on top of the vials. From the perspective that there is a direct opening to the product, there does exist the potential for product contact. The challenge becomes how to approach setting limits for these surfaces. It may be feasible to do a long-term study to evaluate the normally expected residuals on these surfaces and to use this information to establish limits. Another approach that has been used with some success is to develop mathematical models assuming that the residue from a certain size surface area could reach the inner surface of a single vial or bottle. This calculation could in turn be related to the medical dosage of the contaminant and limits could subsequently be established.

13.3.2 Laminar Flow Hoods

Laminar flow hoods are another gray area with regard to product contact. There is the potential for product contact and thus contamination since the product is opened in these areas and air is flowing over open product for a finite period. The normal activity, however, should not result in product contamination unless the surfaces of the laminar flow hood are grossly contaminated, and visual inspection should certainly mitigate against such extensive and obvious contamination. Again, it may be possible to develop models that treat the surfaces as a product-contact surface but use a more generous safety factor for the actual mathematical calculations. For example, a safety factor of 10 to 100 might be readily acceptable for calculation purposes instead of a factor of 1000 for more direct product-contact surfaces.

13.3.3 Ductwork, HVAC, and Air Filters

Ductwork, HVAC, and air filters can directly affect the cleanability and the maintenance of cleaned equipment and facilities. Although facilities should be evaluated individually, the potential for airborne contamination of surfaces is great. Although the direction of air may or may not be monitored and alarmed, there are often temporary glitches or changes in direction caused by ingress or egress of personnel, power surges or outages, and many other types of atypical activity. Many modern drugs are micronized, resulting in particular material that is very difficult to confine and control. Filtering of the air and the maintenance of filtering systems is therefore a critical component of a cleaning program. Although many companies do not sample ductwork, it may be a continuing and long-term source of contamination. The author still remembers sampling secretaries' desks and finding that the white powder was the same material (an active ingredient) weighed out the previous day in the dispensing area of the manufacturing facility. This can represent an Occupational, Safety, and Health Act (OSHA) safety issue as well as a cleaning concern.

13.3.4 Walls, Floors, and Ceilings

Walls, floors, and ceilings also represent a potential source of contamination. Their cleaning may be straightforward, but the establishment of limits represents a stumbling block for many companies. We are approaching a time in which the expectation will be to have an environmental monitoring program for both sterile and nonsterile facilities. The question is "What is clean enough for these nonproduct contact surfaces?" The answer to most of these generic questions is almost always "It depends." Without a doubt, we know that manufacturing environments impact cleaning activities and their validation. We need to study the manufacturing environment regardless of whether the materials are product-contact or non-product-contact in nature. Many materials are particularly absorbent and function as giant sponges, especially ceiling tiles.

14 REGULATORY REALITY CHECK

This section will address a typical FDA 483 observation made regarding a specific cleaning program at an actual pharmaceutical company.

The actual observation made was

The matrix cleaning validation study failed to address and evaluate the following:

1. The cleaning procedures' ability to adequately remove potential product degradents, drug/cleaning agent reaction products and excipients.
2. The ability of analytical methods to detect potential product degradents, drug/cleaning agent reaction products and excipients.
3. Worst case time frames between cleanings. During validation studies cleaning was generally performed within one day of use; however, equipment cleaning procedures lack such stringent cleaning requirements.
4. The most difficult to clean product during cleaning validation studies of product contact, multi-use packaging equipment.
5. Cleaning procedures used to clean product dedicated manufacturing equipment which did not fall into validated equipment grouping categories.

Let's examine these comments in more detail with the objective of identifying how we could correct the problems, and more important, how we could have prevented them from occurring in the first place.

First of all, one might observe that the comments are of such a widespread nature (i.e., from ignoring the chemistry of interaction of cleaning agents and products to not being aware of the importance of hold times) that the 483 document tends to describe symptoms of a larger problem,We might ask ourselves if many of these observations could have been prevented if the company had a comprehensive master plan for cleaning. During the master planning process, we tend to think more globally and try to cover all aspects of cleaning. Cleaning validation is not simply gathering three sets of residue data, signing the approval sheet, and "blessing" the data. The cleaning validation program must be carefully planned prior to implementation. The author knows of many companies that have not addressed the chemistry of interaction of cleaning agents with their specific products and excipients. These programs are in effect small "time bombs" just waiting to be discovered during a regulatory inspection. Now you may be saying that "Gee, you are talking about a monster stability study here if you mean we have to evaluate the potential combinations and permutations of every ingredient in the formulation (including excipients) with the cleaning agent." I think the issue here is evaluating the stability of the active for degradation within the context of the cleaning environment. For example, certain biological and protein molecules will be very unstable in acidic or alkaline pHs. The degradant molecules may be more or less soluble than the parent drug, this needs to be taken into account.

Now let's address the specifics of item 2 in the FDA 483 dealing with analytical issues. The items mentioned should have been identified and addressed by the analytical methods validation package prior to the cleaning validation. Obviously, they weren't addressed in this case and will need to be dealt with in the company's response to the 483. Does this mean the FDA inspector or investigator has some keen insight into the chemistry of your products? From the author's experience, the answer is "absolutely not." It simply means the inspector is saying you haven't proved your case that there is not degradation during cleaning. It would not take a major study to put some product in a solution of cleaning agent at the operational concentration and temperature and evaluate the stability after a normal cleaning cycle.

Item 3 deals with the holding time between the end of manufacturing and the initiation of the cleaning process. Again, this was not factored into the master plan. It would appear that the investigator noted that often the company would allow more than a day to go by before starting cleaning and also noted that the cleaning validation was done for a maximum of a single day's holding time. The company could respond to this comment by simply agreeing to revise the cleaning procedure SOP to require that equipment be cleaned within 24 hr of manufacturing. It is interesting to note that the investigator did not address the other important holding time, and that is the holding time between the end of the cleaning process and the next manufacturing event in that equipment. He or she could probably have found a discrepancy in that regard also.

Item 4 relates to the identification of the most difficult-to-clean representative of each group of products. Many companies have fallen into the trap of assuming that the worst-case product in a group is simply the most potent or toxic product. This isn't always the case. For example, the most potent or toxic product may also be extremely soluble in water and not be nearly as difficult to clean as another "almost as potent" product. The solubility of the products and the inherent difficulty in cleaning should also be evaluated during the selection of the worst-case product. Again, the company would need to re-evaluate the various worst cases in light of these additional factors to determine if the original worst-case product was still a valid choice and submit this evaluation to the FDA.

Finally, item 5 indicates that products manufactured in dedicated equipment should not be ignored with regard to cleaning and cleaning validation. Products may be unstable during the manufacturing conditions and may undergo further degradation if carried over to a subsequent batch of product. For example, proteins are very subject to denaturation and can thus represent a potential contaminate in a subsequent batch of the same product. This issue could be easily addressed in a study that would establish the stability of the materials during the manufacturing process. Again, the challenge to the company is to provide a practical study without progressing to a lengthy and complicated evaluation.

15 WORDS OF WISDOM

Give consideration to cleaning. It is important to prevent cross-contamination of very potent products; it is not just a :"rainy day" project. Good cleaning can prevent recalls and save a company's reputation.

Cleaning processes are considered as critical manufacturing steps, and as such, must be validated.

Most companies will need a cleaning validation master plan to function as a blueprint for all activities. Without a master plan many critical aspects will be overlooked.

Cleaning will usually be a "moving target" because of the ever-changing technology of analytical science.

Don't try to validate procedures that are not validatable. If procedures are filled with vague terms and instructions, they will be open to the interpretation and misinterpretation of the operators doing the cleaning. Take for example the term *hot water.* What does the mean? It will mean different things to different people and is therefore too vague for adequate and consistent cleaning.

Be pragmatic in the approach to cleaning.

Validation must be considered a dynamic exercise as opposed to being static. Cleaning capabilities as well as expectations will continue to develop and will be ever-changing.

Monitoring is an essential and vital component of cleaning validation and validation maintenance.

The success of validation maintenance is contingent upon a comprehensive and far-reaching change management system.

RECOMMENDED READING

General Articles on Cleaning

W.E. Hall, A roadmap to an effective cleaning program, *J Val Tech* 6(1): Nov. 1999.

W.E. Hall, Cleaning in pharmaceutical facilities—So what's the big deal? In: Booth, A. F. (ed.). *Sterilization of Medical Devices.* Buffalo Grove, IL: Interpharm Press, 1999, p 289.

W.E. Hall, Cleaning and validation of cleaning for coated pharmaceutical products. In: Avis, K., Shukla, A. (eds.). *Pharmaceutical Unit Operations: Coating.* Buffalo Grove, IL: Interpharm Press, 1998.

Interview with the FDA: William E. Hall, Ph.D., gets close with the FDA's Robert (Bob) Coleman, national expert drug investigator. *J Val Tech* 5(1), 1998.

W.E. Hall, Cleaning and contamination: Current issues in the pharmaceutical industry. *J Microcontam Det Cont* 1(3), March 1998.

W.E. Hall, Your cleaning program—Is it ready for a pre-approval inspection? *J Val Tech* 4(4): Aug. 1998.

W.E. Hall, Cleaning validation doctor—Diagnosing and treating FDA-483s and warning letters on cleaning. *J Val Tech* 4(3), 1998.

Parental Drug Association. Points to consider for cleaning validation. PDA technical report no. 29. 52(6), 1998.

W.E. Hall, *Validation of Bulk Pharmaceutical Chemicals.* In: Berry, I. R., Harpaz, D. (eds). Interpharm Press, Buffalo Grove, Il: 1997, Cleaning validation for bulk pharmaceutical chemicals (BPCs). pp 335–369.

W.E. Hall, Validation of cleaning processes for bulk pharmaceutical chemical processes. Cleaning *Validation—An Exclusive Publication*, Institute of Validation Technology, 1997.

W.E. Hall, Cleaning validation doctor: Repacking drugs. *J Val Tech* 3(4), 1997.

W.E. Hall, Cleaning validation doctor: Basic cleaning in foreign firm. *J Val Tech* 3(3), 1997.

Articles on Establishing Limits for Cleaning

G. L. Fourman, M V Mullen, *Pharm Tech* April 1993.

D. A. LeBlanc, Establishing scientifically justified acceptance criteria for the cleaning validation of APIS. *Pharm Tech* 24(10): 160–168, Oct. 2000.

D. A. LeBlanc, Establishing scientifically justified acceptance criteria for cleaning validation of APIs. *Proceedings of 15th ICCCS International Symposium and 31st R3-Nordic Symposium on Contamination Control, Copenhagen*, May 14–18, 2000 pp 537–542.

W. E. Hall, Cleaning validation doctor: Residue acceptable levels and contamination. *J Val Tech* 4(1), 1997.

D. A. LeBlanc, Establishing scientifically justified acceptance criteria for cleaning validation of finished drug products. *Pharm Tech* 19(5): 136–148, Oct. 1998 (also reprinted in *Pharm Tech Eur* 11(3): 18–23, March 1999).

Articles on Sampling and Analytical Methods for Cleaning

K. M. Jenkins, A. J. Vanderwielen, *Pharm Tech* April 1994.

H. J. Kaiser, J. F. Tirey, D. A. LeBlanc, Measurement of organic and inorganic residues recovered from surfaces. *J Val Tech* 6(1): 424–436, Nov 1999.

H. Kaiser, D. Klein, E. Kopis, D. LeBlanc, G. McDonnell, J. F. Tirey, Interaction of disinfectant residues on cleanroom surfaces. *J Pharm Sci Tech* 53(4): 177–180, July–Aug. 1999.

D. A. LeBlanc, Rinse sampling for cleaning validation studies. *Pharm Tech* 22(5): 66–74, May 1998.

Articles on Cleaning and Sanitization Agents

D. A. LeBlanc, D. D. Danforth, J. M. Smith. Cleaning technology for pharmaceutical manufacturing. *Pharm Tech* 20(1):84–92, July 1996.

REFERENCES

1. Part 211. Current Good Manufacturing Practice for Finished Pharmaceuticals. Code of Federal Regulations, Food and Drugs, Title 21 U.S. Government Printing Office, Washington, DC: 1978.
2. FDA. Guide to inspections of validation of cleaning processes. Division of field Investigations, Office of Regulatory Affairs. U.S. Government Printing Office, Washington, DC: 1993.
3. Forsyth, R. J., Haynes, D. V., Cleaning validation in a pharmaceutical research facility. *Pharm. Tech.* Sept. 1998.
4. W.E. Hall, Cleaning for active pharmaceutical ingredient manufacturing facilities. In: I.R., Berry, D. Harpaz, (eds.). *Validation of Active Pharmaceutical Ingredients.* 2nd ed. IHS Health Group, 2001.
5. Fourman, G.L., Mullen, M.V., Determining cleaning validation acceptance limits for pharmaceutical manufacturing operations. *Pharm Tech* April 1993
6. LeBlanc, D.A., Establishing scientifically justified acceptance criteria for the cleaning validation of APIs. *Pharm Tech* Oct. 2000.
7. LeBlanc, D.A., Establishing scientifically justified acceptance criteria for cleaning validation of finished drug products. *Pharm. Tech.* Oct. 1998
8. Amer, G., Deshmane, P., Ensuring successful validation—The logical steps to efficient cleaning procedures. *BioPharm* March 2001.
9. Layton, D.W., Mallon, B.J., Rosenblatt, D.H., Small, M.J., Deriving allowable daily intakes for systemic toxicants lacking chronic toxicity data. *J Reg Tox Pharm*, 7(96), 1987.
10. Parenteral Drug Association. Technical report no. 29. Points to consider for cleaning validation. March 30, 1998.
11. Strege, M.A., Stinger, T.L., Farrell, B.T., Lagu, A.L., Total organic carbon analysis of swab samples for the cleaning validation of bioprocess fermentation equipment. *BioPharm* April 1996.

12. Holmes, A.J., Vanderwielen, A.J., Total organic carbon method for aspirin cleaning validation. *PDA J Pharm Sci Tech*, 51(4), July–Aug 1997.
13. Jenkins, K.M., Vanderwielen, A.J., Cleaning validation: An overall perspective. *Pharm Tech*, 18(4), 1994.

8

The Batch Record: A Blueprint for Quality and Compliance

Troy Fugate
Compliance Insight, Inc., Hamilton, Ohio, U.S.A.

The batch record, one of the most critical operations for a manufacturer, can also be one of the most misunderstood or misused. This chapter will evaluate the regulations for batch records, review the life cycle of batch record development, compare European and U.S. batch record philosophy, examine some supporting documentation, and offer a few practical guidelines for batch records.

The primary goal of this chapter is to provide the reader with knowledge of how to compile and create a compliant master production record. The reader will be guided on how to create a complete batch record, detailing operation activities. This batch record should be of sufficient detail and clarity that any third party reviewer could examine the document and have few, if any, questions.

1 GENERAL INTRODUCTION: THE REGULATIONS

Prior to discussing a significant document such as a batch record, one needs to understand the regulations. An effort has been made to simplify the Code of Federal Regulations (CFR) specific to the batch record in order to make it easily understood. To obtain the full wording of the regulations, please reference the actual sections listed in the text below.

One of the first things to understand is the definition of the term batch. As defined in 21 CFR Part 210 § 210.3.(2), *batch means a specific quantity of a drug or other material that is intended to have uniform character and quality, within specified limits, and is produced according to a single manufacturing order during the same cycle of manifacture.*

Breaking this definition down into its subparts, a batch record can be defined as a record or documented proof that

A specific quantity of drug is manufactured.
The product is intended to have uniformity in both characteristics and quality.
The product is manufactured with distinct limits and formalized specifications.
Operations are done within a single manufacturing order or cycle.

A batch record can therefore be construed as a recording of the process utilized to manufacture, package, and label a specific batch of product. The GMP regulations that specifically discuss batch records can be found in 21 CFR 211 § 211.186 and 211.188. A synopsis of these sections follows.

1.1 § 211.186 Master Production and Control Records

This section requires that records assure uniformity from batch to batch, specifically

Master production and control records for each drug product, including each batch size, shall be prepared, dated, and signed (full signature, handwritten) by one person.
They shall be independently checked, dated, and signed by a second person.
The preparation of master production and control records shall be described in a written procedure.

This section of the code details what is required in master production and control records. It states that these records shall include the following:

The name, description of the dosage form, and strength of the product. An example would be: SuperDrug for Injection, United States Pharmacopeia (USP) 10 mg/ml. It is recommended that a single name and dosage be listed on the batch record. If there are multiple names used for the marketed product, then those other names can be referenced at the packaging or labeling stage of operations [e.g., SuperDrug (packaged as ColdRid #1) for Injection, USP 10 mg/ml].

The name and weight of measure of each active ingredient per dosage. A statement could be included on the cover page of the batch record stating this information (e.g., Each tablet will contain 235 mg of Acetaminophen).

A statement of the total weight or measure of any dosage unit. Again, this could be listed on the cover page of the batch record.

A complete list of components for the product. Typically, this list is part of the bill of materials (BOM). This list should be comprehensive, including those raw materials that may not show up in the final dosage form or on the BOM (e.g., purified water, USP).

An accurate statement of the weight or measure of each component.

A statement concerning any calculated excess of component. This statement usually is placed next to the statement of the weight of the component used in excess (e.g., acetaminophen, BP: 1000 g. This includes a 5% excess.)

A statement of theoretical weight or measure at appropriate phases of processing.

A statement of theoretical yield, including an acceptable range. Most processing schemes involve multiple stages of operation. Yields should be calculated at the end of these stages or other appropriate steps. This can determine if any problems exist at an early stage in the production cycle.

A description of the drug product containers, closures, and packaging materials.

A specimen or copy of labeling materials. Verify that the lot or batch number is on the labeling material. It is recommended that specific instructions be given in the batch record detailing when or where to take these samples. It is common practice to take samples at the start of each roll of labels or box of packaging components (e.g., preprinted cartons). At a minimum, samples should be taken and verified at the start of operations, following any interruption in the packaging /labeling process and at the end of operations. Conservative approaches have included samples taken at the splice points of a roll of labels.

Complete manufacturing and control instructions, sampling and testing procedures, specifications, special notations, and precautions to be followed. This is a general state indicating that the batch record should have complete instructions as to how to manufacture the specified product. In-process sampling and testing instructions should be indicated, along with a place to indicate the results of that testing. Limits should be stated in the batch record, including instructions on how to react when those limits are exceeded. Finally, instructions should be given on safety and environmental concerns (e.g., wear safety goggles or room humidity levels should be 40–60%).

1.2 § 211.188 Batch Production and Control Records

Prior to reviewing this section, a critical point should be clearly communicated to all personnel involved with batch records. Batch records should be considered legal documents. Many people view the batch record as merely a "guide" for their use. The batch record is extremely important in support of the verification that the product was produced under cGMP: compliant conditions. This record, along with the signatures that appear on the document, are as legally binding as any formal contract would be.

This section states that batch production and controls records shall be prepared for each batch of drug product produced. These records shall include complete information relating to the production and control of each batch and include

An accurate reproduction of the master production record, checked for accuracy, dated, and signed. An investigator will verify that controls exist to make sure that the batch record issued is reproduced from the correct master copy. If a master record changes, there must be some type of documented "purging" of the files (electronic and/or paper). Additionally, at the time of issuance of a batch record, there must be a documented check for accuracy. This accuracy check should include that all pages issued are complete (no printer or photocopy errors) and that the correct revision of the master was used.

Documentation that each significant step in the manufacture, processing, packing, or holding of the batch was accomplished. Personnel involved with the operations must be knowledgeable of this regulation. There must be a signature and cosigner for each significant step.

To accurately document that each step was performed accordingly, batch records will indicate the following:

Dates. The date and possibly time, if applicable, should be recorded as the operation is performed.

Identity of major equipment and lines used. Equipment codes or line numbers should be unique.

Specific identification of each batch of raw material or component used.

Weights and measures of components used.

In-process and quality control (QC) results.

Inspection of the packaging and labeling area before and after use. This is typically a formal line clearance per Standard Operating Procedure (SOP).

The actual yield and theoretical yield at appropriate steps of processing.

Labeling control records with samples of all labeling used. As stated in the review of the same subject under § 211.168, verify that the lot or batch number is on the labeling material. Specific instructions should be given in the batch record detailing when or where to take these labeling samples.

Description of product containers and closures. Include all items, no matter how minor they are to be (such as vial crimping or coil in a bottle of tablets).

Any sampling performed. Details should be provided as to how much sample to take, how to take or handle the sample (e.g., refrigerate), what test(s) are to be performed on the sample, the specifications for those tests, and what to do if the test results exceed limits.

Identification of the person(s) performing and directly supervising or checking each significant step in the operation. Since most operations involve a significant number of personnel across a broad spectrum of departments or areas, this will be of great benefit to any reviewer. A great deal of time can be wasted trying to determine the identity of a cryptic signature on a batch record that is several years old. There are several options available to comply with this requirement. Signature cards can be maintained on file or a page can be included with each batch record requiring this signature information be recorded for each lot. However it is performed, the minimal amount of information that should be considered is the signature, the handwritten or typed name, and the department in which that person works.

Any investigation performed and the results of that investigation. As with any operation, anomalies will occur. An investigation should clearly state the issue that triggered the report, a detailed account of the investigation, and a documented conclusion. The conclusion sought by many FDA investigators is a concise summary determining the impact on the product and the product's final disposition. Additionally, what the firm intends to do in order to prevent recurrence should be included.

1.3 FDA Guidance Documents/ICH Guidelines

These expectations of a batch record are also reflected in an FDA guideline. The "Guide on Manufacturing, Processing or Holding of APIs" discusses batch production records in Section 6.5. An ICH (International Conference on Harmonization) guide entitled "Good Manufacturing Practice Guide for APIs" also discusses batch records. These two guidelines reflect the same philosophy as indicated in 21 CFR Section 211.186 and Part 188.

Two additional regulatory requirements should be kept in mind about the batch record: (1) the master production record should be maintained and controlled via a change control procedure, and (2) issuance of the batch record should be controlled.

There are several sections of the CFR that address the control of documents. Obviously, documentation supporting each phase of the life cycle of the master production record should be maintained under change control. A change control program at these early stages will verify that appropriate departments are part of the review process and that changes are not made without sufficient data support.

The issuance of the batch record for operational use (either in production or research) is usually a step that is not controlled to sufficient levels. Most organizations have the batch records issued by the quality unit (which is the preferred practice), but little attention is placed on the procedure. Some firms have multiple reproductions of the master production record maintained in a file for easy and quick issuance to operations. This practice can have flaws in that extra safety precautions have to be taken when the master production record is updated. Serious errors can occur if a system is not in place to verify that the current revision is utilized. It is highly recommended that copies *not be maintained ahead of the time of actual issuance due to the liabilities involved in these critical documents.*

2 OVERVIEW OF THE LIFE CYCLE OF A BATCH RECORD

Understanding the critical processes in the development of a master production record will help prevent problems during production activities. It is imperative to document the entire development cycle of the product for future reference. Based upon this premise, the following information should be utilized as a guideline for development of the master production record.

2.1 Design of the Experiment and Scale-Up

As with any product, the beginning is in the research and development stages. Various experiments are usually conducted at this starting stage of the product's life cycle. This experimentation demonstrates what works and what does not. It is at this stage of the product's life that many control parameters are evaluated (e.g., glass container in lieu of stainless steel, temperature controls, optimal range).

At this early stage, the 'batch' record is frequently a simple set of instructions with wide parameters or no parameters at all. Batch records can vary between data recorded in laboratory notebooks to formal, typed documents. Regardless of format, data should be collected in these batch records at processing stages that could readily impact the production of the product. This data can then be utilized to assist in scale-up activities.

Further control parameters for the processing of the product are evaluated or established during the scale-up activities. Document any problems encountered during this stage, as well as the resolutions taken to overcome them. The actions taken to resolve these problems may be indicative of items or data that you should strongly consider including in the master production record. If these parameters are controlled or data are collected, then problems encountered in the future can be more easily resolved. These control parameters are utilized in the next phase of the batch record life cycle—validation.

At this step, the batch record usually takes on the form of a controlled, typed document. Wide parameters are usually given for data collection and processing control. The batch record should serve as a means of data collection during scale-up.

2.2 Validation Protocol Development

At this step, validation of the process serves as the means of setting and/or evaluating control parameters as established during experimentation, development activities, and scale-up. The purpose of validation is to verify or authenticate that the process is capable of reproduction with results within specified parameters. Part of this validation is to collect data over the various processing stages. The validation protocol should be based upon data and information collected during the development activities of the product's life cycle.

2.3 Execution of the Protocol

Once the validation protocol is established and executed, several key parameters will be derived for use in the master production record. It is crucial that these parameters be at least part of the overall objective of the protocol. These parameters are listed and described in further detail below.

2.3.1 Derive Critical Parameters

Critical parameters of the process can be defined as those measures that play an essential role in the production of the product. Be it the quantity of raw material, a reaction time, or a pressure within a vessel, these parameters serve as the basis for the batch record. Any variation from these parameters may compromise the quality and integrity of the product and may constitute a violation of cGMP requirements.

2.3.2 Formalize Optimal Manufacturing Conditions

When obtained, optimal conditions for the process are those control parameters that result in the best-quality product. To optimize manufacturing conditions, you need to vary the process parameters and then evaluate their impact. Many factors should be utilized for this optimization process, including data from the earlier development studies, the scale-up process, or the process involved with determining or evaluating critical parameters. Dependent upon the process or product, the optimal conditions should be placed in the master production record, along with any acceptable variations for those particular manufacturing conditions. These specifications lead into the next parameter for the protocol.

2.3.3 Establish Process Specifications

Process specifications can be established once critical parameters and optimal conditions are determined. This is the critical element of the master production record. As the process is written in a format for the master document, the process controls are already in place. From initialization to the final step of processing, a blueprint for continuous successful production of the product is now established.

2.3.4 Establish Critical Control Points

There should be established points in the process that are used as test and verification points to evaluate the successful execution of that stage of the production operation. Timely indications of a problem early in the process can save time, money, and possibly the lot. Ensure that these "in-process" checkpoints are valid and effectively evaluate the process at critical junctures.

2.4 Develop Master Production Record

The development of a master production record *can* be a monumental task for a company. Trying to determine process parameters and control points from a listing of research notebooks spanning several years is typically a daunting task. If the aforementioned steps are taken into consideration at an early stage, the master production record is usually already in place, in one form or another. The data collected from the protocols should be used to establish a framework for the process. This framework, although the essential guide for making the product, will fall short of being a truly successful framework without some additional guidelines.

2.4.1 Identify Critical Control Points

Ensure these checkpoints are clearly identified in the master production record. Operating personnel must be able to clearly identify the parameters of the test and what steps to take if those parameters are not met. As an example, testing a sample of your product after a key processing test can indicate to you if that key step was performed accurately. Potential problems include results that take too long to obtain. A particular challenge is when the lot has been completely processed or packaged.

2.4.2 Identify Safety Concerns

Many times companies focus a great deal of effort on the successful execution of the process to manufacture an acceptable product but leave out employee safety precautions. Protecting equipment and facility structures from harsh chemicals is very important, but protection of your most critical asset—the employee—should be paramount. If specific personnel protection equipment is needed for particular steps, indicate those requirements in the batch record. Also, evaluate the potential of high-pressure air lines or explosion hazards.

2.4.3 Identify Environmental Concerns

Some products have specific control requirements for the environment, as indicated during development. Relative humidity, temperature, and even light-level restrictions are parameters that may need to be controlled. If there are control requirements, they should be indicated in the batch record. Documenting the actual results of environmental conditions should be a requirement of the batch record. This can be accomplished by either routine monitoring with a calibrated instrument and then recording these results on a log sheet or by attaching data generated by a recorder (e.g., strip chart, circular chart).

2.4.4 Reference Applicable SOPs/Work Instructions

Most batch records cannot include all instructions for processing a product. Many instructions are already included in SOPs (e.g., cleaning of stoppers), which would make their inclusion in the batch record redundant. This reduces the potential for inconsistencies between batch records and SOPs. The batch record would also be very lengthy and user-unfriendly. To resolve this issue, the SOP or work instructions must be referenced in the batch record. It is critical to remember that these referenced documents must be readily available to the operators. Also keep in mind that changes to any referenced document may require an update to the master production record.

2.4.5 An SOP for the Development of the Master Production Record

It is required per 211.188 that a master production record development SOP be established. This will delineate requirements to all departments involved with the process of developing a master production record. An example of an SOP is provided in Table 1.

2.5 Working with the Batch Record

The master production record has been developed, production lots are scheduled, and now everything is ok, right? Almost. There are several other steps that need to be taken to ensure compliance and accurate processing of product. Putting some simple procedures in place can ensure consistent processing and timely release of product.

2.5.1 Training

Production *and* quality assurance (QA) personnel should be trained to use the master production record. Operators should be trained to the batch record prior to its first use. Operators should never be exposed to the batch record just as it is issued for production purposes. Batch record training should include the SOPs and work instructions listed in the batch record. Document all training.

2.5.2 The Reproduction of the Batch Production Record from the Master Production Record

It is now time for the actual batch production record (per 21 CFR Section 211.188) to be reproduced from the master production record. The batch production record will mimic the master in every aspect, from the batch quantity and expected yields to the QC tests needed and the provision for delineating any deviations during production. See Table 1 for details of development.

2.5.3 Completing and Checking the Batch Record During Operations

As operations are performed, appropriate documentation should be made at the time the step is executed. Recording data or verifying a step after the fact may result in either those data being forgotten or a blank space remaining

TABLE 1 A Master Production Record (MPR) Development SOP

Scope: The instructions given in this SOP provide detail as to the development of a master production record to be utilized in routine plant operations.
Applicable departments: QA, QC, microbiology, validation, development, and production.

Procedure

1. It is the responsibility of development to document the development of the early manufacturing process. This would include, but is not limited to, the following:
   a. Initial development of control parameters

      Temperature
      Humidity
      Cycle times
      Processing characteristics

   b. Record problems encountered and the resolutions taken to overcome those problems.
2. The validation department will develop a validation protocol with the assistance of development.
   a. The parameters to evaluate and/or extrapolate will include:

      Critical parameters
      Optimal manufacturing conditions
      Process specifications
      Critical control points

   b. The protocol will be approved by a minimum of QA, QC,
      microbiology, and production.
3. The data collected by validation and development during the validation study will be utilized by production to develop a master production record (MPR).
4. All data and reports pertaining to the development of the MPR will be included in the development report.
5. Laboratory management will review the MPR.

Table 1 (continued)

   a. Verify that all in-process and finished product tests are applicable and capable of being executed.
6. The MPR will, at a minimum, include:
   a. Critical control points
      Limits
      Instructions as to what to do when these limits are exceeded
   b. Safety precaution statements
   c. Environmental controls
      Limits
   d. Measures for controlling the environment

   Reference SOPs and applicable work instructions
7. A BOM trust lists the part numbers, names, and required quantities for each component of the batch. The actual quantities utilized during operations must be recorded.
8. Critical steps must require two signatures: one of the person executing the step and the other of the person verifying the first signature.
9. The assigned lot number for the batch must be required by the MPR and included on every page of the batch production record.
10. Final review by production management.
11. Final review and release by QA.
12. As part of the review and release process for QA, a checklist will be generated. This checklist will include all necessary information required for review and approval *prior* to release of the lot. Items may include the following:
    a. Environmental monitoring results
    b. Water monitoring results
    c. Personal monitoring results
    d. In-process test results
    e. Printouts of monitoring equipment
    f. Cleaning tags
    g. Laboratory slips and work sheets
    h. Transport tickets
    i. Weight slips
    j. Investigation reports

in the batch record. Similarly, recording data on scrap paper to be transcribed to an official document is against regulations. Regardless, a compliance auditor will cite the fact that data are not recorded in a timely manner.

In conjunction with completing the batch record during operations, production management and QA personnel should make periodic checks of the documentation. This ensures that issues impacting compliance to regulations or processing parameters are handled at an early stage.

2.5.4 Review by QA and Production

As indicated above, it is imperative for QA and production to review the documentation of the batch record while in process. This review should be made periodically to verify that all steps are completed accurately and in a timely manner. Online QA verifications can also be performed prior to commencement of activities by production. These verifications can include cleanliness checks of rooms and equipment, confirmation of accurate lot number and expiration date during packaging operations, or verification of raw materials used in the process. An evaluation should be made by QA and production management as to which critical steps should be verified by QA.

For final review and approval of the batch record, QA personnel must take into consideration all documentation having an impact on the lot. This will include deviations, investigations, printouts, and cleaning logs.

2.6 Theoretical Yield Calculations

As stated in the regulations, yields must be stated with acceptable ranges specified at appropriate stages of the process. These yield calculations must be based upon observed data, either with similar processes of other products or on data taken during the scale-up process.

2.7 Deviations

As with any process, there will be situations that require a departure from a predetermined step. A deviation from the process must be documented exactly as it occurred, along with what caused the need for the deviation and how the deviation will impact the quality of the product. The deviation investigation and justification must be approved by both QA and production management, and the root cause of the deviation should be identified if possible. Of critical importance is a statement regarding the possible impact the deviation has on the product.

Deviations should be an infrequent event. A high rate of deviations is indicative of an uncontrolled process that has not been adequately validated. Quality assurance be aware of the overall rate of deviations to batch records and take appropriate actions as necessary.

2.8 Investigations

Investigations are required by 211.188 to explain aberrant events that have taken place during the processing of the product. All investigations must include the following information in order to most benefit someone reviewing the batch record:

The lot number and product name or dosage
The fact that material was quarantined
The reason for the investigation
The investigation process
Impact on other lots
Corrective or preventative actions
Conclusion
Final disposition of material

The investigation should be conducted by production personnel and subsequently reviewed by QA. The best practice is to have production conduct and write the investigation, with QA providing the conclusion relative to the material's final disposition.

Any investigation involving the process of manufacturing a batch of product must be included in the batch record. Part of the review process for release of the product should be to verify that there is an acceptable conclusion to any investigation. Refer to Chap. 13 for an extensive discussion for handling manufacturing deviations.

2.9 Archive and Storage

The batch record is the only documentation a firm has that can demonstrate that the product was manufactured according to specifications. During regulatory audits, the batch record is a primary document for review, and must be easily and quickly accessible. Most firms have two different modes of storage—recently executed batch record storage and long-term archival. It is recommended that a schedule be established to require that the most recent batch records executed (e.g., the last 6 or 12 months) are readily accessible and any batch records older than this be transferred to a long-term archival location.

Most, if not all, investigators respect the fact that not all batch records can be maintained in an immediately accessible location. Most archival locations are off site, and it may take up to a day to retrieve records. Ensure that the investigator(s) know(s) of the archival and storage process at the start of the audit. Accessibility to records may dictate the choice of records picked by the auditor for review. It is recommended that storage and archival procedures be included in an applicable SOP.

Alternate means of archival can also be employed. These alternatives could include microfiche or electronic copying to a CD. Whatever means of archival chosen, it is imperative that the data be easily retrievable.

Ensure that all storage and archival locations are secure. Fire, water damage, and pest infestation have destroyed many data. If the lot is still within expiration and the batch record is destroyed, the firm theoretically has little or no documented evidence of manufacture. This could lead to serious problems or a great deal of time investment to resolve a question related to a batch record.

2.10 Security

Security as it pertains to the batch record is often overlooked. Consider that a firm spends years developing it via specific, detailed steps, only to find that copies of these batch records, thought to be safely discarded, were subsequently retrieved from the trash by the competition. Shredding unauthorized copies and superfluous reproductions of batch records is a wise practice.

2.11 SOP for Handling Batch Record

Requirements for handling and working with batch records can be long and detailed. It is crucial that these requirements be delineated in a formal procedure, such as the SOP in Table 2

3 U.S. VERSUS EUROPEAN BATCH RECORDS

An auditor can expect batch records in the United States to be generally more detailed than those found in Europe. European batch records are more of a guide with extensive references to corresponding SOPs and written work instructions. Either system is acceptable as long as all necessary data are documented and the employees are trained adequately.

4 SUPPORTING DOCUMENTATION

Along with the batch record, there is a wide variety of supplemental documents that further support the manufacture and packaging of a product.

4.1 Bill of Materials

A BOM is a listing of all components to be utilized in the production of a batch. This list included raw materials, packaging material, and filters. The BOM includes the name of the product, a part number or item number, a space for the lot number, the quantity of material to be used, and a space to write the actual quantity used.

TABLE 2 A Batch Record Management SOP

Scope: The instructions given in this SOP provide detail as to the issuance, execution and review, and approval of the batch record.
Applicable departments: QA, QC, microbiology, validation, development, production, and RA.

Procedure

It is the departmental manager's responsibility to verify that any personnel who will be working with the batch record will have documented training on the applicable MPR, SOPs, and work instructions.

1. Upon notification that a product is to be manufactured, production will submit a request to QA.
2. QA will obtain the applicable MPR an generate a copy. The copy will be stamped (on every page) with the lot number.
3. QA will verify that every page was successfully copied, the correct lot number was stamped on every page, and that the current, approved MPR was utilized to make the copy.
   a. QA will sign the document as being issued correctly. The lot number, revision number of the MPR and QA person issuing the document will be recorded in a logbook.
   b. Errors found on any of the copies will require that the page(s) be shredded. New pages will then be issued.
4. The issued batch record will then be delivered to the appropriate production supervisor. The issued records must be adequately controlled to prevent loss during transfer from issuance to operations.
5. As required, production personnel will initiate the use of the appropriate batch record.
6. During processing, if additional pages are needed for the batch record (e.g., pages are damaged while in process), QA must be notified. QA will issue the additional pages requested following instructions outlined in steps 1 through 4. QA should verify that the appropriate version of the MPR has been used for issuance. Production personnel must *never* photocopy pages from issued batch records.
7. During execution of the batch record, applicable checks will be made as required by SOPs or work instructions in the batch record. Attach applicable check sheets to the batch record.

TABLE 2 (continued)

8. All documentation associated with the batch record and operations must be attached to the batch record. These documents include, but are not limited to, the following:
   a. Bill of materials
   b. Cleaning tags
   c. Weight tags
   d. QC work sheets
   e. Investigations or deviations
9. As steps are performed, the applicable data or requested verification should be documented at the time of execution.
10. Upon completion of the batch record by operations, production management should review the documents for completeness and accuracy. The person performing this review will sign on the applicable review section for production. The entire document shall be forwarded to QA.
11. QA will review the batch record for accuracy. Verify that all specifications have been met and that any excursions have been justified with a deviation investigation. Checks should be made to affirm that all applicable documentation associated with the batch record is included. Verification of this check will be documented on the applicable checklist.
12. All issues, deviation, investigations, and so forth must be resolved prior to the final release of the batch record.

The use of the BOM is twofold. First, it provides a list of materials, lot numbers, and quantities for accounting purposes. Second, it provides a means by which production personnel can double check their own work.

4.2 Logbooks

Logbooks are typically used to document ancillary cleaning activities and equipment usage. These logbook entries must consistently correspond to the activities delineated in any given batch record.

4.3 Checklists

Some firms have used a system of checklists to verify such activities as cleaning or assembly of equipment. This documentation may be critical to the proper processing of the product and should be considered for inclusion in the batch record package.

4.4 Printouts for In-Process Checks

Many data are lost due to human error (e.g., forgetting to write down the information or incorrectly transcribing the information from a display). A vast majority of the data can be captured simply by utilizing the print capabilities of the equipment on which the test is performed. Most equipment (pH meters, temperature recorders, scales, etc.) have electronic print ports (typically RS-232) that can generate the data directly. If capable, it is advisable to print out the data, write down the results on the batch record, and then include the printout in the batch record. If this is done, it may be necessary to indicate the lot number and/or step number on the printout.

4.5 Weight Tickets

Many companies stage raw materials ahead of time for operations. This requires the use of weight tickets, which list raw material, quantity of material, item or part number, and lot number. These weight tickets are a key source for checking the proper addition of the correct raw materials to the lot, and should be included in the batch record.

4.6 Cleaning Status Tags

Regulations require that each piece of equipment be identified as to its status. This can be a critical question with sterile operations during batch record review. Additionally, most equipment cleaning programs have time limits between cleaning and use as well as specific hold times and status tags, along with logbooks used to annotate these time frames. Cleaning status tags are important documents that must be included in the batch record.

4.7 Transport Tickets for Mobile Equipment

Most operations have at least one piece of equipment that is capable of movement from one location to another. Many firms use some type of transport ticket to document the movement and cleaning of the equipment. It is essential to include the transport ticket in the batch record to demonstrate the control of equipment from one location to another.

4.8 Campaigned Equipment Logbook

Most firms have specific start and stop points for a batch of product and typically equipment is cleaned between batches. Some firms, such as bulk pharmaceutical chemical companies, conduct campaigns. Campaigning is the act of continuous use of equipment to produce numerous lots of the same or different product. An example would be the use of a blender in which chemicals are continuously added at one end and a specified amount of "finished product" is collected at the other. Data showing the use of equipment utilized for campaigning should be included (or copied) for the batch record.

4.9 Laboratory Work Sheets

Similar to printouts collected during the process, samples are frequently collected for testing in the laboratory that generates laboratory data reporting sheets. Record the results of the laboratory testing in the batch record by attaching the laboratory work sheets.

5 PROBLEMS WITH BATCH RECORDS

The following is a list of common errors associated with batch record development, along with some examples.

5.1 Instructions Open to Interpretation

Using vague words, such as *approximately* or *about.*

Offering a choice of options without giving instructions as to which option to choose. For example, perform steps 3, 4, and 5 or perform steps 6, 7, and 8. Operators usually are confused by such options.

Making note, such as "Make sure the equipment is clean." Does this mean microbial, chemical, both? How clean?

Giving no specified limits; for example, "Test and record data results." Without limits, the operator will not know whether or not the data are acceptable.

"Giving imprecise directions; for example, add five different chemicals and mix for 20 min." Does this mean start the mixer and timer when the first chemical is added to the tank, start the mixer when the first chemical is added to the tank but don't start the timer until all five chemicals are added, or start the mixer and timer after all five chemicals are added? Sometimes simple instructions can result in various interpretations by different operators, causing variability from batch to batch.

5.2 Specifications Too Narrow

No acceptable range is given for the data. Do not give specific instructions "Heat the solution to 50°C," because it is doubtful that the solution will be maintained at exactly 50°C.

5.3 Loss of Data or Information

The instructions to meet a specification without the specific request to record it will result in an incomplete batch record. For example, "Heat the solution to 45°C–55°C" is part of the batch record. The final testing of the lot revealed a problem that could have resulted from the lot being processed at a temperature above 55°C. If the batch record does not specifically call for the information to be recorded, it is likely not to get recorded.

The operator observes something out of the ordinary but does not record the observation.

5.4 Cumbersome or Too Long

The batch record is so long that it is issued in volumes. Unless the process is very complicated, most batch records should not be so detailed as to make the process of issuance or actually following the record an arduous task.

Recording data is great for evaluating the acceptable processing of the batch; however, ensure that the data being recorded are relevant. Recording the start and stop times for each step when the step is not time-dependent may result in operator error and recording irrelevant information.

6 ROLES AND RESPONSIBILITIES

Sometimes interdepartmental roles and responsibilities for batch record development and review can become confused during the "drive" to bring a new product to the manufacturing floor. The information below intends to provide a guide for each department. Keep in mind that individual companies may have differences from what is written here; therefore, it is imperative that these roles and responsibilities are included in SOPs dealing with batch record development, execution, and review. The goal is to make sure that all necessary steps, as listed in this chapter, are fulfilled. This list is by no means all inclusive of every department or of all actions necessary to meet specific internal requirements.

6.1 Quality Assurance

The role of QA should be to serve as a confirmation of compliance to cGMPs and the firm's SOPs. Verification that the aforementioned steps are executed and that all documentation is available is paramount to both releasing finished product and passing an audit. Some additional items for QA to consider include the following:

Maintain the change control system.
Verify training is completed on the batch records.
Review and approve investigations and deviations.
Maintain the master production record.
Issue batch records to operations.
Review completed batch records.
Assign an independent and responsible department for final release of product.
Review and approve validation protocols.

6.2 Operations

Too many firms focus upon the belief that the only responsibility of operations (manufacturing, formulation, packaging, labeling, etc.) is to "make product." The role of operations goes well beyond this notion, and without this department's key involvement, repeatable product quality will be elusive. An overview of operations responsibilities is

Review and approve the master production record.
Train operators to the batch record.
Document operations as they occur.
Perform investigations.
Review and approve executed batch records.
Review and approve validation protocols.

6.3. Validation

The role of validation is to write and execute protocols to collect data and to verify that the process is repeatable and reproducible. It is not its responsibility to "make the process work." The axiom of "bad data in, bad data out" is very pertinent to validation. Also bear in mind that just because the process in the batch record has been validated does not mean that a failure or deviation cannot occur during the execution of the batch record. It is not the FDA's belief that validation will prevent failures; It is its belief that validation will show that a successful process can be repeated when key steps (as should be listed in the batch record) are repeated from batch to batch within a specified variation. Some key responsibilities for the validation group to consider include the following:

Development of protocols prior to performance of a validation
Production of final summary report of the validation efforts
Development of the master production records, based upon the outcome of the validation study

6.4 Research and Development

In relationship to new product development, R&D has a key role in the formation of the batch record. Without proper documentation maintenance at an early state of development, transferring the process from development to operations will be problematic at best. Development personnel should be cognizant of all the information needed by production operators responsible for scale-up and/or technology transfer. Consider the following:

Maintain documentation on any problems encountered during development and steps taken to overcome those problems.
Maintain documentation on critical steps and parameters identified during development.
Record any important processing parameters, such as environmental sensitivities and equipment needs.
Assist in the development of the master production record.

6.5 Regulatory Affairs

Regulatory affairs (RA) has the responsibility of submitting the batch record along with the pertinent application (New Drug Application [NDA], Drug Master File [DMF], etc.). Following initial submission, RA should also be involved with the following:

Review changes to the master production record.
Evaluate changes and report to the regulatory authorities.

6.6 Supervisors and Managers

Supervisors and managers have the responsibility of verifying that all of the steps necessary for the proper development and execution of the batch record are taken. This involves verification of the following:

Proper SOPs are in place, detailing necessary steps.
Appropriate personnel are trained in these SOPs.
Documentation is maintained as necessary.
Crossdepartmental functions are managed and coordinated (i.e., revalidation and RA updates).

7 BATCH RECORD AUDITS

During an audit of a production and/or packaging operation, one of the most important forms of documentation reviewed will be the batch record. Adherence to cGMPs and SOPs, training of employees, investigations of procedures and techniques, and evaluation of product quality trends can all be reflected by an audit of batch records. Other than internal audits by QA, there are typically two different audits groups that will examine the batch records.

7.1 FDA and Other Regulatory Bodies (Occupational Safety and Health Agency [OSHA], Environmental Protection Agency [EPA])

An inspection by a regulatory body will probably be one that receives the most attention and preparation activities. These audits can be focused [such as a preapproval inspection (PAI)] on a specific product general (an overall evaluation of processes, products, and procedures), or a combination of both. These points are some general areas of inspection and should be considered a good basis of preparation, but are not all-inclusive of the questions routinely asked by auditors.

Compare the master production record with the process that was submitted to the agency (e.g., BLA, NDA, ANDA). The two processes should be equivalent.

Evaluate the change control process.

Are changes to the master production record reviewed and evaluated by RA? How are changes to the master production record reviewed in relationship to the regulatory filing? How are changes reported to the applicable government agency? Problems in these areas will reflect seriously on the outcome of the audit.

Are changes to the master production record reflected by data and a development report?

What is the process by which batch records are issued to operations?

Are appropriate documentation practices employed during the execution of the batch record (e.g., steps documented in a timely manner, verification, attached data are present)? Are the data complete and accurate?

Are SOPs or work instructions contradictory to batch records?

Have investigations or deviations been completed in a timely manner? Are the investigations adequate and all-inclusive? Have corrective actions been implemented?

Are the attachments adequate to reflect that the manufacturing operations were consistent and compliant (e.g., cleaning, setup)?

Are personnel who have signed the batch adequately trained (specific, job-related training as well as cGMP training)?

7.2 Customers and Clients

Many firms can use contractors to manufacture product. This can be done for many reasons, such as a sudden demand to increase production or to accommodate new product lines. Active pharmaceutical ingredient (API) firms usually manufacture exclusively for other sites that make finished product. Regardless of the operation, at some point there will be an audit from a client or customer firm.

Many of the items reviewed by a client will be similar to those reviewed by the FDA. There may be specific requirements, however, as dictated by the clients' SOPs or policies or quality agreements with the manufacturer. There will also be some focus on procedures of notifying the client if there is a problem found during stability or with processing. The client will typically be interested in the notification procedures for changes to the master production record. There should be procedures in place to verify that a client's request was at the very least reviewed or implemented.

8 CONTRACT MANUFACTURING AND BATCH RECORDS

It benefits both parties of a contract manufacturing agreement to give specific details on the master production record when a pharmaceutical product will be manufactured by a contractor. Some items that need to be covered by an agreement include the following:

The firm supplying the knowledge for the manufacturing process will assist in training the operators at the contracted site.

Investigation of manufacture deviations will be provided as needed by the contracting firm.

No changes to the batch record can be implemented without the written consent of both parties (contracting and contract manufacturing companies).

As applicable, restrictions to copies of batch records should be made (i.e., control of copies).

9 THE FUTURE ELECTRONIC BATCH RECORDS

The future of the batch record is a paperless system that will record data directly from equipment (start times, weights, temperatures, fill speed, etc.) and use recognition systems such as retinal eye scans for employee "signatures." For some, the future is already here in one form or another, but for the vast majority of companies (especially smaller firms), such a system is just a distant glimmer. The FDA has recognized the use of electronics systems throughout the pharmaceutical industry and has put into place regulations that dictate regulatory expectations. These regulations are detailed in 21 CFR, Part 11.

10 CRITICAL COMPLIANCE ISSUES

For whatever work is done during the course of operations, always keep in mind the overall goal. In this case, the final goal of the batch record is to provide a format for the consistent, successful production of a pharmaceutical product. The axiom that "Success is 99% planning and 1% doing" holds true for the batch record. That said, the following is a quick list of essential compliance considerations for this planning phase.

10.1 Know the Regulations

21 CFR Part 210 and Part 211; specifically sections 211.186, Master Production and Control Records, and 211.188, Batch Production and Control Records.

In Section 6.5, the FDA guidelines on "Manufacturing, Processing or Holding of APIs," discusses batch production records. An ICH guide entitled "Good Manufacturing Practice Guide for APIs" also discusses batch records.

10.2 The Life Cycle of a Batch Record

Design of the experiment and scale-up. It is at this stage of the product's life cycle that many control parameters are evaluated (e.g., glass container in lieu of stainless steel, temperature controls). Information should be collected at critical processing stages and included in the master production record. This information will be used to assist in scale-up activities.

Keep track of any problems. encountered during this stage and the resolutions taken to overcome them.

Development of the validation protocol. Validation of any manufacturing process serves as the means of setting and/or evaluating control parameters established during experimentation and scale-up.

Execution of the validation protocol. Define critical parameters. Critical parameters of the process can be defined as those steps that play an essential role in the successful manufacture of the product. Define these parameters and clearly state them in the master production record and batch production record.

Characterized optimal manufacturing conditions. Once obtained, optimal conditions for the manufacturing process are those control parameters that result in both the most efficient manufacturing process and the best-quality product. Make certain to characterized these optimal manufacturing conditions and note them in the master production and batch records.

Establish process specifications. Once critical parameters and optimal conditions are defined and optimized, the process specifications must be established. The specifications become the targets to be met during batch-to-batch manufacture.

Set critical control points. There should be established points throughout the entire process that are utilized as "stop-and-test" points that evaluate the success or failure of the production process up to that critical point. Critical control points are used as safeguards throughout the process, allowing operators and supervisors to judge whether or not to proceed with the manufacturing process. Make sure these steps are clearly identified in the master production record. Operating personnel must be able to clearly identify the parameters of the test and what steps are required if those parameters are not met.

Develop the master production record. Identify safety precautions. If specific personnel protective equipment is needed for particular steps, indicate those requirements in the master production record.

Identify environmental controls. If there are control requirements, indicate them in the master production record.

Reference applicable SOPs and work instructions. SOP and/or work instructions must be referenced in the master production records. It is critical to remember that these referenced documents must be readily available to the operators.

Working with the batch production record. Training. Production *and* QA personnel should be trained to the batch production record.

Complete and check the batch record during operations. As operations are performed, appropriate documentation should be made concurrently with the step being performed.

Review by QA and production. It is imperative for QA and production to review the executed batch record prior to product release. This review is made to verify that all steps were completed accurately, in a timely manner, and in accordance with the predetermined specifications.

Theoretical yield calculations. There should be stated yields with acceptable ranges specified at appropriate stages of the process.

Manufacturing deviations. A deviation from the process must be investigated and the genesis of the deviation impacts the quality of the product must be explained and fully documented. The investigation must be approved by both QA and production management *prior* to the final release or disposition of the batch.

Investigations. Investigations are a regulatory requirement to explain events and deviations that took place during the processing of the product.

Archival and storage. The batch records should be easily and quickly accessible. Storage should be free of pests, rodents, or environmental assault.

Security. Secure the copying or dissemination of unauthorized batch records.

Associated documents. Make sure all corresponding documentation is maintained as part of the batch record, including

Bill of materials
Checklists
Printouts for in-process checks
Weight tickets
Cleaning status tags
Transport tickets for mobile equipment
Campaigned equipment logbook
Laboratory slips and worksheets

Problems with batch records. The following is a short list of common errors associated with batch record development:

Instructions open to interpretation
Specifications too narrow
Loss of data or information
Record cumbersome or too long
No master for that particular batch production record
Master production record not derived from a validation study
Batch record does not contain sufficient "stop-and-test" points or critical control points

11 REGULATORY REALITY CHECK

With the chapter providing an instructive backdrop, several FDA observations (FD-483s) documented during recent inspections of FDA-regulated facilities are presented below, followed by a strategy for resolution and follow-up corrective and preventative actions.

> FD-483 Citation 1: Unexpected manufacturing deviations are not always documented in the batch record. For example, lot__ had a stainless steel tube get caught in the mixing blade of Tank__. Another example is lot__ into which a stainless steel scoop was dropped. Another example is when a compounder dropped his safety glasses into Tank__ while manufacturing__.

The firm could have easily avoided receiving this citation if it had simply followed the 21 CFR 211.188 regulation requiring that any departure from batch record instructions be thoroughly investigated and evaluated prior to releasing the material in question. In addition to following this fairly clear-cut regulation, it is an industry standard and FDA expectation that FDA-regulated companies install a comprehensive system for investigating manufacturing deviations. The six critical guideposts essential to a manufacturing deviation investigation system are

Identity deviation
Quarantine material
Investigate
Assign probable cause
Install preventive measures
Determine final disposition of material

> FD-483 Citation 2: Documentation in batch records is not always accurate. Additionally, the verification check by a second production employee (usually a supervisor) failed to detect the documentation error.

While it is not essential that the verification check be performed by a supervisor, it is essential that production personnel be absolutely clear regarding FDA's expectations with respect to verification activities. The FDA expects that a verification of calculation and weighing steps be performed concurrently. Additionally, verification does not simply mean initialing the box; it requires a review of the work performed for accuracy and completeness. Operators should be made aware of this during GMP training as well as training specific to the batch record.

> FD-483 Citation 3: Incoming labeled tare weights for containers of active and inactive raw materials used in the manufacture of finished pharmaceuticals are not verified for correct weight or measure. The procedure for tare weight checks is inadequate in that it allows a tare weight range of—for active and inactive raw materials without verifying the labeled tare weight. This would permit for either an insufficient or additional amount of raw material to be used other than the specific amounts approved in a product's formulation.

This observation represents the value of installing and utilizing critical control points throughout the manufacturing process. Had the batch record been designed to incorporate a stop and check immediately after weighing, and prior to proceeding with blending ingredients, the possibility of adding an insufficient or excess amount of raw material would be greatly reduced. The weighing and staging of raw materials is a critical step and a stop and check after this step could save a great deal of resources.

> FD-483 Citation 4: During the walk-through inspection on—it was observed that—product in the fluid bed dryer basket was not protected to prevent contamination as a cursory room cleaning was on going. The cleaning was not recorded in the batch records and was not part of the manufacturing procedure.

This is a perfect example of the firm's not giving enough thought to the essential documentation that must be included in the batch production record. The batch production record SOP should have mandated that all cleaning tags and activities need to be incorporated in the batch record throughout the manufacturing process. Additionally, the design of the batch record should call for verification and documented evidence that cleaning has occurred prior to executing the manufacturing activity.

12 WORDS OF WISDOM

Master production record development must be derived from successfully completed process validation studies that contain information regarding critical parameters, optimal ranges, and specifications.

Batch production records are a reproduction of the master production record. All batch records must have a corresponding master production record.

Batch record management (issuance, review, approval, and release) is the sole responsibility of QA.

Any departures from the prescribed directions in the batch record or from corresponding SOPs and work instructions must be investigated and resolved in conjunction with QA and prior to the release of a raw material, work in progress (WIP), or finished batch.

Prior to launching a manufacturing deviation investigation, it is essential to quarantine the material in question until QA is satisfied with the outcome of the investigation.

Batch records must be user-friendly and conscise with respect to instructions and specifications.

A verification of all weighing and calculation activities must be included in the batch record.

It is essential to build into the batch record several critical control points that allow operators to review the process up to that point. These stop-and-check points can be a simple review of activities or an in-process test that requires waiting for lab results before proceeding with the manufacturing process.

9

Change Management: A Far-Reaching, Comprehensive, and Integrated System

Susan Freeman
Antioch, Illinois, U.S.A.

1 INTRODUCTION

Change is inevitable, exciting, and scary. Change is not always good. Change costs money and is dangerous. Change affects everything surrounding it. Change draws attention and is too much work. Change tickles the very core of our emotional stability. The more things change, the more they stay the same. A small change can make a big difference. Change is all this and much more.

This chapter is all about managing change within FDA-regulated industries. The only way to manage change in our rapidly evolving pharmaceutical, medical device, and biotech environments, is to create it. Adapting to the changes that these rapidly growing industries present can be extremely costly because by the time a company catches up to the change, the competition and innovator of change is way ahead of the race. The FDA decided for us that unplanned and planned changes require a stringent set of controls in order to remain in compliance and maintain product quality, strength, and efficacy; however, good business practices call for precise change management.

Managed change is a positive circumstance. Repeat after me: Managed change is a positive circumstance! Change is an integral part of our lives; it is always going to be here. The better we prepare for it and have a system to manage it, the more positive the impact will be. This is true in business as well. Along with the technology advances that afford us limitless possibilities to acquire, share, and manage information, come changes that affect our business conditions at a record-breaking pace.

Those of us in the health care industry are legally bound to follow the Code of Federal Regulations (CFR). We must stay in compliance with the regulations. To do so effectively, we must be proactive in keeping apprised of pending regulatory changes and anticipating the challenges they present. We must also keep current on inspection trends by being attentive to FDA observations made in other companies throughout the world. Managing regulatory changes must be factored into our cost of doing business or we will be out of business fairly quickly.

We must be keenly tuned in to our customers and their changing expectations for our product. If we cannot satisfy their expectations, our competitors are all too eager to fill the gap.

We must stay current with technology as well as scientific breakthroughs in order to improve our processes and products and to keep our costs down.

Those companies that can respond quickly and systematically to the onslaught of changes from all directions are the ones that will survive in the new millennium.

2 WHY MANAGE CHANGE?

2.1 Compliance to Current Good Manufacturing Practices (cGMP)

Those of us who work in FDA-regulated industries are of course expected to comply with the CFR. The CFR can be quite clear about most aspects of cGMPs, but really does not elaborate on change and how to manage it. The implication and expectation is that each quality system will be established to meet cGMP and that ultimately all these systems will be integrated so they work in harmony.

2.1.1 FDA Regulations Related to Change Management

Whenever a small number of qualified individuals manage an array of critical responsibilities within an organization, it is imperative to develop innovative strategies for managing changes that invariably impact quality and compliance. The constant cycle of change calls for effective tools for managing any change related to pharmaceuticals, medical devices, and biotechnology. Along with other regulatory bodies, FDA understands and accepts the fact that change is a natural part of doing business; nevertheless, the regulators expect these industries to employ effective change management systems that will assure that the integrity of regulated products is not assaulted.

Section 501 (a)(2)(B) of the Food, Drug and Cosmetic Act deems a drug to be adulterated if "the methods used in, or the facilities or controls used for, its manufacture, processing, packing, or holding do not conform to or are not operated or administered in conformity with current good manufacturing practices to assure that such drug meets the requirements of this Act as to safety and has the identity and strength, and meets the quality and purity characteristics, which it purports or is represented to possess."

The cGMP regulations were developed by the Pharmaceutical Manufacturers Association as an industry standard and guideline in the early 1960s. These regulations were later adopted by the FDA as part of the Food, Drug and Cosmetic Act of 1962. The cGMPs specifically require that FDA-regulated industries employ an elaborate change control system to manage any new developments impacting the manufacturer of a pharmaceutical, medical device, or biologic product. The following regulations were codified by the FDA as far back as 1978 for pharmaceuticals and fairly recently (1996) for medical devices. Interestingly, specific regulations governing change control within the biotechnology industry have not been codified; nevertheless, the cGMPs apply to the manufacture of biologic products and, as such, change control is a regulatory expectation and industry standard.

21CFR–Part 211 Subpart F—Production and Process Control

211.100 Written Procedures; deviations:

(b) Written production and process control procedures shall be followed in the execution of the various production and process control functions and shall be documented at the time of performance. Any deviation from the written procedures shall be recorded and justified.

The requirement for change management in a laboratory setting is specifically addressed in the following regulation:

21CFR–Part 211 Subpart I—Laboratory Control

211.160:

(a) The establishment of any specifications, standards, sampling plans, test procedures, or other laboratory control mechanisms

> required by this subpart, including any changes in such specifications, standards, sampling plans, test procedures, or other laboratory control mechanisms, shall be drafted by the appropriate organizational unit and reviewed and approved by the quality control unit. The requirements in this subpart shall be followed and shall be documented at the time of performance. Any deviation from the written specifications, standards, sampling plans, test procedures, or other laboratory control mechanisms shall be recorded and justified.

The quality systems regulations (QSR) that were codified in October 1996 for the medical device industry are more specific than the above regulations, which govern change management for pharmaceutical products.

Part 820 of 21CFR details the QSR that governs the manufacture of medical devices and addresses the subject of change control in the following subparts:

820.30–Design Controls

(i) Design changes

820.40–Document Controls

(b) Document changes

820.70–Production and Process Controls

(c) Production and process changes

820.75–Process Validation

(d) Changes or process deviations

820.90–Nonconforming Product

As evidenced above, the FDA has taken great care in putting forth basic requirements and regulations that will unmistakenly guide FDA-regulated industries in the area of change management. The specific inner workings of a firm's change control system is entirely up to the firm.

Change can come to us from many different sources. Some changes are critical to our operation, and therefore should be considered significant or major. Others have a lesser impact and may be considered minor. Let us explore some of the various ways in which change is likely to present itself, discuss the corresponding quality system(s) that are likely to be impacted by the change, and examine how best to manage the process.

3 WHAT TYPES OF CHANGES DO WE NEED TO MANAGE?

3.1 Sources of Change

While change can come from many different directions, experience has proven that change(s) within a critical compliance and quality system will have the largest impact. The following are some to which you can probably relate some real-life examples:

Regulations
Customer requests/complaints
Personnel
Technology
Testing methodology
Vendors
Facility
Validation
Research
Development

As you can see from Fig. 1, a change in any one area could affect many of your existing systems or require installation of an entirely new system.

Figure 2 depicts an example of how a change that might occur during the development phase could cascade into changes in many related systems. Once development begins toward preapproval, documentation is open for review by the FDA; hence, changes made during the development phase must show evidence of being adequately managed.

Think for a minute about the manufacturing operations associated with the product you sell. Imagine product coming off the finishing line and move backward in the process from that point, taking note of all integrated systems in place and activities that had to occur in order to successfully manufacture a high-quality, functional product. Within the product itself, there are packaging components, labeling, and raw materials. These, of course, are meaningless without all the production equipment to run on, and you may recall the engineering trials and validation protocols needed to get that equipment running efficiently. The product was tested at various stages and perhaps in several different laboratories using many methods and types of testing apparatus. There is a great deal of paperwork—batch records, standard operating procedures (SOPs), specifications, regulatory submissions, and supporting data. Naturally, personnel

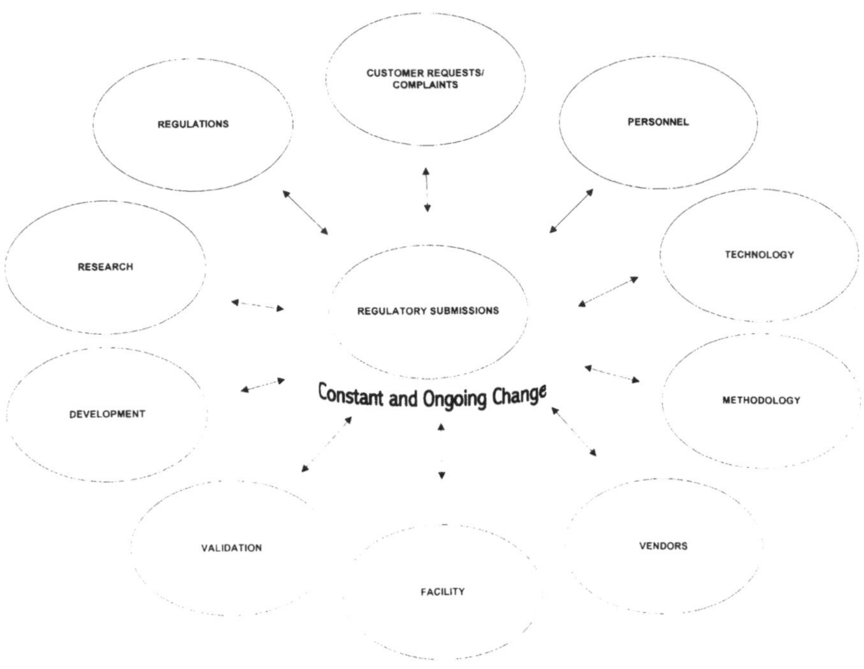

FIGURE 1 Sources of change.

involved throughout the process will require training and retraining, so what you have is an integrated manufacturing process involving myriad functions. Because everything in your process is directly or indirectly related to something else, your change management system must also be integrated. Having isolated procedures to escort change within individual systems may give the appearance of efficiency, but in meeting the real objective of keeping everything synchronized, this approach will not effectively manage change.

3.2 Quality Systems Affected

In keeping with the cGMPs, there are several systems for which you have established controlled procedures in order to keep that particular system compliant. Much effort goes into the execution of each individual system. Because all those systems work in synergy with the manufacture of final product, each one must be covered by the overall change management system.

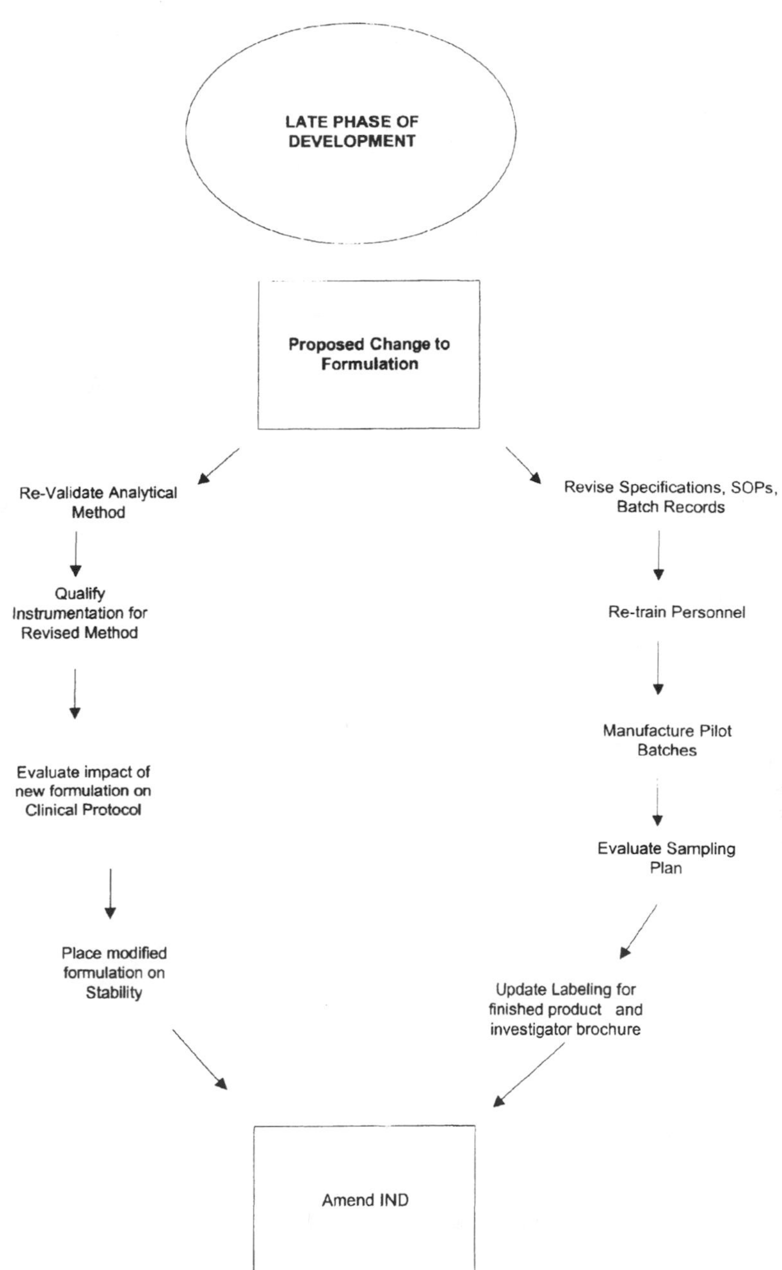

FIGURE 2 Example of cascade of changes.

The quality systems alluded to above include, but are not limited to the following:

Regulatory submissions	Specifications	Validated systems	Training
Procedures (SOPs)	Master/batch production records	Labeling	Purchasing
Third party manufacturers	Clinical trials	Stability program	Environmental monitoring
Sampling Plans	Raw material/ component qualification	Analytical methods	Equipment/ facility qualifications
Complaint handling	Scaling-up program	Design control (medical devices)	Critical support documentation

You can determine which systems need to come under change control by referring to the regulations most applicable to your product(s). All systems addressed in cGMP require some degree of change control. Let's use an example to clarify what all this means.

If you are modern and conscientious in your efforts during product development, the systems cited above will all come into play during the early stages and throughout commercialization of the product. For example, let's say you are producing aspirin.

Perform development work; design clinical trials and stability studies.
Develop and validate analytical methods to measure product quality and conformance.
Write specifications that will align with the product registration.
Submit regulatory filing and seek approval.
Qualify your production equipment to consistently manufacture and meet product specifications.
Validate the manufacturing process for reproducibility and optimization.
Develop labels that are in agreement with the indications and dosage purported.
Write master and batch production work orders and procedures to instruct your personnel on how to manufacture and test the product.
Regularly monitor environmental conditions and cleaning procedures.
Train your personnel in the setup and operation of laboratory and manufacturing equipment and in the manufacturing and testing procedures.

Maintain ancillary supportive documentation, such as development reports, logbooks, and raw data.

All of these aspects of product development are inextricably linked and must be properly coordinated and maintained. After all this work has been accomplished, a change is proposed. A change in any one of the above could necessitate a change in one or all of the other systems. An integrated and comprehensive change management system is needed to effectively and legally carry out proposed changes.

4 HOW DO WE GO ABOUT SETTING UP A SYSTEM?

4.1 Elements of an Integrated, Comprehensive, and Far-Reaching Change Management System

1. *Empowers an organization's personnel by inviting anyone to propose a change.* Everyone in your organization holds unique knowledge and perspective. Make it easy for anyone to suggest a change by having a simple form and an SOP on how to use it.
2. *Affords a way to communicate both vertically and horizontally around issues of change.* With many changes occurring simultaneously, it is very important to have a system that will keep track of them and keep people current as to the status of each pending change.
3. *Provides a viable mechanism for continuous improvement to forward quality.* Change can be a big contributing factor toward continuous improvement. A good system gives you a systematic, documented way to evaluate and incorporate product and process improvements.
4. *Allows for full assessment of a particular change prior to implementation.* What appears to be a good change for one area may in fact have a negative impact on another area. A good system allows for everyone to give input regarding the impact of change.
5. *Provides a systematic and formalized approach to review proposed changes.* A system helps ensure that change justification is documented and the right people are evaluating and approving each change, and identifies other systems impacted.
6. *Allows for a coordinated implementation.* With the potential for many areas to be impacted by a change, all these activities and documents must have a synchronized implementation.
7. *Provides a documented trail and various levels of accountability.* "Not documented, not done." This old rule is still applicable when it comes to change management.

8. *Maintains regulatory filing consistent with plant operations and product specifications.* This is what it's all about—consistency between what is delineated in the submission and what exists in the facility.
9. *Decreases potential for situations of noncompliance.* With many changes occurring, no one could keep track of them all without a good system. Your employees want to do the right thing. Help them with a good system.
10. *Provides an effective matrix to audit against while conducting internal or third-party audits.* Can historically assess what worked or did not work over time.

4.1.1 Change Management System and Its Various Subsystems

The fundamentals of a change management system must be clearly delineated in a corporate quality manual or policy. Current industry standard mandates that a corporation formally design and implement a change management system through a comprehensive SOP with a number of corresponding subsystems that provide specific guidance related to different types of changes.

This comprehensive change management SOP should specify the basic principles and regulations of the overall change management system. This SOP also defines the overall responsibility for identifying, reporting, evaluating, investigating, implementing, and finalizing critical changes within the organization. It is essential that this SOP also define the general categories or types of changes that could be encountered within an organization and more specifically within a manufacturing and laboratory environment. This main procedure will lead the organization toward the appropriate subsystems, depending on the kind of change that is required. For example, a change in a validated system would require using a corresponding SOP related to revalidation or valid maintenance. The objective of having various corresponding subsystems that relate to the overall change management system is to ensure that there is appropriate implementation of any type of change, as well as adequate personnel training and documentation related to the change(s).

4.1.2 Effective Implementation of a Change Management System

Similar to any other critical compliance and quality system, a change management program is only as effective as the policy and procedures upon which it is based, coupled with the consistent adherence to these guidances by management and personnel. It is essential to make everyone in an organization aware of the importance of change management. Personnel at every level must be trained to both the corporate policy and all the procedures related to the change management system. They must be made aware of the importance of documenting and reporting changes in a timely manner before the change has an opportunity to compromise product quality. Often line operators and bench analysts have a tendency to look upon small changes as insignificant and nonreportable. A small change can often make a big difference; additionally, the accumulation of many minor changes can have a cumulative effect over time. This could lead to significant manufacturing deviations as well as serious laboratory failures. A key factor in implementing an effective change management system is consistency by everyone in the organization in the identification, reporting, documentation, review, justification, and ultimate approval of changes.

4.1.3 Personnel Training and Change Management

There are two essential aspects to personnel training relative to a comprehensive change management system. The first is training everyone in the organization to the overall policy and procedures. Just as important is training pertinent personnel to the ongoing changes enacted within an organization, as well as changes imposed by such external parties as regulatory agencies, competitors, and consumers. Training everyone within the organization to the principles and specifics of the change management program should be interactive, using specific scenarios drawn from various operational units to demonstrate the impact a change could have on the overall quality of the product being manufactured. Everyone in the organization should be aware of the seriousness a change in their specific area could potentially have on an FDA-approved submission or license. Employees need to understand that all changes must be evaluated regardless of how inconsequential they seem. Whether a change is planned or not, it must go through the appropriate assessment by the appropriate personnel and quite often quality assurance. It is important for an organization to promote quality through compliance with formalized internal procedures. Change management is a central quality program.

Even the smallest changes will require some training or retraining. An organization must be committed to investing in constant personnel training if its change management system is to be successful and effective.

4.1.4 Regulatory Guideposts for Change Management

Products that have been approved and commercialized must have submitted an application with the FDA prior to approval. There are a number of different types of pharmaceutical applications that can be filed with the agency, such as a new drug application (NDA), or an abbreviated new drug application (ANDA). Changes to the manufacturing and analytical commitments made within these applications must be reported to the FDA. How these changes are reported depends upon the nature of the change and whether or not it requires approval prior to implementation. The reporting requirements for changes made that relate to an approved product could fall into one or more of the following categories.

4.1.5 Preapproval Supplements

There are certain categories of changes that require approval by the agency before they can be implemented. Typically a significant change to the manufacturing process, packaging materials, packaging operations, analytical testing, expiry dating, facility or facility location, and utilities requires submission of a preapproval supplement to the FDA. It is important to note that while implementation of these changes may in fact occur prior to actual FDA approval, any drug product impacted by these aforementioned changes cannot be distributed until such time that FDA provides approval of the supplements.

4.1.6 Change Being Effective (CBE)

In contrast to the preapproval supplement, there is a category of change that requires less stringent review by the agency. The CBE is a relatively new category of change for which the FDA has published guidelines, it has also been added to scale-up and postapproval changes (SUPAC). A change within the CBE category requires the agency be informed about such changes; however, it is not necessary to obtain agency approval prior to implementing the change and distributing products impacted by the change. Typically these changes take the form of tweaking or slightly modifying the manufacturing process, packaging operation, or any number of activities related to operational continuous improvements. It is important to note that these are changes that require an approval level greater than just the company's internal quality assurance unit.

4.1.7 New Drug Approvals (SUPAC, BACPAC)

Scale-Up and Postapproval Changes (SUPAC). Recent FDA guidelines and regulatory changes relative to scale-up and post-approval changes (SUPAC) are a great help in deciding which changes must be reported to the agency postproduct approval. In 1995, SUPAC guidelines were developed by the pharmaceutical industry and academia in conjunction with the FDA in an effort to address the myriad changes occurring within the industry that had created an excessive amount of post-approval supplemental applications for the agency. These guidelines attempt to categorize change for the industry and list what may be considered similar equipment based on comparability in design and operation. The critical question addressed by these guidelines is whether or not a change in some aspect of the manufacturing process will impact critical parameters. The FDA initially published SUPAC guidelines for immediate release (IR) solid oral dosage forms and subsequently published guidelines for nonsterile, semisolid, and solid oral dosage forms with modified release technology. The SUPAC guidelines are limited to changes in the manufacture of the dosage forms referenced earlier; they do not cover the manufacture of bulk active pharmaceutical ingredients; however, SUPAC was recently expanded to include packaging, labeling and analytical methods. There are many similarities as well as differences among these various SUPAC guidelines. For example, the SUPAC guidance for immediate release allows moving product from one facility to another with only a CBE when certain conditions are met. This kind of latitude regarding manufacturing changes allows manufacturers to have better control over the timing of product transfers since they do not have to wait for FDA approval. Batches that are filed with a CBE can be marketed immediately, provided there is stability and manufacturing comparability.

It is important to note that the SUPAC guidances do not supersede the 21 CFR 314.70; they actually complement that section of the regulations. Additionally, the SUPAC guidances do not reduce the scientific and technical data required relative to changes made to the manufacturing site, batch size, component and composition, manufacturing process, and equipment. Prior to SUPAC guidances, the industry employed the 314.70 regulations along with direct input and feedback from the FDA.

Bulk Active, Scale-Up, and Postapproval Changes (BACPAC). A similar guidance document has been developed for bulk active syntheses and manufacture. Active pharmaceutical ingredient manufacturers would benefit greatly from this guidance since the 21 CFR regulations do not directly cover changes to the manufacture of active pharmaceutical ingredients (APIs).

4.1.8 Annual Drug Product Reports

Another category of change that does not require FDA approval comprises relatively minor changes to manufacturing, packaging, testing, holding, and shipping that are inconsequential to the product's, safety, efficacy, potency, and quality. These changes, while not requiring FDA review or prior approval, do require a comprehensive assessment prior to implementation by the company's internal quality assurance unit. Additionally, if these changes correspond to a regulatory filing such as an NDA or ANDA, it is FDA's expectation that companies will categorize and summarize such changes in a document known as the annual drug product report, which is typically filed on the anniversary date of the NDA or ANDA.

4.1.9 Manufacturing Deviation Investigation Procedure

A manufacturing deviation can result from any number of unplanned and sometimes intentional departures from predetermined manufacturing procedures. These deviations may result in the rejection of a batch or the need to reprocess the batch and material. A comprehensive investigation, in accordance with a formalized investigation procedure, will provide insight into the potential source of the failure as well as whether or not the change control system is working adequately. Over time, a significant number of batch rejects or reworks indicates that the procedures or analytical methods have not been appropriately validated or that undocumented changes have occurred. Just as a change control system is a quality program, so is a manufacturing deviation investigation system. History has shown that there is a correlation among quality failures, manufacturing deviations, and inappropriately managed changes.

4.1.10 Annual Product Reviews

FDA is specific with regard to its requirement that companies evaluate their commercialized products on an annual basis; 21 CFR 211.180(e) mandates the preparation of an annual product review summarizing the myriad changes that have impacted product quality for that year. This report typically addresses changes made to the facility, analytical procedures, manufacturing and packaging operations, and utility improvements, as well as any other changes that relate to the compliance status of the organization.

4.1.11 Internal Audit Program

A comprehensive, integrated, and far-reaching internal audit program is not only an FDA expectation and industry standard, but goes a long way toward revealing the effectiveness of a company's change management system. Internal audits should be designed to identify changes and track the documentation and assessment of those changes. Additionally, the audit should assess the effectiveness of those changes over the long term. Frequently changes are made after a thorough evaluation by a number of operating units with the assumption that the change will prove effective without giving any thought to monitoring that change over time. An essential part of change control is to monitor the long-term effect of changes.

4.1.12 Consumer Complaint Handling System

Product complaints provide a great deal of insight and feedback related to potential changes that might be impacting product quality. A series of unmanaged minor changes can frequently have an incremental effect on product quality. Unmanaged major changes in the manufacture, testing, shipping, and storage of a product will certainly lead to quality changes and failures. Similar to manufacturing deviations, there is a strong correlation between an increase in unmanaged changes to a product and the prevalence of consumer complaints related to that product. It is important to evaluate and compare the trends between these critical compliance and quality programs in order to identify inadequately managed changes and prevent product failures or possible product recalls.

4.1.13 Product Development and Change Management

All product commercialization efforts require some level of research and development, new chemical entities more so than generic and over-the-counter (OTC) products. The FDA has become increasingly interested in research and development activities since the advent of the preapproval inspection program in 1987. The generic drug scandal also contributed to FDA's enhanced interest in compliance and quality assurance measures during the research and development phases. As a result of these recent events, the requirement for change management presents itself fairly early in the product development cycle. While the full range of cGMP regulations is not necessarily applicable during early development, change management is. Preparation of a development report, which will be reviewed during a preapproval inspection, is not only a regulatory expectation but also good business practice (Codified regulations requiring a development report do not actually exist for pharmaceuticals or biologic products.) Managing changes during late research and early development is critical since development scientists are typically focused on finalizing the formulation and manufacturing process. As a drug product enters clinical trial activities, change management once again becomes a critical regulatory and quality assurance activities component. Chapter 3 discusses the role of quality assurance through clinical trial and briefly addresses change management issues.

4.1.14 Capturing Changes in the Development Report

The development report pulls together the beginning of the product's development throughout its manufacturing phase for clinical and commercial activities. This report is a direct reflection of all research work, but focuses primarily on development activities, scale-up, and technology transfer efforts.

An FDA investigator will identify both procedural critical control points and critical parameters in formulation that are delineated in this report and proceed to inspect these areas. Appropriate documentation and management of changes to critical parameters and critical controls is essential during development. If there is ever a time that change is welcomed and constant, it is during development.

If the development report does not contain information about changes to the formulation, critical processes, analytical methods and assays, equipment, stability, and so forth, there is a low probability that what has been prepared will be useful to the FDA investigator(s) or the firm during an inspection.

Unlike the annual product review, there are no FDA requirements regarding development reports; nevertheless, it behooves a firm to have a clear, concise, and comprehensive report of its product's development and evolution.

If reformulation has occurred or specifications have been changed, a development report will help the firm identify all the bases that need to be covered.

4.2 Specific Development Report Sections and Change Management

Table 1 represents some of the critical activities that need to occur after late research en route to manufacturing a commercial batch.

Investigators will be looking for evidence that ensures bioequivalence of production size (full-scale) lots to the biobatch. The report must contain

TABLE 1 Research → Production Batch

Formulation
Processes/specifications
Scale-up
Equipment qualification (biobatches, clinical, and full-scale)
Critical and non-critical parameters
Technology transfer/bridge studies
Methods validation
Stability studies
Labeling
QC markers and controls
Process validation

information about the formulation, including justification for any and all changes made in the methods during the development process.

Additionally, the report will include information about the following:

Justification for all ingredients used
Justification for all analytical methods selected
Justification for all of the final manufacturing and analytical processes stated in the application (ANDA or NDA)
Types of equipment used
Manufacturing process (description of evolution of the process)
Scale-up to production
In-process results
Final dosage form test results
Critical parameters of bulk drug substance
Acceptance criteria for critical steps
Conclusions with key variables identified
Stability
Description of pivotal batches

Pivotable batches typically include:

Pilot batch
Scale-up batch
Clinical study material (bioequivalence)
Stability study batch

All of this is fairly simple if your product is a generic version; nevertheless, the information must be contained in a concise report and be available to the FDA during inspection.

If the product is a variation on a theme and there is not an exact product already on the market, the development report really needs to highlight the comparability or bioequivalence of your product to the one seeking commercialization.

The history of the development report should be written on an ongoing basis, concurrent with actual development activities in a chronological manner while everything is fresh in the minds of the scientists.

One strategy is to update the report as each milestone in the drug development process is achieved.

The entire purpose of this report is to point the FDA toward a document that delineates the science and technology that went into making the product and that includes all preliminary studies right up to the regulatory submission stage.

The report is easier to assemble and update for a generic version because the product is simpler to manufacture and justify since it can only vary slightly from the pioneer.

This report must have frequent quality assurance oversight, review, and approval.

Suggested table of contents for development report:

1. Cover sheet

   The manager of research and development (R&D) shall prepare a cover sheet for each development report. The cover sheet shall include the following information:

Company name
Development report member
Product name
Corresponding investigational newdrug (IND)/NDA/ANDA number, if applicable
Name of author
Approval signatures and dates

2. Table of contents

   The manager, R&D, shall prepare a table of contents for each development report.
3. Reason for revision

   List those sections of the development report that have been revised and provide a brief description of the revisions.
4. Introduction

   Briefly discuss the background on the product's intended indication or use. If applicable, complete Table 1, which briefly summarizes the product's clinical programs.
5. Drug substance characterization

   Description of the bulk drug substance, including structural and molecular formula, process impurities/degradants, specifications, rationale for specifications, safety, approved supplier(s), test methods, validation, and stability. Include basic information on synthesis or derivation of the drug substance,or if applicable, reference supplier drug master file.
   List of all applicable reports supporting drug substance characterization (e.g., certificates of analysis, validation reports, raw material specification justification report for drug substance, other technical reports).
6. Formulation/design development

   Quantitative/qualitative formulation (theoretical unit formula per dose).

History and rationale behind the development and selection of the formulation or dose level. Include reasons for excluding other formulations or dose levels.

Description of the role of each excipient (e.g., surfactant, preservative) and cite requirements for the selection. Include discussion on the preservative system.

Discussion of excipient specifications and release parameters. Note any special requirements that are not typical compendial requirements (e.g., particle size). Include comparative evaluation for multiple sources.

List of all applicable reports supporting formulation and design development (e.g., safety reports for excipients, raw material specification justification report for excipients).

7. Manufacturing process development (including in-process controls)

   Brief description of the manufacturing process. Include a flow diagram.

   Description of the history and the rationale behind the development of the manufacturing process (i.e., justify deviations from the established manufacturing procedures that occurred during process development).

   Description of the manufacturing parameters that are important to product performance and the rationale for the selection of these parameters (e.g., processing time, temperature, drying rate, mixer speeds, order of mixing, microbiological control, pressure, spray rate, storage of in-process bulk material).

   Description of rework/reprocessing procedures, or if applicable, state that product will not be reprocessed.

   Summary of the process, cleaning, and sanitization validation or verification studies performed.

   List of all applicable reports supporting manufacturing process development (e.g., process validation protocols and reports, cleaning validation, protocols and reports, and batch production records).

8. Scale-up technology transfer

   Description of the logic behind all pertinent activities that occurred during scale-up, from pilot plant production to phase 3 clinical production. Discuss problems, failures, and so on. Justify the absence of equivalency concerns despite differences in process parameters, equipment, facilities, and systems.

   Description of the logic behind all pertinent activities that occurred during technology transfer from phase 3 clinical

production to the commercial process (determination of full-scale commercial processes, specifications, lot size, etc.). Discuss problems, failures, and so on. Justify the absence of equivalency concerns despite differences in process parameters.
Reference to the report which indicates successful technology transfer (e.g., Validation Report, Verification Report).

9. Manufacturing/packaging equipment
   Summarize information pertaining to equipment used in critical batches and runs in Table 2. (See Sec. 14, "Critical batches and runs")
   Description of equipment designs and functions critical to accommodate product requirements.
   Description of the differences in equipment size, type, and operating parameters between the critical development batches and runs and commercial batches and runs.
   List of all applicable reports supporting equipment (e.g., IQ/OQ/PQ reports, evaluation of equipment comparability between lab/pilot plant and commercial plant, etc.).
10. Finished product testing and results
   Description of the history and rationale for the finished product specifications and release parameters. Note any non-compendial requirements.
   Summarize information pertaining to testing of critical batches/runs in Table 2 (See Section 14.0 "Critical Batches/Runs").
11. List of all applicable reports supporting the development of finished product specifications and release parameters (e.g., finished product specification justification report).
   Analytical/micro method development
   List of all analytical/microbiological methods used for excipients and the finished product.
   History and rationale for the development of all major non-compendial methods for excipients and the finished product.
   List of all applicable reports supporting method development (e.g., method validation reports, technical support documents, etc.).
12. Package/development
   History and rationale for the selection of packaging components, including product/packaging compatibility.
   Description of packaging component specifications.
   History and rationale for the selection of packaging component specifications.

TABLE 2 Potential Changes During Development and Post Commercialization

Formulation
- Composition: Percentage of active and inactive ingredients per unit dose

Active pharmaceutical ingredient (API)
- Alternate manufacture
- Altered impurity profile
- Physical characteristic
- New manufacturing process
- New raw starting materials(s) excipients
- Official grade change
- New raw starting materials(s) excipients
- Official grade change
- Raw material change

Equipment
- Bench scale
- Pilot plant
- Biobatch
- Scale batches
- Commercial size
- Continuous improvement
- Process optimization
- Repairs and maintenance

Cleaning procedures
- Manual to automated
- Equipment configuration changes

Manufacturing process
- Critical parameter changes
- Operating range changes
- Optimal condition changes
- Scale-up
- Technology transfer
- Rework/reprocessing procedure

Environmental controls
- Adjusted to support changes in processing and stability
- Expand in response to in-process control requirements

Analytical method development
- Qualified method
- Validated method

TABLE 2 (*continued*)

Optimize and/or modernize method
Customize a compendial method
Stability profile
Container closure system(s)
Storage conditions
Expiry period
Facilities
Controlled environment requirements
Preventative maintenance
Emergency repairs
Validated classified areas
Structural changes
Cross contamination prevention
General housekeeping and sanitation
Cleaning agents and pest control
Critical utilities
Critical utilities
Clean steam
Potable and purified water systems
HVAC systems
Compressed air system
Dust collection system
Emergency repairs to critical utilities
Changes in PM maintenance schedule
Natural disasters
Facility changes
Utility changes
Validated system changes
Personnel turnover and loss
Critical components
Containers
Closures
Labeling
Packaging materials
Inserts
Desiccants
Vials
Tubes
Changes to any critical parameters of these components must be monitored

List of all applicable reports supporting package development. Include labels for API finished product.

13. Product stability

    Description of the stability of the finished product. Include batch sizes, packaging configurations (including bulk), storage conditions, analytical methodology, specifications analyzed for, and number of batches for which stability data has been generated.

    Summary of the properties of the dosage form or excipients that influence product stability.

    List of all applicable reports supporting product stability (include stability protocols, data, and reports).

14. Critical batches and runs

    A critical batch or run is one that provides primary support for label claims, indications, safety, efficacy, stability, or method development.

    A batch or run listed as critical early in product development may later be determined to be noncritical if the course of development changes. In this case, these batches or runs can be deleted from Table 1 in the next issue of the development report with a brief rationale.

    Complete Table 2, which summarizes the following information pertaining to manufacturing and testing of critical batches and runs:

Formula/design numbers
Product name/label claim (strength)
Batch numbers
Date of manufacture
Batch/run size
Major manufacturing/packaging equipment used

15. Environmental assessment

    Brief summary of the environmental assessment. Address environmental impact and effects.

    List of all applicable reports supporting the environmental assessment.

16. Literature review

    Provide a list of relevant literature references pertaining to development of the drug substance and drug product.

17. Conclusions

Briefly summarize the overall development process. The summary may include a time line that displays the initiation, key intermediate steps, and completion of work in such critical areas as validation of methods, safety studies, clinical studies, scale-up, process validation, and times of regulatory submissions. The summary should also identify key issues to be resolved (future work to be done, if any). Finally, the summary should provide a conclusive statement that links each section of the development report and addresses the equivalency of the clinical/biobatches to the production batches. Include information from in vivo and in vitro studies, as appropriate.

The development report is a tremendously useful tool for capturing and justifying changes during late research and development leading to commercialization.

A number of changes that can occur during development can also occur postcommercialization, and the manufacturer must be prepared for them. Some of those changes are listed in Table 2 .

4.3 Pulling it all Together

Hopefully you are convinced of the necessity and importance of an integrated, comprehensive, and far-reaching system. Here are suggestions on how to begin to build that system in your organization.

1. Get a group together from all key disciplines in your organization, such as regulatory affairs, quality assurance, engineering, manufacturing, materials management, and labeling. Talk through how change should flow through the organization. The end result should be a process flow map of how change is recommended; how notification should occur; which individuals should assess, review, and ultimately approve the change; how change-related activities are tracked; how the change gets implemented; and finally, how to evaluate the change over time.
2. Write a procedure that describes all the information in the process flow map, including responsibilities of individuals, departments, and committees.
3. Develop a change request form. The form is a critical component. You want it to be simple enough that anyone can use it, but it must still capture all the information the approving unit needs to make an intelligent decision about whether or not to move forward with the change request. Included here is a sample of a form that has the minimum guideposts needed. As you can see, the form invites any employee to submit recommendations(s) for change. This is one way an organization can actively honor ideas and recommendations from all employees at any level.

CHANGE REQUEST FORM

Example of Critical Change Control Guideposts

CHANGE REQUEST NO. ____ (a) ____ DATE: ____ (b) ____

CHANGE REQUESTED BY ____ (c) ____
SUPERVISOR APPROVAL ____ (d) ____

Description of Change:

(e)

Justification of Change:

(f)

Change Classification:

(g) Major ____ Minor ____ Departure ____

Related Documents/Activities (itemize or attach listing from cross-referencing system)

(h)

Approved By: (i)	Activity Required: (j)	Ready By: (k)
Manufacturing		
Regulatory		
Engineering		
Materials Management		
Purchasing		
Label Control		
Training		
Other		

Product Hold: (l) Yes ____ (Identify lots) No ____

Change Coordinator: (m)

Date change was implemented ____ Signature ____

CHANGE REQUEST (SAMPLE) FORM

a. *Change request numbers*—Each change request should be assigned a unique number for tracking purposes. By having them sequential and accounting for all numbers, you will not have to explain "gaps" to investigators.
b. *Date*–This should be the date the change request is initiated.
c. *Change requested by*—This should be the name of the individual authoring the Change Request Form.
d. *Supervisor approval*—Many systems advocate supervisory concurrence prior to the change being submitted to the next level.
e. *Description of change*—A brief but descriptive paragraph on the actual change that is being proposed.
f. *Justification of change*—Give good reason why the change is being proposed, include a cost/benefit analysis, identify compliance issues, and explain potential impact on product.
g. *Change classification*—Your SOP should clearly define the difference between major and minor changes so these fields can be consistently checked off. In the case of Departures or Unplanned changes, refer to your SOP for proper procedure. (See Chap. 13.) It does not matter what you call them as long as there is a mechanism for categorizing the level of change in terms of significance.
h. *Related documents*/activities—The initiator should itemize or attach a listing from your cross-referencing database of all documents and activities that could be impacted by this change. All approvers should review this list and add or delete to it as they deem appropriate.
i. *Approved by*—Your approvers should be well-thought-out choices and the reason or responsibility of each approver's signature should be clearly written in a SOP. The number of approvals should be decidedly fewer for minor changes than for major changes. (This is why the "categories of change" must be clearly defined up front.)
j. *Activity required*—Any additional activities not included in the attached listing should be entered here.
k. *Ready by*—If an individual approves the change, he or she should give a date by which he or she can complete his or her activity or document revisions and enter it here. The date farthest out will determine critical path or result in effectivity date negotiations based on project priorities.
l. *Product hold*—This field should be completed by the quality assurance approver, and if lots are to be put on hold, they should be identified here.

m. *Change coordinator*—After the change is approved and all related documents and activities are completed, the change coordinator will determine the implementation date and coordinate the implementation of change-related activities. The change coordinator also maintains the change database (e.g., Excel, Access).

4. Identify relationships between systems. This is where the real work begins. Many of the system relationships are obvious and it is a matter of taking the time to document them in a database. Others will take more thought and will only become evident with time and predictable with extensive experience.
5. Use a manual system or a simple database program (e.g., Excel, ACCESS) to crosslink those relationships in order to illustrate the impact a proposed change will have on multiple established systems and procedures. Included is an illustration of how the database might work.

   Begin by cross-referencing obvious relationships. Perhaps SOPs can be linked directly to specifications, or equipment setup sheets in the batch record can be linked to the asset numbers of equipment. The most obvious relationship will be the easiest to start with.

   a. Cross-reference "knowledge." As you begin to build your database of relationships, talk to employees at all levels of the organization. You will start to undercover the less conspicuous relationships that are critical to the success of your change control system. You will also begin to uncover deficiencies in your systems (e.g., SOPs that need to be written, personnel who need training, or related systems that were overlooked and not evaluated for impact). These could be serious, such as batch records that do not match what was submitted in the regulatory filing or label claims not supported by the clinical protocol and data. If you uncover problems such as these, you are in for a lot of back tracking, but better late than never. You can fix the problems before an investigator discovers them. If you find you uncover many discrepancies in your current mode of operation, highlighting these discrepancies to management may help you get the resources you will need to establish a fully effective change management system.
   b. Add links as they are discovered. It will take awhile to include everything, but do not get discouraged. Just know you are moving closer to an integrated, comprehensive, and far-reaching change control system every day.

c. Maintain the cross-reference system. Establish a means of printing out "links" to accompany change requests as they are routed. This will assist reviewers in identifying the overall impact of change.

6. Appoint a change coordinator and establish a change committee. The change coordinator will be the focal point for all critical, significant, and major pending changes. The committee should consist of representatives from key groups within your organization and should be used to better manage big changes (e.g., capital projects, new product introductions). Lesser changes will be reviewed outside the committee.
7. Define major and minor change procedures. As you are defining your system, you need to keep in mind the speed of doing business and not include nonvalue-added activity. One way to add nonvalue activity is to build in unnecessary approvals. If the change is minor, set the approval requirements accordingly and minimize approvers. Conserve organization resources for the more critical changes that will require quality assurance and executive management input.
8. Address planned and unplanned changes. How do your deviations and departures fall under change review? Unexpected events are going to occur; however, a "planned deviation" is an oxymoron. If you plan on making a change, whether it is temporary or permanent, it must be included in the change control system. If an unexpected event occurs for which you need to write a deviation, you must handle that under a written procedure for handling exceptions. (See Chap. 13.)
9. Annual product reviews should be facilitated by the change management system. The change control system will allow for capturing specific aspects of a particular given time period.
10. The system you establish needs to be able to track pending changes as well as the activities related to each pending change. While there are some software tools available now to help with this process, it is critical that you thoroughly understand how the process works with a manual system before using an automated system.
11. Implementation. When you are ready to implement a change, the first consideration should be product impact. If there is product that has been manufactured incorporating a proposed but incompletely processed change, that product should be placed on hold. Only after all affected areas have had an opportunity to formally evaluate the impact of the change can the change be implemented and product disposition determined. It is also important to be aware of any product

currently in the manufacturing cue, as it may also be impacted by the implemented change. It is easy for a batch that is in the midst of manufacture to be missed.

12. Assessment of specific changes over time. Periodically review (via internal audits) the effectiveness of the changes that have been made. Just because a group of experts supported change, does not mean that it worked over time.
13. Train plant personnel. Everyone in the plant should be aware of the change management SOP and be trained to it. With change constantly occurring, it is important that all employees be trained on how to handle changes affecting their specific functions and respective areas.
14. Documentation. The completed change request forms and all activities surrounding the implementation of each change should be properly documented. These files should be maintained in a state ready to be reviewed by an investigator at any time. These files can also provide good historical data for your ongoing plant operations.

4.4 Benefits of a Successfully Implemented Change Management System

4.4.1 All Systems will be Coordinated and in Compliance with FDA Regulations

Compliance to federal regulations is the law in our industry—it is not optional. A successfully implemented change management system will enable us to follow that law and partner with the FDA to responsibly market products to the general public that are safe and of the highest quality.

4.4.2 Organizational Strengths and Weaknesses will Become More Apparent

Organizations are still made up of human beings. We must have strong systems in place that enable people to readily see what works and what does not work. These systems are a means of benchmarking for continuous improvement. They will withstand the test of time even as the people who execute them move on.

4.4.3 Your Best Practices as Well as Your Weak Links Will Be Identified

Business is more competitive than it has ever been before. Today's technology allows the business process to move very rapidly. It is imperative that we identify what we do well and what we need to improve in order to run our businesses in the most efficient manner. Strong quality systems are major contributors to helping us meet both of these goals.

4.4.4 Internal and External Experts Will Be Called Upon to Make Substantial Contributions to Continuous Improvements

We must make the most of the knowledge base we have access to. By identifying the experts, we can better assure each change is properly evaluated by the best people we have. It will help us to recognize the deficiencies within our ranks so we know when to seek the assistance of external experts.

4.4.5 All Employees Will Feel a Sense of Pride and Ownership

By seeking input from all levels, people will know that their contributions are important to the business and that they are key to its success. By encouraging all employees to recommend change they will feel a sense of responsibility and ownership.

4.4.6 The Organization Never Grows Stale or Stagnant

Change is critical to future business success. It enables us to adapt to the evolving conditions all around us. By having an effective change management system you take the chaos out of the process and can clearly evaluate the impact of each change proposed.

5 WHO IS RESPONSIBLE FOR MANAGING CHANGE?

Just as assuring product quality is everyone's responsibility, managing change is also every employee's responsibility. This is a reality that must be instilled as part of the company culture. As an employee, you have an obligation to recommend change when you have an idea involving a better way. The change control system should encourage and make it easy for employees to suggest their ideas. Many innovative ideas come from those closest to the operation day in and day out!

Recognizing that something has changed is also every employee's responsibility. Employees must be aware of which procedures apply when they notice something is different. Sometimes it is necessary to use change control procedures and sometimes deviation procedures. All deviations do not lead to permanent changes.

Anything that is everybody's job quickly becomes nobody's job. It is essential to assign the function of change management to an individual's role and responsibilities. That person will be the change coordinator. The change coordinator manages the change request system, routes and tracks changes, tracks change-related activities, assures a systematic implementation occurs, and maintains change documentation in a presentable and compliant state.

In addition to a change coordinator, there may be a place for a change management team or committee. These individuals will address major changes that have a critical impact on the operation and that require a significant amount of coordination to assure everything happens within a reasonable time frame and within a clearly defined pathway.

6 SUMMARY

If you are surprised with the magnitude of control and management required for the appropriate implementation of most changes, you are beginning to understand why this system is so vital to an FDA-regulated facility. Change management is complex, but it is the only way to optimize operations through necessary changes without collateral damage. A well-developed system will not only help maintain an organization in compliance, but allow management to make wise decisions because it will have had the opportunity to formally evaluate the risks and benefits of proposed changes. By incorporating a documented review, the full impact of change on all aspects of the organization is realized. Managed change can be a positive experience once an adequate system is installed and followed.

10

The Vendor Qualification Program

Elizabeth M. Troll
Chesapeake Biological Laboratories, Inc., Baltimore, Maryland, U.S.A.
Karen L. Hughes
Guilford Pharmaceuticals, Inc., Baltimore, Maryland, U.S.A.

1 WHAT IS VENDOR QUALIFICATION?

Simply stated, vendor qualification is the combination of activities required to ensure that a vendor will meet the professional and regulatory expectations of the sponsor.

Throughout this chapter, reference will be made to the sponsor, a vendor, and a supplier. The sponsor refers to the company or organization contracting the services or materials. This is the organization performing the qualification.

A vendor is a provider of services, who will in turn provide for all materials needed to meet the requirements of the contract. A vendor is qualified by the sponsors and is also responsible for the qualification of its own suppliers.

A supplier provides the materials required to meet the obligations of a contract. To further illustrate the relationship between these parties, refer to Fig. 1.

The *act* of qualifying a vendor is not so simple. There are different levels of qualifications. Clearly, vendor qualification is not "one size fits all."

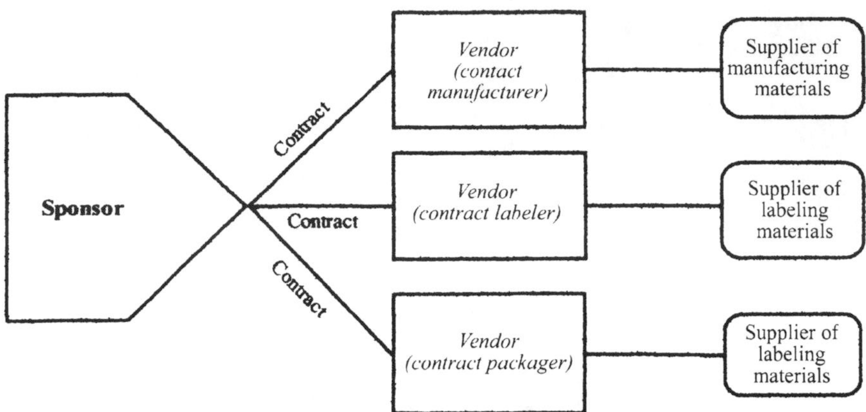

FIGURE 1 Relationship between sponsor, vendor, and supplier.

The extent to which a sponsor will qualify a vendor depends on several factors. The first factor is the type of vendor required to fulfill the needs of a contract. Vendors and suppliers can be grouped as follows:

- Contract vendors
  - Contract laboratory
- Contract manufacturer
  - Active pharmaceutical ingredient (API)
  - Clinical supplies
  - Commercial product
  - Contract packager
  - Suppliers
- Components, containers, and closures suppliers
- Raw materials/excipients
  - Labeling and components
    - Unit labels
    - Unit cartons
    - Package inserts
    - Shipping cartons
- Containers and closures
  - Vials
  - Syringes
  - Stoppers
  - Crimps (seals)
- Manufacturing supplies—catalog items
  - Tubing
    - Filters

Glass carboys
Manufacturing suppliers—custom designed
Stainless steel tanks

The second factor related to the extent of qualification and monitoring required is the risk associated with a particular vendor and the scope of work it is responsible for. The sponsor must evaluate the scope of work to determine the level of monitoring required for a particular vendor.

1.1 Some Examples

If the scope of work is a one-time event or a short-lived project for experimental or developmental manufacturing, there is little risk to the project or the organization. The level of monitoring may therefore be minimal.

If the scope of work is a precursor to a phase I clinical study, enhanced monitoring may be required. The sponsor may want to visit the vendor to ensure that the company is capable of performing to the contract.

If the scope of work involves the manufacture of Phase III Good Manufacturing Practice (GMP) materials, advanced monitoring would be pertinent since this activity could result in the success or failure of a critical path project. The sponsor must be confident that the systems are in place to effectively produce the desired deliverable(s) to complete the project.

The decision to monitor from the "home office" by exchanging information and documentation only or monitoring on-site while each activity is being performed (or a level of monitoring anywhere in between) becomes very complex. Several elements must be considered.

At a minimum, a sponsor should check the vendor's references and financial status. The sponsor wants to ensure that the vendor will be in business over the course of the contract.

Another element of qualification is the quality audit. Will one be required, and if it is, to what extent? A quality audit can be conducted from the sponsor's office through the use of an audit questionnaire. Again, this can be extensive 20-plus page request for information related to the vendor's documentation practices, equipment lists, service capabilities, facility layout, materials control, and personnel credentials, or a three-to-four-page questionnaire identifying such critical requirements as the financial information and industry standards to which the vendor complies. (For an extensive discussion on internal audit programs, see Chap. 12, "The Internal Audit Program.")

Alternatively, the quality audit may be conducted on-site. Again, the audit can be as extensive as a comprehensive GMP, Good Laboratory Practice (GLP), or Good Clinical Practice (GCP) systems audit that reviews all regulatory requirements for strict compliance, or simply a visit to ensure that the vendor is capable of meeting the minimum requirements of the contract. In addition, the quality audit can extend to having a representative at the site each time an activity is conducted. In extreme cases involving maximum risk, the "person-in-the-plant" approach may be essential to a successful outcome.

Establishing the acceptance criteria for the deliverable from the vendor or supplier is a key element in determining the extent of the qualification required. If a representative sample from every lot of every shipment is going to be tested to the certificate of analysis, little monitoring of the vendor may be required. If shipments are going to be accepted on the certificate of analysis alone, however with no confirmation testing, routine monitoring of the vendor is essential to confirm that the vendor is adhering to the contractual agreement.

Finally, there is the issue of trust and reliability. This element is probably the most critical in establishing the level of monitoring desired for each vendor. Vendors are an extension of your own organization. They are not part of your organization, however. They are under different management, have different operating policies, and are separate entities. The sponsor must establish a rapport with its vendors that establishes trust and cooperation. Vendors, similar to sponsors, are in business to make a profit. Partnership in an effort to meet contractual and compliance obligations will benefit both sides.

While it is not the sponsor's responsibility to carry out the work, it is its responsibility to ensure it is done correctly and in accordance with the sponsor's standards and commitments to the regulatory agencies. It is critical to verify that vendors are meeting compliance and contractual obligations by performing an initial qualification audit and periodic monitoring audits.

In a sponsor–vendor dynamic, it is not a good idea to rely on trust and long-standing relationships—ongoing verification is key. There may come a time when it becomes the sponsor's responsibility to assist in getting the job done if it is not being done to its satisfaction. This will increase the need for monitoring and unfortunately require some form of "micromanaging."

While there are several factors to consider in the level of qualification conducted, including the type of vendor or supplier, the scope of work involved in the contract, and the amount of risk associated with the outcome of the project, the evaluation of each of these factors is manageable. The use of a decision matrix is probably the most effective tool in completing this evaluation. An example of a decision matrix is illustrated in Table 1.

The four categories encompass the breadth of the development time line. Some vendors may start as a category 1 and increase in category as the development process moves forward. The opposite is also a possibility;

inexperienced and/or virtual sponsors may underestimate the category of their potential vendors.

2 CHARACTERISTICS OF VENDOR CATEGORIES

2.1 Category 1

Category 1 vendors are generally regarded contracted because of their niche capabilities. They provide services unique to a specific aspect of the development process. They are generally regarded as experts, and data generated from their efforts can be used to define a specific set of characteristics for the project. They are "short-lived" in the development cycle and may be contracted to perform a limited scope of work. Because the risk of them not performing as the "expert" is nominal and the scope of work is limited, minimal monitoring is suggested. An initial audit of the facility and a review of the records generated as they relate to the program may be acceptable.

2.2 Category 2

Category 2 vendors are well-known suppliers of standard containers, closures, raw materials, and excipients. Generally they have process standards in place and are familiar with pharmaceutical development requirements. They may have a wide customer base and produce large volumes of these standard items. They may also be certified to an International Standards Organization (ISO)-9000 quality management system. Historically, they have had an acceptable quality of incoming goods. Enhanced monitoring is suggested. An initial audit of the facility, a review of the records (statistically if possible, with the appropriate action and alert limits defined by historical data), and a standard time frame (suggested at least every 24 months) to reaudit the facility are typically required.

2.3 Category 3

Category 3 vendors may be category 2 vendors who are experiencing quality issues with current incoming inventory or have shown a trend of nonconformance over the last 12 months. They can also be contract laboratory operations that provide routine analysis, sometimes in large quantities. The risk of nonconformance of these vendors is greater than with category 2 vendors. An advanced monitoring program and an annual audit schedule are recommended.

2.4 Category 4

Category 4 vendors are sole-source API manufactures. First-time clinical trial supply manufacturers, offshore single pivotal trial clinical sites, and

TABLE 1

Type of vendor supplier	Financial background	Quality audit			Testing requirements upon receipt			Reliability of vendor		Category of vendor supplier
	Reference financial check conducted	Not required	Question-naire audit	On-site audit annually	On-site audit (during each event)	100% testing upon receipt of deliverable	Minimal testing upon receipt of deliverable	No testing upon receipt of deliverable	No history of quality concerns	History of quality concerns	
Contract laboratory	•			•				•	•		Category 3
Contract Manufacturer—API	•			•			•		•		Category 3
Contract manufacturer—clinical trail supplies	•				•	•			•		Category 4
Contract manufacturer—commercial product	•			•			•		•		Category 3
Contract packager	•				•	•				•	Category 4
Raw material/excipient supplier	•		•				•		•		Category 2
Labeling supplier	•			•		•				•	Category 3
Container/closure supplier	•		•				•		•		Category 2

Supplier of manufacturing supplies—catalog items	•		•		•	•	Category 1
Supplier of customized manufacturing supplies	•	•		•		•	Category 1

Category 1—minimal monitoring. Example: Supplier customizes a formulation tank. Sponsor reviews and approves the blueprints prior to manufacturing. An installation qualification and operational qualification is planned upon receipt to verify acceptability.
Category 2—Enhanced monitoring. Example: Well-known supplier of containers/closures supplies multiple lots per year of vials to the sponsor. The sponsor has no historical quality concerns with the supplier. Testing will be conducted upon receipt to verify acceptability of materials.
Category 3—advanced monitoring. Examples: (1) A supplier provides unique labeling materials that are not readily available through other resources. There is a history of quality concerns with the supplier. Past lots were received containing printing errors. The supplier is audited by the sponsor annually and performs 100% inspection to release incoming labeling materials. (2) A contract laboratory is providing routine analysis for raw material release. There is no history of quality concerns. The sponsor reviews the results upon receipt for acceptability and audits the vendor annually for compliance to applicable regulations.
Category 4—intense monitoring. Examples: (1) A contract manufacture is supplying clinical trials supplies. The supplies are for a phase III clinical trial and the manufacture is scaling up for the first time. There are three compliance batches being manufactured. Since this is the first time that the product is being manufactured at a larger scale, the sponsor has elected to be on site for each event for monitoring and consultation; 100% of the lot will be visually inspected for release upon receipt by the sponsor. (2) A contract packager is packaging material for a commercial product. The sponsor has had quality concerns with the packager in the recent past. Since the contract packager is listed as the approved vendor on the application filed with the FDA, the sponsor is forced to use the packager until the supplement to approve a new vendor is approved. The sponsor will be present during all contract packaging activities and will perform 100% inspection upon receipt for final release.

sole-source bioanalytical laboratories are included in this category. The inability of these types of vendors to deliver a complete and regulatorially appropriate finished work product will have a significant effect on the pharmaceutical development program as a whole. Intense monitoring is in the best interest of the sponsor. Sponsor representatives can have a strong presence at the site (person in the plant) and should retain all release responsibility.

3 REGULATORY REQUIREMENTS

Regulatory requirements will also play a role in determining the frequency of vendor qualification events. A review of the FDA and ICH guidelines as they relate to the scope of the work is recommended. The table below summarizes the subpart references of the Code of Federal Regulations Title 21 as they relate to GLP, GCP, and GMP compliance. They are sorted by system. This is not designed to represent all the current regulations and should be used as a tool for adjusting the focus of vendor qualifications for vendors that operate in more than one regulatory discipline. A more completed description of regulatory requirements can be found in Chap. 2.

4 WHY QUALIFY YOUR VENDORS?

No matter what tool is used to determine if the vendor is appropriate for the scope of work, the real challenge begins with the negotiation of the contract and subsequent initiation of the work. Any vendor or supplier is an extension of the sponsor firm's operations. Ultimately they are partners in the outcome of the project. Supplier partnership is a concept that can be applied to any supplier. Suppliers are viewed as an extension of the customer's overall operation. Critical attributes of a partner relationship are

- Supplier or customer commitment to a long-term relationship
- Information sharing
- Joint agreement on specifications and performance standards
- Performance measurement and feedback
- Customer confidence in the supplier's manufacturing capability, quality, cost, and development

The extent and breadth of these attributes will vary, depending on the status of the supplier–customer relationship. Initiation of a partner relationship generally begins at the qualification stage. The previously mentioned tools of vendor category and regulatory requirements set the foundation for the vendor qualification program.

Regulatory reference system	FDA GLP	FDA GCP	FDA GMP
Organization and personnel requirements	58	50	211(B)
	58.29	54	
	58.31	56	
	58.33	312 (A)	
	58.35	312 (D)	
	58.200	312 (F)	
		314	
Facilities	58.15	211 (C)	
	58.43		
	58.45		
	58.47		
Equipment	58.61	211 (D)	
	58.63		
Control of components	58.83	211 (E)	
	58.90		
Operations, production, and/or process controls	312 (G)	56 (C)	211 (F)
	58.49	312 (E)	
	58.51		
	58.105		
Packaging/labeling	58.113	211 (G)	
Inventory, distribution, destruction	58.51	211 (H)	
	58.107	211 (K)	
Laboratory controls	58.130	211 (I)	
Records and reports	58.81	50.27	211 (J)
	58.120	54.6	
	58.185	56 (D)	
	58.190	312 (B)	
	58.195		
Actions for noncompliance	58.202	56 (E)	211.100
	58.204	312 (C)	211.192
	58.206		
	58.213		
	58.215		
	58.217		
	58.219		

5 SCOPE-OF-WORK DOCUMENTS

Scope-of-work contracts are generally the preliminary starting point in the formal business relationship between the vendor and the sponsor. These are focused on the technical aspects of the project and will contain sections that describe the deliverable from each party involved.

Manufacturing scopes of work (API or finished product) will describe

Raw materials—actives and excipients
Containers, closures, labels
Formulation sequence
Packaging sequence (fill/finish)
Storage and/or distribution criteria
Analytical and microbiological requirements
Stability protocol requirements
Equipment, cleaning, and/or facility validation requirements
Process validation criteria
Documentation practices (electronic and paper raw data, specifications, batch records, shipping records, development reports)
Budget and payment terms

Analytical, bioanalytical, and microbiological method scopes of work will describe the following:

Reagents, reference standards, and prepared materials
Test article characterization requirements
Safety requirements—personnel as well as facility
Protocol requirements
Testing requirements
Data acceptability criteria
Validation criteria
Sample stability criteria
Documentation practices (electronic and paper raw data, specifications, batch records, shipping records, development reports)
Budget and payment terms

Contract research organizations for clinical trial scopes of work will describe

Study start-up
- Protocol and case report (CRF) design and development
- CRF review
- Investigator brochure development and maintenance
- Preparation of informed consent

- Development of study reference manual and monitoring plan
- Site identification selection, recruitment
- Prestudy site visits
- Distribution of study materials (CRFs, protocol, reference manual, etc.)
- Investigator contract negotiation/grant payment administration
- Setup and maintenance of project tracking system
- Regulatory document collection and review
- Setup of trial master files/investigator site files
- Assist are to site with Institutional Review Board (IRB) approval
- Regulatory document submission to appropriate regulatory authorities
- Clinical drug packaging and labeling
- Drug and supply distribution to sites

Clinical monitoring
- Site initiation visits
- Interim site monitoring visits
- Site closeout visits
- Review or resolution of edits and queries

Medical management
- Adverse event management and reporting
- Development and maintenance of study-specific safety database
- Serious adverse event (SAE) reporting to regulatory authorities

Project management
- Provision of regular project updates to sponsor
- Monitoring and management of project resourcing
- Facilitation of regular communications with sponsor
- Management of project issues, study budgets, and time lines
- Regular project team meetings
- Ongoing training of project team

Data management
- Database design and development
- Data entry, editing, verification
- CRF tracking
- Data coding—Adverse events (AE), medications
- Incorporation of local laboratory data into database
- Investigation and resolution of data discrepancies
- Data validation and quality control audit
- Data transfers (test and final)

Biostatistics and medical writing
- Statistical analysis plan
- Review and approval of analysis plan

Statistical analysis
Pharmacokinetics/pharmacodynamic analysis
Data tables and listings
Final integrated clinical/statistical report

6 QUALITY CONTRACTS

Quality contracts or technical agreements are additional tools that can be used to document the responsibilities of each party. These differ from the scope-of-work contracts described previously because they will contain sections that address more global issues and the costs of nonconformance. They are additions to scopes of work and are designed to clearly delineate who is accountable for what during the course of the entire development program, not just a specific scope of work. They can be the foundations for the way the partnership between the vendor and sponsor will be conducted. They define the legal and overall financial obligations of the parties. They contain the following sections to address;

- Regulatory requirements
  - Definitions of what standards are to be adhered to by the vendor
  - Determination of how performance to the requirement is measured
- Technology transfer
  - Who will perform what tasks
  - What tasks will be performed
  - When the tasks will be performed (time frame)
  - Where the tasks will be performed
  - How the tasks will be defined as successful
  - Documentation criteria
  - Approval criteria
- Method transfer
  - Definition of method(s)
  - Transfer design
  - Acceptance criteria
- Information transfer
  - Electronic and paper records
  - Effect of change in ownership of either organization
  - Duration of record retention
- Ownership of intellectual property
- Confidentiality
- Security
- Penalty/cost of compliance

Milestones to be achieved
Time frame for achievement
Cost associated with lack of adherence to milestone/time commitments
Corrective action plans
Definition of roles and responsibilities for corrective action plans
Standard operating procedures governing corrective actions
Auditing frequency and general access to site records
Change control
Systems that assure that the changes made to a current product or process do not affect product outcome
Change management
- Systems that define the ability to document and trace modifications to the current product or process

Release procedures
- Definition of accountability of data authenticity
- Determination of which party will complete the release process

Disaster recovery
- Measures in place to recover from a disaster
- Financial considerations of the disaster
- Determination of who has fanatical responsibility

Recall responsibilities
Project management
Training
Notification of personnel changes

7 COMMON PITFALLS IN VENDOR QUALIFICATION

Once the level of monitoring is determined, the scope-of-work and quality contract requirements have been initiated, and the qualification of the vendor is established, the key responsibility of the sponsor is maintaining the vendor qualified through routine monitoring. Qualification maintenance can prove to be the most challenging aspect of any vendor qualification program. To reduce the extent of this challenge, the sponsor should be aware of some of the common pitfalls and make every attempt to avoid them.

The common pitfalls in establishing and maintaining an appropriate vendor qualification program include

Not establishing the habit of routine update meetings and making sure the meetings have agendas, written minutes, action items within the minutes, and "person accountable" next to each action item

Overestimating the capabilities of the vendor
Having a false sense of confidence that the vendor will maintain itself at the same level throughout the contractual relationship
Underestimating the resource-intensiveness of maintaining a vendor qualified
Using a single-source vendor
Not switching vendors when appropriate
Not dealing with situations of nonconformance in a timely and effective manner
Overburdening the supplier, as though you were its only client
Setting unrealistic time lines
Telling the vendor what to do rather than asking the vendor what is possible
Underestimating the amount of time and effort it takes to initially qualify a vendor, especially if it is a critical supplier
Failing to understand how systems work within the vendor's organization
Failing to periodically authenticate raw data provided by the supplier
Maintaining ownership over intellectual property

The documentation of meeting discussions, decisions, and action items is a key to keeping track of the progress of a vendor on a specific project. Without fundamental record keeping such as meeting minutes, Gantt charts, and progress reports, it is close to impossible to ascertain the status of a given project or issue. Some projects take years to complete. Team members can change, companies can be bought or sold, and details of why a decision was made and by whom, along with who was responsible for implementation, can be lost along the way. Meeting documentation serves as a historical account that can be shared with members of both the sponsor and vendor organizations.

The first visit to the vendor is akin to a first date; everyone is highly aware of their behavior and is trying to make a good first impression. Most people get caught up in the excitement of the possibility of doing business and want to believe that the vendor will do "whatever it takes to make it happen." What the sponsor's representatives tend to forget is that there are myriad other vendor representatives at other vendor sites giving the same presentation, trying to get a portion of the sponsor's business. If every vendor representative got every contract it sought, it is highly probable that it could not meet the demand. An attempt to determine the true capacity of the vendor early in the relationship is a valuable piece of data for the sponsor.

Forgetting that it's your company's money that's being spent is a common practice. Once the relationship is established and trust has a foothold in the mind of the sponsor, there is a tendency to boilerplate like projects. A thorough and comprehensive review of the initial scope of work is expected; however, it may not be realistic to provide the same level of review at every monitoring visit. By the same token, errors may occur, costs of goods may shift, and vendor capacity may change, and this could easily go unnoticed without a periodic comparison of the initial scope of work against the current capabilities of the vendor.

Additionally, contracts may be negotiated for payment or milestones. This challenges the vendor to keep to the time line and complete the tasks in order. Incentives for completing work early can be included. Penalties for nondeliverables or the cost for nonconformance can also be defined. This provides an incentive for the vendor to reach and maintain an appropriate level of compliance.

At the initiation of a project, the choice of vendor may be limited. A specific piece of customized equipment may be needed during manufacturing, a key intermediate may be available in lab-scale quantities only, or a particular investigator may be the only expert in a narrow therapeutic area. An evaluation of how this may impact the overall outcome of the project should be done in parallel with working with the single-source vendor. Does the vendor have the capability to purchase more customized equipment? Can the intermediate be scaled to the tens of kilos? Is there another domestic or international clinical expert available?

The answers to these questions drive the next issue—of when to switch vendors. The needs of the project may drive the sponsor to run one or more parallel programs with other vendors: one campaign to produce materials for first time in human studies, another to produce phase II clinical trail supplies materials, and a third to determine the impact of the different campaigns on the clinical program. Now what started as one sponsor–vendor relationship for one project has turned into three. Multiply this by the number of projects your company is managing at any given time and you will begin to understand the extent of the impact of such a decision.

Loyalty and trust that is built early in the program may become a rate-limiting issue if the needs of the program outweigh the capability of the vendor to supply the contracted service. If the team relationship is working well, why fix it? Something at the vendor site or sponsor site may force a change. Mergers, acquisitions, or limited financial resources may cause one or both parties to redefine the relationship. The risk is that the data that have been generated to date may not be appropriately transferred back to the sponsor. The benefit is that the partnership between the two companies may grow stronger.

Failing to periodically authenticate raw data provided by the supplier during the course of the project can create horrendous compliance problems. Similarly, failure to maintain ownership over intellectual property once the project is completed can have a disastrous effect. This issue is best determined early on and acknowledged in a written agreement between the vendor and sponsor.

There are early warning signs that the partnership between the sponsor and vendor may be changing. A continued inability to meet deadlines, incomplete reports to the sponsor, and modification of the financial terms in the favor of cash flow to the vendor without increased services to the sponsor are examples of such warning signs. When dealt with quickly, these issues generally do not recur. If left to fate, they can destroy the partnership and put projects in jeopardy.

Sponsors set deadlines based on their individual needs. Vendors establish deadlines based on competing priorities and the needs of multiple customers. A sponsor's daily needs do not necessarily have an impact on the vendor's daily prioritization process. Sponsors tend to have the false perception that they are the only ones with a deadline. Vendors are managing deadlines for sometimes hundreds or thousands of clients. Knowing how the vendor manages the preparation, execution, and completion of the scope of work is an important tool for the sponsor. It can help the sponsor help the vendor by assuring that the appropriate issues are being discussed and dealt with at the appropriate time. It behooves the sponsor to become familiar with the vendor's overall operation and prioritization processes. This can only occur if the sponsor takes the time to listen and understand the supplier's concerns and capabilities.

Science drives decisions and shareholder value drives time lines. These issues generally have harmonious outcomes. Sometimes the time line becomes the main focus. This can cause stress and confusion and create "artificial" time constraints on a project. Sponsors can then transfer this new time line to the vendor. Vendors that hurry to meet a deadline and are subsequently asked to delay delivery of the final product (and thus wait to get paid) are unlikely to hurry and be as cooperative the next time. Sponsors need to know what is negotiable at the vendor site and what is not. An accelerated stability protocol that is scheduled to be under test conditions for 12 weeks simply cannot be completed in less time. If the sponsor is truly in need of an expedited event, it behooves the sponsor to have established a history of integrity with the vendor in order to enlist the vendor's full cooperation and understanding; then the vendor knows it truly is a "real" deadline and will generally do anything within reason and regulation to help the sponsor.

When faced with a new time line challenge, the sponsor sometimes forgets the most obvious of potential solutions. Ask the vendor for advice. The sponsor typically brainstorms within its own organization and develops a solution that is best for itself, forgetting that there is another organization involved in bringing the solution to reality. It is very likely that the vendor has other customers that have experienced the same types of challenges, and without breaching confidentiality may be able to suggest a feasible solution for the sponsor. Remember that the vendor is an extension of the sponsor's company and whenever possible should be included in brainstorming sessions and any decision-making process that relates to that vendor–sponsor contract.

Ideally, the sponsor chose the vendor because it offers expertise in a particular field. This is often forgotten when the stress level increases. All vendors prefer to be asked what is possible rather than be told what to do. The politics of the relationship between the vendor and the supplier must be carefully examined and managed by thoughtfully selected individuals within the sponsor organization.

Underestimating the amount of time and effort it takes to monitor the vendor is most common with inexperienced sponsors. In the age of virtual companies, it is understood that the sponsor is purchasing the talents and assistance of the vendor. This means that the sponsor must be an active participant in describing its needs and scope of work and in ensuring long-term conformance with contractual and compliance requirements. The sponsor is more knowledgeable about the product or project than the vendor. Additionally, the sponsor is ultimately liable for any nonconformance to contractual and compliance requirements. As such, the sponsor must be proactive to ensure a successful relationship with and output from the vendor. Sponsors and vendors should take the time and effort to establish and agree upon the rules of the road upfront. Ensure that the partnership includes a formalized and effective change management mechanism. The U.S. Postal Service used to have an advertising pitch during the winter holiday period, "Mail early and mail often. "A successful sponsor–vendor partnership can be characterized as "Communicate early and communicate often."

Failing to understand how systems work within the vendor's organization is almost always guaranteed if a formal vendor qualification visit is not performed by the sponsor's quality organization prior to the start of the initial scope of work. Quality professionals usually have a third-party perspective on how both organizations (sponsor and vendor) typically function. A thorough audit will identify areas of opportunity within the vendor's operation. The timing of the audit will allow for modifications to the scope of work prior to the initiation of activities. This gives the sponsor a chance to assist the vendor in upgrading the quality systems to provide an additional level of compliance, thus ensuring an acceptable level of compliance that meets the sponsor's needs.

8 CGMP REQUIREMENTS FOR VENDOR QUALIFICATION

While there is no specific stipulation in the GMPs for the requirement of a documented procedure defining an organization's vendor qualification program, it is generally understood to be an industry practice and FDA expectation. A comprehensive vendor qualification program should include the purpose, scope, and responsibilities for managing routine audits. Audit management should include the method for tracking the performance of suppliers, maintenance of qualified vendor lists, and corrective action and follow-up requirements, together with defining the frequency of audits, record-keeping requirements for the storage of audit reports, and standards for the performance of quality audits.

Additionally, every vendor qualification program should include provisions for quality planning. This allows the sponsor to plan for audit events across project schedules. Vendor qualification for any project can be divided into two sections, each with three tracks. Each section will contain a track for regulated work products. Good laboratory practice, GCP, and GMP define regulated work products.

The first section is "site qualification." Vendors selected to perform regulated work product will be evaluated for compliance with the appropriate set of regulations. The results of the audit will be reviewed and the need for a "site follow-up visit" will be determined. Site qualification visits are generally performed on a cyclical basis; at least once every 24 months is suggested unless the supplier becomes problematic. A continuous monitoring program is an essential component of a compliant vendor qualification program.

The second section is "site follow-up." If the results of section 1 warrant follow-up, another visit will be made to the vendor during the course of the project. Examples of issues that will usually result in site follow-up include lack of adherence to standard operating procedures, lack of appropriate documentation of training, major renovations to the physical structure of the facility, significant changes to the corporate structure, and inadequate investigation of laboratory and manufacturing deviations. Sponsors may also request follow-up visits if standard operating procedure (SOP) or data integrity questions arise during the course of the study or project. Advanced and intense monitoring programs will integrate these follow-up visits to correspond with appropriate development milestones.

8.1 Section 1 Site Qualification

The *benefit* of performing site qualification is the ability to evaluate the systems the vendor uses to produce regulated work product. Generally, if the systems are well designed the vendor should be capable of delivering regulated work product that meets the appropriate compliance standard. If a systems "gap" is detected in any of the quality systems and the study sponsor needs or wants to use the vendor, the sponsor should request corrective action prior to initiating the scope of work.

The *risk* associated with site qualification is that the systems review is theoretical, not practical. No "real" data can be reviewed prior to initiating the scope of work with a vendor. The systems cannot be adequately tested without "real" data. Nevertheless, the vendor's current operational infrastructure relative to other projects and customers can be evaluated.

8.2 Section 2 Site Follow-Up

Site follow-up visits are scheduled for a variety of reasons.

To monitor site progress with corrective actions identified in site qualification visits
To verify the site is performing in accordance with their SOPs
To verify status of project milestones
To authenticate raw data
To perform technology and analytical methods transfer
To initiate recall or address disaster

The *benefit* of performing site follow-up visits is the ability to evaluate the systems the vendor uses to produce regulated work product in "real time" with data generated for a specific project. By performing these visits in concert with critical program milestones, adjustments that may be needed to bring the regulated work product into compliance can occur in a timely manner.

The *risk* associated with not performing site follow-up visits is that any corrections that may be needed will not occur in a timely manner. Expanding the initial scope of work due to late identification of deficiencies will inevitably delay the project and possibly jeopardize its overall compliance status.

Factors to consider when planning a follow-up visit include the following:

The relationship and experience with vendor
The extent of vendor experience with the sponsor scope of work
The sponsor's regulatory commitments and compliance requirements
The associated risks if project fails (collateral damage)

The quality contract should be reviewed and updated by the sponsor on a routine basis. As projects evolve and the category of the vendor changes, the contract is a dynamic agreement between the vendor and the supplier.

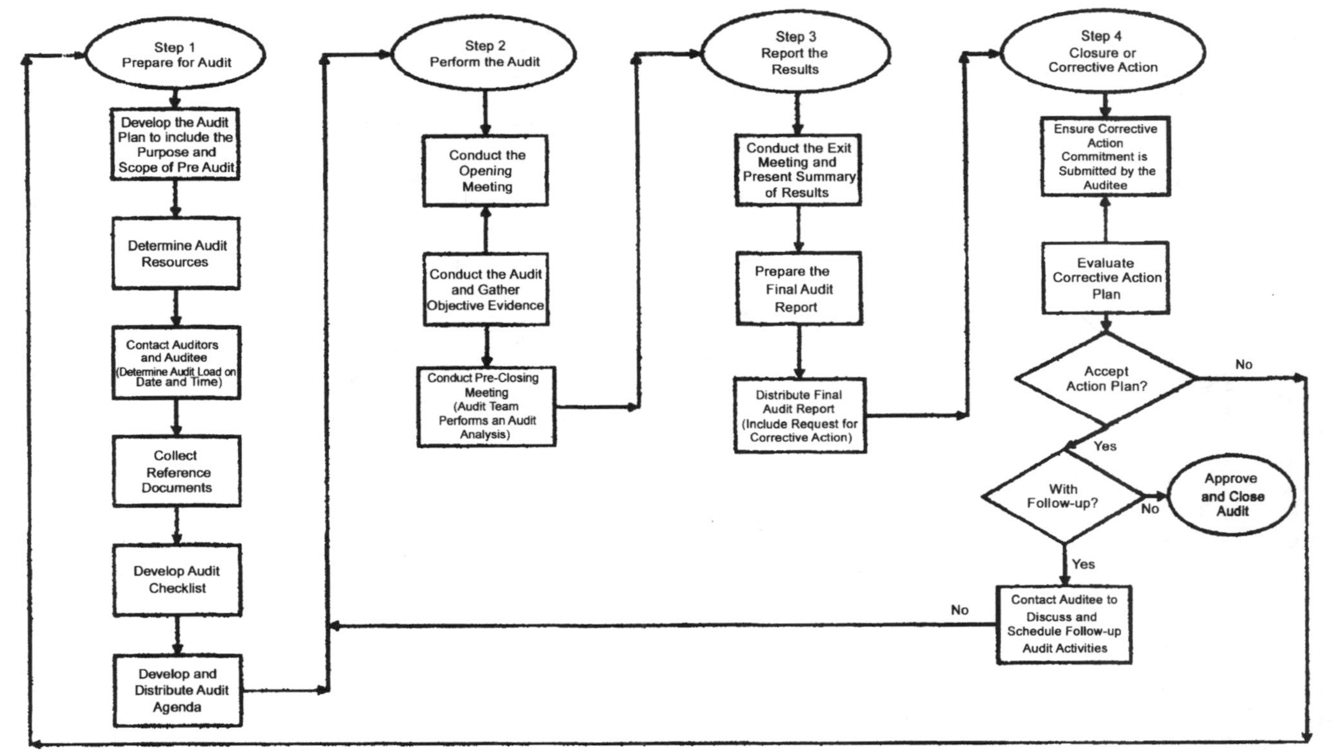

FIGURE 2 The audit process flow diagram.

Standard procedures for conducting the quality audit should also be addressed in the Vendor qualification program. An audit cycle includes the preparation of an audit, performance of the audit, reporting of the results of the audit, and audit closure or follow-up requirements. Figure 2 illustrates the audit process.

8.3 Step1: Preparing for the Audit

The preparation phase of the audit is probably the most critical part of the audit and actually takes longer to complete than performing the actual audit. By putting more time into the preparation phase, the performance of the audit will proceed smoothly. Since this is most likely one of the first sponsor–vendor interactions, a successful audit can begin to form a strong relationship between the two organizations.

An effective audit plan will include the following elements:

Purpose for the audit
Scope of the audit
Resources required of the sponsor to complete the audit
  Number of auditors
  Type of auditor or expertise required
  Assignment of a lead auditor to take responsibility for the audit
Reference documents to be used in planning and performing the audit
  A company organizational chart—to help understand the organizational flow and determine the proper individuals required for interview during the audit
  A floor plan of the organization—to familiarize the audit team with the facility layout and determine what areas to focus on during the audit
  A table of contents for SOPs and/or test methods—to familiarize the audit team with the procedures available for review and plan in advance which procedures to concentrate on during the audit
  Previous FD-483s received by the vendor, along with their corrective action plans. A company dress code for visitors—to determine if there are any special dress needs related to safety or the environment being audited
Checklist(s) to be executed in the course of performing the audit, which may allow the auditor to record the following information:
  The audit response given
  Additional comments and notes
  Reference to substantiating evidence [standard, policy, procedure, or person(s) interviewed]
  Results or verification for activity or work performed

Other information or observations by the auditor
Frequency and timing of activities to be performed during the audit
Audit agendas
Audit agenda—indicate the specific date(s) for the audit, the auditor(s) performing the audit, and the estimated time intervals for each area of the audit, including any time required for lunch or auditor conferences
Opening meeting agenda—introductions, review of audit purpose and scope, review of audit schedule, discussion of audit logistics, including guide for auditors, special clothing requirements, and location for auditors to work
Exit conference agenda—expression of auditor appreciation for the vendor's time, discussion of final report format, and summary of audit findings

Once the audit plan is developed and approved, the audit team leader should formally notify the vendor in writing, typically 2 to 3 weeks prior to conducting the audit.

Formal notification should include a copy of the audit plan, the audit checklist, and a proposed agenda. The audit team leader should also provide a copy of this audit package to the audit team members in preparation for the audit.

8.4 Step 2: Performing the Audit

The performance of an audit begins with an introductory meeting wherein all parties are introduced and the agenda is discussed. It is up to the vendor's senior management to determine the appropriate personnel to be in attendance; however, someone from all management lines being audited must attend. The performance of an audit is the collection of evidence and verification of information through the following means:

Interviews
Examination of documents
Direct observation of activities, processes, and conditions
Review of raw data related to critical documents

It is recommended that the auditors conduct a private conference to collate their findings and prepare a summary of the audit findings. The audit process should entail daily debriefings between the sponsor and vendor as well as a comprehensive exit conference focused on summarizing audit findings and suggested time line for responses of corrective actions. The vendor is then responsible for submitting to the sponsor within the agreed-upon time frame the details and responsible parties for the corrective actions to be implemented.

8.5 Step 3: Reporting the Results

An exit conference is generally conducted to allow the auditors an opportunity to present a summary of the results or findings of the audit. This includes both positive findings and any areas identified for improvement. Providing a written summary prior to leaving the vendor's facility allows the vendor to begin working on the corrective action plans.

The formal audit report is the product of the audit. The lead auditor is responsible for the report content and accuracy and for submitting the formal report in a timely fashion. After completion of the audit, the work on the formal audit report should begin immediately, while audit details are still fresh. In addition, the longer the audit report is put off, the less interested the vendor will be in pursuing corrective actions. It is good practice to complete the final audit report within 2 weeks from the date of the audit.

Each sponsor should set a policy for the circulation of audit reports internally and externally (to the vendor), as well as for the availability of vendor audit reports for review by regulatory agencies. These policies should be defined in their vendor qualification program SOP. While a sponsor must show proof that the vendor audits are being completed with due diligence, there is no formal regulatory requirement to share the audit results with the vendor or the regulatory bodies. Limiting audit report circulation is beneficial, because of the sensitive and confidential nature of vendor operations and the danger of these reports falling into unauthorized hands.

It is important to preclude audit reports from regulatory review during inspections since they fall under the auspices of the internal audit program. During an FDA inspection, it is essential to show evidence of the procedures utilized to conduct the initial qualification audits as well as ongoing monitoring audits of vendor sites. This is the extent of the information that needs to be presented. The actual supplier observations, vendor responses, and final audit report do not need to be presented to the FDA investigator. It is vital that the sponsor's audit component of its vendor qualification program is sufficiently comprehensive. In the event that the audit component is lacking and does not lend credibility that the sponsor has been diligent, the FDA has the right to subpena the vendor audit reports.

8.6 Step 4: Audit Closure, Corrective Action, and Follow-Up

As a result of the audit, the vendor is responsible for developing a corrective action plan to address any weaknesses or deficiencies identified by the sponsor's auditors. It is the sponsor's responsibility to ensure that the corrective actions are effective and implemented in a timely manner. The sponsor must verify all commitments through the course of routine monitoring.

9 REGULATORY REALITY CHECK

With the chapter providing an instructive backdrop, several FDA observations (FD-483s) documented during recent inspections of FDA-regulated facilities are presented below, followed by a strategy for resolution and follow-up corrective and preventative actions.

9.1 FDA Warning Letter Citation

> The firm failed to document how each specific supplier is evaluated in order to assure their products meet specified requirements as required by their Vendor/Supplier Qualification SOP. Additionally, the firm failed to complete adequate purchasing controls in that they fail to ensure that the initial audit of suppliers is conducted prior to the supplier's acceptance, firm fails to assure components meet specifications (physical specifications), and fails to document in-process rejects.

The appropriate response to this observation would be for the firm to revisit its supplier qualification policy and procedures to ensure that they meet current regulatory requirements as well as internal quality standards. Second, the firm will present the agency with a package of new and improved procedures and an umbrella policy related to a state-of-the-art vendor qualification program. These procedures will ensure that all critical suppliers are audited and monitored appropriately in accordance with predetermined criteria and audit protocols. Audit protocols will be designed to meet the unique aspects of every supplier.

The vendor qualification policy will ensure that vendor qualification activities are performed under the auspices of quality assurance. Additionally, a thorough review of the vendor qualification ownership will be performed, and additional resources will be assigned, if needed. An expanded training program will be developed and implemented to ensure that all staff associated with the vendor qualification program are trained on the new and improved policy and procedures. Verification and monitoring of conformance with the company's vendor qualification policy and procedures will be performed periodically through internal company audits.

One way this company could have prevented receiving this citation would have been to assess the adequacy of the company vendor qualification program prior to the inspection. This could have been done through the use of a third-party expert with experience in the area of vendor qualifications or through the company's own quality services group. A formalized vendor qualification enhancement plan that could have been provided to the investigators at the time this deficiency was noted would likely have prevented this observation from becoming a warning letter citation.

9.2 FD-483 Citation

> The vendor qualification program is inadequate in that it does not specify that a container from each lot of active drug substance must be tested to substantiate both lot uniformity and true representation of random sample.

A possible response to this observation would be for the firm to specify the extent of testing to be conducted for all materials received from each supplier. This is achieved by establishing concise specifications and criteria for components, raw materials, services, and supplies up front during the initial identification and selection of the vendor. Additionally, the sponsor must have a mechanism by which it can consistently monitor and assess the conformance of material to the pre-established criteria and specifications. After establishing the vendor's capability to meet the sponsor's contractual and compliance requirements, periodic monitoring audits are necessary to ensure the vendor maintains a qualified status. Testing all incoming materials from all suppliers is covered in the company's receipt and release criteria and specifications procedures. These activities are all delineated in the firm's umbrella policy as well as corresponding procedures related to vendor qualification and ongoing testing.

One way this company could have prevented receiving this citation would have been to have performed an extensive assessment of the vendor qualification program along with other quality systems, such as receipt and release of raw materials, role and responsibilities of quality assurance, and quality control testing of all raw materials. This would have given the firm an opportunity to uncover deficiencies prior to the FDA inspection, along with other deficiencies related to the receipt, testing, and release of all incoming materials.

10 CONCLUSION

In the final analysis, it is apparent that a successful vendor qualification program requires extensive cooperation from both the vendor and the sponsor. Attention to detail and being aware of potential pitfalls will help both parties anticipate and intercept problems before they occur. In summary, you want to assure that these points are covered.

11 WORDS OF WISDOM

Establish the rules of engagement at the initiation of the contract. (This can be done through quality agreements, technical contracts, etc.)
Sponsors and vendors must communicate early and often.
Never overestimate the capabilities of the vendor.
Include the vendor in critical path decision-making processes.
Don't be afraid to switch vendors when contractual and compliance requirements are not being met.
Remember, you are usually not the vendor's only client.
Set realistic time lines and always review time lines with the vendor.
Ask the vendor what is possible rather than tell it what to do.
Attempt to establish a strategic partnership will all critical vendors.
Periodically authenticate raw data provided by the vendor.
Maintain ownership of intellectual property.

11

Handling Laboratory and Manufacturing Deviations

Robert B. Kirsch
R. B. Kirsch Consulting, Arlington Heights, Illinois, U.S.A.

1 BACKGROUND

The occurrence of deviations in pharmaceutical production and analytical testing processes has been a fact of life for as long as materials have been mixed for the purpose of producing a product having medicinal properties. It has only been over the past 40 years or so, however, that both the U.S. pharmaceutical industry and its government regulators have heightened their interest in the accuracy and consistency of production and analytical techniques and technologies used to manufacture and test a drug product, biological product, or medical device prior to its distribution to the public.

Prior to 1960, a lack of sophistication or weaknesses in areas such as the following combined to provide neither regulatory nor judicial motivations to investigate deviations in a careful, prompt, and comprehensive manner:

- Analytical testing techniques and technologies
- Consistent and robust process technologies and formulation methods
- Quality and consistency in regulatory practices and definitions
- Good documentation practices
- Well-defined judicial interpretation of good manufacturing practices (GMPs)
- Compendial positions on GLPs

Moreover, investigations into process and analytical deviations were further impeded by the state of process and analytical techniques and technologies.

After 1960, the ability to detect and resolve manufacturing deviations with some degree of confidence gradually began to improve. To better appreciate the state of current practices regarding investigations into manufacturing and laboratory deviations, a contrast with past technical, regulatory, judicial, legislative, and compendial practices would be instructive. To better understand the extent and manner to which these practices have been refined and strengthened, several processes should be examined.

The first process is the development of new—and refinement of existing—analytical methodology and instrumentation in industry, academic, and government laboratories. The trends in analytical laboratories over the past 40 years have primarily been driven by the need to improve the ability to accurately determine the strength (potency and activity) and purity of the active pharmaceutical ingredient (API) in the drug substance or drug product. Furthermore, a monumental effort has been invested in seeking to reduce the level of analytical detection to enhance analytical system *sensitivity.* Finally, there has been an effort to increase method *selectivity* and *specificity* by filtering out potential interferences normally found in the substance or product matrix that could bias the active's response and compromise the accuracy of the test. Compounds such as impurities, degradation products, excipients, diluents, metals, and extraneous contaminants could potentially interfere with the test results and impact accuracy.

Continual changes in the areas of spectrophotometric optics and electronics, chromatographic methods, instrumentation and columns, thermal analysis and microscopic methods, and especially computerized data acquisition and analysis systems proceeded to drive accurate detection and quantitation of API and related substances to lower and lower levels. As the ability to resolve and accurately quantify active ingredients, impurities, excipients, solvents, soon became possible at increasingly lower levels, the ability to establish more reliable and meaningful specification limits became feasible. As confidence grew in these new specification ranges, industry's ability to investigate, diagnose, and resolve deviations from expected results also improved. In addition to innovation in analytical, microbiological, and physical testing techniques and technologies, the ability to manufacture a drug substance or product more consistently and with higher quality also became more refined through development of new (and optimization of existing) process technology. Examples of improved process technologies are the fluid-bed dryer, the high-shear mixer, automated sterile filling lines, better control over micronization techniques for solid API, and drug products, and better filtration technology and other products for aseptic and terminal processing of parenteral products. Major inroads were also made in the areas of API processing and purification techniques, allowing the production of higher-quality active ingredients and excipients, which resulted in a lower adverse reaction and side-effect spectrum.

1.1 Regulatory and Legislative Processes

The history of regulatory and legislative response to manufacturing and laboratory deviations can be illustrated by examining the evolution of GMPs in production and laboratory environments as well as through technical guidances put forward to address certain key issues in more detail.

The GMPs were first published in the June 1963 *Federal Register* (28 FR 6385) and represented the first genuine attempt by the U.S. government to regulate pharmaceutical production, testing, packaging, labeling, storage, and distribution of pharmaceutical goods, as well as the facility, environmental, and engineering aspects surrounding these manufacturing activities. The prefix *C* (current) has more recently been associated with GMPs to denote the nature of the regulations not as absolute, but as "living" regulations that are evolving and under continual refinement.

The CGMPs are based on the following fundamental *Quality Assurance* concepts [1]:

1. Quality, safety, and effectiveness must be designed and built into a pharmaceutical product.
2. Quality, cannot be inspected and tested into a finished product.
3. Each step of the manufacturing process must be controlled to maximize the likelihood that the finished product will be acceptable.

The *C* in CGMP is significant in that it reflects the periodic need to reassess and revise these GMP regulations to incorporate advances in scientific knowledge and technology and enhance safeguards of the drug manufacturing process. As pharmaceutical science and technology evolves, so does understanding of critical material, equipment, and process variables that must be defined, redefined, and controlled to ensure that finished product uniformity and homogeneity remain within specified limits. The CGMPs must also be reassessed occassionally to identify and eliminate obsolete provisions that are inconsistent with the standards of quality control (QC) and quality assurance that current technology dictates.

The first version of the GMPs contained only the most general and indirect references in responding to deviations found in the manufacturing and testing process. The most pertinent section states that: "A drug will be considered adulterated if it does not comply with established and approved specifications for the product under test."

Revisions to the CGMPs (21 CFR 210 & 211) [2] have been issued through the *Federal Registers* of January 15, 1971 (36 FR 601)[3] September 29, 1978 (43 FR 45014), and most recently and significantly from a laboratory deviation perspective, in January 1, 1995 (60 FR 4087). This last revision contains proposed revisions to 21 CFR 211.192. If the most recent revision is approved, it will clarify and significantly enchance regulatory requirements impacting response to discovery of laboratory results that are out of compliance with approved product specifications. Prior to 1995, deviations were addressed in an indirect and somewhat ambiguous fashion. Although not specifically mentioned in the CGMP regulations, materials used in clinical studies (CTM—clinical trial materials—given to human subjects as part of drug development to assess safety and efficacy) are also expected to meet the requirements of the CGMPs, and thus laboratory results are treated with the same degree of scrutiny as the results associated with products that have already been marketed.

The proposed changes made in the 1995 *Federal Register* were the first to describe in some detail appropriate responses to out-of-specification (OOS) results, as well as within-specification but atypical or unexpected results. This proposal, which at press time has not been incorporated into the most current CGMPs, includes the following relevant passages:

It defines a new term, OOS or out-of-specification result, which refers to a result obtained for the pharmaceutical material or drug product that does not comply with the regulatory specification for the particular test performed. Previously, FDA inspectors would prematurely term OOS laboratory results as product "failures," which clearly was not established without proper laboratory and possibly more extensive investigational activities outside the laboratory environment.

It seeks to update 21 CFR 211.103 to clarify and further reinforce that there must be a written production and control procedure requiring an investigation of any significant, unexplained discrepancy between actual product yield and the percentage of theoretical yield expected for the drug product. The purpose of this clause is to ensure that the source of any potential problem, either in the laboratory or in production, is quickly and accurately identified, investigated, and resolved.

At this point it is worthwhile to digress from regulatory chronology and discuss a specific judicial event and subsequent decision rendered, which arguably has had the single greatest impact on how the pharmaceutical industry and its various regulators (internal and external) respond to a suspected batch failure.

On February 5, 1993, and with further amendment on March 30, 1993, District Court Judge Julius Wolin rendered a decision in the U.S. district court in New Jersey in the case of *United States of America, Plaintiff* v. *Barr Laboratories, Inc., et al., Defendants.* The case, better known as *U.S.* v. *Barr,* resulted in the promulgation of the only definitive judicial decision to date on CGMP issues relating directly to both FDA and industry response to a manufacturing and/or laboratory deviation. Although Judge Wolin's decision is only binding between the U.S. judiciary and Barr Laboratories, its implications have become much more widespread, since its unprecedented analysis of just what CGMP regulations require (and do not require) has been essentially adopted by FDA in various inspection guidelines [4] and further refined in a draft guidance for industry issued in September, 1998 in *Investigating Out-of-Specification (OOS) Test Results for Pharmaceutical Producion* [5].

Some critical points made in the *U.S.* v. *Barr* decision are paraphrased below.

Batch "failure" of the drug product does not necessarily occur when an individual test result does not meet the specifications outlined in the United States Pharmacopeia (USP), manufacturers' new drug application (NDA), or abbreviated new drug application (ANDA). Additionally, an OOS result identified through a thorough laboratory investigation as laboratory error is *not* necessarily a batch failure.

When testing produces a single OOS result, the court required a laboratory investigation composed of more than just retesting. Instead, the analyst and his or her supervisor should take a systematic approach to evaluating all aspects of the analysis, ideally using a checklist to ensure that the investigation is carried out in the most thorough and consistent manner. The results of the laboratory investigation should be preserved in an investigation report and stored in a central file.

Problems more serious than a single OOS result, such as multiple OOS results, product mix-ups, and contamination, require full-scale "formal" investigations involving QC and quality assurance personnel in addition to laboratory and production workers in order to identify exact nonprocess or process-related errors.

Formal investigations extending beyond the laboratory should include the following elements:

Reason for investigation

Summation of events that may have caused the problem and an outline of the corrective actions necessary to save the batch (if possible) or prevent similar recurrence.

List of other batches and products possibly affected, the results of their investigations, and any required corrective actions

The comments and signatures of all production and quality personnel participating in the investigation and who may have approved any reprocessed material following additional testing

Retesting OOS drug product is appropriate only after the laboratory investigation is underway, since the results of the investigation itself in part determine when retesting should occur.

Since it may at times be difficult to pin down the exact cause of the OOS, it is *unrealistic* to expect that the cause of the analyst's error can always be determined and documented.

Retesting the OOS drug product is necessary in cases in which the laboratory investigation indicates that analyst error caused the initial OOS result; retesting is similarly acceptable where a review of the analyst's work is inconclusive.

Whether retesting of drug product is performed at the finished product or blend stage, such testing is to be performed on the same bottle of tablets or capsules in the same drum or mixer, respectively.

If results from the laboratory investigation are inconclusive, the number of retests performed before a firm concludes that an unexplained OOS result is invalid or that the product is unacceptable is a matter of scientific judgment. According to a government witness, such a conclusion cannot be based on three of four or five of six passing results, but possibly seven or eight. The USP accepts the use of retesting for QC purposes, but does not prescribe or recommend the number of individual tests that must be performed in order to reach a definitive conclusion about the quality of the product. The goal of retesting is nonetheless clear; firms must do sufficient retesting to attempt to *isolate and overcome* the OOS result. Retesting cannot continue ad infinitum, however. A firm must have a *predetermined plan* or procedure that stipulates at what point testing ends and a decision is made on the product's disposition. This is the time at which the batch is failed and rejected if the results are not satisfactory. Further testing beyond this point is scientifically unsound and can be interpreted as attempting to "test the product into compliance."

When evaluating retest results from the OOS drug product, it is important to consider them in context of the overall record of the product. Relevant to review are the history of the product, the types of tests performed, and any results obtained for the batch at any other stages of manufacturing and testing. As such, retesting determinations will vary on a case-by-case basis. An inflexible retesting rule designed to be applied in every circumstance thus is *inappropriate*.

With respect to outlier testing, the USP specifically warns against its frequent usage. The USP allows firms to apply outlier testing for microbiological or biological testing due to intrinsic assay variability, but is silent on its use for chemical assays. In the view of the court, the silence of USP on the use of outlier testing to evaluate chemical assay results is *prohibitive*.

Resampling the drug product after the initial test and/or retesting is only appropriate where indicated by USP, as in cases of content uniformity and dissolution testing. Similarly, in limited circumstances in which the laboratory investigation suggests that samples are not representative of the batch under test either because they are adulterated or were obtained using a nonvalidated sampling plan. Evidence, not mere suspicion, must support a resample designed to rule out preparation error in the initial sample producing the OOS result.

Although averaging test data can be a rational and valid approach, as a general rule, firms should avoid the practice, as it may hide variability among individual results. Use of averaging is especially troubling if both passing and OOS results are averaged. Relying on only the average figure without examining and attempting to explain the OOS results is highly misleading and unacceptable.

The elements of a thorough investigation necessarily will vary with the nature of the problem identified. All failure investigations must be performed promptly, however within 30 business days (6 weeks) of the problem's occurrence—and recorded in written investigation or failure reports.

It is clear from these paraphrased *U.S.* v. *Barr* excerpts that Judge Wolin's decision brought far more structure, clarity, and emphasis to issues surrounding manufacturing deviations. Both the industry and the associated regulatory community could now proceed to establish more meaningful processes and procedures in response to manufacturing and laboratory deviations.

In the years following the Wolin decision, the judicial tenets were closely evaluated for reasonableness by placing many of them into practice and judging their practicality. Although many of the principles discussed in the Barr case were shown to be essentially sound when evaluated following implementation, many of the issues required refinement to better accommodate an industry that was now more highly informed and technically capable of dealing effectively with deviation investigations. In other words, the "spirit" of the Wolin decision was upheld by industry practice and FDA review, but the "letter" of the decision was better defined and refined by a more compliance-minded industry.

For example, the number of retests required to isolate an OOS result as invalid or confirm the result as accurate (erroneously ascribed to Judge Wolin as seven passing results out of eight retests) was given a great deal of consideration. First, the seven of eight response was actually given as a "best guess" by an FDA inspector [6] serving as government witness at the trial when queried by Judge Wolin as to the appropriate number of retests needed to isolate the OOS result as invalid (and was presumably not a result of a significant amount of statistical and/or scientific study). Second, when the contributions to a meaningful retest value, such as those derived from analytical and process variability, have been considered using appropriate scientific and statistical tools, the actual number of retests have often been found to be less than suggested in the Wolin decision.

Another example of additional refinement upon reflection has been the use of *outlier* testing in judging the results of chemical testing. While the Wolin decision took the position that USP's silence on this issue with respect to chemical testing is *prohibitory*, upon further consideration, the practice was and is still employed, but far less frequently and only as one piece of information in a more thorough and technically sound OOS investigation. For many individuals involved in making industry policy regarding laboratory or manufacturing deviations, outlier testing remains a useful practice, especially when dealing with a highly homogeneous sample along with test methods demonstrating low analytical variability. Since the outlier test is a statistical evaluation only, however, and thus provides no additional analytical or chemical information, it is more commonly used along with a "preponderance" of other investigational results to identify an OOS result as aberrant and thus invalidate and eliminate the value. Because of its perceived overuse prior to the *U.S.* v. *Barr* case, the statistical test is used only infrequently. According to an industry survey recently conducted by the author, most firms have not entirely banned the use of a statistical test since under the right circumstances it can still be legitimately used to evaluate results leading to final batch disposition.

From the period from 1993 to 1996, the industry prepared and implemented various procedures for investigation and resolution of OOS and other atypical or unexpected results, many of which were diverse in strategy and took different perspectives, but were nonetheless each acceptable in handling and properly resolving these events.

The fact that many aspects of the Wolin decision have been incorporated into FDA's expectations for industry OOS investigations in the QC laboratory can be vividly seen in its *Guide to Inspections of Pharmaceutical*

Quality Control Laboratories [8], published by FDA's Office of Regulatory Affairs (ORA) in July 1993. The guide reinforces such principles as using the term OOS rather than product "failure" to describe a result that does not fall within specified product limits. Investigation techniques are also stressed in the guide, including using a checklist to comprehensively assess the possible causes for the OOS result, following a *predetermined* retesting protocol to assure that the product is not tested into compliance and avoiding the practice of averaging OOS and within-specification results. In the case of microbiological and biological testing in which test variability is usually higher than for chemical testing, however, averaging of OOS results may be a more acceptable practice.

2 RELATIONSHIP BETWEEN THE LABORATORY OOS AND MANUFACTURING DEVIATIONS

In May 1996, the FDA once again proposed to update the CGMPS for drug products by modifying 21 CFR 210 and 211. The proposal appearing in the *Federal Register* [9] contained large portions seeking to amend Sections 211.103 (Calculation of Yield) and 211.192 (Production Record Review) in order to include regulations regarding the proper response to the occurrence of OOS results. The proposed rules would amend Sections 211.103 to codify the need for written production and control procedures to include an investigation for any discrepancies between the actual and theoretical yields of a drug product. The bulk of the relevant changes would be to 211.192 and seeks to amend this section by requiring written procedures, including the following:

1. Procedures for attempting to identify the cause of the failure or discrepancy
2. Criteria for determining whether OOS results were caused by sampling or laboratory error
3. Scientifically sound procedures and criteria for exclusion of any test data found to be invalid due to laboratory or sampling error
4. Scientifically sound procedures and criteria for additional sampling and retesting if necessary during the investigation
5. Procedures for extending the investigation to other batches or products
6. Procedures for review and evaluation of the investigation by the QC unit to ensure a thorough investigation
7. Criteria for final approval or rejection of the batch involved and for other batches potentially involved

The FDA does not intend to issue regulations on specific retesting procedures, but expects that the number of retests performed before a firm concludes that an unexplained OOS result is invalid or that the product is unacceptable be a matter of sound scientific judgment. The proposal does require that the firm have written investigation and retesting procedures in place, applying scientifically sound criteria that limit the amount of retesting permitted, and be specific about the point at which testing ends and the product's integrity is evaluated.

Proposal Section 211.192 requires the written investigation report to include

1. Reason for the investigation
2. Description of the investigation made, including all laboratory tests
3. Results of the investigational testing
4. Scientifically sound and appropriate rationale for excluding any OOS values based on the investigation
5. If laboratory results are found to be invalid, the subsequent laboratory results that support the final products conformity to appropriate specifications for acceptance
6. Conclusions and follow-up (corrective) actions concerning all batches and other products associated with the failure or discrepancy
7. Signatures and dates of the person(s) responsible for approving the record of the investigation
8. Signatures and dates of person(s) responsible for the final decision on the disposition of the batch under study, other related batches, and products involved in the investigation

At press time, the new CGMP proposals have not been officially incorporated into their respective 21 CFR 211 sections.

Finally, in reorganizing that the CGMPs cannot set regulations for OOS investigations in sufficient detail, the FDA has issued a draft, *Guidance for Industry: Investigating Out-of-Specification (OOS) Test Results for Pharmaceutical Production* [10].

As with all other FDA guidances, this document is not binding on industry or the FDA, but represents the agency's current thinking on how to evaluate suspect or OOS test results. The guidance covers laboratory testing performed during the manufacture of API, excipients, and other components, and testing of the finished product to the extent that the CGMP regulations apply. The guidance covers all CGMP testing, including that performed on in-process samples as well as for stability evaluation. Although the guidance is not officially "binding," this, like many other technical guidances put forth by the FDA, takes on the perception of regulatory law, especially in the absence of other FDA-generated regulations relating directly to these issues. The guidance is by far the most specific and comprehensive regulatory document published to date for properly investigating, resolving, and documenting OOS and other unexpected or atypical results.

2.1 FDA OOS Guidance

2.1.1 Critical Compliance Concepts

An investigation must be conducted whenever an OOS result is obtained.

Even if the batch under test is rejected based on an OOS result, an investigation should be conducted to determine if the outcome can be associated with other batches of the same drug product or other products.

To be meaningful, the OOS investigation should be timely, thorough, unbiased, well-documented, and scientifically defensible.

The first phase of the investigation contains the initial assessment of the accuracy of the laboratory data (*before* test solutions are discarded, if possible). Using the original working solutions is an effective way of determining whether analyst error or instrumentation malfunction is the cause of the OOS result. The longer the delay before initiating the investigation the more difficult it becomes to rely on the original test solutions because of their potential instability.

The guidance establishes certain roles and responsibilities for the parties involved in the original analysis and subsequent OOS investigation.

2.1.2 The Analyst

Should be aware of potential problems or subtleties that may occur during analysis that could result in an OOS

Should ensure all instrumentation is calibrated and functioning within acceptable specifications

Should assure that all method system suitability criteria have been met before it is used in the testing

Should not discard and test solutions, vials used for analysis, dilutions, pipets, nor any other analytical tool used during the analysis and which may suggest the cause of an analytical error.

Should immediately inform his or her supervisor of any unexplained OOS result

2.1.3 The Supervisor

Should work along with the analyst to attempt to quickly identify whether the cause of the OOS result is laboratory error or whether a problem has occurred in the manufacturing of the batch under test [i.e., process-related or non-process (operator) -related error]

Should examine all possibilities using a systematic and comprehensive approach (e.g., checklist, method performance history, analyst performance history, trend analysis of product results)

Should compare method performance with validation data to determine if method variability exceeds that obtained under ideal circumstances

Should document and preserve evidence of this assessment

The initial laboratory investigation can include reanalysis of test solutions retained from the original analysis. For certain specialized dosage forms, physical examination of the unit tested may provide evidence that it was damaged prior to the analysis in a way that affected its performance. In addition, further extraction of the dosage unit can determine whether or not the procedure used to remove the active from its matrix was insufficient. The supervisor should be especially alert to developing trends indicating deficiencies in the analytical methodology, inadequate training of the analyst, possible defective instrumentation, or an inability to maintain the system in a controlled, calibrated state (or simply careless work on the part of laboratory personnel).

It should not be assumed that an OOS result was caused by laboratory error without conducting and fully documenting a laboratory investigation. Furthermore, all information gathered during the laboratory investigation phase that concludes that a laboratory error was the source of the OOS, should be taken seriously and addressed in a formal manner. While the product is no longer at risk and a subsequent manufacturing investigation may not be necessary, there is the issue of continuous improvement efforts in the laboratory to respond to laboratory-generated errors and what could potentially turn into a trend. The laboratory is the backbone of any operation, and frequent lab-generated OOS results are a signal that something is amiss in the lab; for example, personnel training may need to be enhanced, laboratory supervisory oversight may need to be increased, laboratory management may need to be examined, equipment and apparatus may need further qualification, the laboratory itself may require more stringent environmental controls, handling of reagents and standards may be inappropriate, and finally and what is frequently the case, method validation may need improvement.

Laboratory investigation reports should be compiled and periodically reviewed by laboratory management and eventually quality assurance if QC management is unsuccessful in reducing the occurrence of laboratory-generated errors.

If the laboratory investigation does not provide a definitive cause for the OOS result, a formal investigation prescribed by a predefined written procedure should commence. The formal investigation should be conducted under the auspices of the quality assurance unit and encompass all other departments that could be "implicated," such as production, process development, engineering, and maintenance.

For the purposes of more clearly differentiating between the laboratory investigation phase and a subsequent formal investigation conducted outside the laboratory, the laboratory investigation will be referred to as *phase I*, while the formal quality assurance-coordinated investigation will be referred to as *phase II*. For the purposes of phase I v. phase II, there is a definite difference, in that QC personnel handles the laboratory functions, while quality assurance personnel monitors and coordinates the overall investigative process.

During the phase II process, all parties (QC, quality assurance, manufacturing, engineering, etc.) will examine their contributions to the batch record and all logs associated with activities carried out during the manufacturing process.

To further investigate the OOS result, the laboratory may (1) retest a portion of the original sample, (2) test one or more specimens taken from a collection of new samples from the batch under study (resample), (3) re-evaluate test data, and (4) use outlier testing to support invalidation of an OOS result. These actions are described below.

2.1.4 Retesting

Ideally, retesting should be performed on units taken from the same homogeneous material that was originally collected from the batch, tested, and yielded the OOS result. Retesting results could indicate an instrumental malfunction or a sample preparation error, such as a weighing or diluting error. Generally, retesting is neither specified nor prohibited by approved applications or the compendium. A decision to retest material from the original collection of laboratory units should be based on the objectives of the testing and guided by sound scientific judgment.

The guidance requires that the retesting be performed by an analyst other than the one who performed the original test. This particular action has been the subject of debate between industry and its regulators. The actual participants in the retesting exercise will be further discussed later in the chapter.

The number of retests to be carried out should be specified in advance (i.e., in a standard operating procedure [SOP] or retest protocol) and may vary, based on the variability of the particular test method employed. The number of retests selected, however, should be based on scientifically sound, supportable principles and should *not* be adjusted based on the results obtained. Statistical factors associated with the analytical method and manufacturing process could play a significant role in assigning a reasonable number of retests. The firm's predetermined testing procedures should stipulate a point of which the testing ends and the product is evaluated. Testing should not continue beyond this point (i.e., testing into compliance is objectionable under the CGMPs). *All* data should be considered in judging whether the OOS result was invalid and the batch is acceptable or whether the OOS is supported by the retest results and the batch must be rejected or held pending further investigation.

For a clearly identified laboratory error (i.e., assignable cause or determinant error found), the retest results would substitute for the original results. The original results, however, should be retained and the explanation recorded in a laboratory investigation report.

In cases in which there is no laboratory error clearly identified, there is no scientific basis for invalidation of the original test results. The guidance recommends that all results, both passing and suspect, be reported and considered in batch release decisions.

2.1.5 Resampling

Following evaluation of all data from the investigation, it may be concluded that the original sample delivered to the lab was prepared improperly and was therefore not representative of the batch. In this case, a resampling of the batch should be performed (i.e., collection of a new sample from the batch) using the same qualified, validated sampling method (or methods) that was used to collect the initial sample. If the investigation determines that the initial sampling method was in error, however, a new, accurate sampling method should be developed, qualified, and documented. Further, if the initial sample given to the laboratory for analysis has become in some way adulterated (e.g., through breakage or exposure to heat, light, or moisture) or was exhausted in the testing process, it may be acceptable to resample the batch under the same constraints as described earlier.

2.1.6 Averaging

Averaging of results can be a valid approach, but its use depends upon the sample and its purpose. If a sample is homogeneous (i.e., an individual sample preparation is homogenized), the use of averages can provide a more accurate result. Because of the inherent variability of the biological test system, the USP prefers an averaging of results for microbiological assays. Since averaging can hide variability among test results, this practice should only be used for chemical tests if specified by the test method or in a formal investigation procedure. In virtually all cases, *all individual values should be reported*. The values could also include the average as an additional piece of information. In some cases, as with evaluating dose uniformity, a statistical treatment (such as standard or relative standard deviation) is also reported.

In the case of replicate measurements taken from a single sample preparation (i.e., replicate injections from a single sample solution), all replicate results should be averaged into a single analytical response. The determination is considered one test and one result. Unexpected variability in replicate determinations, however, should also trigger an investigation followed by documentation of the findings.

The guidance states that it is not an acceptable practice to average OOS results with passing results even though all data are within the variability of the analytical method. The passing results should be given no more credence than the failing results in the absence of documented evidence that an analytical error has occurred. To use averaged results for assay reporting all test results should conform to specifications. One or more OOS results should suggest the possibility that the batch is formulated to contain too little active ingredient. Since CGMPs require the batch to be formulated to not less than 100% of the labeled or established amount of active ingredient [21 CFR 211.101(a)], averaging low-end OOS with within-spec results may be statistically sound action but not acceptable from practical and compliance standpoints.

2.1.7 Concluding the Investigation

To conclude the investigation, the results should be evaluated in light of all information gathered about the batch and analytical testing thus far, the batch quality should be determined, and a release decision should be made. The investigation SOP should describe all activities in detail and should be followed up to this point. Once the batch has been rejected, there is no limit to further testing to determine the cause of the failure. This is critical if a corrective action and preventative measures are to be developed and implemented.

If an OOS result is confirmed by the investigation and no operator-related manufacturing error can be identified, the result should be used in evaluating the quality of the batch. A confirmed OOS result indicates that the batch does not conform to established, predetermined specifications or standards and should result in batch rejection and proper disposal.

For inconclusive investigations in which (1) an assignable cause is not revealed for the OOS result and (2) the OOS result is not confirmed, the OOS result should be retained in the batch record and given full consideration in batch or lot disposition decisions.

In very rare occasions and only if a full investigation has failed to reveal the cause of the OOS result, a statistical analysis (such as an outlier test) may prove valuable as one assessment of the probability of the OOS result as discordant and for providing a perspective on the result in the overall evaluation of the quality of the batch.

Although not specifically stated in the FDA guidance, unconfirmed and unexplained OOS results should trigger a responsible and formalized plan from the quality assurance unit, in conjunction with manufacturing personnel and QC analysts, to increase monitoring of manufacturing and analytical activities for a defined period. Perhaps over the course of the next 3 months or five batches additional vigilance or testing will be imposed. A series of unconfirmed OOS results should lead to a number of quality assurance interventions, the least of which is revisiting the process and method validation activities.

2.1.8 Reporting

Records must be kept of all data derived from tests performed to ensure compliance with established specifications and standards.

For products that are the subject of (approved) applications, the guidance also requires within 3 working days the submission of a field alert report (FAR) to FDA containing information concerning the failure of a distributed batch to meet any of the specifications established in the application.

A formal report containing all information relating to the investigation of the OOS result and the corresponding batch, as well as other batches or lots that have been deemed to be related in this regard, should be reviewed by all investigation participants and approved by the quality assurance unit.

These investigation reports are accessible during FDA inspections and are typically requested in conjunction with the investigator's review of manufacturing batch records. Investigation reports become a central focus during both preapproval inspections and investigator-directed inspections. The OOS investigations should be maintained by the quality assurance unit and periodically trended as part of the overall and ongoing quality assurance program.

3 SOURCES OF OOS AND OTHER UNEXPECTED EVENTS

We've thus far seen how both industry and government interest in the investigation of aberrant results has evolved in the recent past. The next question that could be asked is just how OOS and other unexplained results arise. The following sections will attempt to shed some light on possible causes.

There are myriad sources from which OOS results can be derived. The sources can be found in just about any of the processes associated with drug substance or product production—sampling, testing, packaging, storage, stability assessment, and distribution. Although a comprehensive treatment of all potential sources goes beyond the scope of this chapter, some of the most common sources resulting in OOS and other types of aberrant results are described the following section.

3.1 Selection of Proper Performance Specifications

According to 21 CFR 211.160(b), "Laboratory controls shall include the establishment of *scientifically sound and appropriate specifications*, standards, sampling plans, and test procedures designed to assure that components, drug product containers, closures, in-process materials, labeling, and drug products conform to appropriate standards of identity, strength, quality, and purity."

The specifications selected for a process should permit the determination of whether or not it is in a state of control or more accurately, statistical control. The specifications are values or limits that are represented by the *upper specification limits* (USL) and or *lower specification limits* (LSL) chosen for inclusion in a process control chart. The control chart is a plot on which specific process performance attributes for a given product (i.e., series of batches or lots) are plotted and monitored over time. This continuous charting of critical chemical, physical, and microbiological performance characteristics is conducted to detect developing trends that lead to either (1) a need to change the process to force a realignment tot initial control limits or (2) a need to re-evaluate and perhaps change existing specifications.

This control chart frequently contains a central line representing the desired optimal performance level (e.g., 100% of label claim of the active ingredient), a USL, and/or an LSL. The USL and LSL are based on the desired upper and lower boundaries beyond which a particular measurement should not pass.

A common parameter used by quality professionals to gauge whether or not the manufacturing process and analytical methodology is within the desired state of statistical process control (SPC) is a process capability index [11], such as the C_{pk} value. The C_{pk} value is the value of the tolerance specified for the given performance characteristic divided by the actual process capability.

Most if not all variability associated with a particular performance characteristic, such as *potency, purity, dosage uniformity, or dissolution rate*, is a direct result of the production process variability (produced by variance in process parameters) ***plus*** the precision or variability associated with the analytical method used to measure the characteristic. The only means of controlling the type and extent of overall process variability (i.e., variability associated with production process and analytical testing) is through familiarity with each potential or real source and practical ways to either eliminate or minimize these sources. The more thorough the understanding and origin of each potential cause of variability, the more confident one will be in establishing meaningful specifications.

The C_{pk} value is frequently used during later development phases or postapproval to determine the suitability of the specification value or limits. The value is defined as follows:

C_{pk} = the lesser result of the following two equations:

$$\frac{UCL - \mu}{3\sigma} \text{ or } \frac{\mu - LCL}{3\sigma}$$

where

μ = True mean value of all measurements to be evaluated (estimated by X bar).

σ = Standard deviation of the chosen measurements (estimated by S).

Ideally, the difference between the USL or LSL and the desired result (midline) should approximate 3σ for a normally distributed set of measurements. The 3σ interval encompasses approximately 99% of all possible measurements. The closer the index is to 1.0, the closer the specification approximates the total process variability and the more appropriate it is as a means of controlling this performance characteristic. As the C_{pk} value deviates from 1.0, the limits or specification grow less suitable and appropriate with respect to overall process variability. If $C_{pk} \ll 1.0$, the specification will probably be too stringent for the batch data over time, giving rise to frequent OOS results. These seemingly aberrant results are more a result of misassigned specification limits than a bona fide testing failure. Conversely, if $C_{pk} \gg 1.0$, the specification limits will be too "loose" or tolerant for the process and result in inadequate control of the overall process. Most quality professionals prefer to limit the C_{pk} value to between 1.0 and 1.3. With process capability indices in this range, the chances of OOS results arising from inappropriate specifications are rather remote. Of course the key to successfully setting product specifications is to have gained sufficient familiarity with both process variables and analytical variables during the development phases and early marketing phases to permit a good estimate to be made of overall performance. The specification is only as good as the amount of reliable information available regarding process performance, which basically means the more material produced and tested using consistent process and analytical procedures, the more statistically sound the data used to generate product specifications. If the specifications are invariant and cannot be adjusted based on process performance, the process variability must be reduced by more carefully controlling critical process parameters (e.g., mix time, mix speed, temperature, particle size, compression, and encapsulation rates for solid doses, lyophilization rate, filter size/composition and bulk holding time for parenteral doses, and emulsifying agents for suspensions). Similarly, analytical variability can be reduced through use of automated sample preparation techniques (robotics and better computer data processing) and better control of the measurement technology (e.g., use of an internal standard method for gas chromatography, from packed column to capillary to enhance reproducibility, and better controlled solvent mixing for high performance liquid chromatography [HPLC] gradient systems). This cannot be done using either a single commercial batch or laboratory batches manufactured at too small a scale. It has been the author's experience that establishing specifications that are *not* meaningful and appropriate for the process under study is the single leading cause contributing to OOS results.

3.2 Precision of Analytical Methods

Analytical methodology for testing of drug substances or products should be demonstrated through method validation studies as being suitable for their intended purpose. An integral study in most method validation programs is the assessment of method *precision*. Through the efforts of the International Conference on Harmonisation (ICH), method validation terminology and procedures have been uniformly accepted by regulatory authorities in the United States, the European Union, and Japan. According to ICH guidances on method validation [12,13], three components makeup overall method *precision*: (1) *repeatability*, or the precision under the same operating conditions over a short interval of time (also called interassay precision), (2) *Intermediate precision*, or the measure of within-laboratory variations on different days, with different equipment, by different analysts, and soon (also called ruggedness), and (3) *reproducibility*, or the precision measured between laboratories as used in collaborative studies usually applied to standardization of methodology and technology transfer.

ICH Method Validation—Precision Components

Repeatability	Variability experienced by the same analyst using the same instrumentation, reagents, and supplies in the same laboratory over a short time interval.
Intermediate precision	Within-laboratory variability among analysts, instruments, days, etc.
Reproducibility	Variability measured between laboratories as determined for method/technology transfer or for standardization of methodology via collaborative laboratory studies.

For most analyses, method repeatability and intermediate precision are optimized to within acceptable limits for the particular functional use. Such method performance parameters as precision are evaluated during method validation studies and tested for acceptability, in contrast to acceptance criteria that are developed through familiarity and use of the analytical procedure. In planning the method validation studies, if acceptance criteria for precision studies are too tolerant when compared to the specification limits established for the performance characteristic that is being measured (e.g., potency assay, impurity/degradation product assay, water content, LAL test), OOS results will probably occur frequently (since the validation criteria did not take into account the specification range). If method precision is not compatible with the specified limits for the test attribute, precision should be improved by optimization of the most variable or error-prone steps in the analytical procedure. Usually the single largest source of variability in analytical procedures stems from such sample preparation operations as weighing of small specimens, liquid transfer using improper pipettes, serial dilutions, use of external instead of internal standardization techniques (especially for gas chromatography [GC] applications), solid as well as solvent extractions, and excess heating and sonication. One way of decreasing the variability associated with manual operations is to *automate* one or more of the manual procedures. Analytical precision has successfully been controlled through the use of laboratory robotics and other forms of laboratory automation and computerization. Precision enhancements can also be achieved in the analysis of data or data reduction step(s) by using software that has been optimized for the particular application in use. For chromatographic applications, peak integration settings, optimization of separations through manipulation of flow rate, mobile phase composition, use of peak heights instead of areas, injection volume, analyte concentration, and gradient program control could also contribute to method precision enhancement.

For manufacturing deviations, variability can be reduced by concentrating on *operator-related errors*. Such preventative measures can include but not be limited to

A better understanding of the process and equipment used to carry out production activities
Assurance that CGMPs are well understood and properly implemented
Confirmation of batch record execution accuracy by a second qualified individual who also observes the operations to ensure correctness
Assurance that the equipment is properly qualified (i.e., installation, operational, and performance qualified), properly maintained, and currently within calibrated performance ranges (emphasis on scales and balances)
Assurance that all operations are carried out in a neat and orderly work area
Assurance that equipment and related components are *clean* before usage
Confirmation of proper ingredients and that the amounts are reasonable and consistent with past production

3.3 Quality of Documentation

Proper production process execution and performance of analytical procedures relies on comprehensive, clearly stated, and unambiguous documentation. This is necessary whether the document is used to initiate an activity (e.g., process, method, work instruction) or contains results that will be reviewed by someone else. A master batch record that can be easily misinterpreted by trained operators is a serious source of process variability. Similarly, if the analytical method is not written in a clear, concise, and sufficiently detailed manner, interpretation may be different from analyst to analyst, also resulting in higher than expected variability and perhaps an OOS result. Process and analytical documentation should be written with the user in mind and authored by individuals with intimate familiarity with the task to be carried out. The written procedures should be concise or "crisp," yet contain sufficient technical detail to lead trained operators and analysts through the same set of operations with the outcome being the same when applied to the same material(s).

Usually familiarity with the tasks that are detailed in analytical and manufacturing documentation helps the author prepare a practical and easily interpretable procedure. Moreover, employing a review and approval process that encompasses a technical, editorial, and/or regulatory review by the ultimate user of the procedure (rather than restricting it to functional area management) helps ensure that the document will serve the audience for which it is intended. Also, the review and approval list should contain only those individuals and functions that have a good understanding of the procedure and whose names are mandated by corporate regulatory and compliance policy. Long, never-ending lists of reviewers, many of whom have only a passing (or no) understanding of the processes described, do not add value to the procedure and usually result in a delay in implementation. Delayed implementation of new procedures causes process and laboratory areas to continue to use outdated and obsolete ones. Using outdated or "wrong" procedures is a potential source of OOS results. Nonspecific and imprecise documentation of an analytical method or manufacturing process is yet another potential source contributing to OOS results.

3.4 Training of Personnel

Another major source of aberrant results is improper or inadequate training of the personnel directly involved in the execution of the manufacturing process or analytical testing. There is a simple cause and effect relationship between inadequate training and frequency of errors made in the performance of any complex task. The more complex the task, the stronger and more direct the relationship. According to the GMPs (21 CFR 211.22)

> Each person engaged in manufacturing, processing, packing, or holding of a drug product shall have education, training and experience or any combination thereof, to enable that person to perform the assigned functions. Training shall be in the particular operations that the employee performs and in the current good manufacturing practice as they relate to the employee's functions. Training in cGMPs shall be conducted by *qualified* individuals on a continuing basis and with sufficient frequency to assure that the employees remain familiar with cGMP requirements applicable to them.

Many pharmaceutical firms have implemented operator and analyst qualification programs in which personnel are trained in the specific techniques, technologies, and equipment that they use in their work as well as in general proper implementation of CGMP practices in their work environment. There are still firms not fully familiarizing their staff with all aspects of their assigned tasks before they are asked to carry them out, however. This lack of training is not only dangerous in that a substandard (or adulterated) product can result (with all its financial and logistical ramifications), but the employee may be subjecting himself or herself as well as co-workers to an unsafe working environment. Training should take place for every new hire and whenever the individual is asked to perform a new and unfamiliar task. This training should take place by a qualified "trainer" (i.e., someone who not only is familiar with the task, but with a demonstrated proficiency in communicating technical information to others). There should also be some type of assessment of the effectiveness of the training (e.g., a written or oral quiz, practical exam, or period of observation). A written record should be kept for each employee of all training that has been obtained, including the subject, trainer involved, date trained, procedure trained for, along with the results of the training assessment.

3.5 Equipment/Instrumentation Calibration and Qualification

Aberrant results have been caused by using process equipment or analytical instruments or systems that are not performing according to vendor and user performance specifications.

All equipment and instrumentation used in the manufacture of the drug substance or product as well as in the testing of the resulting material should be qualified before use. Each piece of equipment, instrument, or system should be subjected to a documented installation qualification (IQ), tested for proper functionality during an operation qualification (OQ), and applied to the specific application for which it will be used by conducting a performance qualification (PQ). The qualifications will ensure that the systems perform within requirements established by the equipment manufacturer, user, and regulatory authorities. The qualification documentation should be reviewed and approved by the responsible department as well as the quality assurance organization. The equipment should be placed on a regular preventative maintenance (PM) schedule and be assessed for performance (i.e., calibrated) on a regular basis. Assurance of the system's state of acceptability should be checked and documented prior to use. The PM and calibration program should be maintained by personnel familiar with the operation and performance expectations of the equipment (e.g., metrology group, operator, analyst using equipment, and engineering staff).

3.6 Formulation and Manufacturing Process

Process-related deficiencies are another source of aberrant results. Although the manufacturing process should be optimized and refined during the development phase, such events as the following could all result in manufacturing process problems:

Changes to the process following development, perhaps during scale-up or postapproval
Incomplete process validation (meaning that critical process parameters are not adequately challenged)
Inadequate technology transfer of the process between manufacturing sites
Improper selection of ranges for in-process and final product parameters
Poor development of the process (the precursor to all of the above)

The result is that one or more critical process parameter is not under control and will vary from batch to batch. This problem manifests itself in various ways, the most prevalent of which is poor batch-to-batch uniformity or homogeneity of active ingredient(s) and/or other components. In-process control may not be sufficient or evaluating may not be a requirement to subsequent processing. It is not uncommon to find in-process testing (such as for blend uniformity, filter retention of active and excipients, and the amount of water or solvent in a wet granulations) incomplete before the process is advanced to the next or final steps. All too often and in the interest of rapid throughput, material is manufactured at risk and adjustments that could have been made during critical control points in the process are no longer an option. The decision to proceed with the manufacturing process in spite of incomplete in-process testing is based upon the hope that nothing has gone wrong throughout the process and the knowledge that the material will not be released without appropriate finished product testing and quality assurance review.

3.7 Common Culprits Leading to OOS Results

1. Test specification not properly established
2. Poor or ambiguous documentation
3. Inappropriate use of in-process testing results
4. Lack of historical development documentation
5. Ignorance of development activities that were problematic and were suggestive of future OOS results, lack of familiarity with development history.
6. Incomplete or inadequate method of process validation.
7. Poor training of analysts or process operators
8. Deviation in analytical methods and/or production batch records

4 DEVIATIONS AND THE PRODUCT DEVELOPMENT LIFE CYCLE

Deviations in the pharmaceutical laboratory and manufacturing process areas will probably never be entirely eliminated, but the goal of well-planned and-implemented quality systems is to minimize the frequency of their occurrence. If deviations are to be drastically reduced, one must understand their origins and from what sources they are derived. Several major sources of OOS results were described in the previous section.

If one plans to eliminate or dramatically reduce laboratory and manufacturing deviations, however, the first step must be the careful and comprehensive planning and ongoing execution of each phase of the overall product development life cycle (PDLC). If executed properly, the PDLC can be of enormous value in reducing the number of plant or laboratory deviations and when they *do* occur providing a comprehensive product knowledge base from which to draw information to facilitate a rapid and successful investigation, resolution, and implementation of corrective and preventative measures. Comprehensive, accurate, and easily retrievable knowledge of the overall product and process development can rapidly link an aberrant result to an assignable cause or determinant error in the laboratory or in the manufacturing arena (process- or operator-related source).

It is easy to simply state that a well-executed PDLC will reduce or possibly eliminate future OOS results, since in actuality is process almost always occurs over the course of years and involves tens if not hundreds of technical personnel. Maintaining control over such a long and complex path may seem a daunting task, but if each phase is carefully planned, managed, and controlled through a premeditated and well-documented process:

Methods will be developed, optimized, validated, and documented properly to serve the purpose at each point in the development process.

The specifications for drug substance, drug product, packaging components, excipients, and so on will be established based on sound technical justification, using a continuously expanding experience base that includes information on process, product, analytical methods, stability of drug substance and product(s), packaging components, and so on.

Any deviations experienced with the mature product, process, or analytical procedures can be traced to similar experiences in development and rapidly resolved.

4.1 Product Development Life Cycle

The early steps along the drug development pathway, usually referred to as the "discovery" phase, involve the in vitro and in vivo testing associated with the selection of a viable compound to further develop. The activities involved in this process are numerous and entail such activities as high-throughput screening of literally thousands of compounds using conventional synthetic techniques or state-of-the-art combinatorial or calculational techniques [such as quantitative structure activity relationship (QSAR) computer programs].

During this discovery phase, little time is spent pursuing possible deviations since the emphasis is on lead compound identification, in vitro testing, and testing in animal models to assess efficacy and toxicity. The goal is to quickly decide to either proceed or terminate continued development efforts. Analytical testing during this phase is limited to support for development of laboratory or bench-scale processes, leading to production of relatively small amounts of the active pharmaceutical or biological entity sufficient to conduct these investigational studies. Rapid analyses for potency, purity, moisture/solvent content, intrinsic dissolution, and physical testing are developed for the sole purpose of generating small amounts of relatively impure materials to foster screening for efficacy and toxicity in animals. Analytical procedures and manufacturing processes that are developed at this stage are in a state of flux and must be readily adaptable in order to meet development objectives.

Once the decision is made to proceed with the chemical development or "scale-up" of the drug substance, manufacturing process analytical support must also be provided to the modifications necessary to permit the manufacture of larger pilot- or eventually commercial-scale batches of API. Once again, analytical development must be fluid and able to adjust to variable starting materials, stable and unstable intermediates, synthetic by-products, degradation products, inorganic and organic catalysts, buffers, and testing for compounds used during each step of the synthetic process (including solvents). As the process begins to stabilize and mature, manufacturing control parameters and conditions will be evaluated and established. It is *essential* that sufficient time be spent during this phase to gain a thorough understanding of the process practices, techniques, technologies, and equipment and to accumulate information into a comprehensive knowledge base. The process is eventually sufficiently stable to allow analytical methodology to be developed to the point that the methods can be validated to ensure that they are providing accurate and reliable information about the API produced. Methods to accurately assess the drug or biological substance for various critical chemical, biological, microbiological, and physical attributes may be developed, optimized, validated, and documented to provide a comprehensive characterization of the material.

Although much of the chemical development process occurs well before approval of the investigational new drug (IND) application, sufficient experience has been amassed to permit identification and subsequent investigation of seemingly aberrant or unexpected results. The investigations can be conducted in a less formal manner than required under CGMPs. These observations and investigations, however, lead to a better understanding of the overall API process and thus should be documented in detail for future reference. The manner in which these early investigations are conducted and the information is collected and documented serves as a valuable archive to aid and expedite future investigations that will invariably occur later in development and postcommercialization.

At this point the process should be sufficiently stable and mature to validate methods for API, starting materials, stable intermediates, reaction reagents, buffers, catalysts, synthetic by-products, degradation products, and any other major process component for which control is necessary. The production and testing of the API as well as most, if not all, species used in the preceding steps should now be conducted under CGMPs, including properly accounting for any deviation. This is especially true if the API or biological is used in human clinical trials.

As the process and analytical knowledge base builds due to increasing familiarity, process and analytical *standards* and *specifications* are established to ensure that critical process parameters and conditions are properly controlled and adequately maintained. Analytical testing and limits for the starting materials, intermediates, and final drug or biological substance ensure that the process will continually produce material of acceptable quality and performance characteristics. Once the production of commercial quantities of the bulk drug substance is under control and stable, process validation studies challenge these controls and control specifications to ensure that the process will consistently produce a product with the same characteristics and quality attributes. Further, studies are also conducted ensuring that the final API will remain *stable* throughout its manufacturing, storage, distribution, and use.

Although the API production process may still be slightly modified to accommodate such events as transfer to different manufacturing facilities and change of material supplier(s), process and analytical techniques and technologies should by now be well understood and documented.

A comprehensive development knowledge base and manufacturing history should now exist for the API to permit a scientifically sound investigation of an OOS result to be carried out with confidence and in a timely manner. The key factor in both the reduction or elimination of deviations and the rapid resolution through identification of an assignable cause in the laboratory or process area is the thorough understanding of all control points along the critical production and analytical testing pathway.

As drug product formulators begin to incorporate the API into a suitable drug or biological product matrix or matrices for the clinical testing, analytical methods are needed to provide information on the identity, potency, purity, in vitro dissolution (for solid and semisolid dosage forms) uniformity, stability, moisture content, and physical characteristics of the drug substance in the drug product. In many but not all cases, the analytical procedures developed for the API can be employed for the drug product following minor alterations. As the formulations or delivery systems proceed from simple and preliminary (for human dose ranging, efficacy, and toxicity studies) to the more mature and final formulations, and as the product development life cycle (PDLC) proceeds through phases 2 and 3 of human clinical trials, the analytical methods are relied upon by the formulator to feed back information regarding API and dosage form/unit integrity and stability. The latter is a regulatory requirement since the dosage unit must remain within its predefined chemical, physical, biological, and microbiological specifications throughout its usage in human clinical trials.

Through most of phases 2 and 3, CTM are released to the clinic and assessed for stability characteristics using analytical methods that have been validated as being stability-indicating. The production of the clinical supplies, as well as subsequent testing, should take place using proper CGMP procedures. Since CGMPs are in full force during this process, deviations should be investigated using formal, documented procedures that are appropriately comprehensive and very similar to what will be used for the marketed product.

The analytical methods used to test the drug or biological product(s) used from phase 2 forward should be "validatible"; that is, sufficiently accurate, precise, specific, and robust to permit quality release of the clinical supplies against established specifications, as well as provide information on the stability of the formulation under a variety of different environmental conditions. The conditions simulate the possible climates to be encountered by the dosage unit during storage prior to shipment and throughout its chain of commercial distribution and use. The analytical methods for potency, purity, microbiological, and biological activity must be "specific" or stability-indicating. A stability-indicating method is one in which the compound of interest (usually the active ingredient) can be resolved and quantified accurately in the presence of both identified and potential degradation products. The method must also be selective (i.e., be capable of resolving and quantifying the analyte as well as those species also expected to be present in the drug product matrix).

Just as the API manufacturing process requires a thorough evaluation to establish meaningful controls, specifications, and limits, the drug product formulation process must possess analogously sound controls and corresponding specifications. These controls are for various in-process stages as well as for the final drug or biological product. Such parameters and conditions as the order of material addition, mixing time, speed and temperature, particle size and shape, flowability, compressibility, friability, bulk/tap densities, hardness for dry blends, and wet granulations used for solid dosage forms; the solubility, hold-time stability, filterability, and resistance to microbial growth for nonsterile liquids; and the bioburden, preservative efficacy, pyrogen/endotoxin levels, and sterility issues for parenteral products (solutions and lyophilized) should be well characterized and documented.

Just as with the API, it is crucial to the consistency and ongoing reliability of the process that sufficient time and effort be spent in understanding all critical process parameters and conditions and amassing a comprehensive knowledge base early on for the development of the drug or biological product. Only after sufficient experience with the formulation process and the resulting batch production history is collected and integrated can meaningful controls and specifications be placed on various critical process steps. Statistical process control-trending techniques are frequently used for establishing and maintaining meaningful process specifications that are evaluated using validated analytical methodologies. It is critical that scientifically sound and reasonable specifications and limits be established during this development phase to avoid frequent OOS results following the approval and commercialization of the new drug product. As before with the API process, the caliber and extent of the process knowledge base amassed during these development stages have a direct impact on the manner in which investigations of aberrant results can be conducted post-approval.

In order to assess the state of SPC, a sufficient number of pilot- or commercial-scale batches should be manufactured (at least 10% of anticipated commercial scale) using the same process anticipated for the scaled-up product. When the formulation is finalized, the critical in-process and final product control specifications will be challenged through process validation studies on at least three full- or pilot-scale batches.

5 THE COMPREHENSIVE DEVIATION INVESTIGATION—A REFERENCE TOOL

Up until now, the chapter has focused primarily upon FDA's expectations and the current industry standards related to handling an OOS result within the laboratory environment, with only a small allusion to the handling of manufacturing deviations.

With this information serving as backdrop, what follows is a discussion of how to apply many of these same principles once the OOS investigation has shifted its focus from the laboratory and entered the formulation or production environment.

Since formal OOS investigations involve analytical results from testing materials made in CGMP-compliant environments, this discussion will be limited to the sequence of events associated with QC laboratories that test drug substance or products manufactured for clinical trials or for commercial distribution. A discussion dealing with how these principles relate to results obtained as part of stability studies will also be addressed. The procedure to follow has as its basis the draft FDA guidance for investigation of OOS results described earlier in this chapter. It has been broadened, however, to cover a wider range of technical activities as well as made more detailed to provide guidance in areas that are open to interpretation.

The laboratory and production areas should have procedures in place to properly handle the occurrence of an unexpected or atypical result. Standard operating procedures detailing management's expectations in the event of a deviational event should be available in each operational area. Personnel should be adequately trained to carry out the instructions specified in the SOP. This training should be performed by qualified trainers and repeated on a regular basis.

5.1 Critical Compliance Elements for Handling Manufacturing Departures

1. Timely and prompt identification of departures from prescribed manufacturing procedures.
2. Adequate mechanisms for reporting manufacturing departures, such as batch records that invite operator and supervisor comments and explanations and adequate change control.
3. Immediate quarantine or "hold" of material under investigation.
4. Prompt cross-functional assessment of OOS result.
5. Resampling of material if the original is deemed to be unrepresentative of batch.

6. Alerting of regulatory authorities in event of confirmed OOS on commercialized batch FAR (or clinic if batch is CTM currently in human study).
7. Prompt identification and implementation of corrective actions to eliminate or reduce future occurrences. These may include process changes requiring validation activities necessitating prior FDA approval supplement. (See SUPAC Guidelines [14].)
8. Quality assurance coordination of completion of formal deviation report in a timely manner.

If a deviation occurs, it is usually first observed as an OOS or OOT result in the testing laboratory. An exception to this is the observation during production of an operator-related or process-related problem that can be identified and addressed prior to submission of samples to the laboratory (e.g., through batch record review or operator/supervisor observation).

As described earlier in this chapter, there are now several published guidelines and guidances that effectively outline the laboratory investigation process. These guidances have several steps in common that can be included, in part or completely, in a firm's OOS investigation procedure.

5.1.1 Phase I Investigation: Key Activities

Informal review of all procedural details by the analyst first observing the unexpected result.

Preservation of the analytical work area, including glassware, solutions, sample vials, pipettes, instrumentation, etc.

Immediate notification of laboratory supervision.

Supervisor and analyst discussion of all analytical details. (Use a comprehensive *checklist* as a guide.)

Attempt to identify assignable cause or root cause.

For identified laboratory error or operator-related error, documentation of corrective action(s) required and deadline for completion.

Retesting, if assignable cause is found and documented. Protocol is prepared and executed.

Retesting, if assignable cause cannot be found. Protocol containing justification for number of retests needed to either confirm OOS as real or isolate it as unidentified analytical anomaly. Criteria for retest results are provided.

5.1.2 Phase II Investigation: Key Activities

Initiate formal investigation outside laboratory, if OOS is confirmed.

"Hold" or quarantine material in question.

Resample using acceptable sampling plan if original sample is found to be unrepresentative.
Test resampled material and assess results.
Scientifically evaluate all retest, resampled, and original results and possible use of statistical treatments, such as outlier testing.
Cross-functional assessment of occurrence.
Prepare investigation report containing all results, evaluations, conclusions, and path forward.

The deviation will invariably be detected by generation of an OOS or other type of unexpected analytical test result by the laboratory analyst (chemist, microbiologist, biochemist, etc.). A de facto assumption that laboratory or analytical error is responsible, however, is never acceptable without the support of a scientifically sound and comprehensive investigation.

5.2 Phase I: Laboratory Investigation

Once a result that does not comply with either a regulatory or compendial specification (such as a regulatory limit) or an internal specification (e.g., in-process, alert, warning, or OOT limit) is discovered by the responsible analyst the following events should occur as rapidly as possible:

5.2.1 Step 1

Since the analyst is the most familiar with the analysis just conducted, he or she should revisit all activities associated with generating the aberrant result. The test site should be *preserved*; that is, no solutions, glassware, vials, syringes, flasks, or any other items that could potentially yield information as to the cause of the unexpected result should be removed or disposed. Instrumentation used to produce the original data or further process the raw data should be maintained in its state at the time of the analysis. At the same time, the analyst's supervisor should be notified that an unexpected or anomalous event has occurred and that this fact must be documented. At this point it is up to the supervisor's discretion to notify staff outside the laboratory, such as quality assurance, regulatory affairs, production, and research and development, but a reasonable amount of time should be allowed for laboratory personnel to investigate the aberrant result. The investigation should not, however, proceed for more than 3 days, without formal notification of at least the quality assurance organization.

If the analyst discovers that an error has been made in executing the analytical procedure during the process, the process should *not* be carried through to completion for the sake of determining the impact on the final results. This would create regulatory issues that could compromise the ultimate treatment of the final results.

5.2.2 Step 2

If the analyst has identified the cause of the aberrant result(s), the supervisor should be notified immediately and both analyst and supervisor proceed through a systematic examination of all aspects of the analysis to confirm that the cause that has been assigned is indeed correct. If both scientists agree on the most likely assignable cause of the result or determinant error, this should immediately be documented in a laboratory investigation report form. The original sample should be retested by the same analyst using the analytical procedure and same degree of sample replication as performed initially.

5.2.3 Step 3

If the analyst cannot immediately identify the cause, the supervisor should "debrief" the analyst and all other participants involved in the analysis. A checklist approach to these items would be helpful so that consistency and comprehensiveness is ensured from one investigation to the next. Since there may be aspects of the analysis that are unique to the particular procedure, the checklist should serve only as a guide, and additional possibilities should be discussed and investigated if deemed appropriate. The examination should include but not necessarily be limited to the following:

Method execution. (Was the procedure carried out correctly?)

Whether or not the aberrant result is part of a *trend* observed using this method, analyst, reagents, instrumentation, etc. Records should be kept of the method, analyst, and equipment performance to aid in the investigation and assist in trend identification.

All glassware, solutions, vials, reagents, volumetric solutions, reference standards, columns, etc.

Condition or improper operation of the instrumentation system (result caused by equipment or software malfunction, lack of proper calibration or qualification, misuse by analyst, etc.).

Whether the suitability and/or stability of the system was established before or during the run.

Method appropriateness for test. (Is it the right method for particular test?)

Expected method performance when compared to original method validation results (e.g., higher than expected variability, bias due to interference, or inappropriate quantitation limit).

Solution stability is as expected (per expiration/retest dates).

Calculations appropriate and carried out accurately.

Properly qualified spreadsheet applications for calculations (if used).

Results from other samples in the same analytical run that appear to have passed specification.

Is analyst properly *trained* on method, instrument, analytical techniques, etc.?

Is analyst properly trained in CGMP procedures?

All observations should be documented in the laboratory investigation report form as well as in the analyst's notebook or appropriate test result form(s).

5.2.4 Retesting

Assignable Cause or Determinant Error.

Should the analytical investigation reveal that a determinant laboratory error was the cause of the unexpected result, retesting of the same sample would be in order. Saying the same sample means the "collection" of sample units, powder, liquid, packaging components, and so on that was originally sampled and supplied to the laboratory for analysis. It is presumed that the samples were collected from a larger collection (i.e., batch or lot) using an acceptable, qualified, and documented sampling plan. The samples used by the laboratory for retesting should ideally be from the same homogeneous specimen as was originally sampled by the analyst. For tablets, this could mean the same powder resulting from grinding 20 or more tablets; for capsules, well-mixed capsule contents; for small volume parenterals (SVPs) in vials/ampules, mixed contents of several containers; for large volume parenterals (LVPs) perhaps the contents of a single infusion bag; and so on.

A survey of about 30 pharmaceutical companies performed by the author [15] revealed that laboratories will either select from actual intact dosage units or from the homogenized mixture or solution to obtain samples for retesting. Either is considered acceptable.

Assignment of a definite or probable cause should be simplified by the availability of analytical development reports prepared during various phases of drug substance or product development. If analytical development during drug development was carefully conducted and validated and was well documented, sufficient information should be available to expedite and facilitate the investigation.

If a cause can be assigned to the aberrant result with a high degree of confidence (sufficient evidence exists), then retesting of the same number of replicate samples (or preparations) as was originally tested is generally considered the proper procedure. Retesting should be performed by the same analyst who originally generated the aberrant results. Having the same analyst repeat the testing serves as a valuable training tool not only for the analyst involved, but for co-workers who are qualified to perform the same procedure. As suggested by the FDA Guidelines on OOS Investigations, a second qualified analyst can also perform the retesting or both original analyst and second qualified analyst can both retest the same sample. If both analysts perform the retesting, one may have to allow for the difference in variability by adjusting the acceptance criteria. The supervisor should ensure that adequate training has occurred and has been documented before retesting takes place. All results (original and from retesting) as well as evidence for the assignable cause, corrective actions to be taken to prevent reccurrence, and any deadlines or other commitments should be described in detail in the laboratory investigation report.

If the retest results now pass acceptance criteria, they will replace the original erroneous results in formal analytical records. There should be a note or some other indication in the official record, however, that these results were based on retesting and that a laboratory investigation report containing all details is available. The retest results are considered the official results for reporting purposes. Once the investigation and retesting are complete, laboratory management completes the laboratory investigation report, which is usually filed with the appropriate quality assurance personnel. If the laboratory investigation exceeds 3 working days, quality assurance should be formally notified of both the investigation and the anticipated investigation completion data. The overall laboratory investigation and documentation process should not exceed 30 business days (about 6 calendar weeks) without the knowledge of and approval of laboratory and quality assurance management.

No Assignable Cause Identified. If no definitive cause can be assigned to the aberrant result following a systematic and thorough laboratory investigation this information should be provided to the appropriate personnel outside the laboratory (e.g., quality assurance, regulatory affairs, and production, as defined by the appropriate corporate SOP) and a formal phase II investigation should commence. As the investigation continues in the laboratory, it will be expanded to include the possibility of operator-related (nonprocess) or process-related causes to explain the aberrant result.

The aberrant result may still have been derived from laboratory error; therefore, a different multidisciplinary approach, both within and outside the laboratory must be taken.

If no assignable cause can be identified as a result of the laboratory investigation, adequate retesting of the original sample may be sufficient to overcome the initial aberrant result(s) (i.e., confirm the result as erroneous) and serve to isolate the result as unrepresentative of the batch or lot in question.

If the laboratory investigation proceeds beyond 3 days with no definitive judgment as to the cause of the aberrant result, it may be necessary to inform both the internal and external regulatory authorities. For example, in the United States, the FDA must be informed if the product is already in commercial distribution. Normally, a FAR must be filed within 3 days of a confirmed OOS result on a product that has already been distributed. Detailed procedures should be available to guide the sequence of events should a confirmed OOS result be obtained. If an OOS result is confirmed for a batch that is currently involved in a human clinical trial (e.g., through testing of a stability sample), the clinic must be informed and a decision must be made regarding the continuation of the trial utilizing the "failing" batch. In almost all cases, the trial will end and the CTM will be shipped back to the appropriate clinical development department for further investigation. Not following proper regulatory procedures once a batch failure is confirmed during clinical trials could result not only in severe regulatory consequences for the manufacturer, but have serious legal ramifications as well. (The legality involves exposure of human subjects to unacceptable or "adulterated" clinical samples.)

The number of retests to be performed as well as acceptance criteria to which the results are to be compared are highly controversial topics. Just what is the proper number of retests required to isolate the OOS value as erroneous? Many laboratories have taken their cue from the *U.S.* v. *Barr* decision discussed earlier and used the value offered by FDA investigator Mulligan when he was queried by Judge Wolin as to the "magic" number. Mulligan responded that he believed that seven out of eight retest results would satisfactorily overcome the OOS result. The value is not considered excessive or construed as either testing ad infinitum or an attempt to test the product into compliance. Since the conditions surrounding each investigation are rarely identical, the number of retests should remain flexible and be established on a case-by-case basis. The value of seven of eight retests was an opinion expressed by an experienced FDA investigator, but is it valid? In a recent survey [15] conducted by the author in which OOS investigational practices during stability studies from 36 different pharmaceutical manufacturers were polled, it was apparent that the number of retests used currently by industry during OOS investigations, along with the rationale for these values, were highly *diverse*. The values ranged from two to eight retest results, with various explanations given as to the appropriateness of each choice.

At this time there appears to be no overwhelming consensus within industry as to acceptable retest practice. The recent draft FDA guidance for industry on investigation of OOS test results does not contain a specific number for retesting; it only states that the number must be based on scientifically sound, supportable principles and that it be specified *before* retesting begins. The number of retests may change based on the variability of the analytical methodology and process, but should always be predetermined in such documents as a retest protocol or SOP in which laboratory investigational procedures are well defined. As R. Rutledge of the FDA commented at a recent annual meeting of the Parenteral Drug Association (Annual PDA meeting in Washington, D.C., in November 1998) "don't look to Judge Wolin's decision to determine the number of retests that are necessary, each OOS investigation must be specific to the situation." Mr. Rutledge is also the author of the draft FDA guidance on OOS investigations. Although all laboratories should have an SOP to describe the overall investigation procedure, a separate retest protocol or plan is recommended to specify the required number fo retests for each OOS event. Additionally, the plan should specify the personnel to be involved, the statistical treatment of the results (if any), and the acceptance criterion to be applied to the results. Lastly, the plan should contain any special instructions (such as a data treatment) that are to be used in the particular exercise due to the nature of the investigation. The criterion will be used to either confirm the OOS result as valid or overcome the value as being unrepresentative of the sample under study. The use of a comprehensive investigational SOP rather than specific protocols to dictate the number of retests and criteria is acceptable practice if the testing performed by the laboratory is sufficiently routine, tests are limited in number, and the laboratory utilizes analytical methods of comparable precision attributes.

A *separate protocol* for each retesting campaign that is reviewed and approved by several individuals familiar with the methodology and investigation policy, including laboratory management prior to implementation, is a more effective and efficient way of organizing, executing, and reporting this critical aspect of the investigation, however.

Another factor impacting the quality of the investigation is *personnel*. Just who will carry out the retesting plan? While the FDA guidance on OOS investigations suggests that the retests be conducted by an analyst *other* than the one who generated the OOS result, industry practices with respect to personnel are highly varied. Practices range from having the original analyst execute the entire retest protocol (once aware of corrective action and properly trained), to having two analysts (one being the original) execute the plan, to requiring that only another qualified analyst perform the testing. While all scenarios can be rationalized as scientifically sound, the laboratory should have the flexibility of choosing the retest personnel based both on the qualified resources at hand and on each particular circumstance. In many smaller laboratories, and especially when dealing with complex procedures, there may be only one "qualified" analyst available to perform the retesting. In larger laboratories, it may be more informative and instructive to have the two qualified analysts carry out the retesting in the event that the aberrant result might have been caused by an arcane, systematic error that the original chemist failed to identify. In some cases retesting should be done by a second qualified analyst if he or she is the most familiar with the particular techniques or technologies (subject matter expert). With more than one analyst involved it is important that the retest criteria make allowances for interanalyst variability (taken from an Intermediate precision study in method validation).

As for the actual number of retests, how does one determine this value in a scientifically sound and rational manner? As mentioned earlier, industry practices with respect to the number of retests are highly diverse, and in actuality more than a single approach can be found scientifically sound and valid.

One technique employed to arrive, at an appropriate value has been postulated by L. Torbeck [16], who has taken a statistical and practical approach that in the absence of any other retest rationale can be judged as a technically sound plan. According to Torbeck, the question to be answered—"how big should the sample be?"—is not easily resolved. One answer is that we first need a prior estimate of the inherent variability, the variance, under the same conditions to be used in the investigation. What is needed is an estimate of α risk level (defined as the percentage of probability that a significant difference exists between samples when there is none; what statisticians call a *type 1 error*), the β risk level (β is the probability of concluding that there is no difference between two means when there is one; also known as a *type 2* error) and the size of the difference (between the OOS result and the limit value) to be detected. The formula for the sample size for a difference from the mean is expressed as:

$$N = \frac{(t_\alpha + t_\beta)^2 S^2}{d^2}$$

where

t_α and t_β = One-sided t-distribution values for given α and β risk levels selected (from table)

S^2 = Varaince association with the total product/process/method variability

d = Difference between OOS result and specification value

Of the four values needed to calculated sample size, the α and β values are standardized for most scientific and industrial applications as $\alpha = 0.05$, $\beta = 0.1–0.2$ from the t table for α and β with a given number degrees of freedom (sample size minus 1) of the data used to estimate the variance; S^2 and d^2 are more difficult to obtain. Since methods and processes undergo change, the estimate of variance is limited to methods and products resulting from the most recent changes. Since some products are made only once or twice a year, there may not be sufficient data available to yield a reasonable estimate of overall variance. Moreover, d^2 cannot be determined for future OOS results since this information is not known. This fact represents an inherent and unintended conflict in the FDA position of a predetermined sample size since to use a reliable d^2 value, one must know in advance how far out the OOS result will be. This information cannot be determined until the OOS is actually observed. This is yet another argument for the use of protocols for each retest scenario, since the protocol will be written once the difference between the specification and the OOS result is known.

If certain assumptions are made, however, such as the following:

Intermediate precision of the analytical method is known (from method validation).

A sufficient number of batches have been made to allow a reasonable estimate of process variability.

Assay accuracy or percentage of recovery of the method is represented by a normal distribution centered at 100% and $C_p = 1$ (i.e., 6σ approximates the specification range).

A table of values can be generated using a quasi-simulation from the above formula. (See Table 1.) If one uses the specification of 90.0–110.0% which are frequently used for potency assays of drugs products, and a method intermediate precision (as percentage of RSD) of 2.0 and 2.5%, the values obtained from Table 1 are five and seven samples, respectively. Coincidentally, these values approximate those mentioned in the *U.S.* v. *Barr* decision. It should be mentioned that this method of calculating the number of retests may not work well for some tests, such as LAL bioassays, which produce colony counts that are not normally distributed, but skewed.

In addition to the retest values associated with the OOS sample, samples from lots or batches analyzed along with the aberrant sample (i.e., within the same analytical run) should also be evaluated since they may have been affected by the same errant factor that produced the OOS result. The fact that an accompanying sample passed specification while another failed may be only a result of serendipity. One practice employed by several industrial laboratories is to retest samples that bracketed (surrounded) the OOS sample in the original analytical run. Although the initial results from these

TABLE 1 Sample Size Look-Up Table for OOS

	(%) RSD of intermediate precision															
Spec	0.1%	0.2%	0.5%	1.0%	1.5%	2.0%	2.5%	3.0%	3.5%	4%	5%	10%	15%	20%	25%	30%
99–101 %	3	5	21	79	175											
98–102 %	2	3	7	21	45	79	122	175	238							
95–105 %	1	2	3	5	9	14	21	29	39	51	79					
90–110 %	1	1	2	3	4	5	7	9	11	14	21	79	175			
85–115 %	1	1	2	3	3	4	4	5	6	7	10	36	79	139	216	
75–125 %	1	1	1	2	3	3	3	3	4	4	5	14	29	51	79	113

Note: Cp = 1.0. Assumptions: one sample, measurement is in percentage of recovery, mean at 100%, normal distribution, wish to detect one standard deviation difference in the specification range, there is an estimate of the methods intermediate precision percentage of RSD, samples larger than 250 are not shown, Cp = 1.0; i.e., 6σ equals the specification range.

samples may have passed specifications, the results may have been biased to the same degree as the OOS sample. Only higher initial potency or lower initial impurity or degradant values may have been responsible for passing with equally erroneous results.

Once the retest results have been obtained, how should they be used?

From a laboratory management point of view, should there be a different number of retests associated with each method or product? Do the statistical and technical advantages of different sample sizes and criteria outweigh the need for consistency in order to prevent confusion among the analysts? These are decisions that must be made based on the types of tests performed and the competency level of the laboratory in question.

Averaging of Results. The averaging of results can be a useful technique if the samples under study are derived from a uniform and homogeneous source. Averaging passing retest results and OOS results to make a release decision is not scientifically valid [17] however. The procedure is not valid since the OOS investigation will determine that the result it either (1) valid, in which case the product is rejected or reworked, if permitted) or (2) invalid, in which case it will be released on the basis of the retest results alone.

Averaging passing retest results is valid for homogeneous samples if the objective of the testing is not to evaluate the uniformity of either in-process samples or samples from the finished drug substance or drug product (i.e., blend or content uniformity). If testing is conducted for the purpose of assessing uniformity of the batch, averaging would serve to hide critical homogeneity information. Even in situations in which averaging is acceptable, one should be cautious if the values are near the limit of the allowable range and other data points are outside this range. One should be concerned with product quality if retest results are so close to the specification limit (i.e., within analytical variability) that there is a temptation to average OOS and passing results in order to derive a passing average. If the batch is formulated to 100% of label content of the API, the results should rarely approximate the regulatory limits unless the batch is being evaluated during the later intervals of a stability study. Even if all retest results pass specification limits, it is always best to report *all* results in addition to the average value in official result reports.

5.2.5. Outlier Testing

Outlier testing is a statistical practice used for identifying from an array of data those data that are extreme and unrepresentative of the array.

Using validated analytical methods and well-studied and controlled production processes, a value in rare instances may be obtained, in rare instances, that is markedly different from the other results in a series. Such a result is called a statistical *outlier*. The outlier can be the result of a deviation from prescribed test methods, resulting from variability in the sample, or the result of a parameter that has not yet been identified as impacting either the quality of the manufactured material or the analysis used to evaluate it.

The USP (USP 25) discusses handling statistical outliers in the design and analysis of biological assays section (General Chapter ⟨111⟩). The USP states that for biological assays with exceptionally high variability, elimination of an outlying result may be necessary to prevent biasing of the final results. The USP also states, however, that the "arbitrary rejection or retention of an apparent aberrant response can be a serious source of bias...the rejection of observations solely on the basis of their relative magnitudes is a procedure to be used sparingly."

The USP 24 does not comment on the use of outlier testing as the basis of rejecting results from chemical testing. In the *U.S.* v. *Barr* decision Judge Wolin interpreted this lack of commentary as prohibitive and argued that outlier testing should not be utilized for chemical testing since in most cases method variability is considerably better than is found for biological assays. This opinion has become highly controversial since the original 1993 decision. The FDA draft guidance on OOS investigations states that statistical outlier testing can be only occasionally used for chemical testing under the following conditions:

Use of outlier testing should be determined in advance of using it and should be included in SOPs for data interpretation.

Justification for use of outlier testing should be well documented.

The SOP should specify the outlier test to be applied (e.g., Dixon, Grubbs, American Society for Testing and Materials [ASTM]), with relevant parameters specified in advance.

The SOP should specify the *minimum* number of results required to obtain a statistically significant assessment from the outlier test.

The outlier result should only be used along with a preponderance of other scientifically valid evidence to provide additional support to invalidate an OOS result. It alone cannot identify the source of an extreme result and thus should not be used as the exclusive reason to invalidate.

Based on the survey of industry OOS practices mentioned earlier, many laboratories have begun using outlier testing for OOS results from chemical assays, but the test is being used cautiously and only to support solid analytical evidence that confirms the OOS was truly unrepresentative of the sample. It has been used only to augment evidence that an OOS result is invalid since it sheds no technical light on the possible cause of the aberrant result.

Once the retesting protocol is executed successfully (i.e., no analytical deviations), the results will either support the OOS result and confirm a material failure or render the OOS result invalid and provide ample justification to pass the test specifications. If the OOS result is confirmed, the batch is rejected and the quality assurance department should take appropriate steps for batch disposal. Since confirmation of the OOS result strongly suggests an operator-related or process-related error, the onus is now placed on the formal or phase II investigation participants to identify a cause. If an operator-related error is found and batch rework is possible and prescribed in an approved batch record and approved regulatory submission, this may be the desired course of action. If no rework or reprocessing is possible, the batch is rejected and disposed of in an acceptable manner.

5.2.6 Resampling

As mentioned earlier, retesting, if necessary, should entail the use of the same sample in the laboratory that produced the original aberrant result. This may not always be possible, however. In some cases, the results from original testing or retesting indicate that the sample delivered to the laboratory was not truly representative of the batch or lot under study. The sample may have been improperly prepared prior to laboratory submission (i.e., not performed per predetermined procedures and sampling strategies). Inappropriateness of the sample may be indicated by widely diverse results from samples taken from the same portion or aliquot (with no analytical error being identified). Alternatively, the original sample may have been inadvertently altered by heating, exposure to light, moisture, or cooling: contaminated by quality assurance or laboratory personnel; or completely consumed or destroyed during the analysis. If the laboratory sample is no longer available or representative, another portion or aliquot of the batch may be resampled using the same qualified sampling methods as were used to obtain the original laboratory sample. *Evidence*, not just suspicion, should accompany the need to resample outside the laboratory. FDA inspectors have frequently condemned the ease with which the laboratory obtains a new sample without ample evidence of inappropriate handling of the original sample delivered to the lab. This rationale should be well documented. The resample should be tested using the same validated methods and testing plan as utilized for the *original* sample. If the prescribed plan used to obtain the sample appears to be causing the aberrant results, it should be evaluated and the appropriate corrective action should be taken. The quality assurance unit should coordinate this evaluation in conjunction with production and laboratory personnel. Resampling due to consumption of the laboratory sample cannot be easily justified since the laboratory should be supplied with sufficient samples to cover not only initial testing but the possibility of retesting caused by the generation of an aberrant result.

6 FORMAL OR PHASE II INVESTIGATION

The formal investigation process should be clearly detailed in a corporate or divisional SOP. The procedure should call for the rapid assembly of a team of process participants representing development, quality assurance, regulatory affairs, production, engineering, technical services, and any other discipline(s) that may be involved in the batch production, testing, and release process. The composition of this team may vary, based upon the type of aberrant result obtained and the material needing or requiring investigation. The formal investigation should ensure in parallel with the quarantine of the material in question. The investigation should at least entail the examination of the following items:

Thorough examination of the executed manufacturing batch record for any errors, omissions, discrepancies, inaccuracies, inconsistencies, and so on that could have caused the OOS result.

Interview with the operator(s) involved in the formulation or production process. Seek opinions as to how this type of result could have been obtained. Was anything unusual or out of the ordinary observed during the execution of the master batch formula or record?

Examination of equipment for possible malfunction or misuse. Ensure that all equipment used was properly qualified as required by departmental SOPs and not outside calibration limits over the period of use.

Examination of batch performance history by thorough review of previous product batch records to indicate whether frequent failures point to one or more process variables that may be out of control. Assure that the process has been properly validated and that all control points and limits have been included in the validation trials.

Assure that all critical process control parameters and conditions were utilized per batch record (i.e., the batch record was correctly followed).

Assure that all operators involved in the production were properly trained in the process, use of equipment, and overall CGMPs.

For drug substances, assure that all starting materials, in-process reagents, stable intermediates, solvents, purification columns, and so on are correct for the process and have been properly released by the quality group for their intended use in the process. All items used in the process should have been properly released into production per SOP prior to the start of the process.

Check that the equipment was cleaned properly following previous processing and that acceptable cleanliness was indicated by rinse and/or swab sample results. Ensure that all cleaning agents (e.g., detergents) have been removed from surfaces of the equipment.

For drug product, assure that all API, excipients, solvents, coating materials, capsules, and so on were correct for the process and were properly released by the quality function. Examination of remaining material and corresponding documentation in addition to discussions with vendors may be necessary if performance of an ingredient is suspected.

Compare in-process sample results with final, aberrant results to attempt to localize the cause (e.g., was a subpotent assay result from the drug product predicted by testing on blends taken earlier in the processing?).

Evaluate the physical testing results of the drug substance or product for data that could shed light on the problem.

Examination of product control charts is most useful in trying to distinguish between process-related or non-process-related causes. Trend analysis of key production parameters and attributes could assist in localizing a possible cause of the OOS. For example, if the potency of the product has been trending higher than usual for the last few batches produced (and the OOS resulted from an upper limit failure), this could be indicative of such causations as inaccurate moisture analysis or operator compensation error, error in the batch record, weighing error due to balance or scale bias, change in excipient purity which could impact functional characteristics or failure to maintain and/or calibrate a piece of equipment.

If the results of the formal investigation lead to a nonprocess or operator-related error, the evidence for the event should be fully documented in the formal investigation report. The report should identify at least the following items:

The operator(s) involved

The assignable cause of the error, along with evidence supporting this conclusion

Corrective actions

Preventative measures to be taken to avoid recurrence

The time line for complete implementation of the corrective and preventative actions

The responsible parties required for the implementation

The decision regarding ultimate disposition of the batch

Unless batch rework or reprocessing is permitted and possible, the batch must be rejected at this point and disposed of in a proper manner. Documentation related to final disposition of failed batches is required by the FDA. It is recommendable to have a "witnessed by" signatory on all final disposition documentation if it involved destruction or denaturing of the material.

A more serious investigational finding is a process-related error that implicates the process itself and is not based on any error in process execution. This process problem may have been derived from various sources, such as

Incomplete transfer of the process technology to the manufacturing site
Use of unsuitable process equipment
Incorrect transcription of master batch formula from development to production areas
Inadequate compatibility, optimization, and/or process validation studies conducted during process development or scale-up activities
Flawed process validation; that is, omission of one or more control parameters or conditions so that an out-of-control situation would go unnoticed and still impact batch quality
Inappropriate specifications established for in-process controls
Lack of reoptimization adjustments made following process transfer.

There are other possible causes for process problems, but the most important factors usually result from deficiencies or inadequacies during the product development life cycle in which one or more process events were not accounted for or discovered and resolved prior to transfer. This is why ample time and effort should be spent in various stages of development to permit a comprehensive understanding of API reaction dynamics (chemistries, reaction rates, impurities and degradation products and pathways, process control settings and specifications, etc.) as well as such drug product formulation factors as physical and chemical compatibility, mixing speed and time, influence of moisture and other solvents, and effects of temperature, humidity, and light.

Process-related causes of OOS results will cause the batch to fail and be rejected. Corrective action in this particular case may entail redevelopment or optimization of the process, which usually takes time and effort and may require process revalidation if the change is significant. For drug products, guidelines have recently been made available for postapproval changes and are dose-form-specific. So-called SUPAC (scale-up and postapproval changes) Guidelines have been published by FDA for the purpose of providing procedures for handling changes to the commercial production process.

The guidelines define significant v. insignificant changes to various dosage forms and just what actions are necessary should various types of changes be made. Such actions as process validation studies, specific analytical testing (e.g., dissolution studies), new stability studies, and additional clinical trials are specified based on the type and extent of the change. For API processes, a similar guidance called BACPAC (bulk active compounds post-approval changes) has recently been made available in draft form to guide activities that should accompany changes to synthetic or biological processes.

7 OOS AND OOT RESULTS FROM STABILITY STUDIES

According to the CGMPs (21 CFR 211.166), there shall be a written testing program designed to assess the stability characteristics of drug products. The studies entail subjecting products to various environmental conditions, such as heat, light, and humidity, and assessing those characteristics that could potentially change over time using appropriate validated analytical tests. The procedures to handle OOS or other unexpected results described earlier in this chapter for quality release of material into commerce may also be applied to aberrant results produced while testing stability samples. Stability studies involve batches of drug substance or products that are evaluated for characteristics that may change over time. The analytical methods must be able to distinguish the intact drug molecule from its degradation or decomposition products to determine potency, purity, and physical and microbiological changes over time (i.e., be stability-specific). While quality release testing of a material is performed *once* on a batch, and if all values conform to specifications, not performed on this batch again, stability testing is performed on batch samples repeatedly over a predetermined period of time. This allows for a batch's historical record to be established and continuously updated as the batch ages. The ultimate goal prior to drug product approval is to use the results obtained over time to first predict, using accelerated conditions (about 15○C above anticipated storage conditions), and then confirm, using actual product storage conditions, a meaningful expiration or retest data. For stability studies, an OOS result should usually be preceded by an indication or trend in a particular attribute during one or more prior test interval(s) [unless the OOS is observed at the initial time point (T_o)]. For stability studies, it is also important to detect OOT results (i.e., results that deviate markedly from the trend anticipated or predicted over time but do not fail specifications for the attribute of interest). If an OOT result is observed that deviates markedly from the trend established by either visually or mathematically treating previous data points (e.g., using such tools as regression analysis of trend data or statistical trend analysis), the aberrant result may be in error. Investigation and rapid resolution of OOT results is important since they could be misleading and ultimately bias the calculations for determination of product expiration dating. Establishment of expiration dating and ongoing marketed product evaluation depends on identification of, quantitation of, and/or absence of trends in results. The unresolved OOT result can also lead one to believe that the material under study is unstable and should be redeveloped. The OOT results are more difficult to detect than OOS results, but should be investigated as thoroughly as the latter.

The laboratory investigation should be performed with the same degree of rigor described earlier for OOS results; however, there are other alternative courses available that may be preferable to a formal investigation.

Such alternative actions may be to

1. Increase the frequency of testing (i.e., number of test intervals) to determine whether or not the aberrant result is part of a new trend.
2. Examine other stability batches of the same product or that containg the same materials for similar behavior.
3. Seek an explanation from earlier stability studies performed during development as well as during method validation (e.g., stress studies).

Unlike OOS results, it may not be prudent to retest an OOT result since a retest may unnecessarily lead to an OOS result that would have more serious consequences. In many instances in which an OOT result is obtained, it may be wiser to note and document the event and seek resolution by taking one of the actions described above.

It is common to use so-called *alert* or *warning limits* that are incorporated into the stability protocol to flag one or more results that could signal that a trend is developing. These limits are stricter than the corresponding stability limits and serve to trigger action that could avert an eventual OOS result with all of its ramifications. Also, not complying with alert or warning limits does not carry with it the severity of ignoring an OOS result, but a rapid response to such an event could potentially save the batch or study. One example is the decision to use refrigeration for storage of the product instead of controlled room temperature well before a lengthy study has been completed.

If laboratory investigation of the stability OOT or OOS is not successful in identifying a determinant error, the investigation proceeds outside the laboratory, and as before, a committee convenes consisting of personnel with a strong interest in the status of the stability results. Representation from such critical departments as quality assurance, research and development, and regulatory affairs should participate in investigations for stability studies performed during the development phase prior to approval. If an OOT or OOS is found for a marketed product, cross-functional representation should compose the core investigation team.

It is clear that the quality assurance organization plays a central role in coordinating facilitating, and documenting the formal investigation of an OOS result, whether the affected batch is utilized prior to approval (i.e., clinical trials, process validation) or postapproval as marketed product.

The job of the investigation team is to scrutinize all nonlaboratory aspects of the batch process for a possible cause. For a stability study, the batch record takes on less importance since the executed record has already been audited through the batch release process. The aberrant result may have evolved over time or be the result of sampling, analytical, or process irregularities. Some other possible causes of aberrant results include the following:

Errors may have occurred in executing the sampling plan used during the initial stability set-up.
Design flaws may have been present in the actual sampling plan used to provide samples for testing (i.e., not statistically sound).
Non-uniformity of the batch was not discovered during the limited dose uniformity testing performed during the release process.
Environmental conditions may have exceeded certain limits during the study (e.g., temperature and/or humidity excursions for protracted periods of time).
Analytical method is not stability indicating. Interference of degradation product, excipients, synthetic by-product, and so forth causes positive bias in results. Method validation is deficient.
Samples provided for stability studies were mislabelled (e.g., lower dose mistaken for higher dose of the same product).
If matrixing or bracketing employed, it was not executed properly or else the design was not statistically sound.
The samples were abused prior to delivery or during storage in the laboratory (e.g., exposed to excess heating, moisture, light, shock, and possible contamination).

All of the possible causes listed above can be considered CGMP issues that could have been avoided with careful attention to detail and to the procedures established for production and stability processes.

For a batch that is involved in one or more clinical trials, if an OOS result is confirmed in the laboratory and the manufacturing deviation is confirmed by the phase II investigation, this information must be provided to the clinical lead immediately and the impact on the study must be assessed. Usually the study will cease and the CTM will be recalled by the manufacturer. The impact of the OOS on the substance must also be assessed and corrective actions taken. These actions could include reformulation of the drug product, enhanced purification of the drug substance, a change in the label storage conditions, and/or a change in the container or closure system. If the OOS result ultimately leads to confirmation of a manufacturing deviation for a batch that has been distributed to commerce, a FAR must be filed with regulatory authorities within 3 days of confirmation committing to a recall of the batch from the field. Moreover, other batches of the same product (different strength, container size, etc.) should also be examined for similar stability behavior with the presumption that if the process produced a product that is inherently unstable, other similarly formulated batches could possess the same instabilities. Moreover, products manufactured using the same or similar formulations or processes, equipment (if found to be the source of the deviation), operator or operators (if systematic or common processing errors are observed), or raw material(s) should also be examined for similar aberrational behavior.

8 THE CONCLUSION OF THE PHASE II INVESTIGATION

At the conclusion of the investigation, the results should be evaluated, the batch quality should be determined and a release or recall decision should be made by the formal investigation team. The SOPs should have been implemented to detail every step of the investigational process thus far, including batch disposition following rejection. Once the batch is rejected or the stability study terminated, there is no limit to further testing of the batch to determine the cause of the failure. Continued evaluation of the rejected batch is encouraged since it aids in the assignment and implementation of proper corrective action.

For inconclusive investigations (i.e., the investigation does not reveal a definitive or assignable cause for the OOS result and does not confirm the result) the OOS result and investigational report should be retained in the batch record and be given full consideration in the final batch disposition decision.

It does not end here, however quality assurance must formally increase its monitoring and vigilance of both the laboratory and manufacturing activities related to this product. An active effort should now be mounted to look for trends over time. The FDA expects quality assurance to be proactive and responsible about inconclusive phase II investigations.

9 DOCUMENTATION ASSOCIATED WITH THE INVESTIGATION

Records must be kept of complete data derived from all tests performed to ensure compliance with established specifications and standards (21 CFR

211.194). Once the investigation has concluded, a written laboratory (for laboratory-generated errors) or formal investigation report should be prepared, reviewed, and approved by the proper disciplines. The final approval is usually relegated to the quality assurance department, but should at least be reviewed by all investigation participants for accuracy and completeness. The report should include, as a minimum, the following details:

Accurate and concise identification of the material in question and under quarantine.

Exact time and date of occurrence.

The reason for the investigation.

A running log of all actions taken thus far to investigate and determine an assignable cause to the OOS result. This should include all activities in the laboratory as well as the process and production areas.

The results of the investigation, including *all* routine and additional laboratory results involved in the investigation.

Scientifically sound and appropriate justification for excluding any OOS or OOT results found to be invalid through the investigation.

If the OOS or OOT are found to be invalid, the subsequent results obtained that support the acceptability of the batch and that invalidate the initial OOS results.

The conclusion and subsequent actions concerning not only the affected batch, but all batches and products that may have been associated with the confirmed OOS or discrepancy.

Corrective actions necessary to address the issue at hand and the preventative measures to be implemented to avoid the cause of the failure in the future and the time line for completion.

The signature(s) and date(s) of the person(s) responsible for approving the record of the investigation.

The signature(s) and date(s) of the person(s) responsible for the final decision on the disposition of the batch and other batches or products involved.

Final product or material disposition with a witness.

The investigation report should become a permanent part of the batch record(s) and stored appropriately under the auspices of quality assurance. It is the direct responsibility of the group generating the OOS or process deviation, as well as the quality assurance function, to ensure the corrective action described in the formal investigation report is carried out in its entirely and within the time frame committed to.

10 REGULATORY REALITY CHECK

With the chapter providing an instructive backdrop, several FDA observations (FD-483s) documented during recent inspections of FDA-regulated facilities are presented below, followed by a strategy for resolution and follow-up corrective and preventative actions.

10.1 FD-483 Citation

> There is no formal system to ensure that "corrective and preventative measures" resulting from OOS investigations are carried out. There is no written procedure assigning specific roles and responsibilities to designated personnel. There was no documented evidence that the CAPA (Corrective and Preventative Actions) was actually implemented.

This observation is one of the most common received relative to OOS investigations and deviations. A comprehensive investigation may have been performed and a definitive cause found for the aberrant data, but there is no documented system in place to ensure that the corrective action committed to in the investigation report will be effected in its entirety and in the allotted time frame.

The quality function (quality assurance) should have had an SOP detailing

1. That a corrective action or actions will be taken to prevent a future occurrence of the OOS result and that all concerned are in agreement with these actions and time lines
2. That someone (usually within the quality group) has been assigned the responsibility of tracking and monitoring the progress of these corrective actions
3. That the corrective action has been completed on schedule and adequately documented according to a predetermined format
4. That the formal investigation report has been properly closed out or completed per SOP
5. That the requirement for preventative measures to be put in place is an effort to avoid a recurrence of the incident

The firm could have prevented or avoided receiving this citation by having an OOS procedure that clearly delineates all of the steps required for investigating an aberrant test result. Each step in the investigation would require a concise time line, along with the individual(s) responsible for carrying it out.

10.2 FD-483 Citation

> The firm's investigation of Lot _ related to dissolution failures was inadequate in that it failed to evaluate the potential for and impact of dissolution failures in other lots that were manufactured using the same process, active ingredient and excipients.

This is another common type of observation. Although the batches may have passed the appropriate stage(s) of dissolution specifications, their results may also be erroneous for the same reason found (or not found) for the OOS batch.

The FDA expects the phase I and phase II investigation efforts to proceed beyond the batch in question and to other batches that were also manufactured under the same circumstances leading to failure of the OOS batch. If the investigation has revealed an operator-related or process-related error (e.g., ingredient test failure, equipment malfunction, calibration expiration, irregularity in process execution) and other batches were made using one or more common elements, they should also be investigated for compliance with all acceptance criteria. In the case above, dissolution results should be scrutinized.

Further, if the OOS is found to be caused by a laboratory error, other samples run along with the sample in question should be investigated, irrespective of whether they were acceptable or not.

One example of this type of investigation is to include samples that bracketed the OOS sample in the same analytical run (e.g., chromatographic run overnight). The additional samples or batches evaluated depend on the nature of the identified error. If a dilution error has been found due to improper pipet size, perhaps all samples in which this pipet was used should be investigated. If a chromatographic system error (such as autosampler malfunction) is pinpointed as the assignable cause, there is a chance that all samples in the run were affected and should be investigated as well. If the error is obviously due to misweighing of a single sample, perhaps no other samples should be included in the investigation.

The firm could have avoided receiving this citation by having an OOS procedure that required an assessment of all other lots manufactured using the same process, active ingredient, excipients, reagents, and solutions used for the material in question undergoing an investigation.

10.3 FD-483 Citation

> During Calibration (of the HPLC) the system failed the linearity parameter. The reason given in the investigation report was that the HPLC syringe was changed prior to running the linearity studies

and potentially may have dispensed an inaccurate amount of sample. There was no documentation indicating that this change occurred or verification of any changes to the HPLC system post discovery of the linearity failure.

There are several deficiencies alluded to here. First, during a validation study, an unexpected result was obtained without sufficient evidence from the investigation to confirm the assignable cause as a faulty autosampler syringe. The replacement of the syringe was not tested and confirmed as the source of the error. Evidence, not mere suspicion, must accompany an assignable cause or determinant error during a phase I or phase II investigation. Second, the replacement of the syringe was not documented in any instrumentation log or record that accounts for use and maintenance of the system. While this is not exclusively an OOS issue, it is a CGMP violation in that the system was altered in some way, the modification was not recorded, and its impact on system performance was not evaluated.

The firm could have avoided receiving this citation by ensuring the adequacy of two critical compliance systems:

Handling OOS results
Change management

The procedure for handling OOS results must include assessment of all instrumentation failure and justification for requalification of equipment if it was determined that was the source of the OOS. A state-of-the-art change management program would ensure that any changes made, prospectively or on an emergency basis, would be adequately documented and evaluated prior to implementation. Additionally, there is an evident need for personnel training in the areas of handling OOS results and managing change.

11 WORDS OF WISDOM

Phase I and phase II investigations utilize the same basic principles and strategies to arrive at a final batch determination.

If process and analytical development are systematically performed and well documented, it can provide valuable clues and expedite the resolution and understanding of an unexpected result.

The more time and effort that is spent in understanding a process, establishing and validating sound and robust methodology, and establishing meaningful and realistic specifications, the less chance that an OOS will occur further downstream.

Training is an essential component in the "war" against OOS results, both in the laboratory and production facilities. The more familiar one is with an operation, the less the likelihood of making an error.

All investigations should be expeditiously and comprehensively executed.

Once an OOS is discovered—*"freeze-frame,"* that is the procedures must be in place to ensure that the material in question is quarantined and properly treated during the investigation.

All efforts should be made in the laboratory to identify an assignable cause or determinant error for an OOS result.

Retesting of OOS material should be closely guided by an SOP or protocol in which a *predetermined* number is approved prior to retesting.

The OOS investigations call for corrective actions and preventative actions to avoid future recurrence.

The phase II or formal OOS investigation should take a multidisciplinary approach and include such functional areas as quality, production, research and development, regulatory affairs, compliance, engineering, quality control, and/or any other party that can constructively assist in rapid resolution of the "failure."

REFERENCES

1. Fed Reg 61 (87), 20103–20115, May 3, 1996.
2. 21 CFR 210 and 211 represent the 21st Title of the Code of Federal Regulations, Chap. 1, Sections 210 and 211, which contain the FDA's CGMP for finished pharmaceuticals.
3. Fed Reg 36(61), Jan. 15, 1971.
4. Office of Regulatory Affairs, Food and Drug Administration Guide to Inspection of Pharmaceutical Quality Control Laboratories, July 1993: Website: http://www.fda.gov/ora/inspect_ref/igs/pharm.html.
5. FDA CDER, Guidance for Industry: Investigating Out of Specification (OOS) Test Results for Pharmaceutical Production, Sept. 1998.
6. 812 Federal Supplement, p. 470 (D.N.J. 1993).
7. R.B. Kirsch, Handling of out-of-specification and other unexpected or atypical results, presentation made to AAPS Workshop on Stability Practices, Crystal City, VA, July 1999.
8. Office of Regulatory Affairs, Food and Drug Administration, Guide to Inspections of Pharmaceutical Quality Control Laboratories, July 1993.
9. Fed Reg 61(87): 20104–20115, May 3, 1996.
10. FDA, CDER, Guidance for industry: Investigating Out-of-Specification (OOS) Test Results for Pharmaceutical Production. Rakville, MD, Sept. 1998.
11. American Society for Quality WWW site online glossary: http://www.asq.org/abtquality/glossary.cgi
12. International Conference on Harmonisation, Guidance for Industry: Text on validation of analytical procedures, ICH-Q2A, March 1995.
13. International Conference on Harmonisation, Guidance for industry: Validation of analytical procedures: Methodology, ICH-Q2B, Nov. 1996.

14. SUPAC—Scale-up and postapproval changes guidelines. put forward by FDA (guidance based on type of dosage form).
15. R.B. Kirsch, OOS investigations during stability studies, presented at the AAPS Workshop on Stability Practices in the Pharmaceutical Industry—Current Issues, Arlington, VA, March 29, 1999.
16. L. Torbeck, Determining the number of retests for OOS, presented at Parental Drug Association Course on Handling OOS Results, May 1999.
17. G.E., Gamerman, Do we dump the batch? Resolving OOS results. *J. CGMP Comp*, 2(1), 1997 45–51.

12

The Internal Audit Program: A Quality Assessment

Graham Bunn
Astra Pharmaceuticals LP, Wayne, Pennsylvania, U.S.A.

If you were the chief executive officer of a successful pharmaceutical company would you want to know that your company was meeting and exceeding customer expectations? Do you know who your customers are? Your customers may be consumers of the product, patients and doctors, purchasers (managed care), shareholders, or regulatory authorities. In fact, your customers are all of these. Although their expectations will vary, the company has the common goal of fulfilling all their expectations. There are different ways to determine customer expectations, but without an adequate evaluations tool, a standardized comparison to a common reference point cannot be made. Current regulations for medical devices (21 CFR 820) and good laboratory practices (21 CFR 58 and 40 CFR 160) include requirements for formalized and well-established and, maintained internal auditing programs. Current expectations of most other FDA-regulated industries, such as biotechnology and pharmaceuticals, are no different, although the regulations do not specifically mandate the installation of an internal auditing program. It is only a matter of time before these requirements are more clearly defined and codified in these and other areas.

This chapter will discuss the process of establishing an internal audit program, and should be read in conjunction with Chap. 14 ("The impact of

Total Quality Performance on Compliance"), Chap. 13 ("Preapproval Inspections: The Critical Compliance Path to Success") and Chap 10 ("The Vendor Qualification Program"). Traditional audits by the compliance unit conclude with the issuance of observations. The observation responses are evaluated for acceptability and the file is closed. This chapter presents alternative proactive approaches that can radically change the traditional compliance audit image. The time to eliminate the stigma surrounding internal audits has arrived. Historically, internal audits have carried a very negative connotation because of the adversarial manner in which they have typically been performed. Changing the objective of auditing from identifying deficiencies to assessing and enhancing quality would go a long way in shifting the perspective most companies have of internal audit programs. Quality audits should be viewed as an assessment of compliance to regulatory requirements and as one facet of overall quality.

A quality audit has been defined as "A systematic, independent examination of a manufacturer's quality system that is performed at defined intervals and sufficient frequency to determine whether both quality system activities and the results of such activities comply with quality system procedures that these procedures are implemented effectively, and that these procedures are suitable to achieve quality system objectives" 21 CFR 820.2 (t). Another definition is "Quality Audit: A systematic and independent examination to determine whether quality activities and related results comply with planned arrangements and whether these arrangements are implemented effectively and are suitable to achieve objectives" (International Organization for Standardization/American Society for Quality Control).

The reasons for instituting an internal quality audit program include the following:

- Increase the potential for early identification of regulatory concerns based on FDA interpretations and current compliance focus
- Identify compliance deficiencies and deviations from industry standards and company requirements
- Provide a benchmark of compliance with other companies and regulatory expectations.
- Inform management of compliance status, regulatory risk, and civil liability
- Foster continuous improvement and forward quality
- Provide a tool by which the company can stay ahead of rapidly increasing regulatory demands

The following table shows the types of audits used to perform an assessment of compliance and quality.

Type	Advantage	Disadvantage
Mock pre approval inspection	FDA style with a focus on the product seeking approval. Personnel can communication skills related to regulatory issues, specific documentation, retrieval, and presentation of critical documentation, and answer product-or quality systems-specific questions.	Cannot predict all possible questions or documentation requests. Passing a company audit is no assurance of passing an FDA inspection and may produce a false sense of security.
Product specific from, receiving components to, shipping product	Allows evaluation of several critical departments, multiple systems, and many SOPs.	Need more than one product to identify systemic problems. Product quality trends not easily identified.
Quality systems	In-depth evaluation of critical compliance and quality systems. Auditor gains valuable insight into the system, enhancing auditor knowledge for future audit activities. Good opportunity to identify product quality trends.	Systems focus may not reveal deficiencies related to specific product issues.
Documentation audit (eg., aseptic filling)	Compares actual operations to current SOPs, batch record, policies, etc. Can easily identify areas of non compliance.	This narrow focus does not allow for an assessment of the underlying source or genesis of the deficiency.
"For cause" (eg., manufacturing deviation)	Allows for a very thorough and in-depth assessment of a specific concern and can identify other omissions.	Limited in scope and is not indicative of associated system deficiencies.

Enable the company to compete in an increasingly demanding global marketplace

Essential requirements for a successful internal quality/compliance audit program are as follows:

Executive management total commitment to quality and the quality audit program
An adequate number of qualified personnel well-suited to conduct internal quality and compliance audits
Established standards against which to audit
An independent reporting structure directly responsible to executive management with the authority to allocate resources
The resources (economic, personnel, time) to support the program's and operation's ongoing needs

Historically, FDA-regulated industries have not fared very well when it comes to managing their internal quality audit programs. There are many reasons for this, but the main reason internal audits have not been as successful as they could be is related to an overwhelming lack of adequate follow-up. Many audits are conducted throughout the course of product development and postcommercialization without a real commitment to install the necessary short and long-term corrections required to ensure sustainable compliance. A commitment from executive management with the authority to allocate resources is essential for the success of any internal audit program. The medical device quality systems regulations mandate this commitment from executive management and ensure that comprehensive systems for management review are implemented. The 21 CFR regulations related to pharmaceutical and biologic products have not codified a similar requirement; however, it is FDA's expectation that internal audit issues receive the necessary resources to ensure they are brought to adequate closure.

Some typical reasons internal audit programs have not been particularly successful are as follows:

Meaningful standards are not established or current.
Inappropriate standards are applied.
Auditor skills and expertise in the area being audited are inadequate.
Follow-up and close-out of observations are inadequate.
There is a lack of ownership for the corrective actions within the defined time lines.

There is a lack of accountability for preventative actions to minimize the potential of recurrence of any problems.
There is a lack of implementation of adequate preventative measures.
There is difficulty in measuring compliance status against emerging trends and expectations.
There is a lack of executive management commitment for adequate resources to support the program and corrective actions.
There is non systematic assessment of quality and compliance deficiencies.
There is a lack of identification of the root cause or underlying causes of the observation.
There is a inability to identify the potential impact of observations across multiple processes, departments, and procedures.
The focus is on "quick fixes" without assessing long-term impact.
Benchmarking against other companies typically relies upon consultants.
The staff is comfortable with just meeting the minimum regulatory expectations.
There is a lack of consistent application of a systems assessment to evaluate quality and compliance.
FDA-style failure analysis is not customarily used.
There is insufficient dissection of product and/or system failures.
There is a lack of compliance or quality rating system.

The genesis or root cause could lie in many areas, but is typically due to a substandard system or process or lack of adequate resource allocation.

Traditional auditing by regulatory authorities has included the "bottom-up" technique, which involves requesting data and information concerning manufacturing and quality control operations. Deviations, quality investigation, and out-of-specification logs are typical areas of interest, as this concentration of data will give a rapid overview of the level of compliance and quality in place. The medical device Good Manufacturing Practices (GMP) regulations were revised in June 1997 under 21CFR820 and include the quality system regulation (QSR). The issuance in August 1999 of the FDA's quality system inspection techniques (QSIT) has refocused auditing techniques to examine documentation showing that management has established and is following quality procedures. The document provides guidance to the FDA field staff on a new process for the assessment of a medical device manufacturer's compliance with the QSR and related regulations. Delegation of management's quality responsibilities to the practical level through a quality policy and SOPs is a process of defining the quality role in the company. Management must also be regularly informed of quality issues (audits, complaints, and adverse events) and as a priority where there is a potential product quality issue that may have implications for the patient. The four main areas covered are management control, design control subsystems, corrective and preventative actions, and production and process controls subsystems. While the QSR is a requirement of the medical device industry, the techniques used and standards established can equally be applied to other health care industries.

Quality audit programs must not be designed with the goal of only meeting regulatory expectations; they should be designed to focus on producing a quality product and meeting customer expectations. Meeting or exceeding these objectives should result in acceptable regulatory practices and durable compliance, and ultimately achieve and maintain marketing approval for the product. Audits have historically been seen as a tool for asserting blame for the failure of a system, department, or personnel instead of a means of assessment for quality improvement. Audit reports have been issued and "quick fixes" installed while overlooking the fundamental reason for noncompliance. Auditors were historically given the responsibility of policing noncompliance. They were not the most welcomed employees, since their assignment was to identify deficiencies and report them to management. Employees cooperated only to the extent to whom they would not incriminate themselves or others close to them. Auditors rarely offered advice or answered related questions because it was outside the immediate scope of their responsibilities. The FDA has taken the first steps in the issuance of the QSIT, which contains a radically revised approach to auditing. In support of this paradigm shift the quality assessment program (QAP) from this point of the chapter will replace the traditional internal audit program. In turn, the auditor in this environment would be more correctly identified as an assessor. Underlying causes and the genesis of the deficiencies are more easily identified in a QAP.

Comparison of the attributes related to the two vastly different approaches is shown in the following tables.

Compliance vs. Quality

Reactive	Proactive
Regulatory approach	Continuous improvement
Focused	Reviews entire process
Enforcement	Compliance as a corporate quality mission

Innovative approach using quality assessment	Traditional audits
Employees motivated	Employees fearful
Highlights strengths and weakness in a process or system	Emphasizes deficiencies
Analyzes the entire process or system	Narrow evaluation of noncompliance
Supports global competitiveness	Appeases regulators
Increases quality (six sigma)	Compliance and minimum quality
Promotes best practices	Points a finger at deficient practices
Clearly defines quality standards	Uses the general regulations
Establishes measurable quality and compliance outputs	Relies upon FDA inspection outcomes
Offers a rating system	Uses the FD-483 as measuring stick
Management determines the priorities for correcting deficiencies	Priorities determined by regulatory authorities and resulting actions
Identifies areas for improvement	Assigns blame for failure
Provides process to identify genesis of deficiencies and lack of or need to improve the quality system	Inability to obtain substantial and sustainable improvements
Fosters continuous improvement, total quality management, and forward quality thinking	Focuses on deficiencies and only correcting those deficiencies immediately or easily identified

Innovative approach using quality assessment	Traditional audits
Identifies, installs, and maintains quality standards exceeding regulatory expectations	Focuses on only meeting regulatory expectations
Provides executive management with a clear picture of the overall compliance status of each system or procedure	Identifies only the key deficiencies of the systems and procedures but lacks specifics
Process-gap-analysis rating system provides a quantifiable assessment	Only provides qualitative indicators
Corrective actions presented to bring system back to a 100% compliance status	Responses take a Band-aid approach rather than a comprehensive, far-reaching approach
Ability to identify and correct deficiencies by analyzing root causes	Unable to identify root causes from symptoms

Another nontraditional approach to assessing quality systems is the hazard analysis and critical control point (HACCP). The Pillsbury Company conceived the HACCP in the early 1960s with the cooperation and participation of the National Aeronautic and Space Administration. Essentially, HACCP is a system that identifies and monitors specific foodborne hazards that can potentially affect the safety of food. Some medical device and diagnostics companies are implementing the same principles as the HACCP program.

The principles are
Conduct hazard analysis.
Determine critical control point (CCP).
Establish critical limits.
Monitor each CCP.
Establish corrective actions.
Establish verification procedure.
Establish documentation.
Evaluate long-term effectiveness.

Process flow diagrams of the entire manufacturing process are produced and identification of the CCPs is determined. The assessments include typical hazards and corrective actions. The fact that the thought process and procedures are in place before a crisis occurs offers great benefit. It is a proactive rather than reactive system approach. This same approach can be applied as a tool for the quality assessment of pharmaceutical operations. Examining the systems and procedures using this tool may assist in identifying previously unidentified problems and deficiencies.

The traditional audit may still be useful in that it leads one in the direction of identifying a faulty system or process, and if one chooses one can go further. Using the innovative approach offered by the quality assessment has some limitations in that it is not the current approach used during pharmaceutical and biotech inspections. Mock FDA audits should be used on a routine basis to ensure employees' communication skills are sharpened and that the facility is inspection-ready. Preparation for an FDA inspection should begin during product development. It is a continuous process of the QAP, which the company should implement as early as phase I. No amount of resources in the last weeks before an inspection can make up for lack of adequate planning and preparation.

All FDA-regulated industries should have a formalized procedure addressing the management of regulatory inspections. This SOP should be periodically evaluated against actual practices. Employees may benefit from role-playing during mock FDA inspections and obtaining feedback from the assessors. Specific questions that may not be appropriate to ask during an FDA inspection can be put to the assessors.

All regulatory inspections must be analyzed by the compliance department to determine the main areas for GMP enhancement. An action plan or GMP enhancement master plan should then be designed to correct all deficiencies—and not only those cited in an FD-483. (For a detailed discussion of this subject, refer to Chap. 16, The Compliance Upgrade Master Plan) Subsequent internal quality assessments should initially focus on areas identified during the regulatory inspection and use the FD-483 as a platform from which to launch a comprehensive compliance and quality upgrade master plan.

Regulatory requirements for FDA-regulated industries are defined in the CFR and are open to variations in interpretation. These interpretive variations allow the industry to meet the requirements with some degree of flexibility and creativity for their particular product(s) and company. The FDA has defined the minimum regulatory requirements for internal auditing for the medical device industry but not for the other regulated industries. Regardless of the specific industry, however, internal audits are an industry standard and an FDA expectation. Before any QAP can be established there must be a clear and unquestionable commitment from executive management, including the CEO. A well-defined and implemented program is only as good as the follow-up to the observations made by the assessors and the commitment from management to comprehensively address the findings.

The CEO usually approves the quality policy of the company. The policy is sometimes abbreviated into a quality or mission statement that is in the front of the company literature or proudly displayed in the headquarters' reception area. The policy describes the overall approach to quality and commitment to meet customer (internal and external) expectations. The broad scope of the policy is implemented through the quality manual (a requirement for medical devices), which defines more specific details of responsibilities and processes. Standard operating procedures (SOPs) provide the actual procedures by which the original quality policy is processed and implemented. Quality assessment is one critical component of the quality policy. It is also an integral part of the company's overall continuous quality improvement program. Implementation of the quality policy in the workplace is the responsibility of the quality function. These responsibilities must be defined in a company policy and the details given in an SOP. Management has a responsibility to its customers, patients, employees, and ultimately shareholders to ensure that a quality product is produced. It is management that decides on the allocation and distribution of resources in order to meet the quality policy requirements.

An essential component of the assessment program is the independence of the function performing the assessment. This may not always be practical in small companies but should not be waived at the expense of compromising quality. Management that is fully committed to implementing a state-of-the-art program will provide the resources and give the required authority to the compliance function necessary to maintain an independent function. This function *must* have suitable independent quality management representation at the highest level and ultimately report to the (CEO). Reporting to the CEO, naturally, are the chief financial officer, legal counsel, senior vice-president of marketing/sales, senior vice-president, human resources, and other high-ranking officers. Equally important is a senior staff member representing compliance (e.g., at the senior vice-president level), because this is the key individual who can keep the CEO and other executive management informed of the overall compliance status of the company.

Responsibilities of different quality functions in a company must be defined in an SOP. Generally, quality control covers the testing and evaluation of components, materials, labeling, and products. Components, materials, and in-process materials are released or rejected by this function. This may also include microbiology, which is responsible for sterility testing, bioburdens, and environmental and personnel monitoring. Quality assurance approves procedures, investigations, change control documents, protocols and reports, specifications review, and manufacturing and packaging records to release product for shipping. Although 21 CFR 211.22 covers the responsibility of the quality control unit, the FDA has interpreted this with flexibility to mean the quality function. An independent compliance function is responsible for implementing and maintaining the QAP and reporting the results to executive management. This function will also host FDA and external auditors.

The QAP is a proactive tool that can be used to assess the potential impact of information from one system on another. An ineffective or inefficient customer complaint-handling process is not only a reflection of the company from the customer's perspective but quickly becomes a regulatory concern. This could ultimately result in the recall of product from the market and affect the public image of the company. An effective recall prevention program is closely linked to an effective QAP that consistently identifies deficiencies in customer complaint handling and expeditiously corrects them. (Fig. 1).

Only the quality assurance (QA) department has the distinct advantage and unique opportunity to monitor the overall quality status of the company. All critical documents require some form of approval by a member of QA. These procedures typically include

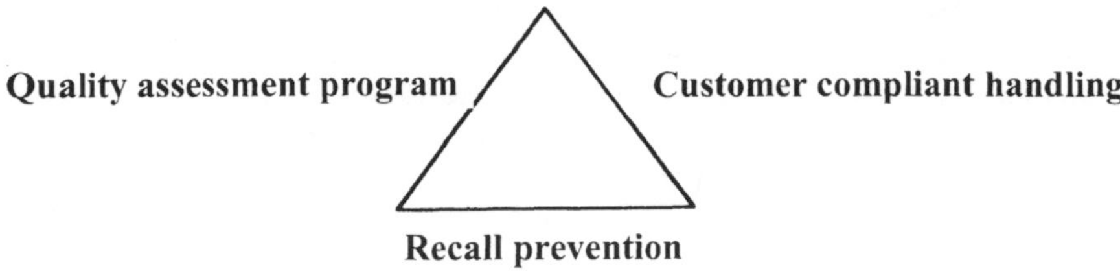

FIGURE 1 Quality triad.

Critical policies and SOPs
Master and production batch records
Deviation and investigation reports
Customer complaints
Change requests
Validation protocols and studies
Facility and equipment qualification reports
Environmental monitoring trends
Product quality during clinical trial activities
Laboratory out-of-specifications and manufacturing investigations
Water excursions

Quality assessments may be delegated to a separate compliance function in the department, but QA should always be apprised of compliance and quality trends. Quality assurance is the "flywheel" at the center of all the company operations (Fig. 2).

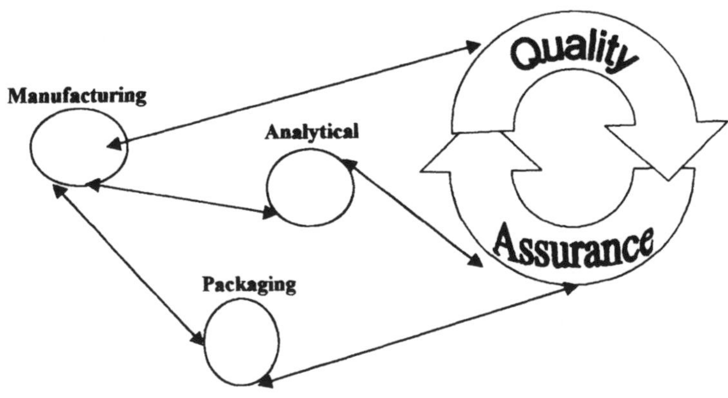

FIGURE 2 Quality assurance feedback loop.

Millions of dollars are spent over many years (10 or more) in developing products that generate supporting documentation and data that are submitted for regulatory review. Obviously the stakes are very high and the result of failure very costly. Even if the product development information and clinical data are sound there may be inadequate documentation to support the submission, which may result in the request for clarification or additional information by the regulatory reviewer(s). Similarly, during the facility inspection there is always the possibility of documentation or observed operations and practices not meeting regulatory expectations, ultimately leading to a recommendation of nonapprovable. Evaluation of the documentation and practices associated with the product seeking approval can easily be accomplished through periodic and consistent audits during product development. Frequent audits enable deficiencies and discrepancies to be identified, investigated, and corrected in a timely manner. Additionally, improvements to the underlying quality systems can be implemented in a timely fashion.

A valuable activity, and one often employed by the industry, is an FDA-style, mock preapproval inspection prior to submission of the regulatory application. It is important to perform this comprehensive audit prior to rather than after submission in an effort to identify and correct discrepancies between the regulatory submission and the actual practices and operations. Observations may be made at this time that could have a significant impact on the outcome of the regulatory inspection. Some of these observations might have been identified earlier had an adequate QAP been implemented and regular assessments performed throughout the product's development. Common systems and procedures that apply to multiple products should require more frequent evaluations to ensure they are meeting their compliance and quality standards. Systems and processes that are employed across the board for more than one product must be constantly monitored, maintained in 100% compliance, and continuously improved since their impact is so far-reaching. Additionally, product-specific data and reports should be assessed on an ongoing basis instead of waiting until the end of product development or shortly before making the regulatory submission. It is wise for management to deploy additional resources around the time of the submission to provide a heightened comfort level relative to inspection readiness; however, inspection preparations should begin at the time of early product development. No amount of testing or retrospective documentation can replace building quality into the process, early on, during the development phase. Procedures and documentation must support the production process and be verifiable during subsequent regulatory reviews some years after the operations were performed. The last thing that executive management wants to receive at the end of a preapproval inspection is a notice that the plant is not ready or that product-specific documentation requirements have not been met. A well-designed, comprehensive, and ongoing QAP would have identified deficiencies in time for corrections to be made and significantly minimize the potential for the failure of an FDA inspection.

1 WHO IS RESPONSIBLE FOR QUALITY AND COMPLIANCE?

Every employee has a responsibility to the company to ensure that his or her responsibilities are met, thus assuring a quality product is produced. Who is responsible for compliance? Everyone. Quality is independent of job title and salary and has no boundaries. Building quality into processes, systems, documentation, and the employee's mindset makes sustainable compliance more achievable. Employees take ownership of the processes and take pride in their work. Ultimately the employees benefit by working in a challenging environment and the operation has a greater chance of meeting and even exceeding industry standards and FDA's expectations. The goal is to have employees be an integral part of the company and know that their ideas and opinions are valued. Retention of valued employees becomes easier and the company can focus on strategic plans for growth and forward quality. Regulatory compliance should not be the goal-quality should be. Building quality into the *entire process* of an operation requires commitment by executive management and the allocation of adequate resources (personnel, facilities, training, etc.).

It is critical to the success of the implementation and continued maintenance of a quality program that management commitment is constantly visible to the company's employees. Management that clearly support quality by way of leadership and adequate resource allocation is more likely to secure and retain committed and enthusiastic employees. Executive management that does not "walk the talk" is likely to be viewed as duplicitous and not genuinely committed to the corporate quality mission. Employees should be encouraged to become involved and take ownership of compliance enhancements and continuous improvements in their specific areas. Individual personnel efforts should be directed by the company's overall program of continuous improvement and forward quality.

1.1 Six Top Drivers to Keep Employees Committed to Quality and Compliance

1. Recognition of the importance of their contributions to the process or system

2. Opportunities for personal growth and professional enhancement
3. A unified program for continuous improvement and compliance enhancement
4. Compliance and quality goals built into the employee's performance appraisal
5. Emphasis on system and process strengths as well as weaknesses
6. Provide a clear and concise road map for measurable outcomes

2 WHAT IS YOUR COMPANY'S CURRENT COMPLIANCE STATUS?

Regulated industries are complex, in a state of constant change, and interdependent across multiple sites in different countries. While 100% compliance may not always be obtainable, the company should always strive toward it. Designing and implementing an effective QAP program is a useful tool for both evaluating the compliance status at any given time and raising the standard of compliance to which the company strives. In addition to enhancing or increasing quality standards within a company, an effective QAP program will help the company assess its inspection-readiness in preparation for regulatory inspections.

3 PARTNERING WITH THE FDA

A partnership dynamic should be established with the local district office and FDA regulators who routinely inspect the facility. Establishing meaningful dialogue and rapport with the FDA during the review phase and field inspection process will go a long way toward enhancing the company's understanding of FDA's expectations. By paying close attention to the guidance provided by the reviewers and field investigators as well as keeping time lines and commitments made in response to noted deficiencies, the company can further establish and maintain credibility with the agency. (For further information on strategic interaction with the FDA see Chap. 15, ("Successfully Managing the FDA Domestic and Foreign Inspection").

3.1 Usefulness of Benchmarking

A company truly committed to meeting or exceeding internal and external customer satisfaction should perform an evaluation in an effort to benchmark each key department. The QA unit would be a good starting point since it is the center of most company activity.

Benchmarking involves performing a measurement and comparing the results with that of the reference standard. An evaluation is made and conclusions formulated. This exercise enables improvements to be made in an effort to meet or exceed industry standards. Establishing an internal standard higher than that which already exists may result from the comparison of one operation, process, or system to another. For example, a benchmarking assessment may ask the following questions of a QA unit:

Does the quality function meet manufacturing and packaging expectations? If not, why not?

What are the best practices and weaknesses of the departments?

Does manufacturing provide the requisite documentation (such as batch records, and quality control data) in a state that facilitates review, approval, and release of product by the quality function?

What improvements can be implemented to ensure the QA department meets FDA expectations and industry standards?

Internal benchmarking involves an evaluation against the best within the same class, whereas competitive benchmarking makes a comparison against the best direct competitors. Comparison with noncompetitors who perform related tasks is functional benchmarking, and generic benchmarking is a direct comparison against the best irrespective of the industry or market. These various evaluation techniques can be applied to FDA-regulated industries because of the wide spectrum and complexity of activities related to product development.

Many companies are only prepared to meet regulatory expectations and provide the minimum requirements. The goal of assessment is not to appease the regulatory agencies but to build quality into the various systems, identify deficiencies early on, identify the genesis of system failures, and ensure continuous improvement. There are myriad resources available from which information can be captured and utilized to assist with the benchmarking initiative.

Conferences present an opportunity to learn what other companies are doing in specific areas and to interface and network with colleagues. The FDA representatives frequently participate in or present conferences that afford the participants an opportunity to learn about FDA's expectations firsthand.

FD-483 trends and warning letters published on the FDA home page or through various industry publications provide tremendous insight into FDA's current areas of critical concern. They should be analyzed within the context of individual company circumstances.

The FDA home page (http/www.FDA.gov) is a good source for currently published and draft compliance guidance.

Consultants offer an expansive knowledge base related to industry-wide practices.

Benchmark partnering provides an opportunity to gauge internal operations against an external standard.

Hosting external customer audits within the company provides an opportunity for information interchange.

Establishment inspection reports are generated by FDA investigators after performing an inspection and delineate all the investigator's findings, observations, and perspectives, as well as FD-483 observations.

Newsletters, journals, and periodicals.

The World Wide Web offers many sites sponsored by consulting companies and contractors disseminating information and articles on current trends.

The actual FDA inspection process is an opportunity to ask the investigator what he or she has seen across the country. Do not attempt to illicit the investigator's opinion because he or she is prohibited from giving it. Ask the investigator to educate the firm as to the various ways in which the industry might handle, for example cleaning validation or personnel training records.

Every company has a wealth of resources as well as a knowledge base within the company that continues to expand as new personnel are added and current personnel are exposed to industry-related training. This reservoir of knowledge should be tapped to enhance the company's competitive advantage.

A formal system should be developed and implemented through which benchmarking information is collected, analyzed, and disseminated to pertinent personnel and executive management.

3.2 The Quality Assessment Program

Figure 3 illustrates a process that can be followed and that should be formally delineated in a corresponding SOP.

The SOP should address the following quality assessment guideposts:

Schedule: Assessments should be scheduled with enough advance notice and frequency to ensure consistent and periodic review.

Preassessment preparation: Involves a review of previous FDA and internal audit findings, industry expectations, and company standards.

Issue agenda: The agenda should clearly detail the type and scope of the audit.

Perform assessment: Ensures that qualified auditors are utilized and that all pertinent personnel are available.

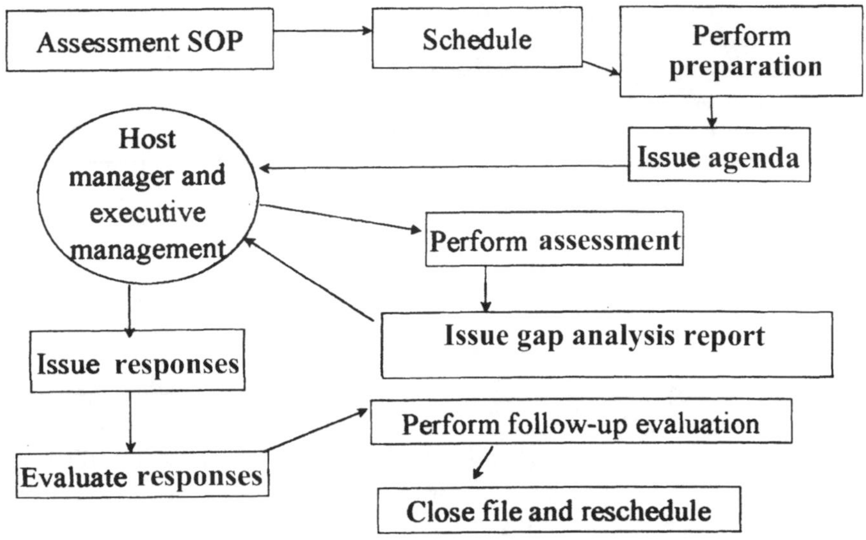

FIGURE 3 A quality assessment program.

Issue report: The report should delineate both the best practices as well as any areas for improvement. Do not turn the audit report into a list of observed deficiencies.

Evaluate responses: Ensure that all issues are comprehensively addressed and that a firm and concrete commitment is made to resolve the findings and any system deficiencies.

Perform follow-up: Confirm that both corrective actions and preventative measures have been implemented.

Close file: If the QA unit is confident that the issue has been adequately resolved, it is acceptable to purge this information upon performing the subsequent audit.

The challenge for the industry has been inadequate follow-up to internal audit findings. Most companies perform the audits with enough frequency, "however, they do not ensure that corrective and preventative measures have been implemented. Another importance aspect of a successful QAP is the ability over time, to evaluate and trend corrective and preventative actions.

3.2.1 Critical Compliance Guideposts for the Quality Assessment Program SOP

The SOP related to assessing quality throughout an organization should be approved by the highest level of site quality management and concur with the company quality policy, mission statement, and any corporate quality requirements.

1. Remember that the objective of the SOP is to plan, execute, report, correct, improve, and follow up in an effort to ensure overall compliance with cGMPs and other applicable standards.
2. Define the scope of the program for the entire company as well as the associated sites providing services and raw materials.
3. Assign authority and responsibility to executive management, QA, and the assessment team.
4. Design an applicable and meaningful rating system for individual processes and systems as well as the overall audit.
5. Define the standards against which the audit will be conducted; for example, cGMPs, ICH guidelines, and six Sigma.
6. Define the skills required for the assessment team; for example, experience, knowledge, judgment, investigative ability, exposure, perspective, and communication skills.
7. Define the criteria for categorizing both the observations and the response time frame; for example, high-risk, low-risk, meets industry standard, above industry standard, best practice, and the required time frame in which to make necessary corrections.
8. Explain the details of the comprehensive audit process, such as planning, scheduling, frequency, areas to be evaluated, and systems to be assessed.
9. Describe the contents of the audit logbook that will be used to confirm the nature and number of internal audits performed.
10. Describe the different kinds of audits that can occur, such as a mock FDA-style audit, quality systems review audit, gap analysis, documentation audit, and preapproval inspection audit.
11. Provide instruction and direction to the assessment team for collecting objective evidence during the audit.
12. Develop a metrics system allowing for quantifiable results wherever possible; for example, use statistical process control charts for manufacturing processes and correlating manufacturing deviations with consumer complaint trends.
13. Prioritize all audit findings into their appropriate risk category and response time frame.
14. Design a user-friendly audit report format. The audit report must ensure that observations are always accompanied by a corrective action and a corresponding time line.
15. Craft a corrective and preventative action plan that addresses all of the audit observations. It is good practice to incorporate such an

action plan into the organization's overall continuous improvement plans.

16. Identify management and personnel to be included in the audit report circulation list. It is a good idea to limit the number of copies distributed as well as track the copies in circulation. All audit reports should be stamped confidential and "for internal use only."
17. Identify the time frame within which executive management must review and respond to the report.
18. Determine an adequate time frame for timely and comprehensive follow-up to the observations.
19. Provide an executive summary to executive management with the power to allocate resources.
20. Ensure that any action items related to the observations go through the appropriate channels prior to implementation. For example, some changes may require formal change control, while other changes may require a supplemental regulatory submission to the agency.
21. Define the record retention policy. It is a good practice to purge internal audit files once they have been completely addressed.

Under 704(e) of the Food, Drug and Cosmetics Act, the FDA has the authority to review and copy all records required under FDA-regulated industries; however, FDA has elected not to review audit reports. Under applicable procedural rules, FDA could request the audits in the event of litigation.

The FDA will require an organization to provide evidence that a QAP does indeed exist. Additionally, it will require proof that periodic internal audits are being performed. It is good practice to maintain a logbook or spreadsheet that contains critical information related to the QAP, such as: the following:

Type of audit performed
Date of audit
Members of assessment team
Areas and systems assessed
Whether observation is pending or closed
Signature of lead assessor

Along with the company policy and SOP, this is the only document that should be presented to the FDA as proof of ongoing audits. If the FDA requires additional confirmation, providing the cover page of various QAP reports should suffice. If the investigator insists on reviewing the entire QAP report, the host should inform the investigator that he or she does not have the authority to release the report and to discuss that request with the company's legal counsel. It should be noted that under 21CFR 820.22 of the QSRs for medical devices, follow-up corrective actions, including reaudits, shall be performed when indicated. The FDA is well within its inspection purvue to request written confirmation from executive management stating that that audits have been performed, along with their specific dates, and that the required corrective actions have been taken (Chapter 17 Quality System Manual). The QSRs for medical devices require planned and periodic audits of the quality system (21 CFR 820.22). Executive mangement is required to review audit reports as part of its review of the suitability and effectivenees of the overall quality system. Executive management review must be formalized, and documented evidence of such reviews must be made available if requested by the FDA.

The FDA's access to documentation supporting the QAP and reports must be defined in an SOP. All corrective actions are open for inspection under the International Standards Organization (ISO) certification. The ISO notified body is not concerned with regulatory follow-up, as is required by the FDA.

3.2.2 Quality Assessor Responsibilities and Qualities

There is an industry trend toward dedicating specific personnel to the function of internal audits and quality assessments. This approach allows the assessor or auditor to focus on the compliance of the company and its vendors and contractors as well as maintain an independent role allowing for impartial assessment of all areas, including the QA unit. Those companies whose assessors and auditors also perform QA responsibilities might seriously consider the regular use of an independent assessment group, such as corporate auditors or an independent consultant. A fresh pair of eyes, coupled with extensive exposure to multiple worldwide operations, makes using external auditors, preferably with some former FDA experience a tremendous asset to any company.

The goal of a quality assessor or auditor is to ensure that the organization succeeds by focusing on quality and not necessarily compliance. Full compliance with applicable regulations is certainly desirable if an organization is committed to forward quality, continuous improvement, and global competitiveness.

Certain attributes are better than others when discussing the qualifications of a quality assessor or auditor. The following characteristics would only serve to enhance the role of any quality assessor or auditor in an organization.

Suitable education
Commitment to and understanding of quality
Understanding of current industry standards and regulatory expectations
Suitable technical knowledge and experience of areas to be assessed
Strong communication skills
Ability to interact with all levels of personnel, including executive management
A person who is approachable, inquisitive, detail-orientated, patient, good listener, assertive, with high ethical standards, and organized
Strong understanding of regulatory requirements and regulatory approval process

Forward-thinking companies will put quality first and build a quality assessment team based on cross-functional knowledge and experience. A lead assessor or auditor should be responsible for coordination of the audit and ultimately responsible for the issuance of the report. Composition of the audit team should depend upon the scope of the audit and utilize personnel with pertinent knowledge. The lead assessor or auditor is responsible for resolving any differences of opinion and presenting a unified executive summary of the audit.

External audits of raw material suppliers and contracted services should be handled in much the same way as the internal QAP. Audit protocols should be customized and tailored to the specific organization to be audited. For example, an assessment of a contract laboratory service should be conducted under the auspices of the internal audit team and follow similar procedures in terms of identifying deficiencies, installing corrective and preventative measures, and reports issuing. It is an advantage to employ the same personnel for both the internal and external audits. This approach assures a standardized and uniform audit procedure across the board.

3.2.3 Assessor Training and Certification

It is not uncommon either in the United States or abroad to find companies using an auditor training and/or a certification program. Many of these training and certification programs rely upon standardized audit protocols for a variety of different operations and systems and also require of their auditors a requisite knowledge of basic compliance requirements. Frequently there is a mentoring aspect to auditor training and certification. The mentor/trainer would certify the assessor after the successful completion of several comprehensive audits. Ongoing training with regard to changing trends and new regulatory requirements is necessary in order to maintain a certified status. All team members must be certified to the same predefined standards. There are a number of independent organizations within the United States that also provide formal auditor training and certification.

Quality assessments are a proactive mechanism for assessing overall operations and individual processes and systems. They should not solely be relied upon to become tools for identifying and highlighting individual problems, but as a strategic approach toward continuous improvement and forward quality. Here are some of the basic compliance and quality elements that should be included in a proactive iternal audit or QAP.

One or more of the following may be used as the reference and standard against which to make a quality assessment.

3.2.4 Quality Assessment Standards and References

Internal	External
Corporate policies	Code of Federal Regulations
Company guidelines	FDA guidance documents
Standard operating procedures	Compliance policy guides
Checklists	Podium policy statements (senior FDA)
Quality standards	Industry trends
Master and production batch records	ISO
Regulatory submissions	International/global trends

Verbal presentations of audit findings to executive management must provide a balanced view of the observations. Significant observations that were corrected during the assessment should be documented as such in the report. Items that are above the regulatory requirements or current industry standards but that the company may benefit from should be verbally presented to executive management before being recommended or committed to in writing. The assessor will evaluate the responses to the observations for completeness and adequacy. The responses should not be limited to the individual observation but rather examine the "bigger picture" and challenge the underlying systems related to the individual observation in order to determine the root cause or causes. In this same vein, an assessment of a particular system should not be limited to a specific batch or product, but rather involve investigations across departments, procedures, and products. If an organization is in the position to respond to a "list of observations" (FD-483), typically issued by an investigator after an FDA inspection, the organization's management should use the FD-483 as a platform from which to launch a more extensive and comprehensive assessment of the company's operations. Each FD-483 should be dissected and it genesis identified in an effort to improve the quality of overall processes and systems. The response to an FD-483 is a "corrective action plan" (CAP), whereas the response to an internal assessment or audit is a "GMP enhancement plan." Both responses would contain comprehensive, integrated responses to observations; however, the CAP would not identify any additional areas not covered by the FD-483 or warning letter from the FDA.

When a company has multiple sites, individual observations must also be evaluated across all sites. Observations noted in one area and not corrected across multiple process areas and additional sites are a reflection of management's shortsightedness and lack of commitment to quality. It would also be typical for a regulatory investigator to identify these issues as recurring systemic problems impacting the overall company. The assessor should discuss any unsatisfacotry responses with the respondent and reach an acceptable solution. Verbally agreed upon changes should be documented and included in the GMP enhancement plan. Time lines must be realistic and reflect the criticality of the observation in relation to public safety, regulatory risk, and compliance liability, and the product quality commitments made by unit personnel and executive management must be binding. Deviations from bilaterally agreed upon commitments should be dealt with by executive management.

Management might consider using "Strategic Quality Contracts" with employees. At least once a year the employee's performance appraisal would include measurable quality and compliance goals. Setting these goals with the employee invites the staff to provide input into what resources they would need to attain these goals. This sort of performance appraisal system must be tied into a larger continuous improvement and forward quality program.

3.2.5 Quality Assessment Report Content

Reports should be written in a positive manner and not solely consist of an extensive list of deficiencies. Typically within multi-faceted organization it is not the people who fail, but the systems upon which the organization is built.

Audit report titles should reflect a more enlightened and proactive approach to auditing; something like the GMP enhancement plan or "quality and compliance improvements" might serve to motivate employees and enlist executive management's support rather than the "cold" audit report.

It is important to list all of the elements that must be present for any given system, process, or operation for it to be considered 100% compliant and meeting industry standards. Whether or not to exceed the industry standard is a matter of executive management's strategic direction and quality commitment.

Reports should highlight strengths as well as weakness within a process, system, or organization. Concise and objective statements should be made relative to the observations. Observations must be based on facts supported by evidence and related to an appropriate predetermined reference standard (or standards). Audit reports need to be very specific and concise with regard to observations and corrective measures. These reports are a tool for executive management to make critical decisions about resource allocation and organizational changes.

A typical audit report should contain the following information:

1. Reason for the audit and type of audit being performed. (e.g. mock FDA audit, documentation audit, systems audit).
2. Scope of audit—Identify all areas assessed, such as batch records and manufacturing processes and areas.
3. Assessors—List of auditors involved.
4. Standards and references against which the audit is performed.
5. Overview—Description of operation(s) and systems reviewed. Overall conclusion—assessment based on predetermined, measurable elements, such as the 211 regulations and industry standards.
6. Specific recommendations—description of strengths and weakness, along with opportunities for improvements.
7. Response and CA—Concise plan for implementing enhancements delineating who, when, how, and why, coupled with corresponding time lines. The lead auditor issues the report, making sure to incorporate information and commitments from area management.

It is important to perform a long-term assessment of any corrections made as a result of an internal audit. If the corrective measure proves to be successful over the long term, the organization can choose to institutionalize the correction. If the corrective action proved not to be as successful as anticipated, the organization must rethink the solution. Corrective and preventative measures must be evaluated within the context of an organization's overall needs and culture in order to adequately determine the potential impact and usefulness of a particular corrective action. There are many different ways to fulfill FDA requirements. As long as the spirit of the regulatory requirement is met, FDA is not as concerned with the specific manner in which a company chooses to comply with its regulations.

3.3 The Rating System

A "rating system" allows the assessor or auditor to establish a basal level of compliance for an organization, measure ongoing improvements, compare sites effectively, and quantify results. Management may also use the system as a key performance indicator to monitor changes and progress. A scale of 1 to 4, in which 1 is the worst situation with few, if any, formalized systems and 4 is the best, might be useful when conducting internal quality audits (See below.)

The following compliance criteria can been applied to the various areas assessed.

3.3.1 Possible Compliance Rating System

Compliance Rating Level.

Level 1: The area, procedure, or activity does not have any system; needs to be installed.

Level 2: Has procedure or system, but not formalized; needs enhanced controls.

Level 3: Has formalized procedure: not state of the art; needs streamlining or has not been fully implemented.

Level 4: No need for modification; current system is formalized and state of the art.

If a process, system, or overall operation receives a rating of level 2, the assessor needs to define what it would take to achieve a rating of level 4. The criteria and performance standards must be predetermined and clearly defined.

Let us take, for example, an assessment of an organization's personnel training program. Column 2 represents a variety of standards against which the organization's personnel training system will be evaluated. Column represents what was observed by the auditor relative to the standards referenced in column.

An assessment that includes the completion of the key components against predefined criteria enables a rating to be assigned. This does not limit the assessor from following the traditional techniques in order to make the assessment. It further enhances the identification of the deficiencies by providing executive management with the complete "picture" of the system, procedure, and so on. In turn, management can allocate the necessary resources required to bring the system into a level of compliance commensurate with the company's quality standards. Traditional internal audits have focused on highlighting the observations and less frequently identify the strengths and means by which to correct the deficiencies. The assessor may provide the manager of the area with a list of the deficiencies with suggestions for improvement. The manager then responds with the corrective actions and

Training Program

Quality standard	Outcome	Comments
Standard operating procedure	Several SOPs governing personnel training	Require minor revisions
Complies with CFR 211.25	Yes	Complies, but not state of the art
Contains CV and job description	24/35 only	Incomplete
Contains Company orientation	34/35	Incomplete
Training matrix defines training requirements	12/15 jobs titles	Incomplete
Required SOPs defined	Yes	Fully accomplished
Required practical training defined	Yes	Fully accomplished
Practical training current	132/150	Incomplete
Process for maintaining training up to date	Maintenance of current GMP requirements	Incomplete
System for tracking effectiveness	Yes	Not utilized to full potential

appropriate time lines. If the responses are acceptable they are sent to executive management to keep it informed of the compliance status of the company.

Each of the items identified in the table above are measurable results and hence can be monitored for improvement. This "process gap analysis" identifies not only what is in place within this process, but also allows the company to see the strengths of the system. Areas of opportunities for any given process or system can be viewed from a different perspective and responded to accordingly.

In the end, all the processes and systems evaluated are placed on a simple bar chart on a scale of 1 to 4. The final chart gives QA and executive management a clear picture of what areas need resource allocation and the overall organizational rating for that assessment period. A mechanism for measuring the effectiveness of the QAP should be designed and implemented. This mechanism might include a metric such as fewer systems with level 1 and 2 ratings. The metric may also want to measure the rates of customer complaints, rejects, and recalls.

3.4 The Future

Multinational companies manufacture products for global markets to attain business goals and meet shareholder expectations. To be competitive and survive in this challenging environment companies must be prepared to establish meaningful and relevant quality standards. This requires forward thinking, creativity, innovation, dedication, resources, and a commitment from executive management to exceed minimum regulatory and industry standards. There is little room for error or failure when the first to gain product approval can gain market placement and prescriber recognition. The rewards of being first to market with a "blockbuster" drug are immense. Every company wants to be the best at what it does, but every organization is not prepared to pioneer an innovative drug. Going from concept to commerce is an arduous process that is complicated by today's evolving compliance demands (Fig. 4).

A Proactive QAP should support the areas in Fig. 5.

A thoughtfully designed and successfully implemented QAP is one of the most effective tools of continuous improvement.

An assessment program is incomplete without the final step of bringing closure to the audit process. This can only be performed after all the quality and compliance enhancements have been implemented and evaluated over a predetermined period of time. This does not necessarily mean that all the action items have been completed, since some corrections may require extensive implementation and assessment. The follow-up assessment, however, must include a review of the previous action

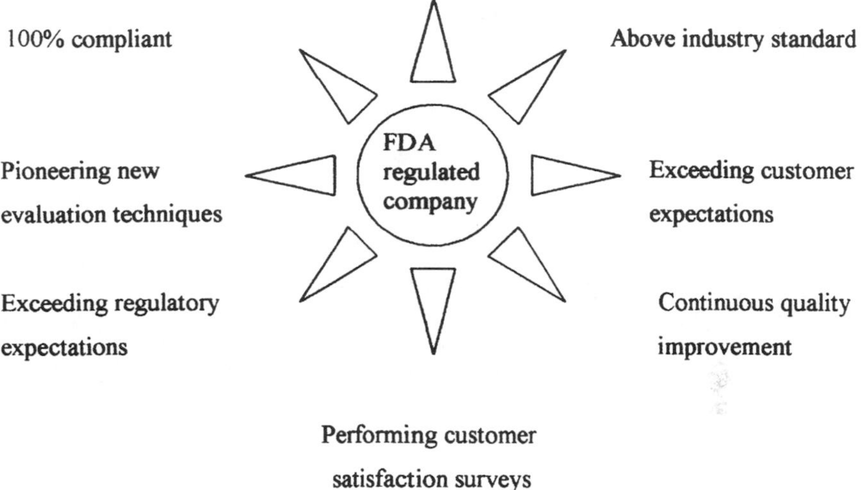

Figure 4 Where is your organization with respect to compliance demands?

Figure 5 Proactive quality assessment program.

items to ensure that they have been appropriately addressed and implemented. It is essential as part of the audit process to confirm the implementation and completion of action items. Management must be informed whenever there is a failure or delay in the completion of a corrective action and a new time line needs to be established.

Executive management must be an integral part of resource allocation and any organizational restructuring that may be necessary. Correcting audit action items and continuous improvement requires a team effort spearheaded by effective leadership.

4 CAN AN ORGANIZATION DO WITHOUT A QUALITY ASSESSMENT PROGRAM?

What is the cost of failing to implement and maintain an effective QAP?

Ethical considerations
Increased external regulatory attention
Potential market recalls
Loss of company credibility with stockholders and the general public
Potential civil and/or federal prosecution due to lack of quality oversight and resulting compliance violations
Decreased quality
Company management unaware of organizational reality
Unmotivated workforce—low morale
Potential injury and or death to consumers and public health

The FDA continues to increase regulatory focus on active pharmaceutical ingredients (APIs), biologic products, and clinical trial materials. The implementation of the FDA Modernization Act of 1997 has increased the agency's inspection latitude relative to raw material suppliers and over-the-counter (OTC) manufacturers. The FDA also expects companies that use external suppliers, contract laboratories and manufactures to monitor them as an extension of their operation. The contracting company must utilize its QA unit to ensure that the supplier or contracted complies with applicable GMPs, regulatory requirements, and contractual agreements. These contractual agreements between the client company or contractor and the contracted or supplier are typically defined in technical or quality agreements. The role and responsibilities of the contractor's QA units is delineated relative to the service or product being provided. The quality of the product or service to be supplied is also delineated very specifically and should address all aspects of purchasing, manufacturing, release, storage, shipping, and any potential recall of the marketed product. Needless to say, a QAP must extend beyond an organization's parameters and monitor supplier relationships. Overall, FDA-regulated industries would do well to respond to the increase in regulatory scrutiny by installing more innovative and useful QAP.

5 REGULATORY REALITY CHECK

How would you answer these citations and what would you do to prevent getting them in the first place?

5.1 FD-483 Observation

> Failure to conduct planned and periodic internal audits of the quality assurance program in accordance with written procedures. For example, no internal audit conducted since 1995. Interestingly, this organization had the foresight to establish a quality assurance program that required periodic internal audits. Unfortunately, they not only violated their own internal policy, but violated FDA's requirement that FDA regulated companies as part of their quality assurance, periodically monitor their operations.

5.2 Preventative Action

One way to have avoided receiving this citation would be for the organization's executive management to have verified that its QA unit was adhering to established corporate policy regarding internal audits. Periodic, formalized executive management review of such quality markers as internal and external audit findings, consumer complaint trends, and manufacturing deficiencies would have revealed a lack of information regarding internal operations. There is no excuse for noninvolvement on behalf of executive management, even where the responsibilities of QA has been delegated to other management personnel. The FDA holds executive management ultimately responsible for organizational shortcomings and lack of quality and compliance initiatives, The agency considers noncompliance to internal policies and procedures as serious as noncompliance to FDA regulations. The fact that this company had a procedure and did not follow it constitutes a violation in itself; however, the fact that the violation relates to negligence regarding internal quality makes it far more egregious.

5.3 Warning Letter Citation

> Company has no internal audit procedures and no internal audits have been performed as required by 21, CFR 820.20 (b). This cit-

ation was levied against a medical device manufacturer. The new medical device regulations (quality systems regulations) mandate that medical device companies install comprehensive internal audit programs as well as perform periodic executive management reviews of quality and compliance data, along with internal audit findings.

5.4 Preventative Action

This citation could have been avoided by establishing a QAP that would evaluate all internal systems and procedures as well as the operations of supplier companies. This citation would be considered a gross deficiency of GMPs, which is why it led to a warning letter citation. The FDA would consider this company's executive management extremely negligent in its duties and could easily require the company to cease and desist all distribution until a comprehensive internal and external quality assessment was performed and all findings adequately addressed.

5.5 Warning Letter Citation

> Company has not conducted periodic quality assurance audits in accordance with written procedures. These procedures call for semi-annual until compliance is achieved and annual audits thereafter. The last quality assurance audit was conducted in 1996. Similar to the FD-483 observation above, this observation rose to the level of a warning letter citation because of the importance and requirement of an FDA-regulated company to employ a formalized quality assessment program.

5.6 Preventative Action

This citation suggests that a baseline audit was conducted, establishing the need for frequent quality and compliance assessments until acceptable compliance was achieved. This company had the foresight to install a QAP, respond to findings by calling for frequent audits, and reach for a level of acceptable compliance; however, its follow-through did not endure. Sustainable, endurable compliance is critical when addressing internal audit findings. Had this company designed a manageable and realistic CAP or GMP enhancement master plan, it might have been in a better position to sustain its quality and compliance efforts, thus avoiding this citation.

5.7 Warning Letter Citation

Failure of management with executive responsibility to conduct reviews of the quality system to determine its suitability and effectiveness at defined intervals as required by 21 CFR 820.20(c). This citation also related to the new medical device regulations, which codified the need for executive management with the authority to allocate resources to periodically review quality systems for their suitability and effectiveness. Good business dictates that executive management be involved at critical levels of a company's operation; however, the FDA found it necessary to mandate this requirement through specific regulations calling for executive management involvement and full awareness of the organizations operations and compliance and quality status.

5.8 Preventative Action

This company could have avoided receiving this very serious citation by establishing formal, periodic management review sessions wherein quality trends are summarized and presented by QA personnel.

5.9 Warning Letter Citation

Failure to perform audits in a timely manner to assure that the quality system is in compliance with the established quality system requirements and to determine the effectiveness of the quality system as required by 21 CFR 820.22. Interestingly, the new medical device regulations have led to a trend in GMP violations related to the lack of executive management review and QAP. As with several of the above Warning Letter Citations, this represents a failure on behalf of a medical device manufacturer to perform quality assessments which would reveal compliance deficiencies and system failures. The effectiveness of an organizations overall quality assurance program can only be measured through periodic assessments of various operations. Identify failures related to manufacturing procedures, analytical methods, or complaint handling, just to name a few, must be brought to the attention of executive management with the power to allocate resources and ensure the necessary corrections are made.

5.10 Preventative Action

In order to adequately address this citation, the company's QA personnel, along with its executive management, must commit to establishing and maintaining a comprehensive QAP that would include periodic executive management review.

The agency requires that all FDA-regulated industries conduct quality assessments. It is up to each particular industry (pharmaceuticals medical devices, biologics, etc.) to comply with this expectation by designing programs that meet the company's quality and compliance standards.

6 WORDS OF WISDOM

Quality assessment programs must foster and support continuous improvement and quality performance.

The traditional audit approach is limited and has historically failed the industry.

The new proactive approach examines the genesis of noncompliance and overhauls systems and processes where needed.

Regulatory compliance should be the minimum quality standard. While not always achievable, 100% compliance is a worthwhile goal.

Executive management must be committed to instituting and maintaining a comprehensive QAP.

A successful QAP is essential to continuous improvement.

A QAP must account for internal operations as well as the external operations of companies providing critical goods and services.

13

Preapproval Inspections: The Critical Compliance Path to Success

Martin D. Hynes III
Eli Lilly and Company, Indianapolis, Indiana, U.S.A.
Carmen Medina
Precision Consultants, Inc., Coronado, California, U.S.A.

1 PREAPPROVAL INSPECTIONS

1.1 Introduction

A preapproval inspection (PAI) is a visit by one or more food and drug investigators to review the adequacy and accuracy of the information provided in a regulatory submission [*The FDA Compliance Program Guidance Manual on Pre-Approval Inspections/Investigations* (Program 7346.832)]. The program was a direct result of the generic drug scandal of the late 1980s. Prior to this time, the Food and Drug Administration (FDA) relied on firms to provide accurate data in support of their submissions. Prior to the implementation of the PAI program, companies essentially operated on an honor system with the FDA. This honor system was effectively terminated with the generic drug scandal. In fact, in 1990 Dr. David Kessler, then commissioner of the FDA, was quoted as saying, "What I learned most from the generic drug scandal is that in the end, the data this agency acts on has to be audited. The honor system is out the window"[1]. As a result, the objectives of the PAI were to:

Ensure that facilities listed in the new drug applications (NDA) have the capabilities to fulfill the commitments to manufacture, process, control, package, and label a drug product following good manufacturing practices (cGMPs).
Ensure adequacy and accuracy of analytical methods (validated methods).
Ensure that the manufacturing process for clinical trial material, bioavailability study material, and stability studies correlates with the filed process.
Ensure that scientific evidence supports full-scale production procedures and controls.
Ensure that firms have submitted factual data.
Ensure protocols are in place to validate the manufacturing process.
Ensure equipment is adequate and suitable for use (equipment qualification).

The PAI program was first implemented in the FDA's mid-Atlantic region. This region has the highest density of pharmaceutical companies in the United States. A 1990 publication authored by Henry Avallone entitled "Mid-Atlantic Region Pharmaceutical Inspection Program" was the first formal notice of the program provided to the pharmaceutical industry. Not long after the publication of the Avallone paper, the FDA issued the first compliance manual entitled *The FDA Compliance Program Guidance Manual on Pre-Approval Inspections/Investigations* (Program 7346.832). The manual outlined a new review step that was being added to the new drug approval process, in which both the Center for Drug Evaluation and Research (CDER) and the district offices would play pivotal roles. For the first time, the FDA district office was to become involved in the NDA approval process. The role of the district office was to ensure compliance with cGMPs, as well as audit the data submitted to the FDA to ensure that it was truthful, adequate, and accurate.

1.1.1 Historical Overview

The historical routes of FDA PAIs can be traced back to the preamble of the cGMP published in the September 1978 *Federal Register* and the draft guideline on the preparation of investigational new drug (IND) products, which was issued in February 1988. The preamble to the cGMP issued in September 1978 contained two statements that are highly relevant to the PAI program.

The relevant section reads as follows:

> The commissioner finds that as stated in 211.1 these cGMP regulations apply to the preparation of any drug product for administration to humans or animals, including those still in investigational stages. It is appropriate that the process by which a drug product is manufactured in the development phase be well documented and controlled in order to assure the reproducibility of the product for further testing and subsequent commercial production. The commissioner is considering proposing additional cGMP regulations specifically designed to cover drugs in research stages [2].

The first concept made explicit in this document is that the FDA has jurisdiction over materials that are used in clinical trials prior to approval and market launch; thus some 12 years prior to the start of PAIs by the FDA it had established an industry standard that compliance to cGMPs, was required during the manufacture of clinical trial materials.

The preamble to the 1978 GMP also contains a key second concept; that is, that the expectations relating to the manufacture of clinical trial materials are different from those of commercial materials. Support for this assertion comes from the statement that the FDA was considering proposing additional cGMPs to cover drugs in the research stage. Despite this suggestion, the FDA has yet to issue cGMPs specifically for clinical trial materials, rather; it has elected to set standards through a variety of different mechanisms, such as compliance programs, inspection guides, and podium policy presentations. There are also ICH guidelines that address myriad compliance issues related to clinical trial activities. Specifically, the ICH guidelines provide direction on the following topics:

For the investigator

Investigator's qualifications and agreements
Communication with IRB/IEC
Compliance with protocol
Randomization procedures and unblinding
Informed consent of trial sutbjects
Safety reporting
Premature termination or suspension of trial

For the sponsor (or company responsible for conducting the trial)

Quality assurance (QA) and quality control aspects of trial
Contract research organization (CRO) responsibilities
Trial design
Investigator selection

Allocation of duties and functions
Notification or submission to regulatory authority (or authorities)

Additionally, the roots of the PAI program can be traced to the 1988 draft guideline on the preparation of IND products. These guidelines clearly stated that cGMPs applied to drugs that were being made for clinical trials involving human subjects.

The need for proper documentation during the drug development process was strongly emphasized in these draft IND guidelines. In addition, control of components, production controls, process controls, equipment identification, packaging, and labeling were specifically addressed. The idea that controls should increase throughout the course of the drug development process as additional experience was outlined in these guidelines.

The FDA had thus firmly established its jurisdiction over clinical trial materials through the cGMPs preamble and the IND guidelines prior to the 1989 generic drug scandal.

The first document issued by the FDA outlining the PAI program was the Mid-Atlantic Region Pharmaceutical Inspection Program. Its goal was to clarify FDA's expectations for the compliance branch of the FDA as well as the pharmaceutical industry. The program was then formalized into two compliance documents: *FDA Compliance Program Guidance Manual on Pre-Approval Inspections/Investigations* (Program 7346.832), issued in October of 1990 and *FDA Compliance Program Guidance Manual on Pre-Approval Inspection of New Animal Drug Applications (NADA)* (Program 7368.001), issued in February of 1991.

1.1.2 Shift in Inspection Trends Over Time

The PAI program has evolved over the past 10 years. This is evidenced by the revision of the FDA's compliance manuals as well as by inspection trends.

Since the first edition of the various compliance manuals referenced above, 7346.832 was revised once, in August 1994. This revision provided improved guidance for all phases of the inspection, sample collection, laboratory evaluation, and assessment of findings. Additionally, this revision outlined roles for CDER and district offices in the inspection process. The FDA also issued the *Guide to the inspection of Dosage Form Drug Manufacturers CGMP* in October 1993 and the *Guide to Inspection of Pharmaceutical Quality Control Laboratories* in July 1993. Both of these guides influenced the evolution of the PAI program into what companies experience today.

The Food and Drug Administration Modernization Act (FDAMA) of 1997, which was implemented in February of 1998, also altered the PAI process. The act significantly altered the balance of authority within the FDA. At the time the inspection program was implemented, FDA field offices were given the primary role in CGMP and data integrity issues. The pre-eminent role was shifted to the reviewing division by the 1997 modernization act. One of the provisions of the act indicated that no action by the NDA review division "may be delayed because of the unavailability of information from or because of action by field personnel." This indicates that the NDA review division may override the concerns of the district office. Additionally, it suggests that the inability of the district office to perform a scheduled inspection may be an inadequate basis for delaying approval. It should be noted that at this writing, the PAI program has not been specifically altered to reflect the changes included in the modernization act.

Additional evidence that supports the change in the inspection program over time comes from looking at the number of inspections that have been conducted in the years since the start of the program as well as the number of approvals that have been withheld as a result of the actual inspection.

The number of PAIs conducted in the years since the program's inception is shown in Fig. 1. As can be seen from Fig. 1, the FDA is conducting slightly fewer inspections over time.

Additionally, the number of inspections that resulted in a delay in approval has decreased over time, as also can be seen in Fig. 1.

This change in the number of approvals withheld is in large measure due to the fact that FDA-regulated industries have invested a great deal of time and energy in preparing for these inspections. This level of improved performance is evidenced by the fact that the FDA's recommendations to

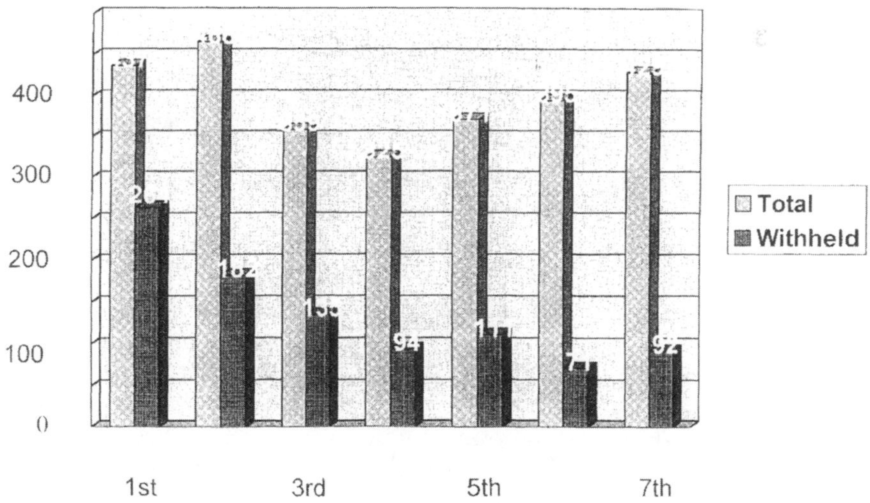

FIGURE 1 Preapproval inspections—7-year summary.

withhold approvals have declined significantly, from a high of 60% in 1990 to less than 30% in 1996.

In the late 1990s, the major reasons that companies were still failing FDA PAIs were CGMP deviations, disparities from commitments made in the application, discrepancies in records, failure to report adverse findings, and suspicion of fraud.

The following is a list of the major CGMP deficiencies that have resulted in failed PAIs [3].

- Standard operationg procedures (SOPs) incomplete, not current, or not available to the operators in the production area
- Batch production records incomplete, not recorded at time of operation, or not specific enough to document significant process steps
- Cleaning procedures not validated or not including all processing equipment and transfer implements (scoops, etc.)
- Failure to establish yields or acceptable levels of rejects for both in-process and finished product
- Failure to conduct stability studies
- Manufacturing equipment not identified and/or qualified
- Inadequate training of employees working in aseptic operations
- Inadequate process change procedures
- Validation protocols that lack acceptance criteria
- Incomplete investigations of laboratory failures
- Failure to follow United States Pharmacopeia (USP) procedures for the bacterial endotoxin test

1.1.3 Difference Between PAI and Other Types of Inspections

The FDA conducts many different types of inspections; some are very tailored to the reason for the inspection. For example, a "for-cause" or "investigator-directed" inspection occurs when the agency has received specific compliants from the public or the trade about a firm's product or practices. This type of inspection could be triggered by a consumer complaint related to an adverse drug experience or misleading information on the product's packaging. What makes this type of inspection along with most others different from a PAI is the FDA's lack of notification to the firm. The investigator arrives unannounced prepared to conduct an assessment of the firm and its product(s) and practices. Whereas the PAI is a scheduled inspection arranged by an FDA PAI manager and a company representative.

The PAI will focus on the firm's commercialization efforts related to the product for which approval is being sought. Other types of inspections are more unpredictable in nature and tend to be more general (i.e., the focus may not necessarily be on one product).

1.2 Trends Over Time: The FDA Modernization Act, Team Biologics, New Medical Device Regulations, and Quality Systems Inspection (QSI)

With the passing of the FDAMA, the agency's inspection latitude was increased and its inspectional responsibility over active pharmaceutical ingredient (API) manufacturers was reiterated. The agency did increase its inspection activity over firms that manufacture APIs and excipients. Additionally, FDA's review of foreign API manufacturers increased slightly. The onus, however, remains with the sponsor company to ensure that the manufacturer of their API is cGMP compliant and the material suitable for use.

1.2.1 Team Biologics

Similar to the FDAMA, team biologics was born of Vice President Gore's Reorganization of Government (REGO) initiative in the mid-1990s. Team biologics was a rational response by the FDA's Office of Regulatory Affairs (ORA) to control various biologic products (vaccines, blood products, invitro diagnostics) more effectively. What resulted from redirecting the manner in which biologics inspections occurred as well as changing which FDA office was responsible for field inspections was an inspection process that more closely mimicked the field inspections for CDER-regulated drugs. This means a far greater assessment of CGMP compliance to biologics' manufacturing and testing, as well as an increase in administrative, regulatory and judicial actions against them. The biologics industry has never been the same since team biologics has imposed its inspectional strategies and techniques on an industry that was unaccustomed to comprehensive compliance-focused inspections by the FDA. The number of warning letters and consent decrees levied against a large number of biologic products' manufacturers since the inception of team biologics approach is unprecedented.

1.2.2 New Medical Device Regulations

The new medical device regulations were codified in 1996, significantly altering the way in which medical device manufacturers were inspected by the FDA. The quality systems inspection technique (QSIT) was launched as part of the new strategy for inspecting the device industry. This inspection technique allowed FDA to move closer to global harmonization guidelines for regulatory auditing. This new inspection approach mandated that investigators assess seven subsystems within a device firm. The seven substances are as follows:

1. Executive management
2. Design controls
3. Material controls
4. Records, documents, and change controls
5. Equipment and facility controls
6. Production and process controls
7. Corrective and preventive actions

This focus helped field investigators conduct more effective, efficient, and comprehensive inspections of medical device manufacturers by evaluating key elements of a firm's overall quality system and compliance status. This new and improved inspection approach also led to an unprecedented number of administrative, regulator, and judicial actions against the medical device industry.

1.2.3 Quality Systems Inspection

Interestingly, along with the changing trends in biologics and medical device inspections, the approach employed by CDER during a pharmaceutical inspection has also changed. There is a trend toward assessing overall quality systems and major compliance categories within a pharmaceutical manufacturing and quality control laboratory from the top down rather than from the bottom up. More emphasis is being placed on management responsibility as well as identifying weaknesses within a particular system, along with the genesis of the weaknesses, versus identifying areas of nonconformance.

2 PREPARING FOR A PREAPPROVAL INSPECTION FROM A CORPORATE PERSPECTIVE

The preparations for PAIs must be incorporated into the overall process of drug development. Preparation efforts must start with the initial formation of the drug development team prior to the start of phase I clinical trials. One of the first tasks the team needs to complete is to author the drug development plan, which integrates the work that is to be performed by the various functions, such as medical, toxicology, metabolism, and process and product development. The work to be conducted by the process and product development groups must follow good development practices that will ensure a successful PAIs. Thus, the preparation work for a PAI needs to begin when the development team first convenes to begin planning the relevant development activities.

Prior to asking development scientists to work on teams and to prepare for a PAI, scientists should be knowledgeable about the FDA PAI strategy, how to prepare for it, and the consequences of a failed inspection. This can be accomplished by sending them to training sessions or relevant industry meetings. Additionally, it is always helpful for scientists who are new to the industry to talk with coworkers who have survived previous FDA inspections.

The chances of passing a PAI are greatly enhanced if the relevant development is carried out in conformance with well-defined quality principles and some applicable cGMPs. These quality principles should cover such topic as batch disposition, stability, process validation, training, deviations, management notification, documentation change, and history of development. These principles have been described in detail in Chapter 2 and by Hynes [3].

If the development work is planned with the goal of passing the PAI and the work complies with CGMP quality principles, there should be only minimal preparations needed for the FDA in the weeks and months just prior to the inspection itself. A number of different methodologies have been developed to help with the short-term preparation efforts. Justice and co-workers at Eli Lilly and Company have described a 10-step process to help guide these preparation efforts [4]. These steps can be found in Table 1.

2.1 The Use of Internal Audits

Many companies have utilized mock FDA inspections to help with short-term preparation efforts[3]. These mock inspections help not only identify issues prior to the inspection but also provide hands-on FDA inspection experience to personnel.

2.2 From an FDA Investigator's Viewpoint

Before an FDA investigator can make any kind of recommendation for approval, he or she must first evaluate several aspects of the firm's operation. Initially the investigator will try to determine whether or not there are any disparities between the information submitted in its application (NDA, abbreviated new drug application [ANDA], biologics license application [BLA], pre-market application [PMA] submissions) and what exists in the facility. Interestingly, disparities almost always exist. For example, the reported method may have been revised or the master or production batch record may have been tweaked, and in either case not reported to the FDA through an application amendment. Change management is a critical

TABLE 1 Ten Steps to a Successful Pre-NDA Approval Inspection

The action plan for a successful pre-NDA approval inspection covers the following 10 steps:

1. Determine overall state of CGMP compliance
2. Compile regulatory documents
3. Prepare regulatory commitment document (RCD)
4. Identify key batch records
5. Compare key batch records to regulatory documents
6. Write analytical methods history review
7. Transfer analytical methods (i.e., site certification)
8. Review analytical raw data
9. Scale-up in preparation for launch materials
10. Write development report (new drug dosage form/bulk drug substance)

quality system post commercialization and equally as important during pre-approval preparation activities. An adequate and comprehensive change control system would ensure that changes made to the process, documentation, analytical methods, facility, and so on, are carefully evaluated prior to implementation and effectively reported to the FDA when needed.

The investigator will also evaluate the various areas used to manufacture, test, package, hold, and ship for which regulatory approval is being sought. Personnel training records will be reviewed as a part of the inspection. Additionally executed manufacturing records will be analyzed, particularly those batch records related to the manufacture of clinical trial material, establishing specifications, bioequivalence, and stability claims. Raw data related to the manufactured material for which data was included in the submission will be verified. Frequently, forensic samples of the material are collected and taken back for analysis and comparison against information submitted in the application.

The process of evaluating the firm's overall operation, such as the laboratories, manufacturing sites, packaging and storage areas, quality systems, and personnel practices is usually a long one and may require an FDA team approach, as discussed above. Having invested in a third-party internal audit that mimics the FDA PAI process is extremely prudent and a smart investment of resources.

2.2.1 Use of Compliance Manual

The preapproval compliance manual referenced above (Program 7346.832) provides the guidelines for the inspection process; however, most seasoned PAI investigators will perform an assessment above and beyond what is delineated therein. The experience the investigators bring is probably the most significant guidance they will follow. Many times the PAI manager will assign the inspection to a particular specialist with expertise in that specific manufacturing process or analytical methodology.

2.3 Role of Previous History

A PAI can become a particularly challenging situation for a firm that has an unfavorable regulatory history with the agency. If a firm has been under a consent decree, a warning letter, or repeated regulatory actions by the FDA, it will have to make an extra effort to counter the agency's perception of it. It may be prudent for such a company to prepare a compliance-centered presentation delineating all the enhancements and quality improvements it has put in place since the last interaction. It is a good exercise to bring the FDA up to date regarding the company's compliance improvements before it begins evaluating anything related to the product seeking approval. This may help to erase any stigma the firm may have and provide a clean slate for the investigators to work from.

3 THE FDA ARRIVES: WHAT THE COMPANY SHOULD DO TO PREPARE

The FDA has the statutory authority to conduct inspections of facilities engaged in the development of new drugs. These inspections can range from for-cause to routine inspections. In the case of a PAI, the inspection itself will occur after the regulatory submission has been made and prior to approval. The firm can therefore estimate the timing of the inspection. The approximate timing can be further refined if the firm works closely with the PAI coordinator in the local FDA office to obtain a more precise time line for the inspection. Even though the firm may have a fairly good idea of when the inspection could take place, the firm needs to be ready for an FDA inspection any time after the regulatory submission has been made.

One important element toward ensuring readiness is having a SOP in place that provides guidance on how to manage an FDA inspection. This SOP should cover the management of the inspection, beginning with the arrival of the FDA to the exit interview and the departure of the investigators from the facility. This SOP should begin with a description of what to do when the FDA investigators arrive at the firm's reception area. The security office should notify the firm's QA group or whatever group has been designated by management to host the inspection. Once QA has been notified, it needs to assign someone from its staff to host the inspection. Shortly thereafter, QA should notify the entire site, including senior management, that the FDA has arrived to conduct a PAI.

The QA staff member who has been assigned to manage the inspection and host the investigators should greet the FDA investigators and check their credentials. Next, the firm's QA representative should verify that it is indeed a PAI and subsequently receive the FD-482 form that serves as the official "notice of inspection." Once these formalities have been completed, the QA representative should lead the FDA official to a conference room on site that has been set aside to house the investigators for the duration of their inspection. Once the FDA investigators have been allowed onto the plant site, they should be accompanied by their QA host at all times. Once situated in a confidential conference room, an overall schedule for the inspection can be developed by QA and the FDA. Additionally, the QA representative should provide and review, with the investigator(s) the firm's SOP for managing FDA inspections.

During a PAI, it is typical for the FDA investigator(s) to want to tour some of the firm's facilities that were and are involved in the development of the drugs, such as the pilot plant, and the formulation facilities, as well as the analytical laboratories and manufacturing areas. The staff should therefore have thought about how they want to deal with this request and documented the preferred approach in their SOPs. As the inspection progresses, a number of FDA requests are bound to arise, such as a request to take pictures during the plant tour and to take test article samples as well as copies of documents. It is extremely important for the firm to have given prior thought to how they will deal with these FDA requests. Will they allow the FDA to take pictures? Will they take pictures of their own? How will they respond to a request for a sample? Will the firm maintain duplicate samples of whatever the investigator collects? Who will retain the FDA receipt for the samples taken? How will documents be given to the FDA? How will they be labeled-confidential, proprietary, or both? These questions need to be addressed well in advance of the actual inspection and documented in the firm's SOPs against which all personnel involved in the FDA inspection should be well trained.

The firm's SOP on the management of the FDA inspection should be reviewed and approved by management.

3.1 How to Manage the Overall Inspection Process

The QA organization or the organization that plays host to the FDA plays a key role in the day-to-day management of the inspection. This ranges from trying to establish a daily schedule through sending out a daily summary (to the firm's management) of the day's key events. The purpose in trying to establish a daily schedule is to help the firm's staff prepare to meet the FDA's request for information and data in a timely manner, as it is in everyone's best interest to limit the duration of these inspections. For example, if the FDA wants to speak directly with some of the firm's development or manufacturing personnel, it helps to have some advance notice to schedule the person into the meeting with the FDA. Additionally, it is most helpful for the firm if it has some advance notice of what documents the FDA wants to review. This allows the firm time to make sure that the FDA has the right to review the documents it has requested (some documents, such as sales and financial data are off limits to the FDA) as well as to make sure that the documents can be retrieved from storage and moved to the inspection site in a timely manner.

A wrap-up session at the end of the day also serves as a debriefing opportunity for the firm. This is an opportune time and forum in which to ask if the investigator has any particular concerns.

As the inspection progresses, the firm's QA host should try to get a read of what observations the investigator has made and the level of concern related to those observations. This begins to provide the firm with some advance indication of what will end up on the FDA-483 form. More important, it gives the firm an opportunity to address these issues prior to their documentation on the FDA-483 and the conclusion of the inspection.

At the completion of the inspection, it is customary to have a wrap-up meeting or exit conference between the officials of the firm and the FDA investigator(s). In general, it is best to keep the number of participants in this meeting small. It should be limited to those who played a key role in the inspection, as well as corporate management from QA, regulatory, and site management.

3.2 What the FDA Will Do

The FDA will present the FDA-483 at this meeting, if one is to be given. In addition to providing the written FDA-483 list of observations, the FDA investigator will read this document and describe his or her findings. It is critical that the staff fully understand the FDA's comments despite the fact that they may not always agree on them. Some investigators are willing to listen to the firm's response to issues and at times may actually remove these findings from the FDA-483 document. This is rare however. The firm needs to decide in advance who the FDA-483 should be issued to. It is important during this exit conference to indicate a response in writing to the findings. It is possible to outline what, if any, corrective action the firm has agreed upon and its corresponding time line. If there is no plan as yet, the firm can state it is committed to correcting it, but has yet to determine a time line. It is also possible to provide a "no comment" response, but a firm should provide some form of a proactive response in writing whenever possible. Remember that the firm's responses to the FDA investigator's individual citations (FDA-483, List of observations), along with the investigator's full report of the inspection (establishment inspection report; EIR), can be obtained by anyone who wishes, through the Freedom of Information Act (FOI). As such, a firm's proactive response to all cited deficiencies goes a long way in depicting the company in a positive and responsible light. At the conclusion of the meeting, the FDA officials should be escorted out of the facility.

3.2.1 Team Versus Individual Approach

It is not uncommon for the FDA to assign more than one investigator to a PAI. Team inspections are most often assigned when the assessment calls for a particular kind of expertise related to the manufacturing process, analytical methods(s), processing or packaging equipment, or critical utility within the facility.

The FDA might partner a "national expert" in vaccine fermentation or microbiological contamination controls with an experienced field investigator well versed in performing quality systems audits. The team members will go in different directions and focus on different areas, allowing them to maximize their time and cover more ground. The firm hosting the team inspection should be prepared to provide a staff member or host to each team member. No FDA investigator should ever be allowed to evaluate or observe an activity of practice without an escort.

Daily debriefings should be conducted with all FDA team members present, as should the concluding exit conference.

4 CONDUCT OF THE INSPECTION: DOCUMENTATION MOST LIKELY TO BE REVIEWED

Preapproval inspections can be very comprehensive in nature. Firms therefore need to be well prepared before the arrival of the FDA investigators. The field investigator will in all likelihood audit development data during the course of the inspection for authenticity as well as for accuracy. Additionally, he or she will determine if the facilities, personnel, equipment, and laboratory methods adequately support the manufacturing process in the NDA. The items likely to be reviewed by the FDA investigator(s) during the course of the inspection are shown in Table 2.

4.1 Good Manufacturing Procedures: New cGMPs

It is critical that an exhaustive data authentication occurs prior to the PAI, first for all reported data and secondarily for all validation data. The data

TABLE 2 The Compliance Program on Preapproval Inspections and Investigations (7346.832)

Items to be reviewed	CDER	District
Biobatch manufacture	X	X
Manufacture of drug substance	X	X
Manufacture of excipients	X	X
Raw materials—cCGMP		X
Raw materials (tests, methods, specifications)	X	X
Finished dosage form	X	X
Container and closure systems	X	X
Labeling and packaging controls		X
Labeling and packaging materials	X	
Laboratory support of methods validation	X	X
Product controls	X	X
Product test methods and specifications	X	X
Product stability	X	X
Comparison of preapproval batches to commercial	X	X
Facilities, personnel, and equipment qualification		X
Process validation	X	X
Reprocessing	X	X
Ancillary facilities	X	X

reported (in the regulatory submission) to support stability, pivotal trials, and bioequivalence and to determine final product specifications should be consistent with all the raw data maintained in the lab or in development notebooks and journals.

4.1.1 Development Records

The development report discussed below is probably the only development document the investigator(s) will focus on; however, it is prudent to prepare all development records in the event the investigator(s) wish to compare original raw data with the information contained in the formal development report. While there is no formal guidance or codified regulation requiring an official development report, since the advent of PAIs it has become both an industry standard and an FDA expectation.

4.1.2 Laboratory Records

Laboratory records will definitely be reviewed and verified by the investigator(s). The records of most interest during the PAI are the ones containing the raw data used to establish both specifications, and expiry dates, in-process, and finished product release. Additionally, out-of-specification and out-of-trend results will be evaluated, along with their corresponding investigations.

4.2 Non-cGMP Documents: The Development History

Although one of the main focuses of the FDA investigator(s) will be ensuring compliance to cGMP, the FDA may also elect to review some documents not covered under the cGMP. The most important of these would be the development report. This report should describe the history of a drug product from preliminary studies through regulatory submission. In general, it is authored by the product development scientist. If properly drafted, it should help the FDA investigator to determine the correlation between the manufacturing process covered in the regulatory submission and the process utilized to make the material used in the clinical trials, the bioavailability studies, and the registration of stability studies. The history contained in the development report thus, delineates all of the scientific evidence in a summarized manner that supports full-scale production procedures and controls.

4.2.1 The Development Report

A suggested table of contents for the development report is offered below.

Cover Sheet. Prepare a cover sheet for each development report that should include the following information:

Company name
Development report—issue number
Product name
Corresponding IND, NDA, ANDA, and BLA number, if applicable
Name of author
Approval signatures and dates

Table of Contents. The product development scientist should prepare a table of contents for each development report.

Reason for Revision. The table of contents lists those sections of the development report that have been revised and provides a brief description of the revisions.

Introduction. Briefly discuss the background on the product's intended indication or use. If applicable, complete a table that will briefly summarize the product's clinical programs.

Drug Substance Characterization.

Description of the bulk drug substance, including structural and molecular formula, process impurities or degradants, specifications, rationale for specifications, safety, approved supplier(s), test methods, validation, and stability. Include basic information on synthesis or derivation of the drug substance, or if applicable, reference supplier drug master file.

List of all applicable reports supporting drug substance characterization (e.g., certificates of analysis, validation reports, raw material specification, justification report for drug substance, other technical reports).

Formulation/Design Development.

Quantitative and qualitative formulation (theoretical unit formula per dose).

History and rationale behind the development and selection of the formulation or dose level. Include reasons for excluding other formulations or dose levels.

Description of the role of each excipient (e.g., surfactant, preservative) and cite requirements for the selection. Include discussion on the preservative system.

Discussion of excipient specifications and release parameters. Note any special requirements that are not typical compendial requirements (e.g., particle size). Include comparative evaluation for multiple sources.

List of all applicable reports supporting formulation and design development (e.g., safety reports and raw material specification justification report for excipients).

Manufacturing Process Development (Including In-Process Controls).

Brief description of the manufacturing process. Include a flow diagram. Description of the history and the rationale behind the development of the manufacturing process (i.e., justify deviations from the established manufacturing procedures that occurred during process development).

Description of the manufacturing parameters that are important to product performance and the rationale for the selection of these parameters (e.g., processing time, temperature, drying rate, mixer speeds, order of mixing, microbiological control, pressure, spray rate, storage of in-process bulk material).

Description of rework procedures, or if applicable, state that product will not be reprocessed.

Summary of the process, cleaning, and sanitization validation or verification studies performed.

List of all applicable reports supporting manufacturing process development (e.g., process validation protocols and reports, cleaning validation protocols and reports, and batch production records).

Scale-Up Technology Transfer.

Describe the rationale behind all pertinent activities that occurred during scale-up, from pilot plant production to phase III clinical production. Discuss problems, failures, and so on. Justify the absence of equivalency concerns despite differences in process parameters, equipment, facilities, and systems.

Describe the logic behind all pertinent activities that occurred during technology transfer from phase III clinical production to the commercial process (determination of full-scale commercial processes, specifications, lot size, etc.). Discuss problems, failures, and so on. Justify the absence of equivalency concerns despite the differences in process parameters. Reference the report that indicates successful technology transfer (e.g., validation report, verification report).

Manufacturing and Packaging Equipment.

Summarize information pertaining to equipment used in critical batches or runs in another table. (See Sec. 14.0, "Critical Batches/Runs.)"

Describe the equipment designs and functions critical to accommodate product requirements.

Describe the differences in equipment size, type, and operating parameters between the critical development batches and commercial batches.

List all applicable reports supporting equipment (e.g., IQ/OQ reports, evaluation of equipment comparability between lab/pilot plant and commercial plant).

Finished Product Testing and Results.

Describe the history and rationale for the finished product specifications and release parameters. Note any noncompendial requirements.

Summarize information pertaining to testing of critical batches and runs in a third table (See Sec. 14.0, "Critical Batches/Runs.")

List all applicable reports supporting the development of finished product specifications and release parameters (e.g., finished product specification justification report).

Analytical or Microbiological Method Development.

List of all analytical or microbiological methods used for excipients and the finished product.
History and rationale for the development of all major noncompendial methods for excipients and the finished product.
List of all applicable reports supporting method development (e.g., method validation reports, technical support documents).

Package Development.

History and rationale for the selection of packaging components, including product–packaging compatibility.
Description of packaging component specifications.
History and rationale for the selection of packaging component specifications.
List of all applicable reports supporting package development.
Labels for API finished product.

Product Stability.

Description of the stability of the finished product. Include batch sizes, packaging configurations (including bulk), storage conditions, analytical methodology, specifications analyzed for, and number of batches for which stability data have been generated.
Summary of the properties of the dosage form or excipients that influence product stability.
List of all applicable reports supporting product stability. (Include stability protocols, data, and reports.)

Critical Batches and Runs. A critical batch or run is one that provides primary support for label claims, indications, safety, efficacy, stability, or method development.

A batch or run listed as critical early in product development may later be determined to be noncritical if the course of development changes. In this case, these batches or runs can be deleted from the appropriate in the next issue of the development report with a brief rationale.

Complete various tables that summarize the following information pertaining to manufacturing and testing critical batches or runs:

Formula/design numbers
Product name/label claim (strength)
Batch numbers
Date of manufacture
Batch/run size
Major manufacturing/packaging equipment used

Environmental Assessment. Give a brief summary of the environmental assessment. Address environmental fate and effects. List all applicable reports supporting the environmental assessment.

Literature Review. Provide a list of the literature relevant to developing the drug substance and drug product.

Conclusions. Briefly summarize the overall development process. The summary may include a time line that displays the initiation, key intermediate steps, and completion of work in such critical areas as validation of methods, safety studies, clinical studies, scale-up, process validation, and times of regulatory submissions. The summary should also identify key issues to be resolved (including future work to be done, if any). Finally, the summary should provide a conclusive statement that links each section of the development report and addresses the equivalency of the clinical biobatches to the production batches. Include information from in vivo and invitro studies as appropriate.

The Equivalent of a Development Report for Medical Devices in the Design History File. This comprehensive file would include critical information about the device's design, changes, review, and design controls. For medical devices, the requirements for the device's history have been codified under 21 CFR 820.30.

4.2.2 Validation Master Plan

The other document that the FDA will probably want to review during the course of the PAI is the validation master plan. It is important to note that at the very least a validation protocol must be in place at the time of the inspection (unless it is a sterile product, then full validation must be completed). The validation master plan should cover the cleaning, environmental monitoring, sterilization (when necessary) process, analytical, and computers. The role of validation in PAI has been reviewed in detail by Nash [5].

Compliance with Part 11. Whether the firm is seeking approval for a pharmaceutical, biologic, or medical device product, compliance with

Part 11 regarding validation of computer software and hardware. Any software used to operate process equipment, generate electronic records, track and transmit laboratory data, automate quality systems, operate critical utilities or software that is itself a medical device must be validated to ensure accuracy, reliability, consistency, and the ability to discern invalid or altered records. Where off-the-shelf software is used for any of the above cGMP activities, it must also be validated for its intended use. For additional information on compliance issues related to computer hardware and software validation refer to Chap. 7 of this book.

5 INTERNATIONAL INSPECTIONS: DIFFERENCE FROM DOMESTIC OR U.S. INSPECTIONS

International inspections have been conducted by the FDA for a number of years. The number of inspections has increased dramatically over the past decade, however. For years, these inspections have focused on determining conformance with cGMP. The FDA does not distinguish between international and domestic facilities when determining conformance with cGMPs. The international PAI process has been thoroughly reviewed by Tetzlaff and Smith [6]. One important similarity is that both domestic and international PAIs are more successful when the firms have prepared and have had prior FDA inspections. For the purpose of this chapter, a few of the major differences between international and domestic inspections will therefore be highlighted (Refer to Ref. 6 for a detailed explanation of these differences.)

The Division of Emergency and Investigational Operations (CEIO) is the focal point of the foreign inspection program.

There are four basic differences between domestic and international inspections.

1. International inspections are scheduled in advance.
2. Language differences pose unique challenges.
3. International inspections are shorter.
4. Reinspections are less frequent.

The firm's ability to work with the FDA to schedule these inspections is very important. It is in everyone's best interest to make sure that all activities they and the FDA consider to be relevant are completed prior to the actual inspection. Given the language difference, it is keenly important that the firm anticipate the needs that English-speaking FDA investigators may have. For example, good judgment would suggest that some important documents be translated into English. Additionally, it might be worthwhile to prepare summary documents in English for key development activities. In addition to having documents in English, it is also important for firms to have

English-speaking personnel to ensure effective communication between the FDA and the personnel at the facility being inspected.

5.1 Critical SOPs During a Foreign Inspection

Standard operating procedures that have been translated for the FDA reviewers are of significant help during a foreign inspection. It would behoove a foreign company to translate the following SOPs into English:

Quality assurance role and responsibility
Change control
Handling out-of-specification results for laboratories and manufacturing
Stability program
Internal audit program
Qualification of suppliers and laboratories
Personnel training program
Environmental control program
Raw material qualification
Annual product reviews
Consumer complaint-handling system
Validation policy (process, cleaning, sterilization, software, methods, etc.)

For more on foreign inspections refer to Ref. [7].

5.2 From a Corporate Perspective

Given that it is difficult for the FDA to conduct reinspections, it is important for firms to correct cGMP deficiencies while the inspection is still in progress. The FDA investigator should be told that the firm is engaged in correcting some, if not all of his or her observations. This should be done during the inspection and again at the exit interview, at which the FDA may issue an FDA-483 list of observations.

5.3 Firm's Conduct During a Foreign Inspection

There is no real difference with respect to a firm's conduct during a PAI in a foreign facility. The investigator(s) will review the same type of documentation, the difference being that the FDA has no jurisdiction under the FD&C act in foreign countries. As such, the investigators do not bother to collect potential evidence toward building a case. Evidence in support of litigation or judicial action cannot be collected; therefore, the focus is on documentation, systems, equipment, process, and practices that support the product seeking distribution in the United States. International firms should make every effort to deal with and nullify any deficiencies cited during the PAI, particularly since the investigator(s) may not be able to return for a follow-up inspection for some time, if at all.

6 INTERACTION WITH MANAGEMENT: QUALITY ASSURANCE INTERACTS WITH ITS CORPORATE SENIOR MANAGEMENT

One of the valuable roles that the QA group can play is to keep corporate management apprised of all aspects of the PAI so that it is not blindsided at the conclusion of the inspection. In addition to keeping management informed about both the time line and daily general activities, QA should make sure that senior management has advance knowledge of any problem areas that could be uncovered during the course of an FDA PAI. This will allow it to assess the potential risk in a timely manner. Once the FDA is on site, QA should keep the management group apprised of the progress of the inspection. This can be accomplished through daily e-mail or voice mail messages, or ideally during the daily debriefing sessions after the FDA has gone.

Whether through daily sessions or communiqués, management should have an idea of what to expect on the FDA-483 documents if one is issued and the investigator's significant comments are concerns.

6.1 Role of the Corporate Legal Group

In addition to management, legal counsel may be needed if the FDA elects to take pictures or issue a subpoena for records the firm is withholding. Legal should not routinely be part of the inspection process; however, its input may be needed if the inspection takes an unexpected turn.

Daily debriefings should continue after the investigators have gone in order to notify management of any significant issues and to strategize for the subsequent day's activities. Many firms retreat to a so-called war room to discuss the apparent direction the inspection is headed in and how best to manage it. It may be necessary to involve the legal department in some of these discussions.

6.2 FDA Investigator Interacts with Management

The investigators will be doing some convening of their own. If there is a team, they may get together to discuss the course of the audit and their respective observations. They may decide upon a collective strategy for the inspection based on their findings and individual expertise. It may also become necessary for the team or any single investigator conducting the PAI to phone the compliance office or management, either for guidance on significant issues or to request additional support. It is not uncommon for an investigator to call his or her resident post or headquarters for additional support if the identified deficiencies are beyond his or her expertise or too significant to dealt with alone.

Whether or not an investigator seeks additional support or is accompanied by a partner, he or she is required to prepare a detailed report of the audit that will accompany the FDA-483 list of observations if one was issued. He or she will also recommend approval or suggest that FDA withhold approval based on the inspection . The district office and headquarters do not necessarily have to adhere to the recommendations; however, they typically do.

7 INSPECTION CLOSE-OUT: ISSUANCE OF AN FDA-483 DOCUMENT

The FDA investigator will document his or her findings of noncompliance to the cGMPs on an FDA-483. These documents can range in length from one observation on one page to multiple observations on multiple pages. Some have been known to exceed 100 pages. The fewer the number of observations on a FDA-483 the better, although it is unlikely that you will complete a PAI without the FDA documenting at least a few observations. The list of observations will be issued during the close-out meeting at the conclusion of the inspection. It is customary for the FDA investigator to read or review the observations with the staff during the course of the exit conference. The FDA-483 is then issued to a member of the firm's management, such as the QA Director or chief executive officer.

7.1 Responding to the FDA

A formal written response to the FDA-483 list of observations is always warranted. Firms should have included as part of their SOP for FDA inspections a clear time line for responding to an investigator's observations. The response should come from the person to whom the FDA-483 was issued, but pulled together collectively by the firm's personnel. The response should either rebut the FDA's observations or describe what corrective action the firm plans to take and institutionalize as a result of the observations, such as changes to specific policies, procedures, and systems. The firm's response needs to be complete, addressing each and every FDA observation. These written responses should be issued by the firm several weeks after the inspection. Obviously, it is in the best interest of the firm to deal with these observations in a timely manner. This is especially true if any of the investigator's observations could significantly delay approval of the application.

The firm can request that a copy of its response be attached to the FDA-483 when it is made available through the Freedom of Information Act.

7.2 Use of Freedom of Information (FOI) Act

The FOI Act can be useful to a firm in two important ways. First, in advance of an FDA inspection, the firm should monitor the FOI results of other PAIs, particularly those conducted within its FDA district, as well as those conducted by the FDA investigators who will be doing their inspection. The information contained in these documents will help a firm understand what the FDA investigators are looking for. The information obtained through the FOI Act can thus be extremely valuable in preparing for the actual FDA inspection.

Some of the information available through the FOI includes FD-483, and EIRs. The EIR is the FDA investigator's report of the entire inspection. The FDA-authored documents will be redacted prior to being photocopied and sent to the requestor. The redaction process is designed to black out any information that the FDA considers confidential or proprietary. It is a good idea for firms to stamp certain documents proprietary and/or confidential prior to the investigators collecting the records.

A firm needs to think carefully about requesting a copy of an EIR and/or the FDA-483 for an inspection conducted at one of its facilities. There are advantages, for example, to knowing what is contained in the EIR, since it is not typically provided to the firm after an inspection. There is, however, a new program in which the FDA sends a copy of the EIR to the firm fairly soon after the inspection and without the firm requesting it. The program has not been implemented consistently across the United States, however. It is important to note that once a copy of a document is requested through the FOI Act, tracking services will pick up the request and include it in their publications. This will tip off other companies that monitor FOI requests that an inspection has been conducted at that firm. Additionally, it will allow them to request their own copies of that firm's EIR. There may be information in the document that should not be in the hands of the competition, even after it has been redacted by the FDA.

7.3 Withholding Product Approval

More firms pass their PAIs than fail them. There are many reasons why a company might fail a PAI. The most obvious would be a lack of compliance with cGMPs, and the lack of compliance is not limited to the product seeking approval. The firm may have manufactured the preapproval batches in conformance with Good Manufacturing Practices (GMPs); however, the facility presents GMP violations. Another reason the FDA may withhold product approval is when there are many disparities between the application submitted and what the investigators actually found during the inspection. This chapter discusses the importance of consistency between what is promised in the regulatory submission and what the investigators observe during the inspection. Another significant reason for withholding approval is a lack of adequate systems for investigating out-of-specification and out-of-trend data as well as investigating manufacturing deviations. This reason for withholding approval has taken on further meaning since Judge Wolin's ruling during the Barr Laboratory incident. The aforementioned reasons are sufficient for failing the inspection, whereas other types of deficiencies, such as limited stability data, incomplete validation, and incomplete personnel training, may only delay approval. An interesting phenomenon that frequently occurs is a PAI turning into a for-cause or investigator-directed inspection. This may occur if during the PAI the investigators observe deficiencies that impact not only the product seeking approval but currently marketed products. The types of problems that could lead to this expanded investigation are deficiencies with qualification of raw materials and components, lack of cleaning validation, lack of adequate personnel training, lack of contamination control and environmental monitoring, and any deficiency that may pose a systemic threat to any product manufactured within that facility. When the focus goes beyond the product for which approval is being sought and begins to target systemic cGMP deficiencies, it is very likely that the firm will suffer regulatory consequences.

7.4 Follow-Up Inspections

In either case, whether the FDA withholds product approval for the product for which approval is being sought or expands the investigation because it has identified broader cGMP deficiencies, it behooves the firm to pursue a corrective action plan that will allow it to host a follow-up inspection in the not too distant future. Follow-up inspections are particularly challenging because the investigators have an impression about the firm's compliance status that will usually taint the investigator's perspective. The firm's executive management must ensure that the follow-up inspection does not identify the same compliance problems that were identified during the first inspection. This repetition of nonconformance or pattern of deficiencies could lead to a warning letter or consent decree. The corrective action plan must be carefully crafted and implemented as to ensure a successful reinspection.

8 CONCLUSION

As this chapter has pointed out, the PAI process is an arduous one with myriad challenges. The words of wisdom below provide a summary of the key compliance components that can lead to a successful PAI. In the end, whether a firm passes or fails the PAI, the result should be a meaningful learning experience that will lead to future successes.

9 REGULATORY REALITY CHECK

> [FDA-483 Citation #1] Firm was not prepared to host Preapproval Inspection, although it had been scheduled with the FDA and the QA Director had indicated the timing was appropriate. Several critical pieces of manufacturing equipment, while in the area, were not fully installed or qualified. Additionally, there were several incomplete validation protocols for the manufacturing and packaging operations.

The firm could have prevented this situation and citation by ensuring that the appropriate person was responsible for communicating with the FDA's PAI manager to effectively schedule the inspection. The appropriate person is someone who has adequate interface with and information from the various departments, units, and personnel that need to be inspection-ready. Ideally, one individual is selected to communicate with various units within the company as well as the FDA. This person could be the PAI project manager or someone in the QA area. It is essential that this individual be apprised of the status of critical inspection activities, such as equipment qualification, process validation, and overall inspection-readiness.

There has been an increase in the number of firms that fail the inspection because they are not ready to be audited when the investigators arrive. Many firms expect that after they have officially made their application submission they will have ample time to make "last-minute" changes to methods, processes, and documentation; however, FDA is responding to inspection mandates much more quickly than ever before and faster than many firms expect. After having submitted its application, it would be unusual for a firm to have to wait beyond 220 days for the PAI. Additionally, in an effort not to waste resources, FDA coordinates the timing of PAIs with the firm in order to ensure that the firm is indeed ready to host a comprehensive PAI. The firm should be ready to host a PAI at the time of submitting its NDA, ANDA, BLA, or PMA and should not schedule the PAI with its district if even the slightest possibility of not being inspection-ready exists.

10 WORDS OF WISDOM

1. Start PAI preparations when new drug development is initiated
2. Author the development report concurrently with all development activities, not after.
3. Safeguard against unexpected weaknesses in critical areas by conducting frequent and effective internal audits.
4. Develop a formalized and comprehensive PAI readiness project plan and assign someone to manage it.
5. Perform a thorough review of all raw data and ensure they are consistent with the reported data.
6. Ensure all pivotal batch records are reviewed for accuracy, completeness, and nonconformances.
7. Ensure there are no disparities between the commitments made in the application and what the investigators find in the facility during the PAI.
8. Develop a strategy for inspection-readiness as well as for hosting the actual PAI.
9. Develop a cross-functional team approach for successful implementation of the inspection-readiness plan.
10. Perform a comprehensive, mock, FDA-style audit of the entire operation prior to the real one, and far enough in advance to correct any deficiencies found.
11. Ensure the timing of the PAI is based upon realistic completion dates for critical activities, such as equipment qualification, personnel training, process of method validation, and data authentication.

REFERENCES

1. D. Kessler, Watchdog or paper tiger? *Medi Adv News*. pp. 23–25, April 1991.
2. U.S. Department of Health and Human Services, Food and Drug Administration, Fed Reg, *Preamble to Good Manufacturing Practices for Human and Veterinary Drugs*, 43 (190):45014–45089, Aug. 1978.
3. M.D. Hynes, III, *Preparing for FDA Pre-Approval Inspections Throughout the Drug Development Process: Gearing Up for Pre-Approval Inspections.* Presented at the Institute for International Research, Miami, FL, February, 1999.
4. R. M. Justice, J.O. Rodriguez, W.J. Chiasson, Ten steps to ensure a successful pre-NDA approval inspection. *J Parenteral Sci Technol* 47: 89–92, 1993 .
5. R.A. Nash, The validation of pharmaceutical processes, Hynes, M.D. III. (ed.) In: *Preparing for FDA Pre-Approval Inspections.* New York, Marcel Dekker 1998; pp. 161–185.

6. R.F. Tetzlaff, P.D. Smith, Food and Drug Administration International Pre-Approval Inspections, Hynes, M.D. III. (ed.) In: *Preparing for FDA Pre-Approval Inspections.* New York, Marcel Dekker, 1998; pp. 31–93.
7. M.D. Hynes, Developing a strategic approach to preparing for a successful pre-NDA approval inspection, Hynes, M.D. III. (ed.) In: *Preparing for FDA Pre-Approval Inspections.* New York, Marcel Dekker 1998; pp. 11–29.

14

The Impact of Total Quality Performance on Compliance

Carmen Medina
Precision Consultants, Inc., Coronado, California, U.S.A.

1 INTRODUCTION

This chapter has been divided into three distinct sections, all of which focus on a forward quality perspective to ensure that the highest level of quality assurance (QA) and durable compliance is maintained throughout the drug development process, during initial product launch, and after the product has been commercialized.

The first section describes essential QA activities during and throughout clinical trial activities. The second section describes critical compliance requirements during the often precarious period of initial launch or commercialization, with a heavy emphasis placed upon the most crucial quality systems required at this challenging time.

The third and final section presents what the author believes to be a necessary culture of quality throughout an organization in order to maintain

With special contributions from Judith Beach, Quintiles Transnational Corp., Research Triangle Park, North Carolina; Valerie Palumbo, Quintiles Transnational Corp., Research Triangle Park, North Carolina; Douglas B. Poucher, KMI/PAREXEL, Atlanta, Georgia.

a commercially competitive global position, as well as durable compliance over several decades.

2 CRITICAL COMPLIANCE ACTIVITIES DURING THE DRUG DEVELOPMENT PROCESS AND THROUGHOUT PHASE I, II, AND III CLINICAL TRIAL ACTIVITIES

Quality can be defined in many ways but is often difficult to measure. In the pharmaceutical industry, the quality of clinical trials (i.e., the studies that are conducted to collect the data necessary for the approval and sale of new drugs) is measured in one way—by how well the trials and the submission of the data from them comply with the good clinical practice (GCP) regulations set forth by the regulatory bodies of the countries in which the new drugs are intended for market. Additionally, the International Conference on Harmonization (ICH) has put forth quality guidelines that address GCPs. These can be found in ICH guidelines (E-6). As such, these form the *standards* upon which the quality of clinical trials is measured. Sponsors and managers of clinical trials are responsible for putting adequate clinical trial development processes and procedures in place to provide compliance with these standards, and hence an assurance of quality. Contract research organizations (CROs) are considered managers of clinical activities and are required to adhere to GCP and ICH standards. Integrating an effective QA program into the clinical trial process to ensure the delivery of a compliant and quality product is essential, but it is not an easy task.

Installing QA throughout clinical trial activities requires a company to do many things, among them: (1) educate staff members to ensure they have a complete understanding of the regulatory requirements for conducting clinical trials and empower staff to provide continuous oversight to ensure compliance against those requirements; (2) provide comprehensive management-approved and-supported procedures and quality systems to ensure consistent and durable quality and compliance; and (3) perform independent assessments of all aspects of clinical trial activities in order to identify and correct noncompliance before the integrity of the study is compromised.

2.1 FDA Good Clinical Practice Regulations

A GCP is an international, ethical, and scientific quality standard for designing, conducting, recording, and reporting trials that involve the participation of human subjects. Compliance with GCP is also intended to provide public assurance that the rights, safety, and well-being of trial subjects are protected.

2.1.1 Good Clinical Practices (GCPs): An Evolving Regulatory Environment

With respect to research involving pharmaceuticals regulated by the Food and Drug Administration (FDA), GCPs are set forth in the twenty first Code of Federal Regulations (21 CFR) Parts 50, 56, and 312. Compliance with these parts is intended to protect the rights and safety of subjects involved in investigations filed with FDA for approval of pharmaceuticals. The agency also expects that sponsors and their agents (such as CROs) have standard operating procedures (SOPs) and other internal guidelines to assist with adherence to GCPs, and that these will be followed. The GCPs apply to studies that meet the definition of *research* as defined in the above CFR references (i.e., for clinical trials conducted under an investigational new drug application (IND), or for approved drugs when new indications and/or a change in labeling is desired).

In the United States in the mid-to late 1980s, pharmaceutical companies were struggling - as they had done a decade earlier with the implementation of the Good Laboratory Practice (GLP) and Good Manufacturing Practice (GMP) regulations – with respect to assuring compliance with the FDA's GCP regulations. These regulations spell out general requirements for the key players in the clinical trial process, namely the sponsors, monitors, investigators, and institutional review boards (IRBs). The application of GCPs and the ICH (E-6) guideline pose an interesting challenge to regulatory authorities because of the evolving nature of regulatory and compliance standards during clinical trial activities. The industry is seeing an increased number of regulatory citations (warning letters and FD-483s) against organizations involved in conducting and managing trials. Table 1 illustrates the key responsibilities of each of these entities. Interestingly, the FDA's GCP regulations do not include the requirement to have a QA function per se, as did the GLP and GMP regulations that preceded them. It was not until the early 1990s that QA both as a group of professionals independent from those involved in the actual clinical development process and as a clinical development tool aimed at assuring compliance—began to take root in the clinical development arena for pharmaceuticals. This makes clinical quality assurance (CQA) a relatively young concept for companies both in the U.S., as well as around the globe.

2.2 Impact of ICH Good Clinical Practice Guidelines on Clinical Trial Activities

In 1996, a guidance called *The International Conference on Harmonization of Technical Requirements for Registration of Pharmaceuticals for Human Use Harmonized Tripartite Guidelines for Good Clinical Practice* (ICH-GCP

TABLE 1 General Responsibilities for Sponsors, Investigators, and Institutional Review Board

Sponsors (defined as the company registering the product, making the submission, and owning the application)	Maintain an effective IND, including making timely submissions of progress and updates for changes to the research study and investigation plan.
	Promptly review all information relevant to the safety of the drug, ensure the reporting of safety information, and take actions on the information received.
	Select qualified investigators and obtain from them a commitment to conduct the investigation study in accordance with the protocol and all regulatory requirements.
	Provide investigators with the information they need to conduct the investigation properly.
	Select qualified monitors and ensure proper monitoring and oversight of the investigation study.
	Ensure that the investigation is conducted in accordance with the study protocol.
	Promptly secure compliance from noncompliant investigators or discontinue use of noncompliant investigators.
	Ensure that all FDA and all participating investigators are promptly informed of significant new adverse events or risks associated with the drug.
	Control the distribution of the investigational drug to only those investigators participating in the investigation and maintain adequate records of receipt, shipment, and disposition of the drug to all sites, as well as drug reconciliation.
	Maintain adequate and accurate records of the research for the required time period.
	Select and qualify CRO if transferring trial management.
	Install QA oversight and SOPs for the management of clinical trial activities.
Investigators	Assure the initial and continuing review and approval of the investigation by an appropriately constituted IRB.
	Promptly report all changes to the research or unanticipated risks related to the investigation to the IRB.

	Ensure the investigation is conducted according to the signed investigator statement, the investigational plan, and all applicable regulations.
	Obtain the informed consent of each human participant in the research.
	Control the investigational drug, and administer it only to person's under the investigator's own supervision or under the supervision of a subinvestigator responsible to the investigator.
	Prepare and maintain adequate and accurate case histories and observations for each individual participating as a subject in the investigation.
	Maintain adequate and accurate records of the investigation for the required period.
	Assure the timely submission of all required progress and safety reports to the sponsor.
	Adhere to the clinical protocol's inclusion and exclusion criteria.
Institutional Review Board (IRB)	Assure its membership meets the requirements specified in 21 CFR 56.107.
	Provide initial and continuing review of research involving human subjects; provide written reports of findings and actions to the investigator and the institution.
	Ensure prompt reporting of unanticipated problems involving risk to human subjects, instances of serious, continuous noncompliance on the part of an investigator, or any suspension or termination of approval is made to the IRB, appropriate institutional officials, and the sponsor.
	Ensure that information given to subjects as part of the informed consent process meets the requirements of 21 CFR Part 50.25.
Contract Research Organization (CRO)	Conduct and manage all aspects of the clinical trial.
	Formalize policies, systems, and SOPs that support these activities.
	Advise sponsor companies of critical issues related to the study protocol, patient safety, investigator compliance, and statistical analysis.
	Adequately monitor investigators.

Table 1 (Continued)

	Train field monitors.
	Install QA unit and oversight.
	Appropriately train investigators to the investigational plan.
	Adequately closeout clinical trials as needed.
	Monitor drug accountability, security, and disposition.
Contract Research Associates (CRA)	Monitor and followup on issues related to clinical trial activities and report to lead CRA.
	Evaluate the conduct of investigators involved with the clinical trial.
	Monitor the drug maintenance and distribution at the various investigator sites.
	Evaluate source data against case report forms at the various investigator sites.
	Identify and report investigator noncompliance.
	Write periodic field reports regarding observations at various investigator sites.
Manufacturer of CTM	Manufacture clinical trial material (CTM).
	Comply with GMPs as needed and when appropriate.
	Comply with the manufacturing, packaging, and labeling (MPL) protocol for the CTM.
	Ship and reconcile CTM to various investigator sites.
	Determine the retest and expiry period for CTM.
	Maintain and retain samples of all CTM.
	Ensure that manufacturing records for CTM production are maintained and available for audit.
	Ensure CTM labeling adequately supports protocol requirements.
	Conduct stability studies for all CTM.
	Ensure equipment is suitable for use for the manufacturer of CTM.
	Ensure testing methods are validated for accuracy and precision.
	Ensure raw materials and components used in the production of CTM have been adequately qualified.

guideline) was accepted and introduced globally into the clinical development process. This guidance was a collaborative effort among the European Union (EU), Japan, and the United States in order to provide a unified standard for the facilitation of the mutual acceptance of clinical data by the regulatory authorities in each jurisdiction. The guidance was developed with consideration given to the current GCP of the EU, Japan, and the United States, as well as GCP requirements in effect in Australia, Canada, the Nordic countries, and the World Health Organization (WHO). This guideline, unlike the FDA regulations, specifically discusses the requirements to have a QA program in place for the generation of clinical trial data that are intended to be submitted to regulatory authorities.

2.2.1 Educate Staff on the Requirements

Many pharmaceutical companies spend significant time and attention on developing and implementing training programs for the clinical monitors they appoint to periodically monitor the investigator sites. It is important not to overlook the importance of training all critical staff involved with clinical trial activities, such as data management and statistical analysis personnel, to the compliance requirements set forth by FDA and ICH.

Sponsors of clinical research must ensure that the staff to whom they delegate responsibilities for conducting the trials are aware of, comprehend, and comply with the requirements applicable to clinical trial activities. Management within the sponsor organization must oversee the development for its staff of appropriate training and education programs that include basic training in the GCP regulations as well as more specific SOP training to ensure the regulations are complied with and quality is installed throughout all phases of clinical trial activities. Management within sponsor organizations should be able to demonstrate the effectiveness of these training programs and periodically assess their alignment with the current regulatory requirements.

It is also important to keep staff focused on the ultimate goal of clinical research and the reason for the regulations in the first place: patient safety. To achieve this, training programs should provide an overview of each employee's specific contribution to the overall picture of the development of new drugs. For example, statisticians should receive training that helps them understand that the statistical analysis plans they develop and the data analyses they perform are not merely exercises in manipulating numbers; they are the key results of a process that the sponsor (and ultimately the FDA) rely upon to determine whether or not further clinical trials should proceed and whether or not the new drug should be approved.

Another training consideration is the detection, reporting, and handling of scientific fraud and misconduct. The emphasis of this training should be on helping staff members identify when a situation could raise suspicions of fraud or misconduct, and instruct them on what actions to take to ensure that the situation is promptly reported to appropriate management within the organization. Training staff on what to watch out for and how to follow up on their hunches and suspicions is the first step toward effectively managing questionable circumstances and ensuring prompt actions are taken to correct them.

2.2.2 Define Roles and Responsibilities Through Formal Processes

As mentioned earlier, the expectation of the FDA is not only that the applicable regulations from 21CFR will be complied with during the course of clinical trials, but also that pharmaceutical companies will develop adequate written policies and procedures to describe how the conduct their clinical research processes in compliance with the regulations. Table 2 provides a list of some procedures that Parts 50, 56, and 312 require in writing.

TABLE 2 Essential Formal Procedures Required During Clinical Trial Activities

Preparation, revision and issue of sponsor SOPs
Preparation, approval and amendment of clinical trial protocols
Investigator recruitment
Regulatory documents
Archival practices
Informed consent
Clinical supplies management
Management of SAE reports
Investigational complaint handling
Quality assurance role and responsibilities
CRO selection & evaluation
CRO monitoring activities
Clinical monitor training
Monitoring clinical trials
Prestudy site visit
Study initiation visit
Close-out visit
Case report form review
Investigational site processing SOPs
Patient subject recruitment
Organizational charts for CRO
Organizational chart for investigational site
Study files and records retention at investigational site
Drug inventory and accountability
Institutional review board SOPs: formation, operation

Written, formalized procedures provide the framework for the various processes that make up clinical trials, including protocol development, selection and monitoring of investigator sites, safety reviews and reporting, and data management and analysis. It is through clearly defined procedures, and responsibilities for all clinical processes—and the critical interfaces among them—that the clinical trial process will be adequately performed and integrated.

2.3 Monitor Compliance to the Regulations and Company Policies and Procedures

There are myriad clinical issues that can benefit from a forward quality perspective and commitment. The entire process—from synthesizing a compound in the laboratory and establishing toxicology in animal studies to subsequent pharmacological and efficacy studies in human trials—offers opportunities for and legitimately requires QA. This is an enormously costly process (typically over $100 million); therefore, embracing a forward quality mindset from the onset of clinical development and throughout clinical trial activities is extremely beneficial.

Many proactive QA measures can and should be installed during all phases of clinical trial activities, such as ensuring that assays used throughout the trials are properly developed and used; clinical trial materials are adequately manufactured, packaged, and labeled; appropriate communication between research personnel and other clinical departments is maintained; and most important, creativity and flexibility are balanced with reasonable drug development timelines.

2.4 The Role of Quality Assurance Personnel During Clinical Trial Activities

In a perfect world, everyone would do his or her job as expected and the results would always be of high quality. In the real world, staff is overworked, underprepared, and oftentimes, can lose sight of the real goal of clinical research.

As QA programs for clinical trial activities developed, so did the role of CQA. The CQA staff is charged with being the independent assessors of the quality and compliance status of many aspects of clinical trial activities. In the realm of clinical trials that can include anything from the trial design (i.e., the investigational protocol), to the review of the statistical analysis plan and the analysis of the data, to the conduct of the trial and the creation of the final study report. It is essential to ensure that QA is an integral part of all clinical trial activities; particularly since firms need to be prepared to respond to FDA's changing regulatory demands.

Some critical attributes and characteristics essential to a CQA professional are delineated as follows:

Proactive mindset—Auditing should not be its primary function. Periodic evaluations focused on compliance and *quality trends* and assessments of personnel practices and adherence to SOPs would serve to alert CQA of potential issues before serious problems arise. Sharing a vision of quality early on is key.

Knowledge transfer and training capabilities—The ability to create learning communities within the various organizations that must interface with one another (such as the sponsor, CRO, investigator sites, and IRBs) can provide a powerful engine driving all entities toward excellence and ensuring long-term success for all aspects of clinical trial activities.

Finesse with the regulatory authorities—Knowing how to host a regulatory inspection and appropriately respond to an auditor's concerns is a true craft requiring experience and finesse. When regulatory inspections are carried out effectively, CQA and other concerned staff can serve to intercept regulatory liabilities and major compliance challenges.

Audit skills—A CQA professional should be quite skilled in the area of auditing and capable of evaluating everything from documentation within the CRO, sponsor, and investigator sites.

Designing a corrective action plan—Along with the ability to audit and evaluate the status of documentation and organizational practices is the need to know how to address and correct deficiencies that are identified. Audit activities should always be paired with the development of a comprehensive enhancement plan that addresses, corrects, and prevents all deficient practices observed during the audit. An audit without this corrective and preventative counterpart is at best useless and at worst quite dangerous if the noted deficiencies go uncorrected.

Submission planning and review skills—It is not uncommon for CQA to become involved with the planning of the regulatory submission for clinical filings, such as the IND, and registration filings, such as the New Drug Application (NDA), Biologics Licence Application (BLA), or Premarket Application (PMA) and 510(k). Clinical Quality Assurance (QA) should assess the acceptability of these filings prior to their submission to the agency.

Building quality throughout all aspects of the clinical trial process and installing QA oversight early on will not only serve to assure the integrity of the study, but will most certainly support the compliance requirements set forth by FDA.

Let us consider some aspects of the clinical development process and CQA's role and responsibilities over each aspect.

1. Adequate study design—The obvious first step in assuring the quality of the clinical phases of development starts with the design of the studies making up each phase of the development. The overall clinical trial plan and design must take into account all of the information already known about the investigational product under study at each progressive stage of clinical development. The design must include appropriate procedures to enable the collection of relevant data to ensure the that objectives of the trial and key endpoints can be evaluated. In order to ensure these goals are met, the clinical trial design team must consist of staff with the requisite knowledge and experience to set the parameters for the trial's conduct and the collection of data. These are the people who assure the overall quality of the design of the trial(s) through the careful planning and consideration of all primary and secondary objectives and endpoints, all procedures and data needed for analysis, and all methods needed for the evaluation and analysis of the data.
2. Design of CRFs—The design of the case report form (typically referred to as the CRF) should have some degree of QA oversight. It is essential that the CRF adequately reflect the investigational issues delineated in the study protocol. The design of the CRF is central to the investigator's ability to adequately capture and document the subject's behavior, compliance level, and reaction to the drug or placebo. Additionally, the design of the CRF should facilitate the monitor's ability to assess investigator compliance against the protocol. Finally, a well-defined CRF will substantially support subsequent statistical analysis of the clinical trial.
3. Validation of computer systems—The role of QA is of particular importance with respect to complying with Part 11, related to electronic records and computer validation. Clinical QA must ensure that any computerized system associated with a study is adequately validated and that the documentation for that validation is concise, complete, and available for audit. Clinical QA must provide the appropriate guidance to both the validation and information technology groups to ensure that clinical studies are not invalidated because of inadequately validated software or hardware.
4. Standardization and transfer of clinical laboratory data—The role of QA must not be overlooked, particularly with respect to data authentication and the transfer of laboratory data from investigator sites to a centralized lab. Clinical QA must ensure that CRAs

are routinely reconciled against available source documents. Additionally, a periodic review by CQA of laboratory data against raw data will significantly safeguard the study.

5. Statistical analysis plan—while it is not essential for QA to have any extensive knowledge of statistical evaluation, it is important for CQA to ensure that the analysts follow the statistical plan and do not perform analysis at inappropriate time intervals, such as an interim statistical evaluation. An interim statistical analysis may serve to invalidate and confound the data, causing the study to be out of compliance and statistically invalid.
6. Accuracy of the final study report—The role of CQA is essential at this phase of trial activity. The final study report must be audited for accuracy and consistency against a broad body of data. If a study is terminated early or extended beyond the protocol period, the final report must reflect all the safeguards employed to ensure both patient safety and data integrity.
7. Selection and qualification of the monitoring group—Clinical QA must ensure that there is a formalized selection and qualification program for all individuals involved with field monitoring of investigators, as well as the supervisory staff responsible for field monitoring staff. The selection and qualification process is a critical one, particularly since these are the individuals who are responsible for evaluating investigator compliance and patient safety. Without well-trained field-monitoring personnel, the statistical data obtained could prove faulty, seriously assaulting the validity of the study. The safety of study subjects is another responsibility of field monitors. Investigators who disregard patients' rights must be identified and disqualified by field staff. CQA must ensure that field personnel understand the protocol requirements, as well as human subject protection laws and how to communicate such violations to the corresponding IRB.
8. Monitoring the monitors—The need for CQA to be involved in the selection and qualification of monitoring staff is a prerequisite for ensuring that qualified monitors interface with clinical investigators and provide meaningful feedback to either the sponsor organization or the CRO. It is not uncommon for the sponsor organization and the CRO to subcontract monitoring activities. In fact, there are organizations that have recently cropped up for the sole purpose of providing clinical monitoring support to sponsors and CROs. The role of CQA becomes particularly important in this case. Clinical QA must ensure that subcontracted monitors are adequately trained by their organizations and that

the documentation of the training is available for audit. Where adequate training has not been provided, CQA will be held responsible for not preparing these monitors before charging them with the role of monitoring a clinical study.

9. CRO/sponsor agreements—This is an area of great interest to both the FDA and the industry. It is essential that these agreements and/or contracts specifically address the roles and responsibilities of both the sponsor organization and the CRO. In the event that the sponsor organization chooses to transfer all clinical trial management to the CRO, the contract must specifically state this and delineate what specific responsibilities have been transferred. The relationship between the CRO and the sponsor must be clearly defined and the scope of work must be concisely delineated. It is helpful to include in the agreement a matrix that delineates all of the various aspects and responsibilities of the trial and differentiates between what is the CRO's function versus the sponsor's. The FDA typically audits these agreements, and where the roles and responsibilities are not clearly assigned, regulatory liability will be.

2.5 Compliance and Quality Issues: Follow-up at the Investigator Site

As briefly mentioned above, staff members designated as "monitors" of a trial by the sponsor or CRO have one of the more crucial roles in the clinical trial process. Monitors are the front line of oversight and convey information about the way the clinical trial is being conducted at the participating investigator sites. Monitors play a key role in the early detection of noncompliance on the part of investigators and in reporting information about the noncompliance to the relevant staff involved in the project. This early detection mechanism allows the sponsor to be kept informed and ready to take action to secure compliance when necessary. Through this early detection and corrective action process, the sponsor ensures that the rights and safety of patients are protected throughout the clinical trial process.

In an effort to layer CQA throughout all aspects of clinical trial monitoring, the following levels of personnel should be considered:

Principal investigator—The individual responsible for conducting the study at the clinical site and selecting additional investigators when needed. The investigator must adhere to the clinical protocol and comply with inclusion and exclusion criteria. Clinical QA must ensure that monitors know how to identify principal investigator noncompliance.

Subinvestigator—This individual is assigned by the principle investigator and is expected to have experience with research studies, as well as have received specific training regarding the study protocol from the principle investigator, sponsor, and/or CRO.

Clinical QA must evaluate the conduct of subinvestigators with the same vigor and expectations as it does the principle. It is not uncommon for subinvestigators to receive little to no training or direction from the lead investigator throughout the course of a study.

Clinical site personnel—These individuals are responsible for the conduct of the study within the context of the daily clinical environment. Clinical QA must ensure that it is aware of such study medication requirements as appropriate storage and security as well as the importance of maintaining the privacy of study subjects and adequate records.

Lead CRA—The lead contract research associate is usually responsible for interfacing with field monitors and addressing investigator noncompliance. They review field reports provided by field staff and determine adequate follow-up to problems that have been identified by the field staff as well as investigators. Clinical QA must ensure that the lead CRA is adequately trained to respond comprehensively and in a timely manner. It must also be formally trained on the notification process required when patient safety or data integrity is in question.

2.5.1 Monitoring of Source Data

The FDA defines *source data* as all information in original records and certified copies of original records of clinical findings, observations, or other activities in a clinical trial necessary for the reconstruction and evaluation of the trial.

Source data are contained in *source documents*, which means original documents and records, including but not limited to hospital records, clinical and office charts, laboratory notes, memoranda, subjects' diaries or evaluation checklists, pharmacy dispensing records, recorded data from automated instruments, copies or transcriptions certified after verification as being accurate and complete, microfiches, photographic negatives, microfilm or magnetic media, X-rays, subject files, and records kept at the pharmacy, at the laboratories, and at medical-technical departments involved in the clinical trial. Source documents should be retained to enable a reconstruction and evaluation of the trial. When original observations are entered directly into a computerized system, the electronic record is the source document. Clinical investigators should retain either the original or a certified copy of all source documents send to a Sponsor or CRO, including query resolution.

Accordingly a *certified copy* is regarded as a source document. Note, however, that carbon copies or "NDC" copies are deemed by FDA to be *original* source documents. As such, CRFs in three-page NDC form are all original source documents. When speaking of certified copies, we are referring to photocopies of original source documents.

A *certified copy* is defined as "a copy of original information that has been verified, as indicated by dated signature, as an exact copy having all the same attributes and information as the original." This means that the person signing and dating the copy has gone through the original and the copy, and verified, signed, and dated that the copy is indeed identical to the original. For example, this would include noting on the copy where the ink color in the original is of a different color than black or the color of the original paper is other than white.

Reviewing these source documents is part of the data authentication CQA must periodically perform during the study.

In some cases, it may be prudent to obtain a certified copy for the subject's clinical trial file. For example, if the records are not owned by the investigator or under his or her control, such as patient medical records held by the hospital, depending on the hospital's archiving procedures it would be advisable to obtain certified copies for the clinical trial file in case the hospital loses, misplaces, or intentionally destroys the documents under an established record retention schedule. This is all part of CQA's oversight role regarding the appropriate transfer and authentication of raw data and source documents.

2.5.2 Safety Reviews and Reporting

The regulations require that the sponsor organization reports any SAE identified during the course of a clinical trial. There are specific reporting time lines for SAEs, depending on whether the trial is being conducted domestically or in a foreign country. The many methods companies use to collect complaint information span the continuum. Many sponsors employ software to electronically capture and trend complaints, while others use elaborate manual systems. Where an electronic system is employed, CQA must assure that it is compliant with Part 11. It is also important for CQA to ensure that SAEs are trended. Of equal importance, CQA should require that nonreportable events are also trended. Trending of nonreportable events can prove to be very valuable during phase IV postmarketing studies, although this is not an FDA requirement.

2.5.3 Key IRB Concerns

An IRB or an independent ethics committee (IEC) is an administrative body that has been formally designated to review and monitor biomedical research involving human subjects to ensure that their rights and welfare are protected. The IRB has the authority to approve, require modifications in or disapprove research that falls within its jurisdiction as specified by the relevant federal regulations and local institutional policy.

The federal regulations apply "to all research involving human subjects conducted, supported, or otherwise subject to regulation by any federal department or agency that has adopted the human subjects regulations."

To have jurisdiction under Health and Human Service's (HHS) regulations (45 CFR Part 46), the activity must involve research and human subjects. Research is defined as a systematic investigation, including research development testing and evaluation designed to develop or contribute to generalization knowledge.

Clinical QA must ensure that any IRB charged with providing clinical trial oversight has formalized procedures for their operation. IRBs come in many shapes and sizes. There are commercial IRBs, as well as IRBs that are associated with a particular institution or hospital. Regardless of their association, they must have SOPs for the conduct of their business. The FDA has increased its inspection activity of IRBs, with a particular focus on the procedures for the protection of human subjects. The FDA has recently cited several IRBs for failure to develop written procedures for conducting periodic review of ongoing research; failure to assure that clinical investigators were adequately trained; lack of procedures to ensure that changes in approved research we are promptly reported; and failure to document IRB voting practices.

2.5.4 Protection of Human Subjects

On November 3, 1999, HHS published proposed regulations (64 Fed. Reg. 59918; 45 C.F.R. Parts 160–164) for "Standards for privacy of individually identifiable health information" to implement the privacy requirements of the administrative simplification subtitle of the Health Insurance Portability and Accountability Act of 1996 (HIPAA). The rule proposes standards to protect the privacy of individually identifiable health information maintained or transmitted in connection with certain administrative and financial transactions. If finalized as proposed, the rule would apply to "covered entities" that is, health plans, health care clearinghouses, and health care providers who transmit health data electronically. The rule proposes standards with respect to the rights of individuals who are the subject of this information, procedures for the exercise of those rights, and the authorized and required uses and disclosure of this information. Pharmaceutical companies (and clinical research organizations) would be required to gain IRB approval for access to individually identifiable patient information for research without patient consent. The board may be an existing IRB or an equivalent "privacy board." HHS' purpose in creating the privacy board is to extend the federal "common rule" human subject protection to the private sphere. Currently the Common Rule applies only to research supported by federal funds or subject to a Multiple project assurance (MPA) with the Office of Protection from Research Risks (OPRR), in which a particular IRB has agreed to apply the common rule uniformly for all research conducted at that institution; Otherwise at present for private research, identifiable information may be used without patient consent and without IRB approval. Under the proposed rule, all researchers seeking to obtain records without patient consent would have to obtain prior certification from either an IRB or a privacy board.

The board would review the project using the following four criteria that were developed to protect individuals' privacy interests for research using existing medical records: (1) Would the research be impracticable without the information? (2) Is the research important enough to outweigh the intrusion into patient privacy? (3) Is there an adequate plan to protect identifying information from improper use and disclosure?; and (4) Is there an adequate plan to destroy the identifiers at the earliest opportunity? [45 C.F.R. 163.510(j)].

In addition, at present the involvement of an IRB is not required for retrospective research studies involving analyzing of data from existing medical records conducted by pharmaceutical and device companies to determine the long-term safety of drugs and devices that have already been approved for marketing generally. Nevertheless, for such studies, CQA should encourage the approval by an IRB or internal privacy board.

2.5.5 The Sponsor–CRO Relationship

Indeed, as the entity seeking approval of new drug products or devices, the sponsor of clinical trials is ultimately held responsible for assuring that the trials and the resulting data have been managed, reported, and analyzed in compliance with the regulations set forth by the FDA. The regulations do permit (Part 312.52) the transfer of obligations to contract organizations. As mentioned above, when the management of clinical trial activities is transferred to a CRO, CQA must ensure that the contract or agreement between the two organizations clearly delineates what responsibilities have been transferred.

2.5.6 Electronic Records

On August 20, 1997, FDA issued regulations regarding the use of records and signatures in an electronic format that are submitted to or appear in records required by FDA (21 CFR Part 11). These regulations describe the technical and procedural requirements that must be met when using electronic records or electronic signatures (ERES) on such documents. The industry is striving to comply with these regulations, which are proving time-consuming and costly. Certain long-standing problems with QA issues should be alleviated once compliance with these regulations has been achieved, however.

Though unclear about what would be expected of a clinical research monitor with respect to monitoring an investigator site for compliance with Part 11, FDA unofficially has stated that it would expect monitoring for particularly egregious noncompliance and considers serious deviations anything that makes it difficult to audit or interpret data or that undermines the integrity of the data. For older electronic systems (i.e., so-called legacy systems) FDA realizes that compliance may take more time, but expects to see a reasonable timetable for corrective measures. In fact, the agency already has issued 483s and warning letters citing Part 11 deviations, such as inadequate password protection and the absence of an audit trail. In view of this information, one would expect that a monitor should routinely make inquiries regarding electronic record keeping of the site and "spot" verify whenever possible. As the first step, during site selection or site initiation, determine whether the site uses ERES for documents that are required to support submissions to FDA, which would include electronic CRF, electronic source documents, scanned paper source documents, and digitized scans. FDA has issued "Guidance for Industry: Use of Computerized Systems in Clinical Trials," from which GCQA may infer that a monitor should make the following inquiries of the clinical trial site:

Whether or not the site is able to generate a complete copy of records in both human-readable (e.g., paper) or electronic form suitable for inspection, review, and copying by FDA

Whether or not each staff member has his own password that is not shared, that is regularly changed and not reused, and that is canceled upon the employee's termination

Whether or not at the start of a computer session individuals can gain access only through entry of their user I.D. and password or through biometric control

Whether or not someone leaves a workstation for a long period, the individual logs off the system manually or there an automatic log-off—either of which is acceptable

Whether or not the computer system has an audit trail that maintains a record of all changes

Whether or not data that are sent over the Internet are encrypted

Whether or not the site has copies of the documentation for the computer system validation and for the installation testing from the vendor

Whether or not there are SOPs and training records for site staff concerning the computer system and programs

Whether or not the site has certified to FDA that it intends all electronic signatures to be equivalent to traditional handwritten signatures

Whether or not during subsequent monitoring visits there are indications of a software error or any issue that would imply noncompliance; if so inquire what steps have been documented to correct the problem.

2.6 Manufacture of Clinical Trial Material

Nowhere is the role of QA more important than in the manufacture of the CTM. During development and early clinical trial activities, these batches typically vary in size and formulation, significantly increasing the role CQA must play to ensure the CTM is properly manufactured, packaged, and labeled. Additionally, CQA must ensure that the batch-to-batch variability does not assault the integrity of the study.

2.6.1 Issues Related to CTM Manufacture

Manufacturing of CTM requires QA oversight, particularly where the CTM is manufactured under the auspices of research and development personnel or a contract manufacturing site. Typically, traditional commercial manufacturing practices and principles do not entirely apply; however, in some cases in which they do apply, they usually need to be more stringent. The very nature of CTM requires flexibility coupled with substantial change management. The manufacturing of CTM must be flexible enough to allow for innovative development, but the changes must be adequately managed so that they do not assault the integrity of the study and so that formulation changes are easily traceable.

Two interesting compliance categories to consider and compare are process validation and stability testing. While full validation of the CTM manufacturing process is not required, it is required for injectable products. Usually it is adequate to have process verification, along with cleaning verification, in place of full validation for CTM. Similarly, stability of CTM does not have to support purported or expected shelf life; however, at the very least, stability data must support the duration of the clinical trial and may also serve to provide expiry data to support the Chemistry, Manufacturing and Controls (CMC) section of the regulatory submission and eventual commercialization.

As the manufacture of CTM progresses from the preliminary formulation(s) and relatively small batch sizes toward the final formulation and a larger batch size (intended to gather CMC information for the purpose of registration), it is essential that CQA maintain a watchful eye over a number of critical aspects. Specifically, QA should be particularly vigilant of the manufacturing procedures, the evolution of critical documentation such as batch records and investigation reports, prevalidation activities, and qualification of raw materials, particularly the active pharmaceuticals ingredients (APIs).

The development report (previously discussed in another chapter of this book) should be well underway at this point and replete with information about preliminary formulations, rationale for changes to everything from raw materials to labeling, pivotal batches made, stability data, and so on.

The entire process of manufacturing, packaging, labeling, and shipping CTM requires extensive QA oversight, as it presents a great deal more

TABLE 3 Critical Compliance Practices When Manufacturing CTM

Proactive QA oversight
Study-specific quality control program
Qualification of Contract CTM manufacturers and laboratories
Zero-defect tolerance level (particularly important since there usually is variation lot to lot and study to study)
Clearly defined sponsor responsibilities
Reasonable adherence to the cGMPs
Equipment suitability and cleaning verification program
Formalized personnel training program (study-specific, customized training)
Characterization of raw materials
Monitor and track evolution of specifications, critical parameters, and processes
Highly vigilant change management program
Establish a cross-functional team (involved with all aspects of trial activities)
Quality control of comparator drug
Drug reconciliation drug manufacture through distribution and stock recovery
Extensive use of in-process controls and assessments where true validation is not yet possible or required (as required for parenteral and sterile products)
Production planning expertise (many variables to be considered)
Reliable formulation and analytical method development (for duration of trial)
Labeling reconciliation
Well controlled manufacturing environment (trained personnel, formalized line clearance, adequate documentation, and verification practices)
Customized and study-specific labeling and packaging procedures
Sponsor-controlled relabeling and product retrieval operations
Limit investigator site manipulation of CTM and labeling

variability than the manufacture of a commercial batch. Nowhere does variability offer a greater challenge than in labeling for randomized and double-blind studies with ascending or descending dose schemes.

For the purpose of registration or filing the regulatory submission, CQA's most important mission is to assure that what is used in phase III trials (or late phase II) is what finds its way into the regulatory submission for regulatory review and approval. Disparities between what was tested during the clinical trials in areas such as the final formulation, manufacturing process, analytical methods, raw materials, release specifications, and container-closure system will result in clinical holds and significant delays.

Lack of QA oversight and guidance during these critical final steps usually results in significant disparities between the material used during clinical trials and the information contained in the regulatory submission's CMC section.

Another compliance activity for CQA to consider throughout the conduct of the trial is adequate CTM accountability and reconciliation. Additionally, at the conclusion of a trial CQA staff should ensure that all CTM has been returned to the sponsor or CRO and adequately disposed of.

Table 3 delineates many of the compliance issues CQA must be sensitive to during the manufacture of CTM.

Table 4 compares the various areas that significantly differ when manufacturing CTM versus manufacturing commercial batches.

3 CRITICAL COMPLIANCE EFFORTS TOWARDS INITIAL COMMERCIALIZATION

Now that the reader has become familiarized with the essential quality issues associated with the full range of clinical trial activities, let us move forward and discuss the compliance challenges typically presented during initial commercialization along with some of the QA practices that can be overlaid to ensure success during this highly visible phase.

3.1 The Development Report

In Chapter 15, section 4.2.1, the author presented the importance and significance of the "Development Report." Without getting too specific or repeating what has already been stated elsewhere in the book, the development report is the tool that pulls all the clinical and early development efforts together in a manner that provides both a real-time and historical perspective. This is the report that is extensively reviewed by FDA investigators during the preapproval inspection. It should concisely detail all critical events and decisions, from concept toward commercialization. In this manner,

TABLE 4 Comparison Between Compliance Requirements for Commercial Versus Clinical Trial Material

Requirement	Commercial product	Clinical trial material
Analytical methods	Fully validated	LOD, accuracy, precision
Manufacturing process	Fully validated	Establish preliminary specs; verify lot to lot (except for sterile processes that must be fully validated)
Cleaning validation	Fully validated	Verification of cleanliness
Batch records[a]	Complies with master record; derivative of the validation study	Follows MPL protocol
Manufacturing area	Full-scale arena	Pilot facility; lab scale
Expiry Dating	Established via stability testing	Preliminary data; retest date
Raw Materials	Qualified suppliers	Proscribed suppliers
Components	Qualified/established component specs and suppliers	Varies according to protocol guide lines; documented and adequately justified
Chemistry and manufacturing controls	Established via validation	Evolving parameters
QC testing	Routine; established	Customized, study-specific
Equipment	D, I, O, and P qualified[b]	Suitable for use; verify cleanliness
Change management	Critical compliance and quality system; QA approval	Critical compliance and quality system; QA approval
Labeling	Adhere to cGMPs	Support study protocol requirements
Indications	Adhere to NDA claims	Adhere to IND claims
Packaging	Established range for reconciliation and accountability	100% reconciliation and accountability
Stability	Support label claim; established	Support study period; evolving
Handling deviations	Formalized QA system	Formalized QA system

Product release	QA Approved and released; complies with specs.	Technical and regulatory green light
Retain samples	cGMP requirement; complaint handling	CTM assessment and critical quality marker
Shipping validation	Part of stability testing	Maintain and verify CTM integrity
Overall Change[c]	Discouraged	Encouraged
Planning	Formalized; fairly consistent	Study specific; unpredictable; requires significant lead times
Operating costs	Very high with profit	Very high without profit
SOPs	Highly evolved system	Abbreviated system
CGMP adherence	Strict; pervasive; rigid enforcement	Flexible interpretation; loosely enforced

[a]The use of formalized batch records during the manufacture of commercialized or soon to become commercialized product is a cGMP requirement. Additionally, the batch record must be a reproduction of a fully evolved master production record. An effective and compliant master production record should be the derivative of the fully executed validation study that typically involves the manufacture of more than one batch of material. In contrast, the manufacture of CTM requires that a study-specific MPL protocol be followed by pilot-scale manufacturing personnel. Similar to the manufacture of commercial product, there is a tremendous need for QC during the manufacture of CTM, particularly with blinded studies in which two or more dosage forms and or placebos and actives share identical packaging and labeling. A mix-up during the distribution, dispensing, and administering of these clinical products would assault the integrity of the study and could potentially injure human subjects/patients.

[b]Design, installation, operation, and performance qualification.

[c]While the industry has devised various mechanisms for adequately managing change, for the most part change during the manufacture of commercialized product is not highly encouraged since it has the potential of disrupting formally validated systems and processes. In contrast, the manufacture of CTM is an evolving scheme that typically requires continuous refinement of specifications, manufacturing processes, release criteria, and method validation. These changes are expected and acceptable, providing the changes are appropriately managed and consistent with early commitments made in the IND. Any changes made that are significantly beyond the promises made in the IND would require a supplement and most likely a newly designed investigational protocol and case report forms.

once product is commercialized, the development report can effectively support trouble shooting activities in both the laboratory and manufacturing arenas.

If it has not already been done during the product's development stage, it is essential that during initial commercialization, executive management and QA jointly examine the firm's overall commitment to compliance and total quality performance. In an FDA-regulated environment, an early commitment to forward quality will set the stage for durable, long-term compliance.

3.1.1 Important Quality Systems During Initial Launch

An early assessment of the firm's perspective of and quality commitment to the following areas can prevent the show-stopping compliance challenges that typically arise during early commercialization:

- Documentation system
- Control of raw/starting materials
- Facility design
- Design and qualification of critical support utilities
- Validation of computer systems, methods, manufacturing, and cleaning processes
- Labeling controls
- Change evaluation and management
- Recall prevention
- Laboratory practices and controls
- Personnel training program
- Environmental monitoring
- Operator self-inspection program
- Supplier and vendor qualification program
- Quality agreements with suppliers and vendors
- Company mission statement
- Executive management commitment and involvement
- Proactive QA unit
- Preapproval inspection preparations
- Complaint-handling system

These systems or programs form a large part of the firm's QA infrastructure and must be fully evolved during the early stages of commercialization. Initial commercialization offers a different set of challenges than continued commercialization. This is the pivotal stage at which a firm establishes its company mission with its employees and with the FDA and other regulatory authorities. This is the phase during which the company's commitment to forward quality and continuous improvement must be repeatedly stated and demonstrated to its employees, shareholders, and the regulatory agencies. Development of a viable product and getting it into the marketplace are preliminary hurdles and just the beginning for an FDA-regulated company. Keeping the product on the market, maintaining quality, and sustaining compliance have become increasingly challenging tasks for a pharmaceutical, medical device, and biologics firm given FDA's evolving regulatory and compliance environment. A number of mature pharmaceutical, medical device, and biologics firms, have floundered during initial commercialization of a new product, and struggled considerably toward sustaining product viability over time. All of the essential compliance systems mentioned above (along with many not listed) are inextricably linked, and a weakness in one will invariably result in a number of weakness throughout the overall operation. Initial commercialization requires setting the stage for continued success by establishing a quality infrastructure that can sustain the company over the long term. A long-term vision is essential during early commercialization; shortsightedness will result in quality systems that do not ensure durable compliance, quality product, and consumer safety.

4 A CULTURE OF QUALITY FOR CONTINUED COMMERCIALIZATION AND SUCCESS

All FDA-regulated industries need to understand the importance, value, and timing of installing all of the essential quality systems. This section will present ideas about which quality systems are critical and when they need to be unveiled in order to most effectively support the continued commercialization of any given product. Successful, continued commercialization is just not possible without an infrastructure of quality and elaborate systems that support QA, QC, manufacturing, and other critical operational activities.

4.1 Establishing Durable Quality Systems: Several Programs That Significantly Contribute to Forward Quality

Early in the drug development process, the role and responsibilities of the firm's QA unit must be defined. As stated, QA oversight during early development and during all aspects of clinical trial activities is essential. Additionally, QA's close management of activities during the product's initial launch is crucial for successful commercialization, in part because this is the time in which the full extent of the firm's quality infrastructure will be put to the test. All systems, from change control and personnel testing to internal audits and annual product reviews, must be evolved and prepared to respond to the many issues that arise the manufacture and distribution of a commercial product. A proactive QA management approach must be taken in order to identify and intercept quality and compliance issues before they threaten public safety and assault product marketability. The days of QA performing one to two comprehensive internal audits are long gone and are no longer useful. It is necessary for a firm's QA management to assume a *prevention-based* perspective and perform targeted and frequent audits. Early interception and prevention are the order in today's fast-paced, FDA-regulated environment. QA must design programs and systems that allow for the prompt detection of problems and ensure the involvement of executive management with resource allocation capabilities. There is no doubt that a ***PQAP*** is a necessary and viable strategy toward sustainable quality and compliance.

In addition to a proactive QA program there are a number of critically important programs that once installed will support long-term, durable compliance and quality. Since the greatest contributor to product variability is usually presented by its raw materials, a far-reaching and well-integrated ***supplier/vendor qualification program*** is vitally important to maintaining product consistency and quality. Experience has shown that outsourcing does not always result in a cost savings. Years of outsourcing has taught us that it is absolutely necessary to partner with suppliers, communicate consistently, and monitor quality levels. Additionally, prior to using any supplier, vendor, or service, a firm or sponsor must adequately qualify its operation and ensure that the supplier can meet the sponsor's standards and quality requirements and that the sponsor is not inheriting any compliance and regulatory liabilities by virtue of the contractual arrangement.

Another especially useful quality system that if designed appropriately can have a great impact on sustaining product quality is a firm's ***complaint-handling system***. A well-designed complaint-handling system will provide the firm with a 360-degree feedback loop that can be used to quickly respond to consumer concerns as well as provide an opportunity for continuous improvements. Forward-thinking complaint handling can also greatly reduce the potential for regulatory and civil liabilities. While FDA only requires the reporting and trending of certain categories of complaints (serious events that are both noted and not noted in the product's labeling), it may behoove a firm to closely monitor all complaints, whether FDA requires it or not. Trending consumer feedback about a product, even if it is not a serious, reportable event, could result in learning something new about the way the product performs throughout the larger population. This information could lead to new, previously unknown product indications. Such was the case with many products that were tested and marketed for a particular indication and later found to provide benefits for another condition. One such case is a drug that was tested and commercialized for hypertension and later discovered to confer hair growth benefits. Monitoring and trending complaints during clinical trials, early commercialization, and as long as the product is marketed are FDA requirements; however, in order for complaint handling to serve as a forward quality system it must go beyond the regulatory requirements and provide a means for continuous improvements.

4.2 Annual Product Reviews

Continuous improvement and forward quality principles are more easily integrated when there is an ***annual product review (APR) system*** in place. When designed and used appropriately, the APR system can serve as an integral part of the firm's continuous improvement and compliance enhancement efforts. The GMP, 21 CFR 211.180(e) regulation requires the following: "Written records required by this part shall be maintained so that data therein can be used for evaluating, at least annually, the quality standards of each drug product to determine the need for changes in drug product specifications or manufacturing or control procedures. Written procedures shall be established and followed for such evaluations and shall include provisions for: A review of a representative number of batches, whether approved or rejected, and where applicable, records associated with the batch. A review of complaints, recalls, returned or salvaged drug products, and investigations conducted under §211.192 for each drug product."

The purpose of this regulation is to provide a reliable system for a manufactures to review the quality standards for each drug product it makes. Every manufacture must customize and establish its own written procedures in order to comply with this mandate for annual evaluations of drug products. It has recently become acceptable for a firm to review a representative number of batches in lieu of examining all records for every batch of a manufactured product. Essentially, FDA requires the annual review of a representative number of batches, whether approved or rejected, along with any corresponding QC records associated with those batches. The stated purpose for this requirement is that firms periodically review an assortment of meaningful quality markers for each product manufactured in an effort to determine the need for changes in its specifications, manufacturing process, QC methods, and process controls. Additionally, the annual review should assist the firm in designing a corrective action plan related to the specific quality issues identified. The minimum sources of information to be reviewed are specified in the regulations. There are a number of other quality systems in addition to manufacturing and control records, such as complaints, recalls, returns, salvages, and failure investigations, all of which are also intended to capture and help correct instances of substandard product. While the GMP regulations require a review of manufactured product on an annual basis, they also require that an investigation of abnormal or unexpected data take place at the time the results occur. As such, the annual product review is really a compilation of information the firm should have been aware of batch to batch throughout the year. Trends and patterns associated with the critical data should also have been compiled and reviewed by QA and manufacturing long before the annual review process begins. Any new information discovered at the time of annual product review indicates that there is a weakness in some other critical quality system designed to capture, evaluate, correct, and prevent product quality problems. APR should be part of a comprehensive and well-integrated quality infrastructure in which primary product quality does not rely exclusively or entirely upon the APR system. APR should function as a safety net underneath other quality systems such as laboratory and manufacturing deviation investigations, establishment of alert and action limits, specification setting, and change management. Ideally, the APR system should provide the firm with an opportunity to correlate and compare information from various quality markers in an effort to identify relationships between trends that may not be apparent when evaluated within the context of one particular quality system. With this basic understanding of the purpose of the APR, a proactive QA system may further enhance the functionality of the APR by extending the data reviewed beyond what is required by the GMPs. A firm can significantly enhance its APR system by utilizing a number of different sources of information and correlating it where possible. For example, a correlation between the frequency of manufacturing deviations for a particular product and the number of complaints received for that product could reveal important information. Another example of a useful correlation is when analytical in-process or release data are correlated with changes in the process, formulation, or raw materials in an effort to identify the variables that contributed to these results. A good APR system should provide the opportunity for continuous quality improvement and sustainable compliance. Additionally, if the APR is well integrated with other quality systems, it provides a platform for a 360-degree feedback loop for all internal and external quality and compliance activities.

It is important that all records required by 21 CFR Part 211 are maintained in a manner conducive to the development of the APR. Data recording, formatting, retention, and retrieval procedures must be designed to provide an efficient means for compiling the APR. Obviously, batch records and corresponding QC reports must themselves be organized in a manner that allows for an efficient APR. Since the regulation requires that a representative number of batches get selected and reviewed for the purpose of APR, it is critical for the firm to predetermine and state within its APR procedure what a "representative" number is. This is usually determined on a statistical basis for all the "passing batches" and it is recommended that all failed batches get reviewed. Consistent with FDA's comments in the *Federal Register* announcement of January 20, 1995, if the review of the representative number of passing batches reveals problems, all batch records should be reviewed.

The question of what constitute a representative number is also addressed in FDA's "Human Drug CGMP Notes" (Vol. 6, Number 1) March, 1998: "The number of batches whose associated records will be reviewed must achieve the purpose of the review. Any reasonable approach to achieve the purpose can be acceptable; the word "representative" was inserted into this regulation in January 1995 to simply confirm that every batch does not necessarily have to be included. Reviewing batches that exhibit varying manufacturing experiences is a critical element in ensuring that a "representative" selection is made. Batches showing different categories of experiences would include those that: (1) have been approved, rejected, and recalled; (2) have unexplained discrepancies; (3) were the subject of FARs (field alert reports) ; and, (4) have any other kind of outcome that may indicate changes are needed."

The "records associated with the batch" to be reviewed are also not definitively described in the cGMPs or the associated background material in the *Federal Register* announcements. The scope of the requirement ("Written records required by this part shall be maintained so that . . . ") however, indicates that the intended scope includes all written records associated with the batch that could provide information indicating that a change is needed to the drug product specifications or manufacturing or control procedures.

FDA's "Human Drug CGMP Notes." Vol. 6, Number 4, December 1998, also addresses this issue: "This section is intended to require firms to perform a systematic review of data relating to product specifications and manufacturing and control procedures to determine if changes are warranted. Data that indicate a need for such changes can be contained within a broad range of records."

The "Human Drug CGMP Notes." Vol. 5, Number 4 states: "Also included in that grouping would be other records that establish product specifications or manufacturing or control procedures. These records are intrinsically relevant to the review because: (1) per CGMP, batches must be made according to those procedures and specifications; and, (2) those very procedures and specifications are what may need to be modified. Because those records may in and of themselves hold indications that changes are needed (e.g., outdated/superseded instructions), a meaningful assessment of the need for change would thus be impossible without at least a minimal review of those records."

21 CFR 211.160(a) —"Any deviation from the written specifications, standards, sampling plans, test procedures, or other laboratory control mechanisms shall be recorded and justified."

21 CFR 211.192—"Any unexplained discrepancy (including a percentage of theoretical yield exceeding the maximum or minimum percentages established in master production and control records) or the failure of a batch or any of its components to meet any of its specifications shall be thoroughly investigated, whether or not the batch has already been distributed. The investigation shall extend to other batches of the same drug product and other drug products that may have been associated with the specific failure or discrepancy. A written record of the investigation shall be made and shall include the conclusions and follow-up."

Complaint files: 21 CFR 211.198(b) —"A written record of each complaint shall be maintained in a file designated for drug product complaints." 21 CFR 211.198 (b) (2) "(2) where an investigation under 211.192 is conducted, the written record shall include the findings of the investigation and follow-up."

Returned drug products: 21 CFR 211.204—"Records of returned drug products shall be maintained and shall include the name and label potency of the drug product dosage form, lot number (or control number or batch number), reason for the return, quantity returned, date of disposition, and ultimate disposition of the returned drug product. If the reason for a drug product being returned implicates associated batches, an appropriate investigation shall be conducted in accordance with the requirements of 211.192."

Drug product salvaging: 21 CFR. 20—"Records including name, lot number, and disposition shall be maintained for drug products subject to this section."

It should be apparent that the APR is a meaningful compilation of myriad critical data, records, and reports intended to provide quality management with a tool to evaluate the quality status of manufactured products. Handling unexpected trends becomes more manageable when there is a comprehensive APR process fully installed and operational. Evaluating trends might also provide an opportunity for process optimization, particularly when specifications seem to be drifting toward the upper or lower limits.

It is unfortunate that for many firms the APR system consists solely of the minimum elements required by the GMPs; that is to say, reviews are performed on an annual basis and only using the quality markers required by the regulations. Since APRs provide a mechanism for management to be advised of quality issues related to manufactured product, the frequency and completeness with which this is done are important factors. Cost reductions are often realized when executive management is apprised of quality issues early on and allowed to allocate resources and redirect priorities in a timely fashion. Sometimes waiting a year to begin the review may not be the most effective timing. The quality systems regulations (QSR) for medical devices require that executive management be apprised of critical quality trends and issues periodically throughout the year. It is important to evaluate how a firm can maximize the APR process and expand it beyond the minimum GMP requirements. A state-of-the-art APR system is inextricably linked to other critical quality systems, such as change control, manufacturing failure investigations, stability, and complaints. Keep in mind that the primary function of the APR is to determine if any changes are necessary in drug product specifications, manufacturing, and control procedures.

4.2.1 Key Elements for Forward-Thinking APR System

Deviations and failure investigation data—The failure of any batch to meet any specification, including batches failing in-process, release, or finished-product (shelf life) specifications is a crucial event. Such events must be reported and captured in the APR system. The identification of the root cause and the determination of corrective actions for it must be evaluated against information reported from other quality systems to confirm that the changes have been completed and were effective in preventing recurrence of the failure without unintended adverse effects.

A common error is to limit the types of deviations reported to and evaluated by the APR system to just deviations from finished-product specifications. All deviations should be evaluated, including deviations from manufacturing procedures, in-process specifications, deviations from raw material specifications, and other expected results. Each of these occurrences could indicate changes are necessary to prevent recurrence. For example, the cause of deviations from manufacturing procedures is frequently evaluated as a lack of training. If there are several of these occurrences by different individuals, however, it is also likely that there may be another root cause, such as unclear or insufficient batch record instructions or inadequately designed or unclear batch record data forms.

4.2.2 Complaint Information and Trends

Complaints related to the quality of the product should also be evaluated. Complaint trends should be reviewed for evidence that recurring problems have been adequately resolved. The complaint trends should be examined for any correlation to the other quality systems. For example, any correlation between complaints and deviations related to the same lot or product line or any-increase (or lack of a decrease) in complaints after a corrective action has been made should be revealed and addressed during the APR process.

4.2.3 Recall Information

All recalls must be comprehensively reviewed during the APR for the subject product. A recall represents a catastrophic failure of one or more parts of the overall quality infrastructure. A review of the recall should focus on identifying the genesis of the failure and future recall prevention. When investigated, most recalls reveal inadequacies in quality oversight or such inadequacies in quality systems as faulty specifications, change control, in-process or finished-product testing, and batch record review. During the APR process, verification of the corrective action and preventative measures installed to prevent future recalls must occur.

4.2.4 Validation Activities

The APR provides an opportunity to review the validation status of products. Over time, numerous small process changes and/or changes evaluated as "equivalent" that may not require requalification have probably occurred. Individually, these changes may not be significant enough to require requalification of a process, but cumulatively sufficient change may have been introduced that requalification has become necessary.

Alternately, should a review of the product quality data show that there have been no significant changes in any analytical test results, stability profiles, SPC data trends, etc., the APR may confirm that these changes did not require revalidation individually or collectively.

4.2.5 Change Control Database

The data contained within the change control system are some of the most important data in terms of accomplishing the goal of the APR. Changes typically fall into two broad categories: a reaction to a problem or an improvement opportunity usually associated with product quality or cost reduction. The first category of change may or may not be effective in solving the problem and preventing the recurrence of the problem, and the second may or may not produce the desired result with or without intended effects. As such, changes must be thoroughly reviewed to evaluate first, if they produced their intended effect, and second, it there were any unintended adverse effects.

Adequate change management can also reveal whether or not the problem was correctly identified and the corrective action was appropriate. A number of changes to the same product or process conducted in succession may indicate that the problem has not been investigated properly and that corrective actions were not appropriate.

The types of changes evaluated should not be limited to changes to the master production and batch record, the raw materials and components, and specifications. Changes to the facilities and equipment used, including equipment cleaning, calibration and maintenance procedures, personnel practices, cleaning procedures, laboratory equipment and analytical procedures, should all be targeted for evaluation during the APR process.

Changes should also be correlated with the statistical data available from stability, inprocess, and finished-product testing. The statistical data should be examined for trends in specification test results. Where trends are found (e.g., increases or decreases in potency, decreases in stability) the change control records should be evaluated to determine whether or not changes have occurred that might be responsible for the shifts in QC trends and analytical results.

4.2.6 Returned Goods

Evaluation of returned goods should be part of the APR process. Product returned from warehouses, distribution centers, consumers, doctor's offices, internal inventory, and of course the market should all be captured in the APR report.

4.2.7 Correlation of Information Between Quality Systems

The benefit derived from correlating significant and compatible information provided by the quality systems become very apparent when a comprehensive APR system has been operational for a while. The relationship between the various quality markers becomes evident and helps to facilitate the development of more concise short and long-term improvements.

With respect to change management and correlation of critical information, the *Human Drug CGMP Notes*, Vol. 6, Number 4, December, 1998, stated "A broad CGMP principle that runs throughout the regulations is the concept of redundant checks and balances to ultimately ensure product quality. The 211.180 (e) periodic review provides that balance with respect to change control because, in addition to having an effective change control system, reviews of records that establish product specifications or manufacturing or control procedures can help firms maintain the currency and accuracy of such things as: (1) cross reference to other documents and standards; (2) materials' specifications (e.g., ensuring they match what's in a firm's most recent NDA submission or USP monograph); (3) equipment references (e.g., too generalized a manufacturing instruction to use a 'suitable mixer' where different types and sizes are at hand would need to be made specific); (4) process step sequencing; (5) process step parameters; and, (6) acceptance criteria. Especially where a firm makes significant or frequent changes, this complex matrix of interrelated instructions and specifications can make change control difficult and may result in operators using outdated or otherwise incorrect procedures or standards to make medicine."

4.2.8 Product Retain Sample Summary

The use of retain samples to routinely examine a statistically sound number of units is a valuable investigative tool. If the retain samples do not reveal any degradation, but other quality markers, such as consumer complaints, manufacturing deviations, and QC profiles, indicate shifts in quality, a problem with the retain sampling method should be considered. On the other hand, if the retain samples reveal problems but the firm's quality systems are not revealing any significant quality problems, there may be a serious deficiency with the quality system infrastructure.

4.2.9 Minimal Compliance Guideposts for APR Procedures

What follows are the minimum elements that must be incorporated in an SOP for the APR process. Anything less than what is suggested here could result in a situation of noncompliance during an FDA inspection. These are critical compliance guideposts that form the basis for a meaningful APR system.

State the purpose

To evaluate the performance of each product manufactured against the approved specifications and expected quality attributes.

To identify trends in product quality or changes in active raw material(s), in-process, or finished-goods performance.

To propose the need for changes in product specifications, manufacturing processes, or control procedures.

To assess the need for validation or revalidation.

To identify product improvement or cost reduction opportunities.

To formally communicate product performance to executive management with the authority to allocate resources.

Delineate the scope

This procedure is intended to describe a state-of-the-art APR system for the evaluation of all products manufactured on an annual basis in an effort to assure that the quality standards of the products are met and maintained.

The APR program will be under the auspices of QA.

The QA unit will coordinate preparation of the APR report and author an executive summary report to be submitted to the reviewers.

The APR committee will consist of the following key personnel: site general manager, executive director of quality operations, and director of manufacturing.

The committee is responsible for review and approval of the executive summary of the APR, and for providing technical and compliance input for the conclusions drawn.

Define the APR schedule: when it begins and ends

Review and approval: Total review time should not exceed 60 calendar days.

Retention and management of reports: Define who is responsible and where the APR will be stored.

Annual product review format: Define it.

For example, cover page must consist of the following information:

Product name and strength

ANDA number

ANDA approval date

Review period

Executive summary should consist of the following information:

The product profile

Final interpretation of the results and analyses

Final conclusion, including the acceptability of the product for continued production

Summary of any corrective or investigative actions necessary

Summary and confirmation of follow-up actions identified as necessary from the previous year's APR

Essential information to include

A brief overview of product performance during the review period. This will consist of the following:

The number of lots produced.

The number of lots rejected.

The number of lots dispositioned.

The number of lots not dispositioned (identify lots being carried over to the next year).

A statement that all sections of the APR have been reviewed and are identified in their respective sections. Any specific trends or changes should be mentioned and compared to previous year, any recommendations, and identified problems and corrective actions taken.

Active raw material information will consist of the following:

Name of manufacturer.

Number of active raw materials utilized.

Number of active raw materials rejected.

Evaluation of assays (loss on drying, bulk density, particle size, and residue on ignition as deemed appropriate for trends, shifts, and outliners).

Comparison of data based on previous year's data.

Comparison of full testing, with lot results compared against their respective certificates of analysis.

Product review will consist of the following activities:

Summary of in-process and finished-product data, including as appropriate, such in-process attributes as weight, thickness, hardness, and total accounted yield; assay, dissolution, and content uniformity.

The summary will consist of an explanation of trends, shifts, and any outliners, if applicable.
Summary of process and method validations and revalidations, when applicable.
QC tools will be used, such as check sheets, histograms, Pareto diagrams, and cause-and-effect diagrams. Scatter diagrams and control charts will be provided, where appropriate, for in-process attributes and finished-product data as an attachment.

Change control will consist of the following:
Summary of changes identifying those that may have had potential product impact.
Identify changes that may have influenced the validated state.

Deviations and laboratory investigations reports (LIR's) will consist of the following:
A review of all product-related investigations and LIRs, identifying any trends (with accompanying Pareto charts, if applicable).

Stability will consist of the following:
A review of all lots in the stability program, identifying any trends for typical or atypical behavior based on regression analysis.
Typical data for evaluation will consist of assay, dissolution, impurity, and hardiness, or whatever critical attributes apply to the product.
Regression analysis will be provided as an attachment.
Specific stability test condition(s) and contents per package in all charts generated.

Complaints will consist of the following:
A summary for all customer complaints based on Customer Complaints Report written by QA, medical affairs, or drug safety.
Pareto chart (description and/or cause) will be provided as an attachment.

Returned goods/recalls will consist of the following:
A summary of returned goods and recalls based on QAs, product release database.
A summary of rationale and corrective measures for all recalls.
A summary of all returned goods and explanation for returns.

Retain sample summary will consist of the following:
A summary of the annual visual inspection of retain samples based on a statistically sound sampling plan.

Product specific internal audit issues will consist of the following:
Resolved, and still pending items identified.

Conclusions and recommendations will contain the following information, at a minimum:

An overall summary of the product performance during a review period.

Recommendations for change or further investigation in order to maintain the quality standards of the product.

Corrective Actions will consist of the following:

A summary of corrective actions, along with the specified target due dates needed for completion, if applicable.

It is important to realize that as a "prevention-based" program, APRs are not the most effective mechanism. While an APR system satisfies regulatory requirements and provides a valuable reference tool, a comprehensive profile of product performance at the end of the year does not provide a useful prevention approach. A comprehensive and well-integrated recall prevention program, coupled with APR, provides a much more effective prevention-based mechanism and system.

4.3 Recall Prevention Program

It has become fairly common for FDA-regulated industries to have a recall procedure in place. A formalized procedure is a cGMP requirement, along with the FDA's recommendation that firms periodically test their recall procedure for effectiveness. A large part of FDA's concern is that firms have a fail-safe mechanism for identifying, tracking, and retrieving their products from commerce if product failure or suspected tampering has occurred.

Beyond these basic requirements of procedural effectiveness and product traceability and retrievability, FDA has not demanded that firms install any kind of recall prevention program per se. The cGMPs inherently demand that firms fulfill certain regulatory expectations, such as the requirement for annual product reviews and laboratory and manufacturing investigations, which provide many safeguards prior to placing the product into commerce. By virtue of these safeguards, recalls can be greatly reduced and prevented.

Since the goal of this chapter is to examine the impact of total quality performance of forward quality on compliance, an interesting and relatively unexplored prospect is a formal *recall prevention program.* In order to begin considering the value of such a program, one must first determine which if any quality indicators can be used to effectively monitor fluctuations in product quality and justify withholding product from commerce in avoidance of a potential recall.

As prescribed by the cGMPs, there are a number of obvious indicators, such as product not meeting release criteria and specifications, a stability failure, and labeling mix-ups that typically alert QA that product should not be released; however, this section will identify quality markers that are not as apparent and that could help a firm prevent a potential recall.

What follows is a preliminary list of potential quality indicators or predicators that can be used as part of a comprehensive recall prevention program.

4.3.1 Manufacturing Deviation Trends and Corrective and Preventative Actions (CAPA)

Needless to say, a manufacturing deviation that results in finished products not meeting its predetermined release criteria should not, from a consumer safety, quality, and regulatory perspective, be introduced into the marketplace. Fortunately for the consumer and public health, failed product does not usually make its way into the market. Such product would be deemed adulterated or contaminated and nonconforming and could be reprocessed if possible, or destined for destruction. This, however, might not be the case for product that did meet its predetermined release criteria but during the course of its manufacture experienced a number of manufacturing deviations. A manufacturing deviation can be defined as *anything that occurs during the course of staging, manufacture, packaging, and labeling that is inconsistent with the batch production record, whether or not what has occurred leads to a true product failure.*

Imagine that a number of departures from the batch production record have occurred; however, upon performing in-process or finished product testing, all specifications have been met. What might this suggest to quality management? Are there certain manufacturing deviations that while not causing a true product failure might strongly preclude product distribution? During the development and validation of the process and its methods, there may have been situations that revealed specific manufacturing deviations that were particularly undesirable while not rendering the material unusable. An example of this might be finished product that met all in-process and release specifications but that was stored in bulk drums for a slightly longer period than validated and under somewhat questionable conditions but that still meets all specifications upon retesting the bulk prior to finished product packaging. From a regulatory perspective, there is no compliance threat or roadblock to releasing the product into commerce; however, from a forward quality perspective there may be legitimate justification for not releasing it, and averting a possible recall.

Another example might be where a specific batch is not in question; but rather a number of consecutive batches across a specific time frame. Imagine a situation in which a number of batches were manufactured. All met in-process and finished-product testing, and they were released to inventory destined for commerce in the months ahead. Upon a routine batch record audit, it was discovered that all of these batches experienced myriad manufacturing deviations. Additionally, it was been noted that there had been some unexpected construction in the facility around the same time in which the batches were manufactured. Environmental monitoring data reveal that there were a number of times during that same period that alert limits were met for several corridors leading to pertinent manufacturing areas.

Once again, while there is no regulatory reason to determine these lots are not marketable, a prudent and forward-thinking QA professional working within the bounds of a recall prevention program might legitimately determine that because of high incidence of manufacturing deviations and the less than optimal environmental conditions during the same period, these lots should not be released into commerce. Another trend the forward-thinking QA manager might consider evaluating is how manufacturing deviations are written. Is there a standardized approach for a deviation investigation? Is the report format uniformly employed? Has management offered some guidelines or de minimus standards for how investigations should be conducted?

Inconsistent investigational approaches coupled with a lack of minimum requirements will invariable lead to some investigations being less comprehensive and effective than others, significantly impacting the quality of the investigations and product release judgments and conclusions.

It is recommended that in an effort to prevent recalls, firms periodically evaluate manufacturing deviation trends within the context of their unique product lines, circumstances, and environments, with the aim of establishing criteria (other than product failure) that would justify not releasing product into market or withholding until further assessment.

In addition to in-process and finished-product specifications that alert QA to possible problems, predetermined criteria could significantly assist with the interception of potentially problematic product destined for commerce.

The industry has come to rely upon in-process and finished-product specifications as the only true reflection of a product's integrity, and for the most part, ignores a number of other important variables that can reveal a great deal about a product's integrity. Since the cGMP regulations require that the industry only perform annual assessments [under 21 CFR 211.180(e)] of their product(s), more frequent assessments of meaningful quality markers, such as manufacturing deviations, could effectively support a recall prevention program. The medical device regulations require multiple assessments throughout the year of a number of quality markers that are intended to alert executive management to potential or existing quality problems and trends.

While the QSR require CAPA periodic assessments of CAPA trending are not required. Similar to reviewing manufacturing deviation trends, assessing CAPA trends could reveal a plethora of quality issues throughout an organization.

When corrective and preventative actions are consistently minimalist and not concise, the results will eventually be reflected in product quality and in the firm's compliance status. Additionally, many problem, can arise when corrective and preventative actions are installed and presumed to be absolute without a short-and long-term assessment of the action. It is important not to assume that the CAPA installed today is necessarily the best practice or solution over the long haul. A long-term review of the CAPA will be the best indicator of its long-term viability.

A periodic evaluation of the long and short-term effects of selected CAPAs or overall CAPA trends can reveal valuable information about a company's quality and compliance status and could provide an indication of product quality for any given period, once again allowing for decisions to be made about future product release activities.

4.3.2 Out-of-Specification (OOS) Trends and Out-of-Trend (OOT) Data

An important piece of information typically contained in an APR is the number of out-of-specification (OOS) results identified over the course of the year for (in-process and finished-product) testing related to any single product. While this is valuable information, it may be of further use to more frequently evaluate the OOS trends relate to high-volume products.

This information could reveal the QC trends related to a particular product for any given time frame. Fluctuations in QC activities, such as an increase in the number of laboratory-generated OOS results, could alert QA to potential current and future problems. An increase in OOS and out-of-trend (OOT) data might suggest that there are such laboratory deficiencies as poor managerial oversight or inadequately qualified laboratory equipment and personnel. A trend of this nature might also point to poorly validated methods. Whatever the genesis of the apparent OOS and OOT patterns, the information strongly suggests unreliable QC practices that could significantly assault the integrity of the products making their way into the market.

In recent years, the industry and FDA have evidenced the impact substandard laboratory practices, rampart OOS data, and lack of forward quality in the laboratory has had on marketed products, attempts to commercialize new products, overall company credibility, and compliance status.

Evaluating OOS and OOT results for the sole purpose of APRs does not serve as a real-time, preventative measure. A more effective mechanism for intercepting questionable material before it is released into commerce would be to periodically assess OOS trends and OOT results within the context of cGMPs in the laboratory and evaluate their impact on manufactured products.

Periodic evaluations of OOT data are a reliable predictor of oncoming trends relative to raw and in-process material and finished product, laboratory practices, and method or assay migration. Evaluating both OOT data and shifts toward the upper or lower points in the range and identifying their origin can go a long way in preventing true OOS results and a decline in product quality. Determining the point at which OOTresults constitutes enough of a quality concern that product should be withheld from commerce until further evaluation is contingent upon the nature of the product, the manufacturing process, its medical necessity, and the speed with which a remediation plan can be implemented.

As is the case with OOS and OOT an effective recall prevention program must establish alert and action limits for the various quality markers results, in such a way that the actions taken—either withholding product or reprocessing it—are timely and part of a continuous improvement effort.

4.3.3 Stability Data and Trends

A failed stability test or trend that points to a fluctuation in the quality profile of a product is an obvious quality indicator that typically results in a field alert, a market withdrawal, and product destruction. Trends in stability data are particularly important when dealing with biologic products, where, for example, a decline in vaccine potency will invariably continue to decline and become a true OOS. For certain categories of products, the observance of a change in the stability profile, even a minor fluctuation, could suggest more serious problems as the product degrades further. Thoroughly understanding the stability profile of the product can serve as a meaningful quality indicator, not just when an OOS occurs, but when the established profile presents unexplained fluctuations that might suggest withholding product from the market.

4.3.4 QC Profiles Related to Raw Materials, In-Process Material (WIP, Bulk) and Finished Products

Similar to unexplained fluctuations in stability profiles and OOS trends, changes to raw materials and in-process and finished product point to any number of problems related to the manufacturing train, process validation, supplier, or environment, and so on. The greatest variability in any product is presented by its raw materials, particularly when provided by an outside supplier. It is essential to establish a system that allows QC to frequently monitor the variability of all critical raw materials. The ability of QC to quickly detect even small shifts in a raw material's performance will allow

QA to anticipate product quality over time and prevent potential recalls. The APR will capture and report shifts in a raw material's QC profile; however, as suggested above, this is a retrospective review, not a prospective, real-time assessment. Another excellent tool for preventing product recalls is ensuring that material is adequately qualified before it is introduced as a raw material or substituted for an approved material.

An evolved supplier qualification program is yet another manner in which a firm can reduce the possibility that unreliable product will be introduced into commerce. Managing changes within a supplier facility, particularly when the facility is thousands of miles away, can be challenging. There are ways, however, to ensure that change management occurs frequently and effectively. It is most effective when the sponsor assumes the proactive role and becomes responsible for periodically checking up on the supplier and inquiring about changes that might affect the quality of the product. While the supplier agreement and purchase orders typically specify the requirements, along with the caveat that should anything change the sponsor must be immediately notified, change management should not be entirely left up to the supplier. A change that the sponsor would consider significant may not be deemed so by the supplier.

Firms might consider installing a mechanism that allows the sponsor to periodically and frequently e-mail or fax a short questionnaire to the supplier regarding changes over a specified period. This will keep the sponsor apprised of changes that have occurred and permit the sponsor to determine what impact, if any, the change(s) may have on the product.

Tracking, trending, and evaluating shifts in the specifications and quality of raw materials, in-process or bulk materials, and finished products will provide a fail-safe alert system and mechanism with which to monitor product before it is introduced into the marketplace.

With respect to raw materials provided by a third-party supplier, it is essential to monitor, track, and evaluate the performance of the suppliers involved. Qualification of materials, along with extensive qualification of the supplier or vendor, is a critical activity for maintaining compliance and quality. Recurring supplier issues can significantly assault product quality. Supplier profiles, coupled with adequate monitoring and trending of supplied materials, are useful indicators that can effectively support a recall prevention program.

4.3.5 Specification Drift and Device Feature Creep

For medical devices, there is a phenomenon recognized as "feature creep", in which the specifications for the design features of a device begin to drift toward a specification limit because inadequate design reviews, output verification and change control. This same sort of variability occurs with pharmaceutical and biological products, and without adequate change control, validation maintenance, revalidation, and frequent monitoring, the specifications for any given product can also shift or drift toward a specification limit very quickly.

Product variability can be reduced through constant monitoring, either by SPC once enough batches have been manufactured, or through batch-to-batch comparisons of data. Where a specification range has been established, whether for an analytical method or a manufacturing process, it is very important to observe fluctuations and drifts in any direction. This is an easy and practical indicator that enables QA to anticipate potential product quality issues before they actually arise.

Inadequate change control and regulatory reporting of changes will eventually result in uncontrolled and unapproved product finding its way into commerce. It is not uncommon for proposed changes to receive incomplete and inadequate assessments prior to the implementation of the change. Additionally, when changes are made to a product's critical parameters (e.g., such as its specifications, packaging, and labeling) without the appropriate notification to the applicable regulatory agency, the product is considered in violation of what was approved. This would constitute reason for a recall, cease and desist order, and possible fines from the FDA. Specification drifts and changes in a product's approved parameters can easily occur without adequate change management. This is precisely why the use of SPCs, such as the ones delineated in the APR section are helpful in preventing the release of substandard product.

4.3.6 Complaint Trends

Along with medical trends revealed by consumer complaints, quality defects are indicative of potentially serious issues related to the product. Drug safety information and events are typically trended in an effort to evaluate if the reported product complaints reveal quality and medical issues. Some medical issues, such as ineffective product, could be traced back to any number of quality issues, such as raw material potency, the weighing step (too little active ingredient added) , the manufacturing process, product stability, or the storage of raw materials or finished product. A firm must have a formal system for evaluating adverse reactions and medical trends to ensure that any possible deficiencies within quality operations that could negatively impact the product's safety are adequately addressed.

Quality assurance must also have a formal system in place to thoroughly investigate product complaints that are referred to quality operations for an assessment. Criteria and guidelines should be developed that provide a standardized framework for comprehensive complaint investigations. Additionally, periodic trending of all quality investigations should be conducted by QA in an effort to identify recurring quality themes and promptly address them. Trending data that reveal deficiencies in the manner in which a product is manufactured, tested, stored, or shipped can serve as a very effective recall prevention mechanism. These trends can alert QA of quality problems that could lead to a product failure in the field, thereby preventing distribution.

4.3.7 Internal Audit Findings

Similar to complaint trends, internal audit information that reveals recurring noncompliance patterns related to such product-specific activities as manufacturing, process controls, stability testing, and operator performance should alert QA of the possibility this may have on released product. The key is to intercept the product before it is released and a recall is demanded.

Lack of adequate follow-up to product-related audit findings is another clue that there may be a need to hold off on releasing product until further investigation occurs. One recent case of recurring cleaning validation deficiencies at a firm, coupled with a change in the dosing schedule for a drug product led to a significant increase in the number of complaints received for that product. QA correlated the fact that the cleaning validation deficiencies had yet to be corrected with the information received by drug safety personnel and quickly moved to place an internal hold on all lots of that product still in inventory and within distribution centers and company warehouses. A major recall was averted.

A great failure routinely exhibited by many internal audit programs is the lack of an internal rating system. The results of every audit should receive a rating, and each audit should be measured against the previous audit(s) in an effort to determine whether or not the firm is making the necessary improvements. Evidence of continued digression or lack of any improvement audit to audit would serve as a very useful quality indicator and real-time alert mechanism for QA to use as part of its product quality assessment and future release decisions.

4.3.8 Sterility Failures and Trends (Retest Rates)

A sterility failure is sufficient reason not to release product and may eventually lead to the need to destroy it. A periodic assessment of stability test failures and retest trends would alert QA to investigate any possible risks associated with the continued distribution of that particular product.

4.3.9 Environmental Monitoring Data

Similar to establishing alert and action limits for the number of OOS and OOT results allowed before the release of a particular product is suspended, establishing these limits as part of a comprehensive environmental monitoring program is essential, pending further investigation (discussed in Sec. 4.3.2) . These alert and action data can also form part of a product's release criteria and provide a safeguard against releasing product that may be contaminated or that requires further testing.

4.3.10 Water Data

Water system qualification and requalification data, along with periodic water testing, will reveal any critical excursions and alert QA to the need to withhold product from the market. Establishing alert and action limits for this crucial raw material, particularly since it is used pervasively, can go a long way in supporting QA in recall prevention.

4.3.11 Returned Goods

Periodic tracking and trending of returned goods whether or not re-entered into inventory, along with product that required rework or reprocessing, is another useful indicator that can serve to alert QA of potential quality problems with specific products.

4.3.12 Label Reconciliation Practices, Results, and Trends

Adequate label reconciliation is a regulatory requirement, as well as a critical part of QA for all products. Studies have shown that the majority of recalls are attributable to a label mix-up. QA should ensure that as part of the firm's internal audit assessment, a review occurs of its labeling practices, label storage and security, reconciliation, and line clearance. A full understanding of how this activity is controlled is in and of itself a significant recall prevention measure. Additionally, if internal audits reveal that there are deficiencies in this area, QA can assess the risk related to product destined for distribution. Recurring label reconciliation problems must be investigated and evaluated against the risk of releasing product.

4.3.13 Formal Line Clearance Practices, Results, and Trends

Similar to label reconciliation, inconsistencies related to formal line clearance procedures can also result in shipping product that may later need to be recalled. It is important for QA to ensure that periodic assessments of this activity occur and that any deficiencies uncovered are immediately addressed.

4.3.14 Quality Assurance Philosophy

An interesting component in preventing recalls is the philosophy and mindset of the QA managers within a firm. Are they proactive or are they reactive?

Do they make an effort to install quality systems that support "prevention-based" activities or do they identify problems after the fact and frequently too late? Is there a culture of prevention or one of correction? The perspective embraced by a firm's QA management plays an important role in recall prevention. QA can significantly reduce the number of risky lots released by performing periodic evaluations of any number of quality indicators.

4.3.15 Correlation of Quality Markers

As stated in the APR section above, the benefits derived from comparing information from one quality system, such as manufacturing deviations, against the frequency of consumer complaints or laboratory investigations for that same product can be quite revealing and useful.

Comparisons and a meaningful correlation between compatible quality markers will invariably yield information that can server to assist QA in its risk-based analysis prior to product distribution.

4.3.16 Annual Product Reviews and Outcomes

As stated earlier, an APR system is a regulatory requirement and a useful quality tool. Knowing how to develop a comprehensive APR system is just the beginning of having the APR consistently support quality activities. APR information must be used routinely in order to adequately monitor product performance.

Many companies prepare an APR but do not take full advantage of this tool as a recall prevention mechanism. An annual review may not offer the timeliness required to prevent the release and distribution of substandard product. Real-time monitoring of changes that could impact product integrity is an essential component of recall prevention.

4.3.17 Personnel Component

There is no denying the personnel component of product quality. Issues such as training, personnel turnover, staff's core capabilities and knowledge, job expectations, personnel motivation, and leadership all contribute to overall quality and product integrity.

QA must periodically evaluate these factors and determine whether or not they have a negative impact on product destined for the market. Frequent internal audits that examine the ebbs and flows of personnel practices and issues can serve to reveal the potential impact personnel patterns might have on product quality. Significant findings with regard to personnel activities might suggest to QA that withholding certain products from the market would be prudent.

4.4 Organizational Enrichment and Personnel Training

Among the myriad challenges confronting FDA-regulated organizations today, particularly in the face of FDA's evolving regulatory environment, is how to make the best use of the company's talent and employee knowledge. Corporations must learn how to tap into the wisdom within their organization and support professional learning communities rather than provide employees with dry and uncreative training. A great deal has been written on how to cultivate professional learning communities at work and how to define the best practices for enhancing student achievement. This perspective of education and learning places emphasis on the whole person and values the social element quintessentially important to education. Additionally, this perspective emphasizes that learning is a process of participation in *communities of practice*, participation that is at first legitimately peripheral, but that increases gradually in engagement and complexity. A community of practice or a community of professional learning is defined as groups that share a common concern, set of problems, or passion about a topic and choose a mutual venue in order to expand their knowledge and expertise in this area by interacting on an ongoing basis. It has become increasingly easier to create and sustain these global communities for transferring and exchanging knowledge through the use of the World Wide Web; however, without a formal infrastructure these professional learning communities cannot sustain themselves. Building a community of practice in an industrial society and within a highly regulated environment requires some planning and effort.

Communities of practice presents a theory of learning that starts with the assumption that engaging in social practice, or a common goal is the fundamental process by which we learn, as well as become partnered and cooperative. This is essentially how teams are built.

Professional learning communities is a term wherein each word has been chosen purposefully. Let us begin with the word *professional.* She is someone with expertise in a specialized field, usually with advanced training, and frequently required to maintain her expertise through continued education. The word *learning* suggests a mode of action and intellectual curiosity. The word *community* refers to a group united by a common interest or goal. It is easy to see how personnel within the structure of an FDA-regulated organization can become involved in professional learning communities. It is important to create an environment that fosters mutual cooperation and team work in order to achieve what could not be accomplished individually. Since the theme of this book is FDA compliance and this chapter has focused on forward quality, let us examine the essential building blocks for creating a professional learning community within an FDA-regulated organization.

The mission statement: A powerful driving engine for any organization seeking excellence and long-term success is the statement of a common goal and vision shared throughout the organization. In the case of the FDA-regulated company, sharing a common goal and vision should not be difficult. Such a company usually exists to manufacture products that benefit humankind, and for the most part, FDA regulations mandate the minimum requirements for this kind of business. This immediately clarifies for all personnel what the business priorities are along with the organization framework. Mission statements are not new to FDA-regulated industries; Johnson and Johnson established one over 10 years ago. Establishing a mission statement for the organization is a prerequisite to another important building block-a common vision for the organization.

Company vision: How does the company plan to accomplish its mission? This requires some thought, because while the mission establishes the company's purpose, the vision determines the direction in which it will go. The vision should present a realistic and concise picture of the organization's future, so that all employees are motivated and inspired to work toward it. A collective vision is an essential part of a professional learning community. A shared vision means that employees are going in the same direction, embrace the same principals, share the same value system, are committed to learning, and are prepared to work together towards the same goal. Executive management must not be afraid to adjust the mission statement and company vision in order to keep pace with evolving regulations, marketing demands, changing standards, and philosophical shifts. The mission statement and company vision play an important role throughout the company's life cycle; therefore they should not be static. It is inevitable that an organization will need to be redirected and transformed, and the first step should always be redefining the mission statement and company vision as needed.

Leadership: Much has been written about leadership, whether on the battlefield or in the business world. With respect to creating professional learning communities, executive and quality management must have a clear-cut plan for marshaling the organization in a way that is consistent with the direction delineated in the company's vision. Shared values and a shared mission will languish without the appropriate leadership. Quality management must define its expectations and unambiguously communicate to all personnel what their roles and responsibilities are with regard to the common goal. Effective leadership must also recognize individual talents and diversity within its workforce. It must support positive relationships within the organization, both horizontally and vertically. It must encourage the organization to approach situations with an open mind, commit to continuous improvement, and remain constantly aware of the organization's mission and vision. It must promote an environment in which personnel feel safe expressing their ideas and recommending changes. Most important, effective leadership must provide the blueprint for organizational success. Objectives and goals must be clearly defined and communicated, along with specific tasks and time lines for accomplishing them. Measurable milestones must be apparent to all involved in order to assess progress and sustain motivation. Oftentimes such activities as a preapproval inspection preparation project or responding to a serious regulatory action such as a warning letter or consent decree can bring an organization together and result in benefits far greater than expected. The GMP enhancement master plan discussed below represents a journey rather than a destination. It is an improvement process that fosters enormous individual accountability and collective responsibility. Leaders must recognize that continuous improvement requires continuous learning.

Opportunities for learning and knowledge transfer: It is important for executive management to provide ongoing learning opportunities for its employees. A professional learning community requires a great deal from its participants. It requires that they are open to learning alternative ways of looking at things. It requires a great deal of collaboration and commitment to respecting one another's ideas. The FDA requires ongoing GMP training for all employees; however, executive management must ensure that personnel training goes well beyond what is required by the GMPs. Executive management must promote the value of continued education and facilitate professional learning communities. Achievements and contributions must be recognized and celebrated. A system of rewards, personnel recognition, and celebration are an important part of a professional learning community.

Personnel must have an opportunity to explore external educational venues in which perspective and culture different from their own company might be presented. In-house educational venues benefit greatly by including several levels of personnel, from operators to managers, within the same session. This will allow employees to discover how differently their colleagues might view a particular situation. It is important to provide employees with a balance of both internal and external learning opportunities. This will help keep the organization supplied with fresh perspectives and avoid becoming stale. Create collaborative opportunities with a focus on knowledge transfer and learning.

Develop a global E-learning community: Participation in monthly teleconference calls with quality management from other companies may prove valuable. Establishing a knowledge transfer Website that invites global participants provides easy access to those wishing to participate in a professional learning community. Professional learning communities can take many forms. For effective learning to occur, it is not always necessary for people to meet face to face. These communities of practice vary widely in style within different organizations and throughout the world. Knowing what these variations are is important because it will allow people to recognize a professional learning community despite a different name or style. Some professional learning communities are very large, perhaps taking the form of an international organization with frequent conventions and meetings, while others may take the form of smaller venues. The common denominator is that both communities encourage their participants to contribute and share their ideas. Creating a Website allows for common ground and common identity. A well-defined domain will lend credence to the professional learning community by having a name, affirming its purpose, and stating its mission and value to members and other stakeholders. Officially sanctioning a Website as a formal professional learning community will inspire participation and facilitate a true exchange of ideas and knowledge. An official Website might benefit greatly by presenting periodic opportunities for people to meet face to face either through teleconferencing or actual meetings.

If corporate management is concerned about proprietary information being shared, it can develop a charter that provides the framework and boundaries for the professional learning community. Informal learning communities will continue to exist whether or not corporate management wants them; therefore, it may prove more beneficial to establish a formal platform for the exchange of knowledge. Executive management may wish to provide a more private opportunity for the exchange of ideas and create a community within the organization that supports formal and informal personnel training.

Monitoring: Executive management must be committed to periodic monitoring of its professional learning community and associated personnel training. It must remain aware of ineffectual or counterproductive training practices and be committed to effecting change when needed. It must design tools that enables it to measure the effectiveness of the organization's personnel training practices, knowledge base, and personnel expertise. A mechanism for systematically reviewing the status of ongoing education and learning opportunities should be developed.

4.5 Process Analytical Technologies (PAT), Applications, and Benefits

Scientific and technological advances in the area of process analytical chemistry, engineering, and multivariate data analysis offer new opportunities for improving the overall efficiencies of drug development, manufacturing, and associated regulatory processes, and lend themselves nicely to forward quality. While the pharmaceutical community has long recognized the need for improvements in these areas, little progress has been made.

Process analytical technologies (PAT) is quickly becoming a cutting-edge quality philosophy within FDA-regulated industries. It has the capability of changing a firm's manufacturing environment for the better, and is a new phase in the evolution of pharmaceutical manufacturing control that has fortunately been examined and promoted by the FDA.

What is PAT? It is the implementation of technology tools to monitor and control the quality parameters of pharmaceutical products while in process as they are being manufactured. Examples of PAT tools are

NIR
LIF
Raman
Mass spectroscopy
Remote hyperspectral imaging
Acoustically optical tunable filters
Multidirectional flow injection sensor technologies

Examples of pharmaceutical processes that can benefit from PAT are

Blending
Formulations
Dissolution testing
Drying
Tableting
Raw material testing and release

4.5.1 Compatibility of PAT with Traditional Validation Approaches

PAT can be used concurrently with traditional validation approaches. Within a pharmaceutical environment, PAT can be applied during product and process development, analytical method development, and scale-up. It can reduce manufacturing cycle times while maintaining quality. In some cases it may even improve product quality and support recall prevention in that in-process results are rapidly obtained. Incorporating the availability of information and rapid feedback during in-process control can go a long way in preventing substandard product from being released and introduced into commerce.

Obtaining critical product quality information while the product is still being manufactured can lead to improved patient safety as well as increased product quality. PAT can also help establish causal links between process variables and product performance. The key is to develop suitable methodologies specific to the manufacturing process and product. It is also important to develop validation strategies that allow for the incorporation of PAT.

4.5.2 Regulatory Implications Related to PAT

PAT have significantly impacted the regulatory paradigm under which the industry has historically operated. In some cases when PAT has been used as a quality control measure FDA has allowed the number and type of required drug product specifications to be reduced. Some companies have actually been allowed to skip selected tests in the final drug product specification because of PAT QC. The FDA is cooperating with a number of large pharmaceutical companies to create a regulatory environment conducive to the deployment of PAT. Among other things, it is attempting to define the conditions under which PAT may replace current end product release testing. The goal is to expedite new product development and commercialization. Shortening time to market can occur if PAT is employed concurrently with acceptable validation methods. If PAT proves to be a successful in-process quality control tool, three validation batches would still be required, but future batch release might come to rely solely on PAT testing. While PAT is distinctly useful for new product development and process improvements, it can also serve to improve the process of products already marketed, particularly those processes that have shown manufacturing problems.

4.5.3 PAT as a Continuous Improvement Strategy

PAT is a system for continuous analysis and control of manufacturing processes based on expeditious, real-time measurements during processing. They can measure quality and performance attributes of raw and in-process materials as well as processes to ensure acceptable finished product quality. What makes PAT forward-quality is that it can occur while the process is underway, as opposed to the more traditional approach of measuring quality at the end of the process. Clearly, this significantly supports the recall prevention program discussed earlier. Additionally, PAT measurements are typically nondestructive to the ongoing process. The process is not consumed or contaminated by these PATs. Additionally, the need to prepare samples are eliminated and timely testing turn around times can be avoided. PAT is designed to measure critical quality parameters and performance attributes, thereby yielding quality-specific information. They also contribute to a forward quality manufacturing environment in that they support the quick resolution of process-related problems and decrease protracted production cycle times. In-process QC testing can often take days (frequently exacerbated by OOS investigations), subjecting the WIP to potential environmental contamination or excessive holding times. PATs are capable of addressing this common problem. These advances in PATs can provide high-quality drugs to the public and hopefully at a faster pace. They also provide an opportunity to move away from the current in-process and finished product testing paradigm to substantiate quality towards a built-in (continuous QA) paradigm that ensures a higher level of quality. Online measurements of product performance are always preferable to finish product testing. It provides greater insight in a timely manner, allowing for appropriate manufacturing and QA intervention. The industry must be prepared to justify the use of PAT during a regulatory inspection since the agency is not yet totally accustomed to and comfortable with these technologies. It is essential for firms to include in their regulatory submissions their intention to utilize PAT in conjunction with traditional validation. It would not be prudent for a firm to assume that PAT will be acceptable during product development or scale-up since this forward quality approach is relatively new. It is important not to position the use of PAT as a means to increase capacity, but to emphasize its usefulness as a QA tool.

4.6 Risk Management Through a Comprehensive GMP Enhancement Master Plan

The FDA recently launched a significant new initiative to help ensure that cGMPs are applied in a more consistent manner and reflective of technological advancements within FDA-regulated industries. The initiative has been called "Pharmaceutical cGMPs for the 21st Century: A Risk-Based Approach." It is essentially a risk-based approach intended to enhance the focus of the agency's GMP requirements by identifying and targeting those manufacturing operations that pose the greater risk on public health and focusing their efforts accordingly. Additionally, FDA has committed to enforcing pharmaceutical product standards while not impeding innovation or the introduction of new manufacturing technologies in the industry. This initiative promises to enhance the consistency and predictability of FDA's approach to monitoring and assuring production quality and product safety. While FDA-regulated industries must continue to fully comply with existing cGMP regulations, the agency has promised to evaluate how it might improve its operation through a risk management-based approach.

The FDA's cGMP regulations were initially issued in 1963, according to a *Federal Register* announcement, with changes in 1971, 1978, and 1995. The 1995 revision clarified certain requirements and was "intended to allow drug manufacturers more flexibility and discretion in manufacturing drug products" (60 FR 4087). Since FDA-regulated industries have experienced such remarkable advances in the biologics, drugs, and veterinary medicine areas, it is necessary for the agency to re-evaluate its current approach. The industry's increased reliance upon automation and computerization and the increasing role of biotechnology and products regulated by the FDA mandates that the agency reposition itself to confront these advances. Additionally, advances in such QA tools as PAT and new manufacturing technologies provide an opportunity to evaluate how these advances can be applied to pharmaceutical manufacturing. This initiative will integrate the most current quality systems and risk management approaches and will encourage the adoption of modern and innovative manufacturing technology. In addition, it will better integrate its current inspection program with the review of drug quality that is performed as part of the preapproval process. The initiative will also use existing and emerging science and analysis to ensure that limited resources are best targeted to address important manufacturing quality issues, especially those associated with predicted or identifiable health risks. It is expected that this initiative will strengthen public health protection achieved by FDA's regulation of pharmaceutical manufacturing. It will also allow FDA to enhance the scientific underpinnings of the regulation of pharmaceutical quality and to facilitate the latest innovations in pharmaceutical engineering.

4.6.1 A Rational Response to the New FDA Initiative: The GEM Plan

In response to FDA's risk management orientation, FDA-regulated industries might begin to consider ways in which to adopt a similar orientation. In fact, why not stay a few steps ahead of the agency and perform an organizational assessment that will lead to a comprehensive GMP enhancement master plan (GEM plan). A GEM plan is a proactive initiative that has the potential to significantly improve an organization's overall compliance and quality status. A GEM plan could use an FDA-issued, list of observations (FD-483) as a platform from which to launch the initiative. The FDA inspection does not preclude the need for company self-assessment, however. The GEM plan is a compliance-centered and forward quality initiative that is

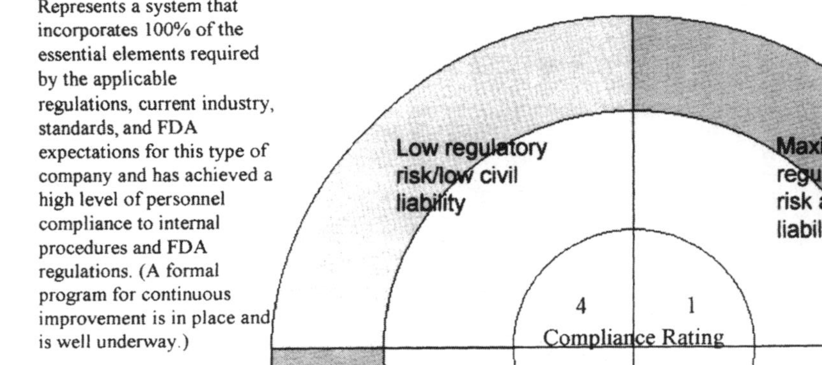

FIGURE 1 Compliance profile rating for a biologics or pharmaceutical firm.

marshaled by the company's executive and quality management. The GEM plan allows a company to evaluate its weaknesses and install the necessary improvements before the FDA mandates such an action. This is why a GEM plan should be undertaken before the FDA arrives to perform an inspection. A GEM plan is a risk management tool in that it can be provided to the FDA in response to any deficiencies unearthed during the inspection. The GEM plan's objectives and goals must be far-reaching and comprehensive in order to convince the agency that the company can launch such an initiative successfully and ensure that it addresses the deficiencies revealed during the inspection. The GEM plan is a strategic initiative for managing crisis before it strikes.

Additionally, a GEM plan can help enlist the district's cooperation during implementation of the plan because FDA will recognize the firm's proactive, forward quality efforts. A GEM plan can integrate compliance issues cited on an FD-483 with continuous improvement activities already underway. This will allow the firm to restore its credibility with the agency, the public, and other stakeholders.

An important part of a GEM plan is the ability to rate the company's compliance and quality status once the assessment has been completed. There are many possible rating systems that can be employed as part of a GEM plan. Figure 1 is one possible rating system that will enable executive management and QA to establish a baseline during this forward quality initiative.

This is an example of a rating system for a biologics or pharmaceutical manufacturing operation. The comprehensive assessment should cover all critical compliance systems (process validation, personnel training, change management system, etc.) and determine whether or not they meet FDA expectations and current industry standards.

4.6.2 Role of Executive and Quality Management

Another important aspect of the GEM plan is the role and responsibilities of executive management. A GEM plan reflects forward thinking on behalf of the firm's executive management. It demonstrates that it is effectively communicating and interfacing with QA management and are committed to a crisis management strategy before crisis actually strikes. A highly visible executive and quality management team with the authority to allocate resources will ensure success during the implementation of the GEM plan. Figure 2 represents essential elements for a successful GEM plan.

Mutually agreed upon objectives and goals must be spelled out in the GEM plan. The alliance of and commitment from both divisions (executive management and QA) will invariably create companywide enthusiasm and motivation. It will also demonstrate to the FDA that there is internal collaboration and executive management commitment to support a long-term,

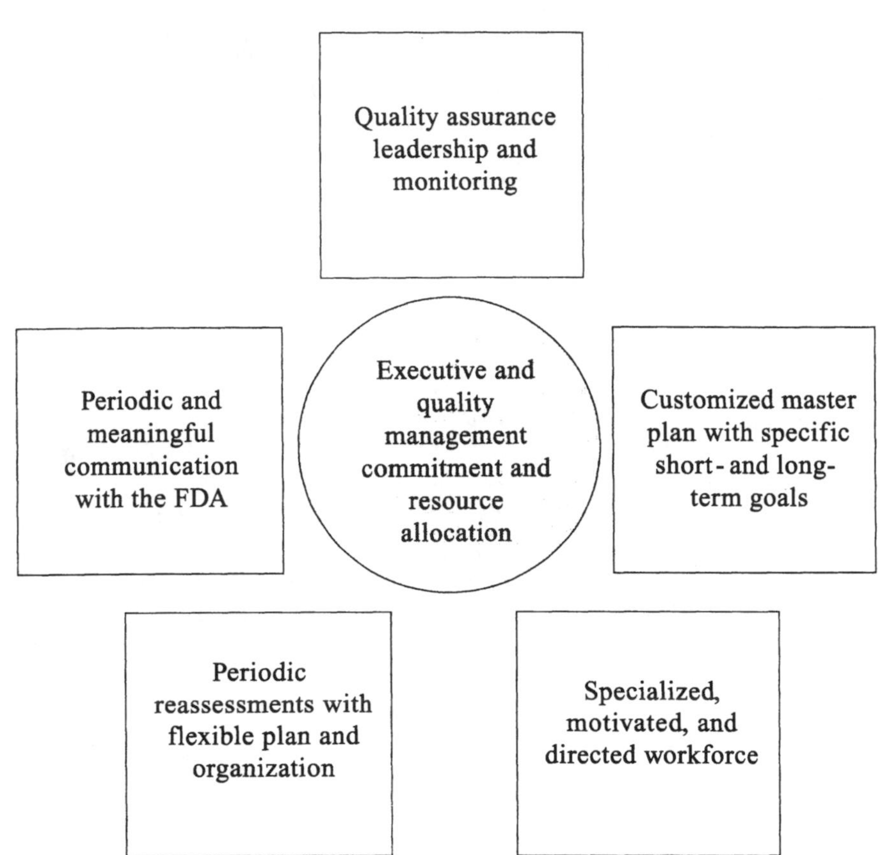

FIGURE 2 Elements of a successful GEM plan.

comprehensive, forward quality improvement plan. After all, FDA shares the same objective—effective risk management that will ensure high-quality, safe, and effective products.

There may be a need to change the GEM plan as it unfolds. Executive management and QA must ensure that the organization remains flexible and realistic. Management's commitment and rapid response to reorganizing, redirecting, and allocating resources where needed is the key to success and full compliance.

5 WORDS OF WISDOM

Embrace a forward quality perspective during the early stages of product development.

Install a quality assurance infrastructure during development and throughout all clinical trial activities.

Become familiar with the sorts of quality assurance issues that could assault the integrity of clinical trial activities.

Periodically evaluate the quality and compliance status of the organization, regardless of the stage of product development or commercialization it is in.

Communicate the firm's commitment to forward quality to all employees, stakeholders, and regulatory agencies.

Commit to employing innovative quality control tools, such as process analytical technologies.

Install a recall prevention program, in conjunction with a meaningful annual product review system.

Create professional learning communities for knowledge transfer and employee training.

Develop a GMP enhancement master plan (GEM plan) before FDA inspects the firm and imposes it own corrective action plan.

15

International Compliance Issues and Trends

Alan G. Minsk
Arnall Golden & Gregory, Atlanta, Georgia, U.S.A.

Globalization, harmonization, standardization, uniformity, and mutual recognition agreements are all prevailing trends that significantly affect compliance in the pharmaceutical, medical device, and biotechnology industries.

This chapter will focus on some, but not all, of the areas in which the U.S. Food and Drug Administration (FDA) and the European Union (EU) regulatory authorities have attempted to coordinate their efforts to provide uniform rules and standards for the pharmaceutical industry. Specifically, we will review the efforts to harmonize approaches relating to inspections (including public disclosure of confidential information) and product approval or authorization (including clinical trials). While space limitations do not provide sufficient opportunity to describe each regulatory authority's system or the harmonization attempts in detail, it is our hope to provide some background of where the efforts are now, where the efforts are intended to go, and what we believe will be the results of these efforts. In addition, the author is much more familiar with the U.S. system than the EU system because of his experience and daily exposure with FDA, this chapter will focus more on the U.S. structure.

1 INSPECTIONS

1.1 Overview of FDA's Inspectional Authority

To understand the ongoing harmonization efforts concerning pharmaceutical inspections, specifically as they relate to uniform quality systems standards, we must first describe the current inspectional authorities that FDA and the EU possess.

FDA conducts inspections of pharmaceutical companies for many reasons, such as a directed inspection for a specific reason (e.g., notice of a complaint about a drug product), a routine audit to ensure company compliance with current good manufacturing practice (cGMP) requirements, a reinspection after a warning letter or other enforcement action, a recall effectiveness check, a preapproval inspection (PAI), or because the company either has bid to supply products or is a supplier to U.S. government agencies. According to the Federal Food, Drug, and Cosmetic Act (FDC act), the agency is authorized to inspect the premises and all pertinent equipment, finished and unfinished materials, containers, and labeling either within the establishment or on any vehicles in which drugs are manufactured, processed, packed, held, or transported [1]. The inspection, which is to occur at reasonable times, within reasonable limits, and in a reasonable manner, "shall extend to all things therein (including records, files, papers, processes, controls, and facilities)" bearing on whether the products comply with the FDC act and FDA's implementing regulations. The FDA investigators pay particular attention to process validation, laboratory operations, bulk pharmaceuticals, and microbial contamination. The FDA's authority to inspect drug establishments applies to both prescription and over-the-counter (OTC) products [3]. The agency may also review and copy all records of common carriers and those receiving products in interstate commerce (i.e., cross state lines), showing the movement of such products in interstate commerce or holding of the products after movement in interstate commerce [4].

In certain cases, FDA may inspect foreign manufacturing sites. Typically, FDA must be invited to inspect by the company, and at times by the foreign government. If FDA should not receive the invitation, however, it can detain or refuse admission of products to the United States that "appear" to be noncompliant with U.S. law. In addition, FDA can refuse to approve pending marketing applications before the agency if it cannot verify compliance with certain requirements as part of the new drug approval process, which might involve manufacturing activities at a foreign site. Some international inspections include bioresearch inspections, which cover clinical trials, preclinical trials, and other activities that are used to support a marketing application.

Most foreign inspections conducted by FDA are in European countries, although many are in Japan and other countries in different regions of the world. Typically, a foreign inspection trip lasts 3 to 4 weeks, with two to five inspections per trip and more than one country covered. A foreign inspection often involves a review of administrative information, raw materials (e.g., handling, storage, controls), production operations (e.g., standard operating procedures (SOPs), validation, production records, packaging and labeling, facilities, equipment, and maintenance), and product testing (e.g., procedures and methods). The agency may deny the importation of drug products that "appear" to be violative [5]. FDA may also reject a marketing application if it considers the company to be in violation of the law.

Firms should be aware that FDA investigators are *not* entitled to review everything they want. The FDC act makes it clear that the following documents are *not* subject to review during an inspection (although the agency might attempt to obtain a search warrant from a court):

Financial data.
Sales data (other than shipment data).
Pricing data.
Personnel data (other than data as to qualifications of technical and professional personnel performing FDA-related functions).
Research data (other than data relating to new drugs and antibiotic drugs and subject to reporting and inspection) [6].
Although not specified in the FDC Act, internal audit data as it relates to compliance with cGMP requirements, are typically not subject to FDA review, but the agency might seek confirmation that audits were performed. FDA has described situations, such as a for-cause inspection or during litigation, in which it might seek a review of this type of information.

1.2 EU Inspection

Several laws and guidelines describe inspections EU [7]. The EU consists of Austria, Belgium, Denmark, Finland, France, Germany, Greece, Ireland, Italy, Luxembourg, the Netherlands, Portugal, Spain, Sweden, and the United Kingdom. The EU membership can, and is expected to, expand. Within the EU, the European Commission is responsible for the harmonization of inspection procedures and technical matters. The European Medicines Evaluation Agency (EMEA) coordinates national inspections and pharmacoviligance, and the "supervisory authorities" in the member states conduct inspections of manufacturers within their respective countries [8]. The inspection will be conducted by someone from the responsible country in which the facility is located, but the inspection is done on the EU's behalf, not only for the member state [9].

1.3 FDA vs. EU Inspections

The FDA and the EU regulatory authorities seek the same objective when conducting an inspection—to determine whether or not products manufactured at a particular site comply with applicable regulatory requirements so that products distributed to consumers are lawfully produced. There are some general differences between the two inspection approaches, however. (Of course, there may be exceptions to these general observations.)

Inspections conducted by EU authorities focus primarily on cGMP-type compliance. In contrast, FDA's current focus is on PAIs, although cGMP compliance is part of the PAI. The two styles seem to be reversing roles, however. The EU has recently begun reviewing PAIs more closely, while FDA's new foreign drug inspection plan calls for a shift of enforcement emphasis from preapproval product evaluation to postapproval cGMP compliance, sometimes referred to as a "risk-based" strategy, although some at FDA might say that its inspection focus remains application-driven.

Another general distinction between FDA and EU inspections relates to the disclosure of certain information obtained during an inspection. In the United States, anyone can submit a written request to FDA, pursuant to the Freedom of Information Act (FOIA), to obtain a copy of an FD-483 issued to a company (i.e., a listing of observations by an FDA investigator of a facility's potential noncompliance), the company's response, the establishment inspection report (i.e., the investigator's diary of the inspection), and any resulting enforcement action, such as the issuance of a warning letter.

Typically, FDA will comply with the FOIA request. There are exceptions in which the agency may not disclose certain information, however, such as confidential, trade secret data, and when FDA is considering further law enforcement-related actions against the company. Separately, FDA's regulations permit communications between the agency and foreign government officials on certain matters, although there are certain conditions that must be met [10]. In contrast, the EU appears more concerned about the releaseability of company or product information. Member states' laws vary, and only a few have an FOIA-type law. In general, according to the EU, public access to information is not a right.

In the mid-1990s, FDA and the EU began discussions to harmonize cGMP-type inspections. Specifically, in 1997, Congress enacted the Food and Drug Administration Modernization Act (FDAMA), which amended the FDC Act, to encourage FDA to support the Office of the U.S. Trade Representative, in consultation with the Secretary of Commerce, in promoting harmonization of regulatory requirements concerning FDA-related products through mutual recognition agreements (MRAs) [11]. According to FDAMA, FDA was required to publicize a framework for achieving that MRA of cGMP inspections no later than 180 days after FDAMA's enactment date (i.e., by May 21, 1998). The FDA met the deadline [12].

While the passage of FDAMA was significant, the concept of harmonization was not new. Harmonization is generally understood to mean the adoption and application of a common approach to regulatory activities, and the United States has numerous regulatory cooperation agreements with foreign countries on drug- or device-related issues, including agreements on imports/exports, product approval, labeling, and compliance. Separate from MRAs, some agreements may be called "memoranda of understanding," "memoranda of cooperation," and "cooperative arrangements."

Several reasons led to the MRA with the EU on quality systems. First, funding for FDA inspection has been decreasing over the years, and harmonization with the EU on foreign inspections would save FDA significant financial and personnel resources [13]. Second, during the 1990s, there were efforts to reform FDA, including the passage of FDAMA and export reform.

Third, many members of Congress expressed concern that foreign firms were not held to the same standards as U.S. firms, which was reflected in an April 1998 General Accounting Office (GAO) report entitled, "Improvements Needed in the Foreign Drug Inspection Program," which stated that only one-third of foreign firms that had informed FDA of intentions to ship drug products to the United States had been scheduled to be inspected by the agency. Furthermore, the GAO report, which had been commissioned by Congress, found that 85% of foreign inspections in fiscal year (FY) 1996 revealed cGMP deficiencies sufficiently serious to merit a formal response from the firm. The number of warning letters went down in FY 1996 compared to FY 1995, however. Despite FDA conducting nearly as many foreign inspections in 1997 as in 1996, the number of warning letters issued to foreign drug manufacturers for cGMP deficiencies declined by more than 50% during FY 1997 compared to FY 1996. Meanwhile, during that same period, there was an increase in warning letters issued to domestic manufacturers on cGMP-type issues [14].

The GAO noted that two-thirds of foreign inspections in FY 1997 related to PAIs (due in large part to user fee funding), and only one-third to risk assessment issues. The GAO suggested that in FY 1997 FDA headquarters frequently downgraded inspections in which field investigators recommended enforcement action. As a result, there were fewer reinspections [15]. The GAO report said that FDA should ensure that serious manufacturing deficiencies were promptly identified and enforcement actions initiated. In addition, GAO was concerned that FDA was taking too long to issue warning letters to foreign drug firms for serious cGMP deficiencies and allowing companies to continue exporting products to the United States despite manufacturing problems.

With this background, the MRA between FDA and the EU on inspections was formally signed on May 18, 1998, with two sectoral annexes on drugs and medical devices. According to an FDA Talk Paper issued on June 16, 1997, it is anticipated that "both Partners to the agreement [FDA and EU] could streamline their processes and save considerable resources while enhancing their public health standards" [16]. The stated purpose of the pharmaceutical cGMP annex is to "govern the exchange between the parties and normal endorsement by the receiving regulatory authority of official good manufacturing practices (cGMP's) inspection reports after a transitional period aimed at determination of the equivalence of the regulatory systems of the parties" [17]. The general idea is that FDA might be able to defer to the EU on GMP-type inspections of sites in the EU so that FDA does not have to conduct them and vice versa. (For the purposes of this chapter, we will only discuss the pharmaceutical-related issues of the MRA, and not those associated with medical devices.)

The annex covers preapproval and postapproval cGMP inspections and describes systems under which FDA and participating regulatory authorities of EU member states will exchange information about products and processes subject to the annex [18]. According to FDA, however, the cGMP sectoral annex does not affect FDA's *current* cGMP regulations. The products subject to the annex include biological products for human use, active pharmaceutical ingredients, drugs for human or animal use, and intermediates and starting materials [19]. The following products are not covered: human blood, human plasma, human tissues and organs, and veterinary immunologicals. In addition, human plasma derivatives, investigational medicinal products/new drugs, human radiopharmaceuticals, and medicinal gases are excluded during the transition period (to be discussed), but these products' coverage may be reconsidered at the end of the transition period [20].

The pharmaceutical cGMP annex provides for a 3-year transition period in which FDA and EU were to review the "equivalence" of one another's cGMP-type inspection programs to determine whether they provide the same level of consumer protection as their own systems [21]. "Equivalence" of the regulatory systems means

> the systems are sufficiently comparable to assure that the process of inspection and the ensuing inspection reports will provide adequate information to determine whether respective statutory and regulatory requirements of the authorities have been fulfilled. "Equiva-

lence" does not require that the respective regulatory systems have identical procedures [22].

The transition period began in December 1998. During the transition period, FDA and EU authorities were to participate in such "confidence-building activities" as the exchange of inspection reports and cooperate on joint inspections and training [23]. These activities would enable a party to assess "equivalence" of its counterpart regulatory authority.

As described in FDA's regulation, the following are the criteria to be used by FDA and the EU to access equivalence for post- and preapproval cGMP-type inspections:

- Legal/regulatory authority and structures and procedures providing for post- and preapproval
  - Appropriate statutory mandate and jurisdiction
  - Ability to issue and update binding requirements on cGMPs and guidance documents
  - Authority to make inspections, review and copy documents, and take samples and collect other evidence
  - Ability to enforce requirements and to remove products found in violation of such requirements from the market
  - Substansive manufacturing requirements
  - Accountability of the regulatory authority
  - Inventory of current products and manufacturers
  - System for maintaining or accessing inspection reports, samples and other analytical data, and other firm/product information
- Mechanisms in place to assure appropriate professional standards and avoidance of conflicts of interest
- Administration of the regulatory authority
  - Standards of education/qualification and training
  - Effective quality assurance systems measures to ensure adequate job performance
  - Appropriate staffing and resources to enforce laws and regulations
- Conduct of inspections
  - Adequate preinspection preparation, including appropriate expertise of investigator/team, review of firm/product and databases, and availability of appropriate inspection equipment
  - Adequate conduct of inspection, including statutory access to facilities, effective response to refusals, depth and competence of evaluation of operations, systems and documentation; collection of evidence; appropriate duration of inspection and completeness of written report of observations to firm management

Adequate postinspection activities, including completeness of inspectors' report, inspection report review where appropriate, and conduct of follow-up inspections and other activities where appropriate, assurance of preservation and retrieval of records

Execution of regulatory enforcement actions to achieve corrections designed to prevent future violations and to remove products found in violation of requirements from the market

Effective use of surveillance systems
- Sampling and analysis
- Recall monitoring
- Product defect reporting system
- Routine surveillance inspections
- Verification of approved manufacturing process changes to marketing authorizations/approved applications [24]

At the end of the 3-year transition period, FDA and the EU regulatory authorities that participated in the activities will assess the information obtained during the transition period [25]. Representatives from FDA and the EU will then meet in the Joint Sectoral Committee to jointly determine which regulatory authorities are equivalent, using the criteria described in the annex [26]. Regulatory authorities not listed as equivalent at the end of the transition period may apply for reconsideration at a later date, once the necessary corrective measures have been taken or additional experience is obtained. After equivalence determinations have been completed, the operational period will begin.

The operational period follows the transition period and applies to pharmaceutical inspection reports "generated by authorities listed as equivalent for the inspections performed in their territory" [27]. Reports provided during the operational period will represent inspections performed in member states whose authorities are listed as equivalent. If certain conditions are met, inspection reports provided during the operational period may also include inspections carried out in EU member states whose authorities are not listed as equivalent [28]. Inspection repots will be exchanged during the transition period, but only to get a sense of what is in them for the purpose of determining equivalence. During the operational period, those reports from authorities found to be equivalent will be exchanged and "normally endorsed" (to be explained shortly).

The regulatory body that conducts a postapproval GMP inspection will submit its report to the authority of the importing country within 60 calendar days of the request [29]. If a new inspection is necessary, the inspecting body will provide its report within 90 calendar days of the request [30]. Reports of PAIs will be sent to the requesting authority within 45 calendar days of the request unless an "exceptional" case arises [31].

For preapproval GMP inspection reports, equivalent regulatory authorities will give preliminary notification that an inspection may need to take place [32]. Within 15 calendar days of the notification, the regulatory authority requested to perform an inspection will acknowledge receipt of the notice and will confirm its ability to perform the inspection [33]. If the authority performs the inspection, the resulting report will be sent to the requesting authority within 45 calendar days of the request [34]. If in an "exceptional case" an authority requests a report to be transmitted in a shorter time, it must describe the exceptional circumstances in the request [35].

According to the MRA, each party retains the right to conduct its own inspection if it considers it necessary, and may suspend or detain the product distribution to protect human or animal health [36]. Similarly, if a regulatory authority is unable to perform the inspection as requested, the authority making the request will have the right to conduct the inspection itself. According to an FDA Talk Paper dated June 16, 1997, the "regulatory authorities and bodies of the exporting countries will measure manufacturers' compliance according to the requirements of the importing country." It is thus important to remember that each party retains full responsibility for products marketed in its own country and "has a right to fulfill its legal responsibilities by taking actions necessary to ensure the protection of human and animal health at the level of protection it deems appropriate" [37]. The importing country may request reinspection by the exporting country and may conduct its own inspections at will [38].

Once an equivalent authority receives an inspection report from another equivalent authority (post- or preapproval reports), the receiving authority will "normally endorse" the report, except under specific and delineated circumstances (e.g., material inconsistencies or inadequacies in inspection report, quality defects identified in postmarket surveillance, and specific evidence of serious concern in relation to product quality or consumer safety [39]. In these exceptional cases, the importing country's regulatory authority may request clarification from the exporting country, which could result in a request for reinspection. In addition, the importing country might conduct its own inspection of the production facility if attempts at clarification are not successful [40].

Normal endorsement will be based on findings in the inspection report as they are measured against the importing country's own laws and "based on the determination of equivalence in light of the experience gained." FDA expects that it will be able to normally endorse inspection reports received from authorities listed as equivalent.

Any party to the MRA may contest in writing the equivalence of a regulatory authority [41]. The Joint Sectoral Committee composed of FDA and EU officials will discuss the matter and determine whether verification of equivalence is required [42]. The Joint Sectoral Committee will try to reach unanimous consent on the appropriate action [43]. An authority may be suspended immediately by the committee if there is agreement, or the matter may be referred to the Joint Committee, which establishes the Joint Sectoral Committees and which is also composed of U.S. and EU representatives [44]. If this committee cannot achieve unanimous consent on the issue within 30 days, the contested authority will be suspended [45]. Upon the suspension of authority previously listed as equivalent, a party is no longer obligated to normally endorse the inspection reports of the suspended authority. A party must continue to normally endorse the inspection reports of that authority prior to suspension unless the authority of the receiving party decides otherwise based on health or safety considerations.

The suspension will remain in effect until unanimous consent has been reached by the parties on the future status of that authority [46].

In addition to its status as an arbiter of disputes, the Joint Sectoral Committee monitors the activities under both the transitional and operational phases. The Joint Sectoral Committee's functions include

- Making a joint assessment (which must be agreed upon by both parties) of the equivalence of the respective authorities
- Developing and maintaining the list of equivalent authorities, including any limitation in terms of inspecting type or products, and communicating the list to all authorities and the joint committee
- Providing a forum to discuss issues relating to the MRA, including concerns that an authority may no longer be equivalent, and an opportunity to review product coverage
- Considering the issue of suspension [47]

The Joint Sectoral Committee meets at the request of either party (i.e., FDA or EU) and, unless the cochairs of the committee agree otherwise, at least once each year. The Joint Committee is kept updated on the meeting agenda and the conclusions reached at these meetings.

The MRA also provides for FDA and the EU to establish an early warning system to exchange information on postmarketing problems with a drug, and the agreement includes a section on maintaining the confidentiality of, and providing public access to, certain information about an inspected company [48]. The following are the criteria to be considered by FDA and the EU in developing a two-way alert system:

- Documentation
  - Definition of a crisis/emergency and under what circumstances an alert is required

Standard operating procedures
Mechanism of health hazards evaluation and classification
Language of communication and transmission of information
Crisis management system
Crisis analysis and communication mechanism
Establishment of contact points
Reporting mechanisms
Enforcement procedures
Follow-up mechanisms
Corrective action procedures
Quality assurance system
Pharmacovigilance program
Surveillance/monitoring of implementation of corrective action [49]

1.4 Implication of the Mutual Recognition Agreement

What is the implication of the MRA between FDA and EU on GMP-type inspections? The potential downside is uncertainty, because both FDA and the EU as well as industry, have much data to gather and interpret. Until there is final resolution there will be status quo to some extent, during which FDA and the EU will continue to conduct their own inspections, although they will both be working toward harmonizing their efforts. In addition, there is concern that neither party, particularly FDA, will change its enforcement approach, despite congressional pressures to do so. It is also unclear whether the MRA will indeed bring consistency in enforcement approaches when many in the pharmaceutical industry complain that there is a lack of consistency today with current inspections and investigator observations.

On the positive side, it is hoped that, after the parties can reach agreement, there will be uniformity and harmonization concerning cGMP-type inspections. In addition, with a streamlined inspection process, it is expected that drug products will be approved at an accelerated pace because more PAIs will be conducted in a more expeditious manner. Finally, the goal is that, if the MRA objectives are met, industry will have a better understanding of what to expect during an inspection and the cGMP standard bar will have been raised.

As of this writing, there are a number of outstanding issues that remain unresolved and that could threaten the viability of the MRA. One of FDA's biggest concerns is the unharmonized definition of GMP and the EU member states' interpretation and enforcement of quality system standards. Another potential problem is the EU's reluctance to conduct inspections of APIs, despite their inclusion in the pharmaceutical GMP annex.

FDA has also expressed frustration that the EU member states have not always provided inspection reports and other relevant information describing EU regulatory systems in English, thereby delaying the "equivalence" evaluation process. In addition, the agency has said that the EU has not been too responsive in completing certain joint activities during the transition period, such as establishing a comprehensive alert system and assurance of confidentiality.

The EU member states also have their concerns about FDA's efforts. For example, FDA assessments will not likely be finished by the end of the transition period. The agency estimates that it will complete equivalence evaluation of only one member state, with the others being in varying degrees of completion. FDA wants to determine equivalency of all member states while the EU wants one state to represent all. The EU has expressed frustration with FDA's progress, as to its equivalence assessment of all EU Member States. The EU noted that its "evaluation of the FDA is on target to be completed within the three years of the transition period." In addition, the EU has suggested that FDA's review of each member state's regulatory system appears to go beyond the "spirit" of the MRA. Lack of resources is also another obstacle. Finally, the EU is concerned that FDA does not intend simply to rely on member states' regulatory conclusions concerning GMP compliance, although the agency will conduct a comprehensive review of each member state's regulatory system.

As of this writing, the MRA is not officially dead, but it is stagnant. To date, FDA provided the EU with copies of virtually every document describing what the agency does (e.g., Code of Federal Regulations, Compliance Policy Guide, Regulatory Procedures Manual). Some of the EU member states have reciprocated. EU representatives have visited FDA's headquarters to learn more about the agency and its operations. In addition, EU officials have observed FDA investigators and analysts conducting inspections of U.S. firms. Similarly, FDA representatives visited Medicines Control Agency in the United Kingdom to perform an equivalence assessment. FDA officials likewise observed several inspections of UK firms by MCA inspectors.

To the author's understanding, Ireland was scheduled to be the next country for FDA to observe. However, the process is on hold.

It is vital to the industry to stay apprised of international initiatives to harmonize inspection efforts [50]. In addition, industry should keep updated on new developments and monitor enforcement trends. Finally, firms should coordinate efforts between domestic and international operations, including third-party distributors, so that all parties are on the same page.

2 PRODUCT APPROVAL

2.1 Overview of FDA's Drug Approval Process

The manufacturing and marketing of "new" pharmaceutical products in the United States requires the approval of FDA. Failure to obtain premarket approval might cause a drug product to become "unapproved," in violation of the FDC Act. Noncompliance with applicable requirements can result in fines, issuance of a warning letter, and other judicially imposed sanctions, including product seizures, injunction actions, and criminal prosecutions [51]. Similar approvals by comparable agencies are required in foreign countries.

The FDA has established mandatory procedures and safety and efficacy standards that apply to the clinical testing, manufacture, labeling, storage, recordkeeping, marketing, advertising, and promotion of pharmaceutical and biotechnology products. Obtaining FDA approval for a new therapeutic product may take several years and involve the expenditure of substantial resources.

A "new drug" is one that is not generally recognized among scientifically qualified experts as safe and effective (typically referred to as "GRAS/E") for use under the conditions stated in its labeling [52]. In other words, if a product is GRAS/E, it is not a new drug. Furthermore, even if the product is not GRAS/E, it might not be a new drug if it was on the market prior to 1938. If a drug was marketed prior to 1938, premarket approval is not required, as long as no changes to the composition or labeling of the drug have occurred. A drug may also be a new drug if it has not been used outside clinical investigations "to a material extent or for a material time under [labeled] conditions" [53]. [We will not discuss here the drug efficacy study implementation (DESI) review or the "paper NDA" policy because, while important from a historical context, these issues relate to older drugs on the market or older policies.]

FDA's regulations state, in relevent part, that a drug may be new because of

- The newness for drug use of any substance which composes such drug, in whole or in part, whether it be an active substance or a menstruum, excipient, carrier, coating, or other component
- The newness for a drug use of a combination of two or more substances, none of which is a new drug
- The newness for drug use of the proportion of a substance in a combination, even though such combination containing such substance in other proportion is not a new drug

- The newness of use of such drug in diagnosing, curing, mitigating, treating, or preventing a disease, or to affect a structure or function of the body, even though such drug is not a new drug when used in another disease or to affect another structure or function of the body
- The newness of a dosage, or method or duration of administration or application, or other condition of use prescribed, recommended, or suggested in the labeling of such drug, even though such drug when used in other dosage, or other method or duration of administration or application, or different condition, is not a new drug [54].

Pharmaceutical products under development are required to undergo several phases of testing before receiving approval for marketing. The first step involves preclinical testing, which includes laboratory evaluation of product chemistry and animal studies, if appropriate, to assess the safety and stability of the drug substance and its formulation. The results of the preclinical tests, together with manufacturing information and analytical data, are submitted to FDA as part of an investigational new drug (IND) application. An IND must become effective before human clinical trials may commence. Moreover, once trials have commenced, FDA may stop either the trials or particular types of trials by placing a "clinical hold" on such trials because of concerns about, for example, the safety of the product being tested. Such holds can cause substantial delay and, in some cases, might require abandonment of a product or a particular trial.

Clinical trials involve the administration of the investigational pharmaceutical product to healthy volunteers or to patients identified as having the condition for which the pharmaceutical agent is being tested. The pharmaceutical product is administered under the supervision of a qualified principal investigator consistent with an informed consent. Clinical trials are conducted in accordance with good clinical practice (GCP) requirements and protocols previously submitted to FDA (as part of the IND application) that detail the objectives of the study, the parameters used to monitor safety, and the efficacy criteria evaluated. Each clinical study must be reviewed and approved by an independent institutional review board (IRB) at the institution at which the study is conducted. The IRB considers, among other things, the design of the study, ethical factors, the safety of the human subjects, and the possible liability risk for the institution [55]. Government regulation of both IRBs and informed consent along with the resulting enforcement action, appear to be on the rise, with several high-visibility institutions known for clinical testing coming under regulatory scrutiny for their clinical testing practices.

Clinical trials for new products are typically conducted in three sequential phases that may overlap. In phase I, the introduction of the pharmaceutical into healthy human volunteers, the emphasis is on testing for safety (adverse effects), dosage tolerance, metabolism, distribution, excretion, and clinical pharmacology. Phase II clinical trials usually involve studies in a limited patient population to determine the initial efficacy of the pharmaceutical of specific, targeted indications, to determine dosage tolerance and optimal dosage, and to identify possible adverse side effects and safety risks. Once a compound is found to be effective and to have an acceptable safety profile in phase II evaluations, phase III trials are undertaken to more fully evaluate clinical outcomes. Phase III clinical trials generally further evaluate efficacy and test further for safety within an expanded patient population and at multiple clinical sites. FDA reviews both the clinical plans and the results of the trials and may require the study to be discontinued at any time if there are significant safety issues. In certain cases, FDA may request so-called phase IV studies, which occur after product approval and may be set as a condition for product approval. These studies can be designed to obtain additional safety data or efficacy data, detect new uses for or abuses of a drug, or determine effectiveness for labeled indications under conditions of widespread usage. These studies can involve significant additional expense.

The results of the preclinical and clinical trials and all manufacturing, chemistry, quality control, and test methods data are submitted to FDA in the form of a new drug application (NDA) or a biologics license application (BLA) for marketing approval.

There are seven broad categories in which the required data fall.

Preclinical data, such as animal and in vitro studies, evaluating the drug's pharmacology and toxicology

Human pharmacokinetic and bioavailability data

Clinical data (i.e., data obtained from administering the drug to humans, which must include "adequate tests" to demonstrate that the drug is safe for use under the proposed conditions for use, as well as "substantial evidence" that the drug is effective under the proposed conditions)

A description of the proposed methods by which the drug will be manufactured, processed, and packed

A description of the drug product and drug substance

A list of each patent claiming the drug, drug product, or method of use, or a statement that there are no relevant patents making such claims

A financial certification or disclosure statement (or both) by clinical investigators, where applicable

The drug's proposed labeling [56]

An NDA must also contain a certification that the applicant has not and will not use the services of any person who has been debarred by the Secretary of Department of Health and Human Services because of a felony conviction for conduct related to drug approval, or for conspiring, aiding, or abetting with respect to such offense [57].

In addition to the aforementioned requirements, the NDA applicant must provide a summary "in enough detail that the reader may gain a good general understanding of the data and information in the application, including an understanding of the quantitative aspects of the data" [58]. The summary must conclude with a presentation of both the risks and benefits of the new drug [59].

There are three types of premarket applications for new drugs. The most onerous is the "full" NDA, submitted under section 505(b)(1) of the FDC act [60]. A full NDA , in particular, requires extensive clinical data to prove the drug's safety and efficacy. FDA usually requires two adequate and well-controlled clinical studies to support approval [61]. The type of information that FDA will require for the NDA submission is described in the agency's regulations [62]. The law permits FDA to approve an NDA based on only one adequate and well-controlled clinical investigation and confirmatory evidence if certain conditions are met, however [63].

The NDA (including the conduct of clinical trials) usually takes several years to prepare and file and is very costly (which can include the payment of "user fees" to FDA to review the application). The FDA review period usually takes 1 to 2 years and the outcome is not certain. The approval process can be affected by a number of factors, including the severity of the side effects and the risks and benefits demonstrated in clinical trials. Additional animal studies or clinical trials may be requested during the FDA review process and may delay marketing approval.

A company may also submit a supplemental NDA(sNDA) for certain types of modifications to a drug product for which the company is the NDA holder. A firm may submit an sNDA to market a drug product for which it already has approval but, for example, in a different strength or dosage form or for a new indication. Typically FDA will only ask for information in the sNDA, which could include data from further clinical trials, to support the safety and efficacy of the change or modification.

Another type of NDA is established by section 505(b)(2) of the FDC Act (a variant of the old "paper NDA"). Specifically, a 505(b)(2) application is defined as

> an application submitted under section 505(b)(1) of the act for a drug for which the investigations described in section 505(b)(1)(A) of the act and relied upon by the applicant for approval of the application were not conducted by or for the applicant and for which the

applicant has not obtained a right of reference or use form the person by or for whom the investigations were conducted [64].

FDA has stated that section 505(b)(2) NDAs apply to the following applications:

> [those] that contain investigations relied upon by the applicant to provide full reports of safety and effectiveness where the investigations were not conducted by or for the applicant and the applicant has not obtained a right of reference or use from the person who conducted the investigations. Thus, section 505(b)(2) of the act is not restricted to literature-supported NDA's for duplicates of approved drugs; it covers all NDA's for drug products that rely on studies not conducted by or for the applicant or for which the applicant does not have a right of reference [65].

A 505(b)(2) NDA usually requires published studies or similarly available information. FDA also expressly recommends an application under section 505(b)(2) for a modification, such as a new dosage form, of a previously approved drug that requires more than only bioequivalence data [66]. Some advantages of the 505(b)(2) NDA route are that, if certain conditions are met the applicant may obtain marketing exclusivity (discussed below) and can rely on studies not performed by it for approval, thereby saving the applicant significant money and time. A notable disadvantage is that a 505(b)(2) NDA may be subject to certain patent and exclusivity restrictions imposed by law, to the benefit of a 505(b)(1) NDA holder, thereby delaying approval of the 505(b)(2) NDA.

While FDA has discussed the types of applications it considers appropriate for filing under section 505(b)(2), there is little guidance in the FDC Act or its legislative history. In 1999, however, the agency issued a document entitled "Guidance for Industry: Applications Covered by Section 505(b)(2)" in an effort to provide some assistance on the type of information needed to support a 505(b)(2) NDA.

The little FDA guidance, the inconsistency at times shown by FDA reviewers in determining the information required to support these NDAs, and the agency's general lack of familiarity and experience with 505(b)(2) NDAs are significant barriers and limitations to this route of marketing.

The least burdensome application is the abbreviated NDA (ANDA), which may apply to a new drug that is bioequivalent to a "reference listed drug" (a new drug approved by FDA for safety and effectiveness and designated as the reference product for approval.) [67]. An ANDA must contain the "same" active ingredient as the brand name drug and have essentially the same labeling [68]. As background, Title I of the Drug Price Competition and

Patent Term Restoration Act (commonly referred to as the Waxman–Hatch Amendments or the 1984 Amendments) amended the FDC act by establishing a statutory ANDA procedure for duplicate and related versions of human drugs approved under section 505(b) of the FDA Act [69].

There is no requirement that ANDA applicants conduct complete clinical studies for safety and effectiveness. Instead, for drugs that contain the same active ingredient as drugs already approved for use in the United States, FDA typically requires only bioavailability data demonstrating that the generic drug formulation is, within an acceptable range, bioequivalent to a previously approved drug [70].

FDA will consider an ANDA drug product to be bioequivalent if

> The rate and extent of absorption of the drug does not show a significant difference from the extent of absorption of the listed drug when administered at the same molar dose of the therapeutic ingredient under similar experimental conditions in either a single dose or multiple doses and the difference from the listed drug in the rate of absorption of the drug is intentional, is reflected in its proposed labeling, is not essential to the attainment of effective body drug concentrations on chronic use, and is considered medically insignificant for the drug [71].

The statutory conditions to demonstrate bioequivalence, as described in the FDC act, are not exclusive and do not preclude other means of establishing bioequivalence.

It is important to note that generic drug products can be developed and approved more quickly than innovator drugs, because clinical trials are not typically required. A 505(b)(1) or 505(b)(2) NDA holder may delay approval of an ANDA due to patent or exclusivity considerations, however. These issues, however, while described briefly later, are beyond the scope of this chapter.

If a manufacturer wants to submit an ANDA for a drug that differs from a listed drug in the identity of active ingredient(s), route of administration, dosage form, or strength, the manufacturer must first file a petition to FDA requesting permission to do so [73]. This type of submission is typically referred to as a "suitability petition." The petition should include

- A description of the action required, which should specify the differences between the ANDA product and the listed drug on which it seeks to rely
- A statement of the grounds of that action, which addresses the basis for the petitioner's conclusion that the changes proposed in the petition meet the statutory criteria for acceptance

An environmental impact analysis or claim for categorical exclusion from the requirement to prepare such an analysis

A certification that the petition contains not only all information on which it relies but also representative data and information known to the petitioner that is unfavorable to the petition

The petition must provide information that the active ingredient of the proposed drug product is of the same pharmacological or therapeutic class as the reference listed drug [74]. In addition, the petitioner must demonstrate that the drug product can be expected to have the same therapeutic effect as the reference listed drug when administered to patients for each indication of use in the reference listed drug's labeling for which the applicant seeks approval [75]. The petition should also include a copy of the proposed labeling for the drug product that is the subject of the petition and a copy of the approved labeling for the listed drug [76]. A suitability petition must be granted, unless, among other things, FDA finds that safety or effectiveness studies are needed for approval of the drug [77]. The FDA must act on the petition within 90 days of submission [78].

The FDA may disapprove a suitability petition if it determines, in relevant part, that

Investigations must be conducted to show the safety and effectiveness of the drug product or any of its active ingredients, its route of administration, dosage form, or strength that differs from the reference listed drug [79].

Any of the proposed changes from the listed drug would jeopardize the safe or effective use of the product so as to necessitate significant labeling changes to address the newly introduced safety or effectiveness problem [80].

If the agency approves the suitability petition, it may describe any additional information required for ANDA approval [81].

The FDC Act does not differentiate between prescription and OTC drugs with respect to new drug status [82]. An OTC new drug requires premarket approval. FDA has adopted an administrative process, however, the OTC drug review, to determine which active ingredients and indications are GRAS/E for use in OTC drugs [83]. With the aid of independent expert advisory review panels, FDA is developing final rules, referred to as monographs, that define categories of GRAS/E OTC drugs.

Once a monograph is final, any drug within the category may be marketed only in compliance with the monograph or under an approved NDA [84]. FDA does provide for an abbreviated form of NDA in which the drug would deviate in some respect from the monograph [85]. This so-called NDA deviation need include only information pertinent to the deviation [86].

2.2 Rx-to-OTC Switch Petitions

To provide a context for our discussion of Rx-to-OTC switches, we will describe briefly the classifications of drugs that must be limited to prescription use. Specifically, the following types of drugs must be sold for prescription use only: (1) drugs not safe for use except under the supervision of a licensed practitioner because of toxicity or other potential for harmful effect, method of use, or the collateral measures necessary for use; and (2) drugs limited to prescription under an approved NDA [87]. FDA may remove by regulation a drug subject to the premarket approval requirements from prescription status "when such requirements are not necessary for the public health" [88].

Neither the FDC Act nor FDA's implementing regulations define the term *switch,* which typically refers to moving a drug product from Rx to OTC status. FDA will allow a switch when the prescription requirements are no longer necessary to protect the public health and the drug is safe and effective for use in self-medication as directed in the proposed OTC product labeling [89].

There are three ways to accomplish a switch: (1) an NDA holder submits a supplemental application for the switch, bearing the burden of proof to demonstrate that the prescription status is no longer needed for the safe and effective use of the drug; (2) FDA, on its own, makes the switch by creating or amending an OTC monograph; or (3) a manufacturer petitions FDA to make the switch, typically by creating or amending an OTC monograph to include the prescription drug [90]. Frequently, FDA might require as part of the switch process additional warnings to be provided on the product to ensure the safe and effective use of the drug [91].

The switch issue received a great deal of attention recently when, in response to a citizen petition filed by a third part, FDA discussed the possibility of unilaterally reclassifying prescription antihistamine products, such as Zyrtec, Allegra, and Claritin, from Rx to OTC status. Several legal issues have been raised as to whether FDA has this authority without obtaining the manufacturer's permission and, to date the agency has not made a final decision. In the case of Claritin, the manufacturer submitted an sNDA for the switch.

The most common way to switch a product is for a manufacturer to submit an sNDA, because the NDA owner might obtain 3 years of market exclusivity if the supplement is supported by new clinical investigations (other than bioavailability studies), conducted or sponsored by the applicant, that are essential to approval of the application or supplement [92]; that is, FDA will not approve (but will accept the filing of) a 505(b)(2) NDA or an ANDA before the expiration of 3 years from the date of the approval of the first applicant's NDA or sNDA. A second manufacturer, however, may still obtain approval of its version of the same drug through the submission of a 505(b)(1) NDA and/or where the NDA does not refer to the first manufacturer's product or to any investigations conducted by or for the first manufacturer [93]. In addition, FDA will not reject a 505(b)(2) NDA or an ANDA merely because it has approved a supplemental indication for the pioneer drug upon which the new generic is based within the preceding 3 years; FDA will not approve the generic version for the supplemental indication, but may approve it for others [94]. (Refer to Chap. 1 for a detailed discussion on regulatory submissions.)

2.3 Fast Track and Accelerated Approval Provisions

While the benefits of FDA's fast track approval and accelerated approval procedures are described in Chapter 1, it should be noted that a company seeking to use this option must also expedite its preparation for preapproval inspection and GMP compliance. Because FDA signoff of the manufacturing facility will likely be required for NDA approval and such signoff is to be accelerated, a firm must recognize that expedited compliance is necessary for this expedited review and have its own manufacturing house in order.

2.4 Overview of the EU Drug Approval Process

The EU drug approval process is working toward a harmonized FDA-type system. The transformation is not yet complete, however, and currently the EU has two drug approval systems in place.

The efforts to develop a harmonized system concerning direct safety, efficacy, and quality go back to 1965, when the first harmonized directive was issued [95]. Ten years later, the Committee for Proprietary Medicinal Products (CPMP) was established [96]. In 1989, the International Conference on Harmonization (ICH) was founded, and the EMEA began operation on January 1, 1995 [97]. The EMEA serves as an advisory board, but is responsible for coordinating the approval, manufacturing, and inspection of medical products between the CPMP and member states' regulatory bodies [98].

Until only a few years ago, a company intending to market a new drug product in the EU would submit a "national" NDA in each country in which authorization was sought. While national procedures may still be used in limited situations, such as for approval of product line extensions, a firm must choose the "mutual recognition" or "centralized" system. (We will not discuss the approval process for generic products.)

According to the mutual recognition approach, a company may first apply for approval in one EU member state, such as the Medicines Control Agency in the United Kingdom or the Agence du Medicament in France

[99]. That particular country will review and make a decision on the marketing application and issue an assessment report [100]. After that decision, the firm may apply to other EU countries for approval [101]. "Recognition" of the other country's approval is not mandatory, however, and if the countries cannot agree to recognize the approval within 90 days and the applicant does not withdraw, the application is referred to the CPMP for arbitration and a binding decision [102].

A firm intending to market a medicinal product in the EU should review "The Rules Governing Medicinal Products for Human Use in the European Community" [103].

The second approval system is centralized, but at this time is limited to biotechnology drugs, new active substances, new blood products, and high technology products. Specifically, a product application may be sent directly to the EMEA, which consists of the CPMP, a secretariat, an executive director, and a management board comprising representatives from the EU member states, the EU Commission, and the European Parliament [104]. The EMEA uses experts from two countries who are assigned to serve as rapporteurs to review the marketing dossier, and who then report their findings to the CPMP. The CPMP consults with its standing committee on medicinal products for human use [105]. The CPMP has 210 days (and this is very strictly kept) to review the application, and makes a recommendation to the European Commission in Brussels whether or not to approve the drug application. The application may be rejected if quality, safety, and efficacy are not "adequately" shown [106]. If the CPMP recommends approval, the opinion is provided to the European Commission, all EU member states, and the applicant [107]. The European Commission prepares a draft opinion, which is sent to the standing committee on medicinal products for human use [108]. If this committee affirms the draft decision, approval is made final, and is valid in all of the EU states [109]. If the standing committee rejects the proposal, the European Commission must act within 90 days or the proposed rejection is automatically overridden, with the CPMP draft decision becoming final [110]. Each member state's national legislature is not required to accept the European Commission's decision.

The EU drug approval system is *moving* toward harmonization. A system that permits individual states to reject the centralized body's recommendation is a significant obstacle to this goal, however. One author has written

> The inability to bind the member nations continues to be serious hindrance. The success of this endeavor, as measured by reduced inspections or a decrease in time to market, remains subject to the pleasure of each member state's legislatures. The conflicting nationalistic attitudes delaying formation of the agency, may yet prove a substantial barrier to its final success. Viewed in this light, integra-

tion into a single, multinational, central control mechanism as proposed by the ICH seems destined to continue to meet with difficulty if not impossibility [111].

There are proposals to reduce the time it takes to approve or authorize a new drug from approximately 18 months to 9, significantly less than FDA's approval rate [112]. As a result, more companies could shift research and development (R&D) efforts to Europe; R&D investments have doubled in Europe in the past 10 years, while the increase has been fivefold in the United States [113].

Other proposed changes to the drug approval process would include increased centralization of drug approvals, with more new products being submitted to the EMEA, "fast-track" approvals for drugs intended for poorly treated diseases, and "compassionate use" of unapproved drugs where no alternative exists [114].

The EU may also ease restrictions on direct-to-consumer advertising for drug products, such as permitting companies to release information on treatments of AIDS, asthma, and diabetes in response to specific requests from patients. The proposed reforms would require agreement by both the European Parliament and the EU governments, which could take some time [115].

3 CONCLUSION

There is no question that FDA and the EU have taken significant steps in coordinating their efforts on inspections and drug approvals. It is far from clear, however, whether these efforts can be sustained or if international harmonization can be achieved. Like any family, there are internal squabbles within FDA, among EU member states, and between FDA and the EU. This is not to say that such debates and challenges cannot be overcome but, as lawyers frequently say, it depends—on whether the key players in the harmonization efforts can reach consensus on important definitions, standards, and techniques. If this consensus can be reached, the pharmaceutical industry, as well as the regulatory authorities, can greatly benefit from uniformity and consistency. If the current problems persist and are not resolved, however, we will be forced to continue to operate under disparate regulatory systems and approval processes.

4 TIPS FOR MINIMIZING COMPLIANCE/REGULATORY CITATIONS RELATED TO INTERNATIONAL COMPLIANCE TRENDS

Instead of the "regulatory reality check" found in most of the other chapters, this author will provide some tips to help minimize the risk of receiving any regulatory or compliance citations during an inspection.

Consider potential differences between different countries pharmacopeias (e.g., USP, SP, EP), because firms that are multinational and producing drugs for different countries would be smart to ensure that specifications for finished product cover all bases.

Develop an inspection readiness plan and a formalized SOP for handling both overseas and domestic inspections.

Retain outside consultants and regulatory counsel to review SOPs and ensure conformance with FDA's regulatory requirements.

Adhere to SOPs whether the firm is foreign or domestic.

If there is a need to deviate from an SOP, document the reason and have proper sign-off by appropriate personnel.

Thoroughly investigate complaints and out of specification and out of trend results.

Conduct periodic audits of laboratory and manufacturing facilities, utilizing internal and external resources.

Ensure coordination among quality assurance, manufacturing and regulatory affairs departments concerning cGMP issues.

If your company has foreign manufacturing sites, make sure that this unit is informed of prevailing company policies.

Consistently train all employees company policies as well as on foreign and domestic FDA requirements, and share information within the company; there is no "I" in "team."

5 WORDS OF WISDOM

Know your rights concerning foreign and domestic regulatory authorities inspectional jurisdiction.

No affidavits, no photos— have corporate policies on this in advance.

Typically a company should not go above an investigator's head unless it is really necessary and only after much internal discussion and, where appropriate, with outside counsel.

Go to the district director if the investigator is disrespectful, mean, unreasonable, or exceeding legal authority.

Review FD 483 (and any subsequently issued warning letter) for accuracy, clarity, completeness, and foundation.

Submit a timely written response.

Prepare for follow-up inspection, especially if a warning letter has been received.

Convey commitment to compliance, including obtaining senior management support.

Determine whether or not it is useful to bring the matter to FDA headquarters, particularly where scientific or technical issues are involved.
Remain courteous and professional.
Have a crisis management plan that can be implemented if FDA pursues further enforcement action.
Remember that the response is disclosable under the Freedom of Information Act (so be careful about what you say and how you say it).
See the big picture. Make certain that you understand the government's concerns and respond accordingly with the right people and the right answers.
No excuse; fix the problem and prevent recurrence.

REFERENCES

1. 21 USC § 374.
2. 21 USC § 374.
3. 21 USC § 374.
4. 21 USC § 373.
5. 21 USC § 381.
6. 21 USC § 374.
7. See, for example, Directive 75/319/EEC, as modified by Directive 89/341/EEC; Directive 91/356/EEC; regulation 93/2309/EEC, July 22, 1993; Compilation of Community Procedures on Administrative Collaboration and Harmonization of Inspections. III/94/5698/EN. Jan. 1995.
8. K. Bredal Jensen, Good manufacturing practice inspection in Europe, Drug Info J 29: 1211–1216, 1995.
9. P. Meyer, The future cGMP inspection system, Drug Info J 28: 977, 1994.
10. See 21 CFR § 20.89.
11. See 21 USC § 383.
12. FDA's final rule, issued on Nov. 6, 1998, provides for an agency's monitoring of the equivalence assessment, including reviews of inspection reports, joint inspections, common training-building exercises, the development of alert systems, and the means for exchanging information regarding adverse reports, corrections, recalls, rejected import consignments, and various other enforcement issues related to products subject to the annex; 63 Fed Reg 60122; see also 21 CFR Part 26 ("Mutual Recognition of Pharmaceutical Good Manufacturing Practice Reports, Medical Device Quality System Audit Reports, and Certain Medical Device Product Evaluation Reports: United States and the European Community")
13. It is estimated that the yearly cost of FDA inspections, both domestic and foreign, is $3 billion. D. V. Eakin, The International Conference on Harmonization of Pharmaceutical Regulations: Progress or stagnation. Tulsa J Comp Int L 6:221,222, spring 1999.

14. Based on informal discussions with FDA officials, FDA counters that inspections of manufacturers that result in serious findings might lead to satisfactory corrective actions implemented by the firm, and in turn, negate the need for a warning letter or other action.
15. Again, FDA counters that investigators might make recommendations on what to do about a firm, and the firm may make satisfactory corrections after the inspection report is submitted to FDA but before any compliance decision has been made.
16. FDA negotiations with EU. FDA Talk Paper. June 16, 1997.
17. 21 CFR § 26.2.
18. 21 CFR §§ 26.3(a) and 26.6.
19. 21 CFR § 26.4(a).
20. 21 CFR § 26.4(b). Products regulated by FDA's Center for Biologics Evaluation and Research as devices are not covered by the annex.
21. 21 CFR § 26.5.
22. 21 CFR § 26.1(b); Article 1 of the Sectoral Annex for Pharmaceutical GMPs.
23. 21 CFR § 26.6.
24. 21 CFR Part 26, Subpart A, App. D.
25. 21 CFR § 26.6.
26. 21 CFR § 26.17.
27. 21 CFR § 26.11(a).
28. 21 CFR § 26.11(b).
29. 21 CFR § 26.13.
30. 21 CFR § 26.13.
31. 21 CFR § 26.14.
32. 21 CFR § 26.14(a).
33. 21 CFR § 26.14(b).
34. 21 CFR § 26.14(c).
35. 21 CFR § 26.14(c).
36. 21 CFR § 26.21
37. 21 CFR § 26.21.
38. 21 CFR § 26.12.
39. 21 CFR § 26.12(a).
40. 21 CFR § 26.12(b).
41. 21 CFR § 26.16(a).
42. 21 CFR § 26.16(b).
43. 21 CFR § 26.16(c).
44. 21 CFR § 26.16(c).
45. 21 CFR § 26.16(c).
46. 21 CFR § 26.16(d).
47. 21 CFR § 26.17.
48. 21 CFR § 26.19, 26.20, 26.76(a).
49. 21 CFR Part 26, Subpart A, App. E.
50. EMEA/MRA/US/95/00. July 5, 2000; http://www.fda.gov/oia/jscjune00.htm.
51. 21 USC § 331–334.

52. See, for example, 21 USC § 321(p)(1).
53. 21 USC § 321(p)(1).
54. 21 USC § 321(p)(2).
55. 21 CFR § 310.3(h).
56. Currently there are efforts to harmonize GCPs. Specifically, the ICH is working to establish a uniform standard for designing, conducting, recording, and reporting trials that involve human subjects. Sec 60 Fed Reg 42948, Aug. 17, 1995. The ICH guideline includes elements of FDA's regulations and the European GCP requirements.
57. See 21 USC § 355(b), (d); 21 CFR § 314.50.
58. 21 USC § 335a(k).
59. 21 CFR § 314.50(c).
60. 21 CFR § 314.50(c)(2)(ix).
61. 21 USC § 355.
62. 21 CFR § 314.50(c). 21 CFR § 314.50. FDAMA provides that when appropriate and when based on relevant science, the "substantial evidence" of efficacy required for approval of an NDA may consist of data from one adequate and well-controlled clinical investigation and confirmatory evidence (obtained prior to or after such investigation).
63. See 21 CFR § 314.50.
64. 21 USC § 355(d).
65. 21 USC § 355(b)(2). The 505(b)(2) NDA is a hybrid between a full NDA (requiring full reports of investigations that demonstrate whether or not the drug is safe and effective) and an ANDA, to be discussed below.
66. 21 CFR § 314.3(b).
67. 57 Fed Reg 17950, 17952, April 28, 1992.
68. 21 CFR § 314.54.
69. 21 USC § 355(j); see also 21 CFR § 320.21.
70. 21 USC § 355(j)(2)(A). "Bioavailability" is defined in the FDC Act as "the rate and extent to which the active ingredient or therapeutic ingredient is absorbed from a drug and becomes available at the site of drug action." 21 USC § 355(j)(8)(A).
71. 57 Fed Reg 17951, April 28, 1992.
72. Reference deleted.
73. 21 USC § 355(j)(8)(B); see also 21 CFR § 320.1(e) (FDA's implementing regulation that defines "bioequivalence"). An applicant may qualify for a waiver of the in vivo bioavailability or bioequivalence requirement if certain prescribed conditions are met. See 21 CFR § 320.22.
74. 21 USC § 355(j)(2)(C);21 CFR § 314.93. An ANDA may not be submitted for approval of new indications, and other changes from the listed drug that would require safety or effectiveness data; these changes may be approved through a 505(b)(1) NDA or a 505(b)(2) NDA.
75. 21 CFR §§ 10.30(b) and 314.93(c).
76. 21 CFR § 314.93(d)(1).
77. 21 CFR § 314.94(d)(2).
78. 21 CFR § 314.93(d).

79. 21 CFR § 314.93(e).
80. 21 CFR § 314.93(e).
81. FDA defines "investigations must be conducted" as information derived from animal or clinical studies necessary to show that the drug product is safe or effective. Such information may be contained in published or unpublished reports. 21 CFR § 314.93(e)(2); see also 21 CFR § 314.93(e)(1).
82. 21 CFR § 314.93(e)(1)(i), (iv).
83. 21 CFR § 314.93(e)(3).
84. See 21 USC §§ 321(p) and 353(b)(1).
85. 21 CFR § 330.1.
86. 21 CFR § 330.1
87. 21 CFR § 330.11.
88. 21 CFR § 330.11.
89. See 21 USC § 353(b)(1)(A)-(B).
90. 21 USC § 353(b)(3).
91. See 21 CFR § 310.200(b).
92. See, for example, 21 CFR §§ 314.70–71, 310.200(b), and Part 330. According to 21 CFR § 310.200(b), FDA may exempt any drug limited to prescription use from the prescription-dispensing requirements if the agency concludes such requirements are not necessary for the protection of the public health by reason of the drug's toxicity or other potentiality for harmful effect, or the method of its use, or the collateral measures necessary to its use, and the FDA Commissioner finds that the drug is safe and effective for use in self-medication as directed in proposed labeling.
93. See, for example, 21 CFR §§ 310.201(a)(16)(viii)(b) (triaminoheptane), 310.201(a)(26)(methoxyphenamine hydrochloride), 369.20 (pheylpropanolamine hydrochloride).
94. 21 USC §§ 355(c)(3)(D)(iii)–(iv) and 355(j)(5)(D)(iii)–(iv).
95. See Burroughs Wellcome Co. v. Bowen, 630 F. Supp. 787 (E.D.N.C. 1986).
96. Bristol-Myers Squibb Co. v. Shalala, 91 F. 3d 1493, 1496 (D.C. Cir. 1996).
97. Council Directive 65/65/EEC (Approximation of Provisions Laid Down by Law, Regulation, or Administrative Action Relating to Proprietary Medicinal Products)"laid down the principle... that [authorizations] for medicinal products and all Member States should be granted on scientific grounds of quality, safety, and efficacy, without regard to socio-economic considerations." See 1965 OJ (22) 269.
98. Council decision of May 20, 1975, setting up a pharmaceutical committee, 75/320/EEC. Off J Eur Commun no. L. 147/23, 9.6.75 207. The CPMP's purpose was to function as a central celaringhouse for drug approvals submitted to any single European state by any one of the 12 member states of the European Economic Union. Once approval was sought in any single state, application could be made to as many as five states within the union. Those states were required to consider the approval in the initial state when conducting their own reviews. By retaining broad authority to raise objections, each state could decide to reject a drug even though approved by the initial state of submission. This caused substantial uncertainty and effectively added another layer of approval

without any apparent benefit in expediting market access. Essentially all submissions under this system resulted in objections that precluded their general approval. See Ref. 13, p. 224. (Citations omitted.)

99. Council Regulation no. 2309/93/EEC of July 22, 1993. Off J Eur Commun no.L.214/1; Euro Agency for Drug Evaluation sets up shop in London's East End. Nature 371:6, 1994.
100. See Ref. 13, p. 225.
101. T. M., Moore, S. A., Cullen, Impact of global pharmaceutical regulations on U.S. products liability exposure. *DeF CoUNS J,* J 66:101–103, 1999.
102. T. M., Moore, S. A., Cullen, impact of global pharmaceutical regulations on U.S. products liability exposure. *DeF CoUNS J,* J 66:101–103, 1999.
103. T. M., Moore, S. A., Cullen, impact of global pharmaceutical regulations on U.S. products liability exposure. *DeF CoUNS J,* J 66:101–103, 1999.
104. T. M., Moore, Cullen, S. A., impact of global pharmaceutical regulations on U.S. products liability exposure. *DeF CoUNS J,* J 66:101–103, 1999.
105. Commission of the European Union. The Rules Governing Medicinal Products for Human Use in the European Community, 1998; see also 65/65/EEC and 79/319/EEC, amended by 93/39/EEC,93/40/EEC, and 93/41/EEC.
106. Commission of the European Union. The Rules Governing Medicinal Products for Human Use in the European Community, 1998.
107. See Ref. 13, p. 225.
108. See Ref. 13, p. 225.
109. See Ref. 13, p. 225.
110. See Ref. 13, p. 225.
111. See Ref. 13, p. 225.
112. M., Mann, D., Firn, A. Michaels, Europe set for fast track approval of new drugs on www.FT.com, July 18, 2001.
113. M., Mann, D., Firn, A. Michaels, Europe set for fast track approval of new drugs . www.FT.com, July 18, 2001.
114. M., Mann, D., Firn, A. Michaels, Europe set for fast track approval of new drugs . www.FT.com, July 18, 2001.
115. M., Mann, D., Firn, A. Michaels, Europe set for fast track approval of new drugs . www.FT.com, July 18, 2001.

16

Strategic Planning for Compliance and Regulatory Defensiveness

Ron Johnson
Quintiles Consulting, Rockville, Maryland, U.S.A.

1 INTRODUCTION

I was recently talking with the CEO of one of the world's largest pharmaceutical companies and he asked me, "How did this happen?" The "this" was an FDA injunction, potentially a corporatewide injunction. My answer to the question required considerable diplomacy, because in looking at the company's history, all the signs of impending regulatory doom were clearly apparent. This very sophisticated company and its most senior manager, however, were unaware of what got them where they were.

Not knowing where they had arrived or how they got there are symptomatic of a company's failure to develop a corporate culture of quality and regulatory compliance. This culture must permeate the entire organization, from the production worker to the receptionist to the accounts receivable clerk to the director of operations. Anything less results in chinks in the armor, and thus regulatory vulnerability. As they say, "The best defense is a strong offense." The best defense against FDA enforcement action is an effective, far-reaching quality system. The backbone of an effective quality system is a quality culture driven and overseen by the company's top management.

Establishment of strong quality systems is fundamental to regulatory compliance, but should be complemented with both an awareness of specific FDA concerns, along with an action plan to react to them.

This chapter discusses management's role in creating an effective overall quality operation and provides some insights into and techniques to address specific regulatory threats.

2 GOOD COMPLIANCE MANAGEMENT SYSTEM

The Food, Drug and Cosmetic Act and several precedent-setting court decisions have generally defined management's role and responsibility in assuring compliance. The FDA has not developed much specific guidance on how it expects management to execute these responsibilities, however, at least not until the medical device quality system regulation was published a few years ago. In this regulation, FDA for the first time has specifically identified and defined the responsibilities of management. While this regulation applies only to medical devices it does reflect FDA's current thinking and can be used to identify the elements of a good compliance management system for other FDA-regulated industries.

The quality system regulation also gives some interesting insights as to how FDA defines management from its own perspective. As the director of compliance for CDRH, I was responsible for developing and publishing the quality system regulation as a proposed regulation. There were a number of what we considered to be improvements to the existing GMP regulation that needed to be included in this new regulation. One area we felt strongly about was the need for the regulation to specifically address management responsibility. I was involved in developing and processing enforcement actions for most of my FDA tenure. One common feature of virtually all of the legal actions pursued by FDA was what we often referred to as *detached management*; that is, senior-level management who simply did not put a priority on regulatory compliance and made little effort in leading the company's compliance efforts. It is detached management that asks the question "How did this happen?" It seemed clear to us that detached management was a primary cause for most of the regulatory situations requiring FDA intervention. The management we were concerned about was at the highest levels of the company, so in crafting the quality system regulation we included specific duties and responsibilities for what we defined as executive management (the most senior levels of the company; e.g., the CEO or president). These individuals had the executive authority to allocate resources where needed to address regulatory and compliance challenges. Ultimately, the final regulation was modified to refer to "management with executive responsibility," but essentially did not change FDA's perspective. It is senior management that is pivotal to a company's compliance status.

Establishment of a good compliance management system enables senior management to effectively carry out the management responsibilities assigned to it. The elements of a good compliance management system are the following:

Compliance plan
Commitment
Compliance training
Organizational structure
Resource allocation
Periodic reviews and commitment tracking

2.1 Compliance Plan

Senior management (e.g., the CEO, president, business group responsible head) should establish a compliance plan that defines the quality standards and practices, resources, and activities relevant to its operations. A core element and guiding principle should be the *quality policy* or *mission statement.* This policy or mission statement should reflect senior management's quality objectives for the company. It represents the foundation upon which all other quality systems are built.

2.2 Commitment

There must be a recognition by the employees of the company that management is indeed committed to compliance and adheres to the company's quality policy. Management needs to visibly demonstrate this commitment. It is more than just posting a corporate policy or mission statement on the wall of the lunchroom; the daily decisions and actions made by management must reflect this commitment. Demonstrating this commitment is not always easy; balancing the company's need to get a new product to market in time to contribute to fourth-quarter revenue while assuring that it is properly designed and manufactured frequently puts such commitment to the test. Such tough decisions test management's resolve and reveal its priorities to quality and its employees, as well as to the FDA.

2.3 Compliance Training

Management must not only develop a compliance plan reflective of the overall quality policy or mission, it must assure that the elements of these are clearly known by all employees. This requires training, but more important, the training must be effective, as the quality system regulation requires. Going through the motions will not assure the company of success unless the employees have absorbed the training and made it part of their culture. Any training program needs to have an element that assesses effectiveness of the instruction given. The FDA's investigators can be expected to ask rank-and-file employees to describe the company's quality policy. Employees aren't expected to have the policy memorized, but should be able to articulate its principles. Failure to do so leaves a negative impression with the investigator, spurring a more intense investigation into the company's quality systems and overall practices.

2.4 Organizational Structure

Most companies recognize the need for an independent quality assurance unit. There needs to be a direct link between this unit and executive management to assure awareness of the status of the organization's quality systems. The FDA expects careful consideration of an organization's management structure. The basis of noncompliance the potential regulatory crisis is frequently due to an inadequate management structure and the lack of an appropriate and proactive quality assurance presence. The interface between executive management and quality assurance must be carefully defined and formalized. The quality system regulation requires that quality assurance periodically presents a series of quality markers such as the following to executive management:

Consumer complaint trends
Manufacturing deviations
Corrective and preventative actions
Design controls and changes
Out-of-specification and out-of-trend laboratory results
Internal audits

2.5 Resource Allocation

As with most things, if adequate people and monetary resources are not provided, even the best-laid plans will not be successful. The manner in which resources are allocated reflects management's priorities and can predetermine the success of the activity or unit. Executive management's choice of resource allocation will reveal where its genuine commitment lies. If quality assurance and quality control resources are substantially less than those allocated to marketing and production there is no question where the priorities of executive management lie. When considering resource allocation, executive management must ensure an appropriate balance inter- and intradepartmentally, particularly in the areas of quality and regulatory compliance. The FDA will carefully assess management's choices with regard to allocating resources in its overall appraisal of a company's commitment to compliance.

2.6 Periodic Reviews and Commitment Tracking

Management must establish a mechanism to be kept apprised of the state of the operations' quality system. This amounts to a great deal more than just quality assurance performing internal audits. By design, an effective quality system provides a 360-degree feedback loop that can and should be used for continuous improvement. Management must regularly review quality system data in conjunction with the results of the internal audits. This should include feedback from the corrective and preventive system, complaint system, service data, and other sources of quality markers. Management reviews must focus on identifying and correcting underlying system defects. Frequent management reviews allow an organization and its executive management to anticipate potential product failures, compliance deficiencies, regulatory departures, and areas for quality improvement. Management reviews must look comprehensively beyond just corrective actions to include anticipatory preventive actions. More important, management should ensure that it is aware of all FDA inspectional activities, the findings of those activities (e.g., FDA-483's), actions taken to correct any deficiencies, and any communications with FDA. There should never be a time in which a company's executive management is unaware of daily quality concerns and trends, let alone any serious FDA regulatory threat.

Even though a good compliance management system can go a long way in avoiding or minimizing product quality problems, there are no absolute guarantees. The best systems fail occasionally and there are factors beyond your control that can adversely affect your company. Problems can surface from the company's own internal auditing systems, from customer complaints, or as the result of an FDA inspection. When they do, a company must respond in a strategic manner, focusing not on the specific failure, but on the system(s) that caused it. Many well-intended companies conscientiously correct specific problems, but fail to address their genesis or underlying source. Invariably these companies find themselves fixing individual problems over and over again as FDA points them out during repeated inspections. Eventually the company loses credibility with the FDA, FDA loses patience, and regulatory crisis strikes.

3 STRATEGIES FOR RESPONDING TO THE FDA-483 AND WARNING LETTER

Most people in the industry are familiar with the FDA-483 (inspectional observations) that is left by the investigator at the conclusion of an FDA inspection. Whether responding to an FDA-483 or to an internal quality audit finding, the company's approach should be the same; the systems that caused the problem should be the focus of any corrective action.

Before responding to the FDA-483, there are some preliminary facts that should be understood about this regulatory document. Such an understanding will help in reacting and responding to an FDA-483.

The observations listed on an FDA-483 represent what the investigator believes to be objectionable. It is important to realize that frequently an investigator's "observations" do not represent the violation of a specific regulation or statute enforced by the FDA. This fact is widely known and bemoaned by FDA compliance officers, the individuals who review and follow up on the FDA-483 citations issued by field investigators. As such, when company management has the opportunity, either during the daily debriefing sessions or during the exit conference, to discuss specific FDA-483 citations with the investigator, all efforts should be made to identify the applicable statute or regulation related to the citation. If the investigator is unable to link his or her observation to a specific statute or regulation, he or she may be inclined to remove it from the FDA-483.

Once the investigator issues the FDA-483 and leaves the premises following an inspection, the FDA-483 cannot be changed even if it contains errors. This is particularly important when considering that the FDA-483 is available in its entirety under the Freedom of Information Act (FOIA) to anyone in the public arena who requests it. Even if the company is subsequently successful in convincing FDA that a particular observation is inaccurate or inappropriate, once the FDA-483 has been issued, it will be provided without qualification to anyone who requests it under the FOIA. Unless the individual requesting the FDA-483 also requests the EIR (establishment inspection report) and any supporting correspondence and documentation generated in response to the FDA-483 it will not be provided. Additionally, unless the company that was inspected and issued an FDA-483 specifically requests that all of its response documentation be included in any FOIA request, it will not be automatically provided. It thus behooves the company to ensure that the investigator as well as the district is provided with all the relevant information that may appropriately counter the investigator's observation.

Investigators are encouraged by FDA not to include insignificant observations on an FDA-483. Unfortunately, in their zeal to demonstrate their effectiveness, some investigators will insist upon issuing an FDA-483 even for minor or insignificant violations. Sometimes there is a "piling-on" phenomenon, in which insignificant observations are included along with a large number of significant ones. It does not hurt to ask the investigator to prioritize the observations. This will help in setting corrective action priorities and may result in the least significant observations being deleted.

The investigator is generally not permitted to include any observations on the FDA-483 relating to labels, promotional materials, product classifications (new drug application [NDA], premarket application [PMA], etc.), or registration requirements. It is not inappropriate to ask why such observations are included on an FDA-483. It could be an error that could result in either removal of the observation or clarification that headquarters has approved its inclusion.

In spite of the current "industry-friendly" attitude at FDA, many investigators still take considerable pride in issuing a lengthy *list of observations* (FDA-483s). To them (and to many in the agency), it is an indication that they have done an effective job. It is a measure where there are few metrics available to quantify an investigator's efforts in assuring compliance within regulated industries. Those investigators who seldom or never issue FDA-483s are regarded as ineffective by their peers and frequently by their supervisors and managers. Up until a few years ago, an investigator's performance was actually evaluated using the numbers of "violative" inspections conducted, violative meaning issuance of an FDA-483 containing serious deviations from the regulations or statute resulting in at a minimum issuance of a warning letter. Even though FDA has removed this element from performance appraisals of investigators, the agency's culture continues to recognize and reward those who find and document extensive violations.

The FDA-483 opens the floodgates to virtually any enforcement process available to the FDA, thus it must be taken seriously no matter how innocuous or even erroneous it may appear.

3.1 During the Inspection

The best time to respond to an FDA-483 is during the inspection *before it is issued*. The importance of this timing cannot be stressed too strongly. Most companies make efforts to learn of investigator concerns in a timely manner. A first step is obtaining a commitment from the investigator that perceived problems will be pointed out as they are encountered, or at minimum, in a daily briefing. It is essential, however, that there be a clear understanding of the problem by the investigator and the company. By doing so, the investigator may discover that the observation is not as significant as he or she originally thought or may not represent an observation that warrants inclusion on the FDA-483. Moreover, the company will have a clear understanding of the investigator's concern in order to properly initiate corrective action. Too many companies spend their energy and resources solving what they thought was the problem only to find out that it wasn't really at the heart of the investigator's concern.

Another important reason to make sure that the investigator understands what he or she is seeing and has been provided with all pertinent information is to prevent his or her from forming an erroneous hypothesis about the company or its products. For example, if in reviewing complaint files the investigator concludes from the information reviewed that the company's product has caused a death, he or she will focus the remaining inspection on pursuing evidence to support that belief. It will invariably lead to further investigation at some of the company's customer sites. If the company is able to ascertain an investigator's perspective in a timely manner, it can provide the investigator with a comprehensive and accurate summary. If there was a causal relationship between the product and the death, the company can share the results of its investigation and corrective actions. If the company has information that its product was entirely unrelated to the death, however, it can share that auspicious information and avoid the rigors and corporate embarrassment of a "for-cause" investigation.

A more subtle reason to be assertive and proactive in making sure FDA gets all of the information it needs to assess a particular situation is that it can very well derail the snowball effect. Once FDA believes that it has discovered a violation, especially one it believes constitutes a public health threat, the dynamic between FDA and the targeted company dramatically changes. Communication is usually one of the first casualties. The FDA's naturally distrustful attitude is energized; "Guilty until proven innocent" becomes its approach. The company's perspective becomes less and less credible to FDA as it attempts to prove its hypothesis. Even if mitigating information is obtained by FDA, the snowball effect may have moved the action too far along for it to back off. In this situation, FDA will frequently pursue whatever action it can, even a less serious regulatory action than it initially had planned. The snowball effect starts with the investigator and is augmented by the district office. By the time it is submitted to FDA headquarters, considerable investments have been made and the issue has an emotional charge that cannot be ignored by headquarters reviewers. Headquarters personnel have been criticized and ridiculed by congressional oversight committees for declining to approve enforcement actions submitted by the field offices. Consequently, there is some timidity by headquarters to independently evaluate field enforcement recommendations when faced with the snowball. A common refrain from headquarters' compliance personnel is that "It is easier to approve an enforcement action than to turn in down!" I once heard a senior compliance official who had just been brutalized during a congressional oversight hearing say, "The message is clear. Do not turn actions down."

The time to stop the snowball effect is during the course of the inspection, long before issues mount and become a lengthy FDA-483 and potential warning letter.

3.2 At the Close of the Inspection

The next best time to respond to an FDA-483 is at the conclusion of the inspection when the investigator issues it. Efforts made during the inspection may have successfully prevented some items from appearing on the FDA-483, and for those that do appear, the company has one last opportunity to present its perspective. The investigator is required to discuss his or her observations and record any response made by the company, particularly corrective actions that have been completed or promised. This is the only time that the FDA-483 itself can be changed. If the investigator can be convinced that the observation is incorrect or no longer significant in light of additional information provided by the company, he or she can remove it from the FDA-483. Current agency initiatives also require the investigator to annotate the FDA-483 observations with corrections made or promised by the company. A corrected or annotated FDA-483 more accurately reflects the findings of the inspection to whomever may obtain it under the FOIA.

If the company's efforts in communicating with the investigator during the inspection have been successful, the FDA-483 will contain only confirmed observations and the company will have already conducted necessary investigations and developed acceptable resolutions that can be provided to the investigator. In some cases it may be appropriate to visually demonstrate to the investigator the corrective measures that have been implemented (i.e., documents or physical plant/equipment modifications). The company should solicit a reaction from the investigator to obtain some sense as to whether or not the response is regarded as adequate. The investigator may be reluctant to commit, but frequently will give some indication that his or her concerns have been addressed.

It was suggested earlier that responses to FDA should focus on the systems that underlie the observation or deficiency. During the inspection, there may not be time to do so, so it is acceptable during the course of the inspection for the company to offer the short-term corrective measure and subsequently provide the more comprehensive solution. The investigator may note the specific deficiency in an effort to display a larger systemic problem; therefore every effort should be made by the company to effectively counter or invalidate each citation wherever legitimately possible. After the inspection, all responses and corrective actions must focus on systems rather than just address the specific observations.

Unfortunately, many companies have a policy that all FDA-483 items can only be effectively responded to in writing and curtail comments or responses during the conclusion of the inspection. There are some advantages to this approach in that subsequent responses presumably will be well-organized and reflective of the company's collective mindset and official position. Regrettably, however, it also communicates to FDA that the company does not have confidence in the ability of its employees to adequately respond. This raises concerns about the ability of these same employees to carry out their assigned responsibilities. Worse, it raises suspicious in the investigator's mind that a "corporate spin" will be placed on the response. Company representatives who are relegated to listening to the investigator's list of objectionable conditions and who are restricted to a simple "No comment; a formal response will be made later" quickly lose credibility with the FDA.

3.3 After the Inspection

Nowadays, most companies wisely provide an extensive written response to the FDA-483. This typically takes the shape of a reiteration of inspectional findings accompanied by the firm's collective response regarding corrective measures. This comprehensive response should be signed by a senior management official of the company and be appropriately directed to the district director with a copy to the investigator. This response should include compelling, corroborative documentation that may serve to intercept further regulatory action on behalf of the agency.

One of the major benefits of satisfactorily responding to the issues raised during, at the conclusion of, or after the inspection is the chance that a more serious communication from FDA such as a warning letter may be averted. There currently is a pilot program underway in FDA that will allow companies to avoid a warning letter if a satisfactory response to the FDA-483 is made. Even before this is an official agency policy, many districts as a matter of practice have not issued warning letters when the deficiencies were adequately resolved during or immediately following the inspection. The current pilot is reflective of the agency's enforcement philosophy of the past several years. Anyone who has been around the industry or within the agency knows that the enforcement pendulum regularly swings between aggressive enforcement and voluntary compliance. Even though we are experiencing a voluntary compliance respite, the agency, particularly the field offices, continues to take pride in the issuance of a warning letter. Just as investigators continue to see the issuance of an FDA-483 as a tangible measure of their effectiveness, the district office likewise sees the warning letter as a palpable indicator of its efforts to protect the public health. Even in this instance, however, workload pressures and "common sense" may result in the district office foregoing issuance of a warning letter if the company has made a comprehensive and satisfactory response.

3.4 The Warning Letter

If a warning letter is received it should be taken seriously. While it technically represents the institutional position of the agency, it will more often than not echo the investigator's observations. Reviews of the investigator's FDA-483s and EIRs are frequently performed by compliance officers in the district office who may not have the technical skills and knowledge base to adequately evaluate the significance of the observations. Interestingly, most warning letters are not reviewed or approved by FDA headquarters for medical devices, which may increase the district's reliance upon the investigator's observations and interpretations. From a regulatory perspective, the warning letter does go a step further than the FDA-483 by identifying those observations regarded as serious violations and aligning them with specific statutes and regulatory requirements. The FDA views the warning letter as a potent regulatory action toward placing company management on notice of serious regulatory violations. Warning letters will be directed to the most senior management of the company, or at minimum, will copy most senior management. The purpose for initially targeting specific senior management officials is to enable FDA to later demonstrate that every attempt was made to solicit voluntary correction from executive management with the means to allocate resources. This notification becomes a predictor of which company officials will be named as defendants if the situation escalates to an FDA enforcement action, such as an injunction or prosecution.

3.5 Warning Letter Recommendations

The following are recommendations relative to warning letter responses:

Meet the 15-day response deadline.
Only make commitments that the firm is absolutely capable of meeting.
Ensure that the corrective action goes beyond the specific citations and addresses the underlying systemic deficiency.

Offer solutions that address system improvements related to the specific citation.

Noteworthy is the fact that the FDA's threshold for issuing a warning letter is its conclusion that the company failed to assess the specific citation within the context of the larger system deficiency.

Do not use the response as an opportunity to chastise the agency for the number of days and work hours expended by the company in hosting the inspection. The FDA really is not interested in the extent of resources extended by the firm to host an inspection. Raising such an issue seriously detracts from the message the firm should be sending, specifically that the issues have been seriously discussed and that comprehensive and far-reaching corrective measures will be implemented in a timely manner. Moreover, it characterizes the company as being more concerned about its finances than the serious compliance and quality deficiencies officially noted by the agency.

Do not use the warning letter to complain about the investigator's conduct. Any legitimate concerns in that regard should have been raised during the course of the inspection, not after a warning letter has been issued.

Do not attempt to bury FDA with superfluous and exhaustive documentation. The FDA does not appreciate receiving volumes of documents in the firm's initial response to the warning letter. The agency may perceive this as a maneuver on behalf of the company to intentionally overwhelm the district with documentation. On the other hand, it certainly benefits the company to augment the warning letter response with substantive and appropriate supporting documentation. Should the agency pursue regulatory actions beyond the warning letter, excessive material initially provided to the agency as part of the warning letter response could be used to support its case against the company. As such, the amount of supporting documentation that is included in the response should be limited unless the company has a legal department that thoroughly examines and collaborates in the development of the response package.

The company should make every effort to meet with the district director to present an overall plan intended to address issues that are complex or particularly serious. A face-to-face meeting will provide the company with a firsthand impression of district's perspective related to the warning letter, as well as affirming the company's commitment to compliance.

Make every effort to partner with the district director. Clearly, if there are issues for which resolution cannot be reached with the district or the firm believes the district is wrong in its interpretation of the regulations, the firm can and should seek higher-level adjudication. In doing so, follow protocol by advising the district director that you are seeking higher-level resolution and invite him or her to participate.

It bears repeating that an effective response to an FDA-483 or a warning letter must strategically address underlying system problems. In crafting a response, a simple matrix can be helpful. Placing each observation within its specific quality system will provide an indication of systems that require significant augmentation and improvement. Utilizing a systems approach to address any observation will go a long way in reestablishing the firm's credibility with the agency, as well as providing a roadmap for an integrated, continuous improvement plan. In the midst of presenting the firm's approach, every effort should be made to enlist the district as a partner, from the standpoint that it buys into what the firm has proposed and agrees to the proposed timelines. This quasi-partnership requires keeping FDA informed of the firm's progress or lack thereof and maintaining a high level of credibility with the agency throughout the entire process. This approach will elicit greater cooperation and accommodation from the agency.

It is important to note that after issuance of a warning letter FDA will usually monitor the company to assure a response is received and to schedule and conduct a follow-up inspection. This is done differently by the various FDA offices and usually involves some form of a compliance tracking system. It may be automated or simply paper-based. The purpose of these various systems is to assure that "violative" companies are consistently monitored and tracked. In today's kinder and friendlier atmosphere some offices do not aggressively follow up. Sometimes a follow-up inspection is not required when the violation is adequately resolved with documentation (i.e., a validation study is submitted). When a follow-up inspection is required, it should occur within 6 months following the warning letter. Unfortunately, due to limited resources within the agency, many reinspections do not occur within this prescribed time frame. Nevertheless, a firm that has received a warning letter should be proactive and begin preparations for the follow-up inspection immediately. Sometimes the follow-up will take the form of a comprehensive FDA inspection and other times it will just focus on the issues highlighted in the warning letter. In either case, a company must be prepared for what may turn out to be a comprehensive, full-scale inspection. If the reinspection uncovers continuing problems, the company should expect escalated enforcement action from the FDA.

4 PRODUCT SEIZURE

The Food, Drug and Cosmetic Act provides for the seizure of foods, drugs, and devices that are adulterated or misbranded. While FDA has used this provision sparingly in the past several years, it continues to be a formidable enforcement tool that can have devastating effects for a company.

The statutory authority to seize goods was intended to permit the government to remove adulterated and/or misbranded products from commerce before they reached the consumer. The reality is that it takes FDA so long to develop and process a seizure action that the offensive goods are usually long distributed and consumed by the time the seizure action has been administratively processed. Consequently, FDA has come to rely upon voluntary recalls as the primary means of removing violative products from commerce. Seizure is still used, but now more frequently as part of an escalating enforcement strategy. As FDA attempts to interrupt a company's violative conduct, it employs increasingly more threatening regulatory tools until correction is obtained. Such a strategy usually begins with the issuance of an FDA-483 followed by a warning letter, then seizure, and ultimately an injunction. When one of these enforcement tools is not effective in bringing about compliance, a more forceful action will be pursued. Occasionally the FDA may leap several levels without necessarily imposing gradual regulatory actions. For example, if a reinspection reveals repeat violations, and the warning letter was not effective in eliciting the desired outcome, the agency may consider a seizure or move directly to an injunction. Product in the possession of the company or at its clients' premises can be seized. The purpose of the seizure, when used as part of such a strategy, is really not to remove product, although that will be its immediate effect; its primary objective is to turn the heat up under the company in an effort to obtain corrective actions. Remember the orange juice seizure in the early 1990s? This seizure action was initiated to let the industry know FDA was serious about regulating claims of freshness. Clearly it removed some product from the market, but its major impact was alerting the industry to the ramifications of false and misleading claims. Even seizure of a small quantity of product can be disruptive, embarrassing, and publicly damaging to a firm. The FDA hopes that generating negative publicity about a company will provide the necessary leverage to mobilize the firm's management as well as the industry to commit resources to compliance.

An expanded form of seizure is the mass seizure; a draconian action that can and has brought companies to their knees. Mass seizures vary in scope, but the basic premise is that all goods at a certain location are misbranded and/or adulterated. Mass seizures and frequently made at the company's manufacturing facilities and include all raw materials and in-process and finished product. Such an action essentially serves to paralyze the operation.

There are some telltale signs that can serve to warn a company that the FDA may be considering using the seizure tool. Individually these may simply reflect "normal" evidence gathering by the investigator. Collectively they may forewarn the firm of the agency's intent to initiate a seizure action. The following FDA activities are prerequisites to a seizure:

The investigator collects a physical sample of a product and requests shipping records documenting its distribution. The investigator will leave a receipt for any physical samples collected.

The investigator marks his or her initials and the date on selected cases of a lot of goods in the warehouse. This is for purposes of identifying the goods when the seizure is executed.

The investigator obtains production documents relating to a specific lot or batch and requests copies of labels for that lot or batch. This is referred to as a "documentary" sample since it is not actual product but represents the product.

The investigator asks for an inventory of specific products or physically makes a count of certain products in the warehouse. The FDA has minimum quantity thresholds for a seizure action.

The investigator asks for records identifying interstate consignees of specific products. These consignees may be visited by FDA investigators for the purpose of documenting receipt and location of the entire breadth of goods to be seized.

The investigator asks a company official to sign an affidavit or other document attesting to the interstate source or destination of a particular product. This basically represents an admission by the company that FDA has jurisdiction over the product or that the company has shipped a product believed by FDA to be in violation of interstate commerce (an act prohibited by the statute).

The investigator visits one of the company's customers and collects a sample of the company's product. This sample may be used to initiate seizure of the goods at the customer's premises.

If any of these activities are initiated by an investigator, the firm should be everything it can to ascertain the investigator's intentions. Unfortunately, the investigator is not always aware of whether or not a seizure strategy has been initiated by the district. Nevertheless, the investigator should always be questioned about his or her concerns about the product and the extent to which the district may be aware of these concerns. If the investigator is aware of the district's intentions, he or she will not usually communicate this information to the firm for fear that actions may be taken to thwart the district's efforts. As soon as a firm has confirmed that a seizure action is imminent, the company should consider the following:

Contact the district office to explore FDA's concerns about the product in an effort to be responsive to those concerns as a means of making seizure unnecessary.

Quarantine all suspect product within the company and in distribution centers and inform the district that the firm has proactively ceased any kind of distribution of the product.

Voluntarily recall the product, depending upon the significance of the problem and the level of FDA interest. Recalls should always be considered by the firm if the product pose a health hazard, or if it is reasonably expected that FDA may request or encourage a voluntary recall on behalf of the firm.

Destroy suspect product in the presence of FDA.

Any consideration of these actions should also involve communications with the local district office and should address any other products or lots of the same product that may have similar problems. This approach accomplishes several things. First, it may serve to intercept an FDA seizure; second, it will allow the firm to gauge the degree of FDA's concern. It may be that FDA was not really pursuing seizure, just seeking a responsible action on behalf of the firm's management.

Some companies have gone so far to avoid a seizure that they physically moved the suspect products to a different warehouse or prematurely shipped them to a customer. This is a dangerous strategy because it results in distribution of product that may be harmful, and it may provide evidence that the company knowingly shipped adulterated or misbranded product in interstate commerce—a prosecutable act. In my experience as a regional director, I remember a specific drug manufacturer with serious GMP violations that had been uncovered during the course of several FDA inspections. The company learned that FDA was considering a mass seizure of goods at the company's principal production facility. To avoid seizure, the company leased a number of tractor trailers, loaded them with goods from the facility, and parked them in multiple locations throughout the city. When the FDA seizure was attempted, there was nothing to seize at the firm's facility. The FDA's suspicions were confirmed after a thorough investigation and several company officials were subsequently prosecuted for obstruction of justice.

Convincing the FDA that seizure action is not required is not enough; it is necessary to fully understand the extent of FDA's concerns and comprehensively and adequately address them. If these concerns are not fully understood and resolved, a subsequent confrontation could result in a more serious action against the firm, such as an injunction.

5 THE INJUNCTION

The injunction is used to stop a company's violative conduct. It is most frequently used when FDA has exhausted all of it administrative tools to obtain the company's voluntary compliance. If the company's actions pose a public health hazard, FDA will seek an immediate temporary restraining order as a prelude to an injunction. Alternatively, FDA seeks an injunction when it has been unsuccessful in utilizing the gradual enforcement action scheme. The injunction simply asks a judge to order the company to comply. The complaint filed in court by FDA usually will cite a lengthy history of FDA's efforts to elicit voluntary compliance. This history will include FDA-483s, letters, meetings, and enforcement actions that FDA has initiated to no avail. By this time the company has lost most of its credibility with FDA and any promises at this point are not seriously considered. In fact, FDA is now in the hot seat because the record it must lay before the court calls into question its own effectiveness and provides fodder for its critics; so by the time FDA decides to seek an injunction, the firm can expect that the agency is determined to obtain it.

The following criteria are prerequisites for an injunction:

Multiple inspections in which the same kinds of problems have been identified. Although the company may have corrected the specific findings resulting from each inspection, the repetitiveness and similarity of the violations indicates that the company has failed to identify and correct the underlying systemic deficiency.

The company has repeatedly promised but failed to correct deficiencies.

The company has failed to meet proposed or imposed time frames or deadlines.

The company continues to have problems that result in recalls or field corrections, suggesting an ongoing quality concern.

The company is incapable of addressing its problems.

The company appears to be unwilling to correct the problems.

When seeking an injunction, FDA will as a matter of policy specifically name company officials as individual defendants. These are usually senior-level managers, but occasionally will include lower-level middle managers. A traditional feature of an FDA-sought injunction is what is internally referred to as a "shutdown" provision. This requires the facility or facilities subject to the injunction to cease operations until the company comes into compliance, secures a third party to certify that substantial compliance has been achieved, and receives confirmation by FDA through reinspection that compliance has been achieved. This shutdown can apply for an extended period of time; companies can be shut down for a year or more. Obviously an injunction can seriously impact the viability of a company.

There are additional monetary features of most FDA injunction. These include the defendant's payment of all costs associated with FDA's monitoring of compliance with the injunction, such as inspections and sample analyses. The FDA frequently attempts to recover its significant expenditures related to the investigations and enforcement actions that led up to the injunction. Sometimes this includes what amounts to a fine, even though the language of the injunction will call it something else. A large IVD manufacturer recently paid a whopping "fine" of $100 million. Not uncommonly, there are also penalty provisions if the company fails to meet its imposed time frames or deadlines. All of these costs, along with fees associated with attorneys and consultants and lost revenues, can easily total in the millions of dollars.

Even after the firm is permitted to resume operations, the company is perpetually enjoined by order of the court. This means that any violation of the order constitutes contempt of court. If FDA encounters such a breech, it can file a civil or criminal contempt action, which can result in additional monetary penalties or even jail time for some of the individual defendants.

5.1 The Corporate Injunction

During my tenure with the Center for Medical Devices, I was part of several injunctions that were referred to as corporate or corporatewide injunctions. These actions were intended to address what the agency believed to be a corporate culture that permitted, and possibly encouraged, company noncompliance. Corporate injunctions were invoked when compliance problems were found in different facilities of the same company. A pattern of noncompliance and negligence on behalf of the parent company led the agency to consider a corporate injunction as the regulatory enforcement tool. Corporate injunctions typically include the corporate parent, its senior officers, and all of its facilities around the world. It usually requires that some of those facilities cease operations for some period of time until compliance is achieved. Its characteristics are similar to the traditional injunction, with the exception that this particular enforcement targets the corporate umbrella.

5.2 What to Do When Faced with an Impending Injunction

Once FDA develops a strategy to enjoin a company, it is very difficult to derail the process. This process traditionally begins with a recommendation from the local district office and subsequently is reviewed and approved by the appropriate center. Thereafter, it is submitted as an approved recommendation to FDA's lawyers in the chief counsel's office. This is where the first real legal review will be made of the facts and the violative conduct. Of course the snowball effect is alive and well and has impact on the lawyers reviewing the case file. If concurred by the chief counsel's office, the Department of Justice must also approve the injunction before it can be forwarded to the local U.S. attorney for filing. The injunction development and review process is long and arduous and ultimately relies upon the willingness of the local U.S. attorney to file and argue the case. In an effort to streamline the process, FDA has taken the approach of calling the company in to sign a consent decree of permanent injunction before it actually leaves the hands of the FDA. Approval from the Justice Department will be obtained, and may or may not be involved in the negotiation of the consent. As indicated above, multiple warnings and interactions from FDA usually precede an injunction action, so at this point it is very difficult to change FDA's mind. The only way it can be done is if the company undertakes actions that unquestionably eliminate the violations about which FDA is concerned and convinces FDA that there will be no recidivism.

The fact that this is a long and arduous process can work to the advantage of the targeted company. The FDA recognizes that when seeking an injunction its evidence to support alleged offenses needs to be timely. Ideally, it attempts to have inspectional evidence no older than 30 days when it files its case. In order to meet this objective, a last-minute inspection of the defendant's facilities is usually made just before filing the case. If this reinspection finds that significant improvements have been made, the government's case is seriously weakened and may lead the FDA to reconsider. A company thus has one final opportunity to change the course of FDA's intentions. The improvement must be dramatic, compelling, and impressive enough to overcome FDA's skepticism and sufficient enough to dilute FDA's representation to the Justice Department. Even if corrections appear to have been made, FDA may not fold. A major corporatewide injunction that hinged on the findings of a final inspection of a particular facility come to mind. Initial inspection findings revealed no significant violations; however, a second, more experienced investigator was dispatched to ensure violations were uncovered. This second inspection found serious violations of the type complained of in the injunction and were sufficient to convince the company to sign a consent decree.

Once again, this importance of an effective management review system that provides continuous, up-to-date, comprehensive, and concise information about the company's overall compliance status is clear in light of the possible regulatory ramifications. In the event that a company has failed to institute a review system that provides for such up-to-the minute compliance information, it may almost certainly be surprised when serious enforcement actions are called for within the agency. When and if a company becomes aware of a potential injunction or some other regulatory action, it behooves the firm to act rapidly. Most important, success will be contingent upon the extent to which executive management becomes involved with and committed to making improvements.

Executive management should be prepared to present its plan to the agency, as well as make such significant concessions as

Plant closure for some period of time.
Recall of suspect products.
Cease-and-desist distribution of suspect products.
Employment of a third-party consultant.
Commitment to and execution of a comprehensive compliance upgrade master plan. (For an extensive discussion on the compliance upgrade master plan, see Chap. 16, "The Compliance Upgrade Master Plan.")

Often these proactive actions on behalf of the firm are not sufficient to ward off an injunction. The company may need to decide whether to negotiate a consent decree or take its chances in court.

6 SUMMARY

Preventive measures such as developing a proactive and comprehensive compliance upgrade initiative go a long way in preventing serious regulatory enforcement actions by the agency. A firm's credibility depends entirely upon its management's commitment to quality and compliance, allocation of resources necessary to support that commitment, and consistent follow-through and communication with the FDA.

7 REGULATORY REALITY CHECK

With the chapter providing an instructive backdrop, several FDA observations (FD-483s) documented during recent inspections of FDA-regulated facilities are presented below, followed by a strategy for resolution and follow-up corrective and preventive actions.

7.1 Warning Letter Citation

"The firm failed to install a mature and comprehensive quality assurance department."

7.2 Response

This is one of the most damaging observations the FDA could make during an inspection. It suggests that deficiencies are widespread and insidious to the point of calling into question all aspects of the firm's operations. The only response to this serious violation of the Current Good Manufacturing Practices (cGMPS) would be for the company to voluntarily cease and desist all activities and immediately begin the design and implementation of a well-integrated and expansive quality assurance program. Anything less than a wholehearted commitment to identify and retain experienced quality assurance leadership, coupled with the installment of state-of-the-art quality assurance programs, would result in a possible corporate injunction or consent decree by the FDA.

7.3 Preventive Action

In this day and age any company that attempts to operate an FDA-regulated facility without a fully operational quality assurance department is asking for severe regulatory actions by the agency. There is no justification for even the smallest FDA-regulated firm to operate without some degree of quality assurance and quality control. Not only is this a regulatory requirement, it is smart business sense.

8 WORDS OF WISDOM

Use the executive management review aspect of the quality system regulation as a quality guidepost within the pharmaceutical and biotech industries.

Be aware of discernable compliance warning indicators to predict that a regulatory crisis may be approaching.

The best defense against an FDA regulatory crisis are effective, far-reaching, and integrated quality systems.

The backbone of effective quality systems is a company culture driven by its top management with a commitment to compliance and quality.

Executive management is responsible for the creation and maintenance of a quality mission statement that permeates all levels of the company.

17

Unique and Unprecedented Compliance Challenges in the Biologics Arena

Anne P. Hoppe
Serologicals, Inc., Clarkston, Georgia, U.S.A.

Curtis L. Scribner
Biomedicines, Inc., Emeryville, California, U.S.A.

What is unique about the biologics arena? How does this impact the compliance challenges faced by biologic product manufacturers?

The mission of the Center for Biologic Evaluation and Research (CBER) is to protect and enhance the public health through the regulation of biological and related products. The CBER's Vision is "to advance the public health ... through high quality science-based regulation to ensure safe and effective products reach the public as rapidly as possible." As one of several FDA centers, CBER shares many things in common with those other centers, particularly the Center for Drugs Evaluation and Review (CDER) and the Center for Devices and Radiological Health (CDRH).

The 1990s witnessed expanded efforts to harmonize administrative procedures among the FDA centers, and recorded significant progress in reducing differences in the way in which compliance is assessed and enforced. Despite a common application form for new drug and biologic product approvals (FDA 356h) and reforms in regulatory procedure that eliminate some historic differences in operational perspective, there are unique features that set biologic products apart from all other FDA-regulated products.

1 WHAT IS A BIOLOGIC PRODUCT?

The CBER regulates a wide range of products, including allergenic extracts, monoclonal antibodies, therapeutics derived through biotechnology, somatic cell and gene therapy, xenotransplantation, tissues, whole blood, blood components and derivatives, vaccines and medical devices related to blood bank use, including in vitro diagnostic tests and related equipment, and software. The CBER repertoire thus encompasses drugs, devices, and biologics. Most recently, the responsibility for reviewing biologic's applications for products, other than vaccines and blood, was removed from CBER and placed under the purview of CDER.

The CBER utilizes all of the mechanics of oversight and licensing of investigational new drug application (IND), biologics license application (BLA), new drug application (NDA), and the abbreviated new drug application (ANDA), and medical device clearances [510(k), investigational device exemption (IDE), and premarket approval (PMA)].

This has led to considerable confusion when trying to decide how a product is to be approved. One example of the CBER regulatory schematic demonstrates the interplay of the alternative processes as follows: an empty container for blood collection is cleared as a medical device by 510(k) or PMA; if anticoagulant or preservative is added to the container, it is approved as a new drug (NDA), if blood is then collected in the container with anticoagulant, the blood or blood component requires a biological product license before it can be shipped interstate.

Another interesting demonstration of the multiple regulatory mechanisms that complicate the biologics area is the way in which tissue is regulated under the Code of Federal Regulations (CFR), Title 21 (21 CFR 1270) [1] Since FDA chose to regulate tissues under section 361 of the Public Health Service (PHS) Act as opposed to section 351 (the basis for licensing any biologic product in interstate commerce) the focus is primarily prevention of infectious disease. The 21 CFR 1270 regulations, however, apply only to minimally manipulated or processed bone (including demineralized/with glycerol), ligaments, tendons, fascia, cartilage, corneas, skin, veins (except umbilical), and pericardium. Title 21 CFR 1270 does not apply to whole organs, semen, milk, bone marrow, or autologous tissue.

Some of the other tissues regulated as traditional biologics, devices, or drugs include:

- Biologics: blood components (CP 7342.001 [2])

- Devices: heart valves, corneal lenticules, umbilical veins, interactive wound dressings and dura mater (CP 7382.830 [3]) and
- Drugs: autologous, allogeneic, xenogeneic cells with ex vivo altered biological characteristics, and gene therapy products (CP 7356.002 [4]).

The unique perspective of biologics is also borne out in approval of products by different centers (CBER vs. CDRH), depending upon their intended use. Cell separators are cleared by CDRH if their labeled use is therapeutic apheresis. The CBER requires a second review or approval if the same equipment is used to prepare blood components, such as platelets or plasma; likewise, CDRH clears in vitro diagnostic device (IVD) tests for syphilis for general diagnostic use. If a blood establishment chooses to perform the test (required for blood donor screening) with a particular reagent, however, the IVD manufacturer must include a specific claim for blood donor screening and undergo a second review by CBER. The result is that some serologic tests for syphilis on the market have more than one test method described in the package insert; some of these methods can be used for blood donors and others cannot be used for blood donor testing, although they continue to be used in infectious disease screening clinics. While FDA is making considerable effort to clarify cross-center jurisdiction issues, confusion still reigns.

This chapter will focus primarily on the compliance issues associated with the subset of biological products that include injectable products for human use. There will be no attempt to cover such biologic devices as blood bank software, in vitro diagnostic reagents designated for blood bank use, or equipment designed for performing these tests. There are several aspects of biologics regulation that create unique compliance challenges that will be elucidated in this chapter. Among these factors are

The inescapable variability of the source material and potential threat to safety due to unrecognized infectious agents [transmission of AIDS by blood products and the undefined potential for either human or bovine source materials to spread Creutzfeld–Jakob disease (CJD)/transmissible or bovine spongiform encephalopathy (TSE, BSE)]

The difficulty in establishing surrogate end points because in vitro tests often fail to correlate with meaningful clinical outcomes (e.g., a chromatograph or electrophoretic profile of a protein does not assure stability, reactivity or specificity of an immune globulin);

The dual application of PHS Act requirements (licensing of products and establishments in accordance with 42 U.S.C. 262, PHS Act section 351) as well as the pharmaceutical current good manufacturing practice (cGMP) regulations derived from the Food, Drug and Cosmetic Act (FD&C Act)(21 U.S.C. 374)

The outdated regulations that fail to keep pace with scientific advances and technological changes (resulting in many "requirements" applied only through the licensing process and multiple other sources of guidance that complete the regulatory puzzle and confound an understanding of the regulatory expectations, particularly in the allergenics area, in which manufacturers have brought suit against FDA for the plethora of informal requirements or the blood area, in which memoranda to blood establishments compose the bulk of current requirements and whole books are written to guide manufacturers through the maze [5–7]).

The explosion of cell-derived products and methods for manufacturing both therapeutic and clinical diagnostic products using living cells or recombinant DNA/RNA technology

There is an often repeated, potentially apocryphal story about the Office of Regulatory Affairs (ORA) field investigator finishing his inspection at a major pharmaceutical drug manufacturer and walking to another part of the complex only to have his way blocked by his escort. "I am sorry," says the escort as they start the other way. "That building is off limits to you. That is where we make the biologicals, you know." The investigator glances over his shoulder one more time wondering what really goes on there before proceeding to the conference room to write up his current list of 483 items.

All of this changed when an alert pharmacist in Kansas reported that a patient in his hospital became septic after receiving albumin. Efficient and insightful epidemiology from the pharmacist and later the Centers for Disease Control (CDC) led to isolation of the same organism from both the patient and the albumin, identification of other incidents, and the isolation of multiple organisms from sealed bottles of albumin. FDA field investigators were called to the hospital as well as the manufacturing site, where they found that a pallet of albumin in glass bottles had been dropped, then hosed off using city water. Each bottle was then visually inspected for damage. If no damage was found, the bottle was released. Unfortunately, several bottles were contaminated through small cracks, and the era in which the manufacture of biological products remained the province of "product experts" at the CBER ended abruptly.

History tells us that is not really new. Biologic products were initially regulated, as with all other drugs, because of a disaster. In 1901, 10 St. Louis children died after receiving horse antidiphtheria antitoxin that was contaminated that tetanus toxin. The investigation demonstrated many lapses in optimal manufacturing practice, including mixing of multiple lots, lack of a responsible head to oversee and guarantee production, and inadequate safety testing of the final product. The result of this disaster and the subsequent investigation, after a few days' debate in Congress, was the Biologics Control Act of 1902, which initiated the regulatory differentiation between drugs and biologics and still forms, along with the FD&C, the foundation for biological product regulation today.

Biologic products are defined in 21 CFR 600.3 as [8]:

> (h) any virus, therapeutic serum, toxin, anti-toxin or analogous product applicable to the prevention, treatment, or cure of diseases or injuries of man: (1) A virus is interpreted to be a product containing the minute living cause of infectious disease and includes but is not limited to filterable viruses, bacteria, rickettsia, fungi, and protozoa. (2) A therapeutic serum is a product obtained from blood by removing the clot or clot components and the blood cells . . . (4) An antitoxin is a product containing the soluble substances in serum or other body fluid of an immunized animal that specifically neutralizes the toxin against which it is immune. (5) A product is analogous...(ii) to a therapeutic serum if composed of whole blood or plasma or containing some organic constituent of product other than a hormone or an amino acid, derived from whole blood, plasma, or serum

Biologic products therefore present many problems that are unknown or unusual in drug manufacturing. For example, the source material is often in short supply, poorly controlled, potentially infectious, and active only when maintained within narrow physiologic ranges. The biologic product itself is often delicate and easily denatured. Moreover, biologic products, especially if used as a replacement in human deficiency states such as hemophilia, can themselves induce an immunogenic response in the host, resulting in further bleeding because of the inactivating antibodies. The result is that FDA's CBER had declared that regulation of biological products starts at the earliest possible spot, whether it is the human or animal supplying the plasma, the cell line used to make a recombinant protein, or the infectious agent, and that at a minimum, the manufacturing site must be inspected every 2 years. Transfer of partially processed intermediate products is allowed, but only from one licensable site to another licensable site, or for export if the finished products will not return to the U.S.

The "new" biological products that present increasingly unique manufacturing control issues are challenging the regulatory objectives almost daily. For example, the many biotechnology companies that are rich in science but poor in capital and experience question the need for such tight manufacturing controls, yet at the same time, vaccine and protein production in live plants presents both significant concerns for the control of manufacturing at the field stage and especially a completely new set of contaminants (e.g., pesticides, insecticides) requiring unique control measures.

In fact, there are few fundamentally new compliance issues in regulating the new biological products. The dependence on tight control of the entire manufacturing process defines the strengths and pitfalls of biological product control. It also forms the significant regulatory and compliance conundrum facing the biologics industry as it moves to implement "drug quality" cGMPs today.

2 VARIABILITY OF SOURCE MATERIAL

The regulation of biologic products presents the highest level of challenge for FDA *and* the manufacturer because there is no such thing as a generic biologic product. The inherent variability of the starting materials always leaves an element of doubt concerning whether brand A = brand B, whether lot C = lot D, and even whether potency X "today" = potency X "past or future." The theoretical concerns are magnified by the fact that the testing that can be done "in process" or "prerelease" often does not adequately reflect the key reactivity desired. For example, levels (concentration/titer) of antibodies in an immune globulin preparation can be accurately measured, and with modern tools, the proteins can be exquisitely characterized, but none of the in vitro testing can tell with certainty whether the antibodies are stable throughout a designated shelf life and will enhance immunity or be protective when challenged in vivo by exposure to an infectious agent. Further, even minor changes in tertiary or quaternary structure can result in significant activity or immunogenicity concerns.

Control of manufacturing from the earliest stages forms the basis for biologics regulation, but purity and control of source material has always been a significant problem for biological products. Attenuated or killed infectious organisms have formed the basis for vaccines for years. Impure cultures or the contamination of attenuated strains with wild type or virulent bacteria or viruses can be disastrous. Likewise, the source material for many lifesaving plasma protein products—human plasma donors—is imperfect at best. Significant infectious diseases have been passed from plasma or blood donors to recipients, including hepatitis B, hepatitis C, and HIV. The source for allergenics can be even more problematic, as the source of cat dander or dust may be difficult to document and standardize.

For vaccines the starting material problem was addressed early, and methods were developed to control the identity and purity of the starting material. These studies often were completed with the aid of, and in the laboratories of, scientists at CBER with the conditions mutually agreed upon at the time of approval. Specifically, the development of master cell bank (MCB) and manufacturers working cell bank (MWCB) concepts helped stabilize starting material. A single stock of multiple vials of the starting material is created and then "laid down," usually in the frozen state. This single source of starting material, or MCB, is exhaustively analyzed and expected to form the only starting material for the manufacture of the product for the life of the product.

In order to make the MCB last longer, a MWCB is used. Usually a single vial of the MCB is expanded in culture to form many vials with homogenous content. The MWCB is again exhaustively characterized and used as the starting material for the production of the virus or bacteria for the vaccine. Control and validation of the storage and stability conditions are required because of the fragility of the cells and the need to have an identical start for each manufacturing campaign.

The same concept of MCB and MWCB has been adopted for the cell lines used to manufacture recombinant products, and forms the basis for one of the most fundamental and difficult areas of recombinant product manufacturing. As new processes are used or as new manufacturing risks have been identified, it is not unusual that the initial cell line needs to be adapted to the new growth requirements. This often includes limiting dilution growth patterns to ensure clonality and the development and validation of new MCBs and MWCBs. This presents problems later in development since the final material used in Phase III most often needs to be made from the new clone, but if it takes 6 to 9 months to create and validate the new MCB followed by the MWCB, significant delays can be introduced in going toward approval. As outlined later, the decision to use product made from the new cell line or even by new process changes is made after an exhaustive physicochemical evaluation of the "new" and "old" product to make sure that significant changes have not been introduced. This comparison must be made in a data-rich environment.

For plasma-derived products the problem is slightly different. The source of the plasma (humans or animals) cannot be cloned or be closely controlled (especially in humans), and the concept of a MCB cannot be introduced. Plasma collected under controlled conditions at licensed collection sites and distributed as FDA licensed product (Source Plasma), however, significantly mitigates potential infectious disease complications. Donors are questioned carefully about activities that may put them at high risk of having infectious diseases, given physical examinations, and tested repeatedly for signs of infections. Potential problems, of course, are numerous. People forget or lie about exposures or activities. Even the minimal remuneration given for plasma donation or the burden of the unreimbursed costs of phlebotomy for people with polycythemia vera may be enough of an incentive for donors to "forget" about their risk behavior and put recipients of the plasma or blood at risk.

Whole blood and blood components (e.g., platelets, red blood cells, granulocytes, and peripheral blood stems cells) present similar source product control issues and rely on similar control measures. The difference, of course, is that while a unit of blood is given to only one or possibly a few people, a single contaminated unit of source plasma could present significantly more widespread problems because it is blended with hundreds or thousands of plasma units to extract the ultrasmall quantities of plasma proteins in each unit.

Controlling of source material is also a problem of acute interest with the use of blood (and now cells or whole organs) from nonhuman species. Horse serum has been the source of antisera for years. As noted above, the contamination of antidiphtheria antiserum with tetanus toxin formed the nidus of biological regulation. Likewise, the recent transspecies jump of prion-related disease into cattle in the United Kingdom and Europe and the epidemiological association between consumption of food products potentially exposed to BSE (also known as "mad cow disease") and the development of "new variant" Creutzfeld–Jacob disease (nvCJD) had led the U.S. FDA to strongly recommend against (or ban) the use of European-derived bovine products in the manufacture of biological products. The FDA has gone as far as to defer people from donating blood who have lived for more than 6 months (during the years 1980 to 1996) in the United Kingdom, even though the direct transmission of BSE or CJD through blood, plasma, or biological products has never been demonstrated and, through the use of animal models, is thought to be *extremely unlikely* [9].

More recently, cells, tissues, and even whole organs from swine, primates, and sheep have been transplanted into humans. Oftentimes this transplantation across presumedly unacceptable immunological differences has occurred with significant immunosuppression of the recipient. This has raised considerable concern about the potential transmission of xenogeneic infectious agents, especially retroviruses, into humans. This concern is greatly heightened by the severely immune-compromised patients who do not have the normal infection control mechanism needed to contain the infection. Species jumps have occurred. The FDA has responded to this concern with requirements to ensure that the source material for the cells, tissues and organs is tightly controlled and tested before transplantation [10,11].

3 SHORT SUPPLY

This concern about source material control, however, has led to unexpected applications and problems. For example, the control applied through the "short supply" regulations (21 CFR 601.22) has led FDA to insist that final product manufacturers specify in detail their biologic source material (preferably licensed!) and to license the monoclonal antibodies intended for use in the further manufacture of licensed final products.

Short supply is a very old provision unique to the biologics area, but it is often misunderstood. Short supply was initiated to provide a mechanism that permitted interstate shipment of biologic materials without a license. It was recognized that although the PHS Act (section 351) would apply even to the suppliers of biologic source materials (and hence force them to obtain licenses), this would be unduly burdensome in many situations and could result in shortages of critically needed biologic products for which there was no substitute. The only remaining short supply materials recognized by CBER are snake and hymenoptera venoms and the classic example, Recovered Plasma.

Recovered Plasma to this day is the most widely used application of the short supply provisions. In the 1940s, when it was recognized that lifesaving plasma derivatives could be prepared from human blood, the primary source of such plasma was in-date or outdated whole blood stored in glass bottles. Because of the difficulties in aseptically extracting the plasma from glass bottles that had been filled under vacuum (with a 4-hr limit on the use of the red blood cells [RBCs] for transfusion due to concerns for bacterial contamination), it was almost essential that the plasma recovery would occur in the hospital blood banks after it was determined that the blood would be given as packed RBC. It was completely impractical for FDA to license every little blood bank with plasma to contribute for plasma derivative fractionation, and so the acceptance of unlicensed recovered plasma under short supply was an ingenious solution.

Today all blood for transfusion is collected in plastic bags, and most of the plasma is immediately separated by the collector. In addition, source plasma collected by apheresis as a licensed product provides approximately 80% of the U.S. plasma supply [12]. The FDA still permits unlicensed recovered plasma without FDA-defined specifications to move interstate under short supply, however. This solution to the supply dilemma requires the final product manufacturers to assume the responsibility for assuring that their source materials are appropriate and effective in their manufacturing process.

The licensed manufacture of the final product has to institute controls that assure the FDA requirements are met. Although some manufacturers inspect their suppliers, generally, the control exercised is primarily a written agreement specifying that all FDA-required tests are performed, records are kept, separation is aseptically performed by the supplier before shipment, and so on. This agreement does not, however, release the primary manufacturer from responsibility for appropriate control of the source material. Specifically, 21 CFR 601.22(b) requires that

> A biologics license issued to a manufacturer and covering all locations of manufacture shall authorize persons other than such manufacturer to conduct at places other than such locations the initial, and partial manufacturing of a product for shipment solely to such manufacturer only to the extent that the names of such persons and places are registered with the Commissioner of Food and Drugs and it is found upon application of such manufacturer, that the product is in short supply due either to the peculiar growth requirements of the organism involved or to the scarcity of the animal required for manufacturing purposes, and such manufacturer has established with respect to such persons and places such procedures, inspections, tests or other arrangements as will ensure full compliance with the applicable regulations of this subchapter related to continued safety, purity, and potency. Such persons and places shall be subject to all regulations of this subchapter except Secs. 601.2 to 601.6, 601.9, 601.10, 601.20, 601.21 to 601.33, and 610.60 to 610.65 of this chapter [13].

A manufacturer wishes to implement short supply arrangements therefore must remember that

- The final product manufacturer retains all responsibility for assuring FDA requirements are met.
- A product to be shipped under short supply requires appropriate labeling with the intended use indicated ("Caution: For Use in Manufacturing . . . [name of product] . . . Only").
- The written agreement must be filed with CBER as a BLA supplement–and must be periodically updated to reflect any new requirements, changes in authorities of the signatories, etc.
- The agreement must be signed by both parties *before* the product referenced is collected. One cannot, for example, collect and store a lot of Recovered Plasma and then later find a short supply buyer for it.
- A product collected under short supply can only be shipped once—and only to the final product manufacturer. If the buyer does not use it as specified in the agreement, it will need special permission from FDA to divert to another purpose.
- The agreement must be between the final product manufacturer and the collector. Brokers can play a role in facilitating these arrangements, but they cannot sign the agreements. Any broker that takes

physical possession of a blood product must register with FDA in accordance with the provisions of 21 CFR 607.

Although short supply technically permits import or export as well as other interstate shipment, there are some additional requirements that may apply. A U.S. licensed fractionator, for example, could not obtain unlicensed foreign recovered plasma under a short supply agreement because FDA wants the added assurance of knowing that it came from an FDA registered establishment [14].

4 CONTROL OF MANUFACTURE

The degree of difficulty in assessing biologic effectiveness is also a harbinger of the additional complexity of compliance issues in the biologics area. The assessors of quality/good manufacturing practice (GMP) throughout the world agree that control of critical starting materials is an essential element in assuring consistent production of a finished product that meets all requirements [15–19]. Nowhere is this borne out more dramatically than in the biologics arena. However, when one considers the special challenges evolving from very small volumes of identical material (single donor) such as unknown safety risks, biologic variability of the impure, complex milieu from which the active ingredient is extracted, and the often unknown effects of silent tag-along compounds or micro-organisms, it seems miraculous that any products reach market and perform reliably. This complexity forces one to control as effectively as possible every variable that can be controlled for biologic products, and yet all would agree that every starting pool of biologic materials is absolutely unique, whether it be human plasma for plasma derivatives, fetal kidneys or pooled/random donor urine for streptokinase, RBCs for transfusion or protoporphyrin (Hematin™) production, viral cultures for influenza virus vaccine, house dust laden with mites for allergenic extracts, or horseshoe crab blood harvested on the beaches of North Carolina for Limulus Amebocyte Lysate. The decisions concerning how one defines GMPs (e.g., how much control is sufficient and what safeguards should be in place) are complicated by the lack of control and inherent variability and heterogeneity of the basic starting material, yet one must achieve a safe and effective final product.

The control of source material is imperative to assure the same starting material every time. The importance of controlling the manufacturing process completely is even more critical. Most biologic products are made up of proteins, as purified proteins, living organisms, or whole cells. Under normal conditions, these cells survive in very controlled conditions of temperature, pH, tonicity, and nutrients. Changes outside these narrowly controlled conditions are not tolerated as the protein is easily inactivated, the culture is contaminated, or in some other way made useless or even dangerous.

The FDA has been forced to require a very rigid, tightly controlled manufacturing plan because its extensive experience (usually painful and disastrous) teaches that even what were considered to be relatively minor changes in manufacture have sometimes resulted in serious consequences. Unfortunately, examples of the potentially serious nature of minor changes are many. In the spring of 1955, cases of polio were traced to the use of the Cutter Laboratories polio vaccine. Very slight changes in the manufacturing process resulted in virulent live polio virus surviving the formaldehyde treatment and causing disease. The very small change in pH was thought to be negligible, but obviously in this case it proved to be clinically significant; likewise has been the unexpected immunogenicity of factor VIII subjected to dry heat sterilization in Europe as well as the new onset of hepatitis C (HCV) infections from plasma-derived immune globulin products after it was mandated that only plasma from HCV-negative donors could be used for plasma products. In the first case, the development of inhibitors was detected only in humans. In the second, the removal of anti-HCV antibodies effectively removed the protective neutralizing antibody levels that could prevent resident undetected hepatitis C virus from transmitting infection.

In each case, relatively minor process changes or major safety "improvements" caused significant human disease. As a result, the FDA became hypervigilant, and believing that the control of the manufacturing process was so important that both a product license application (PLA) and a separate establishment license application (ELA) were required in order to review both the product *and* the manufacturing site information were thoroughly reviewed before licenses were approved. Further, it declared that each manufacturing site must be inspected at least every 2 years. Only recently, with the development of the biotechnology industry and prodding from Congress and the commissioner, has CBER implemented the BLA to replace the PLA and ELA, but the biennial inspection requirement remains.

The major manufacturing process for plasma-derived products is the fractionation of human plasma, the liquid part of blood, to remove the minute amounts of plasma proteins present in each unit of plasma. To make the process commercially successful, very large quantities of plasma are mixed together and then fractionated. Dr. Elias Cohn developed the initial process in the early 1940s at Harvard University using differential cold alcohol precipitation. Essentially the same process (as modified by Oncley) is used today, with the addition of more rigorous viral inactivation techniques to increase safety. The conditions have been set to both efficiently fractionate the protein and to maximize viral partition, inactivation, or removal. Considering the serious nature of prior product contaminations with hepatitis and HIV viruses, once the manufacturing process has been set and validated, there can be little or no deviation from that process for the life of the product.

Likewise, each unit of blood or blood component remains highly controlled through control of manufacturing, often more to control or prevent infectious disease transmission than to protect the red blood cells or platelet products from degradation. Just as plasma or vaccine products are controlled from the very start of manufacture or collection of source material, so is blood for transfusion or further manufacture.

5 COMPARABILITY PROTOCOLS

The development of recombinant products has led to some progress in loosening FDA's tight control of manufacture. Driven primarily by a need to help the biotechnology industry, CBER developed and adopted a "comparability" theory to help facilitate rapid changes in manufacture of recombinant products by relying on biochemical and biophysical comparisons of original product and "new" product [20]. Depending on the degree of identity between the originator and new product, changes in manufacture could ostensibly be made much more quickly and inexpensively than revalidating the entire manufacturing process and repeating clinical trails.

Unfortunately, while the comparability paradigm sounds good, implementation has been difficult. All comparisons depend on the use of validated methods to compare the physicochemical characteristics of the product before and after the change. However, because of the continual fear that minor changes in manufacture can result in significant changes to activity or immunogenicity that might not be seen under comparability testing conditions and thereby put recipients at potentially life-threatening risk, FDA has gone so far as to require site-specific stability testing and comparison. Further, the innate variability of biological assays often approaches 50% at best, so creating this data-rich comparison is not as easy to use as originally thought.

The "invisible difference" type of problem is exemplified not only in biologic products, but in an amino acid supplement. While nutritional supplements are now protected from FDA oversight by the Dietary Supplement Health and Education Act of 1994, there was an epidemiologic association between the sudden occurrence of a relatively new clinical entity, eosinophilic myalgia and the consumption of tryptophan as a dietary supplement. Unbeknownst to the FDA, a major manufacturer of tryptophan in Japan used recombinant technology to increase the yield of tryptophan from its bacterial cultures. While the gross amount of tryptophan produced was significantly increased, extremely small amounts of 1,1′-ethylidene-bis-L-tryptophan contaminated the source amino acid and was implicated by National Institutes of Health (NIH) as the cause of the epidemic eosinophilic myalgia. The FDA has clearly taken this lesson to heart and remains extremely vigilant—and rigid—in controlling the manufacturing process.

At the same time however, there is increasing pressure from the biotechnology and biologics industry to modify and implement comparability testing to add significant flexibility to manufacturing process development. The meeting sponsored by the U.S. Pharmacopoeia (USP), CBER, and the International Association for Biological Standards (IABS), Biologics 2000, Comparability of Biotechnology products: A Global Approach to Accelerating Development and Availability, addressed in great depth industry and regulatory experience in this area. This very high level conference ended with a recommendation that comparability be taken up as a topic by the International Conference on Harmonization (ICH). This meeting and these recommendations should assist in developing a more flexible process, while clearly bolstering arguments on both sides of the "generic" biologics disagreement.

6 CIRCLE THE WAGONS

Until comparability changes come, however, either through ICH or the regulatory authorities, the tight control of manufacture and the serious problems posed in trying to change the process continue to pose a major liability, especially for the development of new plasma-derived products. The FDA/CBER decreed early on that each change in a product should be considered to yield a new product subject to new manufacturing process validation, new clinical studies, and new licensing. Further, CBER has never entertained the concept of a generic biologic product, primarily based on the importance of the process, which is as critical as the final product testing. Even as standardized as the manufacture and safe use of albumin has been over more than five decades, it is not treated as a generic product. This conservative approach is extremely costly and appears to have resulted in a serious lag in improved product purity or ease of manufacture for plasma-derived products. Because FDA essentially said that it will penalize any changes by their demands for data, changes have been made only slowly or when forced on the plasma product industry by FDA.

An example of the problem is the FDA "recommendation" that the manufacturing process for all plasma products incorporate new viral inactivation or segregation steps to reduce the risk of transmitting enveloped viruses such as HIV and Hepatitis B and C. However, the FDA also required the plasma fractionators to compare old product to new product in large pharmacokinetic (PK) studies and pharmacodynamic studies. For many products, this was the first PK evaluation that had been conducted with the product since its introduction in the 1960s and 70s. The studies actually required three to five times as many patients for the comparison study as were included in the entire original licensing study. While there was little if any justification given for this highly conservative FDA approach other than "I am concerned" and "It is the best science," it also graphically displays why plasma fractionation using 1940s and 1950s technology remains little changed today.

Likewise, the static manufacturing process reviewed only by CBER investigators led to an isolation and stagnation in the manufacture of blood- and plasma-based products. Products that were approved in the 1960s and 1970s using then state-of-the-art techniques were made the same way year after year, even though GMP advances for drugs continued to evolve. Any proposed change from industry was met with very expensive negative reactions from FDA requiring significant clinical studies. With the development of Team Biologics to "bring the biologics industry into the 20th century" and impose by fiat drug cGMPs on the blood, biotechnology, plasma, allergenic extracts, and vaccine industries, there was a great deal of "circle the wagons" mentality. The industry has said, "But we have made these products this way using techniques that have been specifically reviewed and approved by CBER for 30+ years. There have not been problems before. If we are to change our manufacturing to conform to cGMPs, that will be a violation of our license." The CBER has replied, "These are the same reasonable manufacturing practices we have been enforcing for years." The field/ORA interrupted with, "This entire industry is out of control and using totally unvalidated processes and should be shut down immediately as a threat to the public health." However, not having product is an even greater threat to the public health. Major portions of U.S. licensed fractionation capacity now operate under consent decrees or permanent injunctions. The FDA-induced shortages of albumin and immune globulin may have significantly jeopardized the health of their users.

The basis of these problems are complex. The FDA/CBER said you cannot change, industry said we cannot afford to change, and ORA defined itself as the arbiter of biologic GMP. Unfortunately, the lessons of the larger drug industry cGMPs were never allowed into the equation. The discussion above reveals some of the historical reasons for the inordinate level of concern for controlling starting materials and manufacturing processes. The agency's awareness of these issues, however, did not always result in optimal control in manufacturing operations, and some threats to product safety occasionally surfaced. The FDA responded to these problems with changes in the way compliance for biologics is evaluated and enforced.

7 INSPECTION PERSPECTIVES—TEAM BIOLOGICS

Historically, biologic products were regulated under section 351 of the PHS Act, and inspections performed by the headquarters experts familiar with the product license tended to be more from a "collegial scientist" perspective. From 1955 to 1972, the inspection staff came directly from the division of biologics standards in the NIH where they also reviewed license applications and did related research [21]. The central tenet of the PHS Act is that biological products cannot be shipped interstate without a license, and it is rarely feasible economically to operate a manufacturing facility for products precluded from interstate use. The exquisite simplicity of the biologics regulatory scheme evolves from the fact that no complex administrative procedures are required to suspend or revoke a license.

Although it is true that current biologics operational principles mimic those of the larger FDA in almost every way (including *Federal Register* publication of notice concerning the opportunity for a formal hearing before license revocation), it remains convenient that license suspension is among the most simple of FDA enforcement actions. After the July 1972 incorporation of biologics review and inspection programs under the Bureau of Biologics, FDA (later CBER), a new compliance perspective emerged. The industry cringed as the GMP emphasis traditionally applied by FDA field investigators experienced in applying the FD&C Act to drug inspections also began to take precedence in difficult biologic compliance issues [22]. This long-standing difference of investigator perspective was a source of tension throughout the next two decades of CBER's regulation of the biologics area, but it was not until the late 1990s that the agency formally addressed harmonization in a new program known as "Team Biologics" [23].

An important thing to remember when dealing with the FDA, is that the regulation of all products by nature occurs in an atmosphere of distrust. The FDA does not trust its regulated industries, physicians, or clinical investigators. Moreover, employees from one Center do not necessarily trust or share the same compliance perspective with their colleagues from the other centers. It is also clear that the General Accounting Office, the Institute of Medicine, the HHS inspector general, and especially Congress, do not trust FDA either, especially in safeguarding the public health after a series of inspection problems.

The reassignment of biologic inspectional duties from CBER to the ORA "field" began in the late 1980s when CBER staff were no longer able to inspect all blood banks in a timely manner and significant concerns about the control of the Red Cross blood supply came to light. While the CBER staff was made up of product experts, frequently with special blood bank certification, the field staff was not. The CBER staff understood blood bank practice and could troubleshoot problems on site. Field staff could look at overall compliance trends and systems but rarely addressed specific scientific questions. At the insistence of Congress, expertise was developed in the field force, and blood bank and plasma center inspections became the province of investigators who might be inspecting airline food service one day, a blood bank the next, and a pharmaceutical company on the third day.

Team Biologics was developed to specifically leverage and synergize the cGMP expertise and training of FDA's field staff and the scientific product experts at CBER. Product experts were developed and cross-trained, often to leverage prior expertise in, for example, small-volume parenteral injectable products, and formed into core teams. By having CBER and field experts working closely, both drug cGMPs and the exigencies of biologic product manufacture could be accommodated.

Initially, whole blood and plasma collection inspections were transferred to the field investigators, and Team Biologics inspections for fractionated blood products followed shortly thereafter. Team Biologics then added licensed in vitro diagnostics on April 1, 1998, biotechnology and allergenic products in October 1998, and vaccines and all other products in October 1999. At the time of this writing, 100% implementation has not been achieved in all districts because of the agency's need to divert many of its resources to manage the consent decrees that have resulted from the Team Biologics initiative.

While the teams are made up of both field and CBER members, the core team is clearly under the control of the field staff. The leader of the inspection is almost always from the field. The field leader chooses the team members. Compliance issues are under the control of the ORA compliance officer, and conflicts over potential health hazards due to the deficiencies are not resolved by the center, but by the associate commissioner for operations.

The important point is that both drugs and biologics are being held to the same compliance standard and requirements for product, manufacturing, and testing controls. Control must be shown through the use of standard operating procedures (SOPs) with attendant training and proficiency testing, validation of process, in-process testing, and strict lot release and final specification testing. Team Biologics operates under a dual inspectional mode. Overall compliance systems, along with manufacturing and testing controls, are the field investigator's expertise; while product characteristics and testing are in CBER staff's area of expertise.

This has led to increasingly more difficult and complex compliance challenges in the biologics community. How does one change a decades-old process to comply with the drug cGMP requirements that the rest of the pharmaceutical industry accepted and implemented on an evolutionary basis? How do manufacturers assure that the relatively delicate proteins of biological products are not significantly changed by the overwhelming demands made by FDA field investigators using traditional pharmaceutical inspection models? How do statisticians appropriately power the clinical equivalence or noninferiority trials when the data needed to adequately control and power the studies are not available because of the sudden revolutionary implementation of drugs-style requirements? Most important, how can a firm finance the physical upgrades, clinical trails, and long-term immunogenicity trails in light of typically thin margins and increasing compliance demands?

In response to these escalating compliance demands, several companies, especially in the blood bank and fractionated plasma product industry, have either gone out of business or signed consent decrees of permanent injunction that effectively allow FDA to control and oversee the function and improvement of the company. Specific plans, time lines, and personnel issues must be filed with, accepted, and overseen by FDA and supervised by the federal courts. Violations of the "negotiated" (i. e., dictated by FDA) conditions constitute a contempt of court with increasingly serious penalties. With not only the company but also the chief executive officers being named as defendants in the case, appropriate attention is paid to the plan at all levels of the company.*

All of these changes have resulted in the biologics industry's increased need for a more thoughtful and complete product manufacturing development plan. In addition, preparation for preapproval inspections, Team Biologics inspections, and expanded internal audit programs have become essential.

The problem is most acute with the small biotechnology companies, in which there are significant conflicts among optimal development and management and shareholder focus, resource allocations, and time lines. Most often after preclinical testing has been completed the major focus of the company turns to clinical testing. It is expensive, it is definitive, and it sells stock. With the great emphasis on company valuation, in the typically underfunded biotechnology company any emphasis on validation is usually secondary, if given any credence at all.

8 IRRATIONAL OUTCOMES—DUALITY DILEMMAS

Compliance issues are further complicated by the analogies the field would like to apply from the pharmaceutical or medical device environments. That

*The interesting exception being that Elizabeth Dole was not named in the American Red Cross action, even though she was ARC president.

"blood is a drug" was established by the courts in the 1960s and there has never been any serious challenge as to whether 21 CFR Parts 210, 211 (cGMP—Pharmaceuticals) are applicable, as well as 21 CFR Part 600 (600, 601, 606 [cGMP—Blood], 607, 610, 640 etc.) [24]. However, when application of pharmaceutical GMP is taken to extremes, bizarre compliance issues arise because of the unique difficulties associated with collecting the raw material and the inability to apply efficiencies of batch testing when "every donor is a different lot."

The German Drug Law, for example, is applicable to all pharmaceutical and therapeutic products, including plasma derivatives, and has created enormous expense for U.S. Source Plasma collectors to resolve compliance deficiencies (cited as failure to meet GMP requirements) as interpreted by this law [25,26]. One example derived from this experience related to the facility requirements recently imposed on plasma that is exported to Germany. Since the beginning of plasmapheresis in the early 1960s, the donor center facilities often consisted of one large room with donor screening partially segregated at one end and a counter at the other end where filled plasma bottles were processed, pilot tubes filled, and containers sealed before placing in the adjacent freezer. This "one room" handling never seemed objectionable since the process was essentially "closed" except for the momentary exposure of the sterile needle as it is uncapped for insertion in the donor's arm. However, German interpretation of GMP has forced full segregation of the various activities, and the filled plasma bottle can no longer be processed in the same room in which it was extracted from the donor, and the donor cannot remain near his product when he rests after the donation. The expense of constructing walls between these functions and providing supervision in each area has escalated cost and complicated compliance. Similarly, the generation of biomedical waste occurs primarily in the donor room. However, once the box is filled and sealed (including double leakproof liners) it must be stored in its own locked room (Nothing else permitted!), and it cannot pass back through the donor room as it exits the building with a licensed medical waste hauler; so although the underlying principles are sound, the increasing compliance demands on what used to be a relatively simple operation clearly drive the cost of operations through the roof.

9 CRITICAL DESIGN ELEMENTS

The lack of critical attention to the physical development of the biologic product is problematic; not only do sufficient experience and data need to be collected to support licensure, the same data set is needed to justify changes and maintain comparability to the manufacturing process.

Process development is never truly complete with biotechnology products. Changes to the cell bank, manufacturing process, formulation, and regulatory submission occur constantly. The development of a sufficiently deep product characterization and manufacturing database is required to support the contention that the changes in process do not necessarily result in significant changes in the finished product. The larger the database on product characterization, the more closely the "new" product can be compared to the original product. The science of exacting specifications relies upon complete familiarizations with the manufacturing process. On the other hand, the fewer full process runs that are executed, the less likely it is that appropriate specifications will be established. Further, while it is optimal to have the entire manufacturing process "locked down" before the start of the phase III trial, process changes are often being introduced right up to the time of BLA submission. Without sufficient characterization and process data, the potential for a "non-comparable" decision from FDA goes up quickly.

At the same time that significant data are being recorded for the product, it is also important to assure that the entire process is critically reviewed and defensible. For example, the transfer of a biotechnology product from an academic laboratory into a commercial environment is usually more difficult than anticipated. The FDA will require data on the origin and evaluation of the cell line. The conditions under which the hybridoma or recombinant cell line was developed, cloned, grown, passaged, and controlled are often lost through the passage of time, no matter how good the lab books seem to be. At one time significant concern was raised about the potential for contamination of the cell lines with prionlike material from bovine fetal calf serum (FCS). The type, grade, quantity, and vendor of that FCS can probably not be known. Similarly, the cloning strategy, selection strategy, and storage conditions are unknown, so it is imperative that when a cell line is imported into a manufacturing facility a critical evaluation of its origin, growth characteristics, and clonality be established.

Another problem area is understanding and justifying each step in the manufacturing and formulation process. Just because the same process and formulation as the originator uses is used, one should not assume that that particular process has been optimized. Instead, every manufacturer must evaluate each step of the process to understand its purpose and characteristics. What is the purpose of the column? What are its load and flow characteristics and the outer parameters of its use? How often can it be used and what is its maximum capacity? These questions must be addressed before licensure. Each step must be tested to ensure adequate justification for operating parameters and ranges.

Finally, the same questions should be asked in a critical manner of the in-process and final acceptance tests for raw materials, in-process intermediates, and final product. Imported and research tests often do not make good quality control laboratory tests. The quality control tests need to be robust, meaningful, validated, and reproducible. Many research assays are narrow in scope, unstable, user-dependent, and complex, and they also yield wide-ranging results. These assay characteristics are exactly the opposite of what is needed in quality control; therefore, all analytical methods and assays must be critically evaluated for relevance and validated for use.

The same stringent assessment needs to be made at each step of the process, such as the:

Raw materials and acceptance testing
Master cell bank and MWCB
Manufacturing process and control
In-process and final product testing and specifications.
Formulation
Regulatory submission
Stability testing program

Demonstration of adequate controls in the manufacturing process includes critically understanding, justifying, and validating each step in the process. The most important part of the control process is ensuring that the product is pure, potent, safe, and efficacious. More important, the firm should have sufficient data on hand to evaluate and justify any and all changes in the manufacturing process to assure that the products are comparable and acceptable and that the process will pass the next FDA inspection. (For an extensive overview of compliance elements related to change control, refer to Chap. 9, "Change Management.")

10 CURRENT PRODUCT INSPECTIONS

The FDA's inspections and regulatory reviews are challenging under the best of circumstances. Given FDA's awesome responsibility to protect the public's health and the intensity of the current Team Biologics inspections, new product and routine compliance inspections are particularly demanding. However, with comprehensive planning and critical evaluation, coupled with using the FDA inspections as a learning experience, all inspections can prove successful.

As noted in the previous section, a critical evaluation of each step of the optimized manufacturing process is imperative. Data, which are best collected under worse case conditions and usually during process development, are needed to support the control and robustness of each step and each conclusion. All methods and manufacturing processes must be validated using scientific justification and sound compliance judgment.

Develop a plan, establish critical pathways and time lines, and ensure every critical step in the manufacturing process has been adequately validated. See Appendix A for a synopsis of FDA's inspection guide for assistance in performing internal audits.

11 QUALITY SYSTEM CONTROLS

For any company, whether new or old, implementation of quality systems is the first step toward meeting the FDA's current expectations of the biologics industry. The quality systems approach was developed and has been implemented by CDRH for use in the development, testing, and control of devices. The regulations are codified in 21 CFR 820 and make numerous valuable suggestions that are applicable across all FDA-regulated industries, particularly the biologics arena.

For example, the use of a product design and control history file is useful in understanding what has been done and why. Each step of product design and research is documented. Changes to the product, process, or formulation are noted, justified, and saved. It is no longer necessary to rummage through old lab books or call someone who retired 10 years ago to figure out why early design decision were made and implemented.

Similarly, the use of a failure analysis approach for unexpected changes in final product or in process controls markedly help both the analysis and documentation requirements at each step. Finally, the use of a written and continually revised process development document is most helpful. The early identification of data requirements, special studies, and comparability decisions for use in process planning is important, especially with budget constraints.

12 PREVENTATIVE MEASURES

There are several additional critical steps to be taken and techniques to be applied in preparation for an inspection and to assure that the current manufacturing process is both "in control" and of the highest quality.

One of the first steps is to develop a reasonable and rational initial product development plan. Hasty or "bare-bones" process development frequently causes significant delay during the clinical study phase and during preparation of the BLA [27]. Included in the firm's overall planning should be a validation master plan, a top-to-bottom validation effort that would include everything from facilities qualification, SOP development, implementation, and training to environmental controls. Developing the validation master plan early and then updating it regularly as the process or facilities change helps maintain constant control over the multiple aspects of the operation. The validation master plan will assure that critical interdepartmental interfaces are established and maintained, and products and processes are controlled through the utilization of quality systems.

At the same time, and especially as the manufacturing process matures, it is also important to perform a reality check before FDA walks through the door. While the FDA investigator represents a potentially harsh reality check, preparation for that visit should be an ongoing process and should begin during product development. Preparation for a Team Biologics inspection frequently entails repackaging early development work and subsequent clinical and commercial data so that it is reviewer-friendly and is easily retrievable, manageable, and transparent during the inspection process. Even after a firm receives approval during a preapproval inspection for a new product, subsequent inspections may not be as successful if the firm loses its commitment to inspection-readiness.

Benchmarking the facility and its various operations against other companies is also very useful. The secrecy of proprietary manufacturing schemes and closed plants can make benchmarking a daunting process, but the use of outside third-party experts often allows the comparison to occur. One of the most fruitful, if painful, strategies is to invite a "gap" analysis (i.e., determining the difference between an existing process and current cGMP expectations, by an expert third-party reviewer). As in many areas, chemistry, manufacturing and controls (CMC) staff are often too close to the process to see everything, so these former FDA or industry reviewers can help identify problems. Most important, they should not be restricted to just one area. Manufacturing is an integrated process, so SOPs, training, documentation, data collection, reports, failure control, and analysis should also be reviewed before FDA comes through the door.

These inspections can be either informal or formal. The informal approach is to have a complete in-depth review of the entire process with a written report. The formal approach is to have a mock FDA inspection, which is extremely useful in identifying system and process failures at all levels. The latter offers practice in both preparing for and interacting with an FDA investigator as well as testing the breadth and depth of documentation that would be needed for the FDA inspection process. In all cases, the use of a third-party experts allows for a more objective evaluation of the entire process from a fresh pair of eyes and from someone who can offer an objective comparison against current industry standards.

For facilities currently in production, the use of benchmarking and outside experts is also useful in developing the compliance upgrade master plan or GMP enhancement master plan (GEM Plan). As with the development of a Validation Master Plan, the quality and control review process needs to be dynamic; cGMPs change on a regular basis and manufacturing engineering advances create opportunities for further process development. Through the development of an upgrade or enhancement master plan, changes in processes can be handled in a controlled manner.

These plans take into account advanced planning and budgeting and revalidation of key processes, as well as identification of potential problems both up- and down-stream of the key process change and product supply issues. They are somewhat similar in many ways to the corrective action plans that are required as a response to an official FDA communication, such as the issuance of an FDA-483 (notice of inspectional observations) or a warning letter. The difference is that the compliance upgrade and GEM plans are proactive plans and are developed well before an FDA inspection occurs. It is also a tool that the firm can have in its back pocket should a lengthy FDA-483 or warning letter be issued. The GEM plan illustrates that the company has acted proactively and responsively in identifying issues and is working diligently in resolving them in a carefully controlled manner. Whether it is a GEM plan or a corrective action plan in response to some form of FDA regulatory action, firms should establish realistic time frames for completion and subsequently monitor to assure the commitments have been met. (For further discussion about the GEM Plan, see Chap. 14, "The Compliance Upgrade Master Plan."

13 WORKING WITH FDA

Another useful source during the product development phase is the FDA, which probably knows as much or more about the proposed product then you do since it has both the scientific expertise and prior knowledge of everyone else's similar product. If the CBER reviewers are engaged at a scientific level with data-driven discussions and proposals, they can help not only with the successes, but also the failures. It behooves a firm to establish meaningful dialogue and long-term relationships with the assigned reviewers at CBER. The firm should take advantage of presubmission meetings as much as possible, while taking care not to become over reliant on the reviewers. With an increasingly large workload being handled by an increasingly smaller and younger cadre of reviewers, CBER is putting more and more restrictions on sponsor contacts with agency staff. When used judiciously, CBER can be one of the most effective sounding boards and evaluators of what the firm will need to ensure product approval.

The same can be said of the field investigators. In spite of the inherent difficulties between the investigators and the firm they are investigating and the recognized variability among investigators, today's Team Biologics investigators are highly trained professionals. They often have specific expertise in product areas and understand that it is their mission to ensure public safety through their thorough assessment of facilities and manufacturing controls. Their main focus during an inspection is to evaluate the qualification of facilities and equipment related to the manufacture of product seeking approval. It behooves a firm to pay special attention to all investigator comments during the inspection process and it is particularly important for key personnel to consistently interact with the investigators and ask questions about their concerns. If the FDA inspection proves too chaotic to engage the investigators in protracted discussions about their concerns and perspectives, the exit conference is the last opportunity for a firm to elicit meaningful feedback about their processes and overall operations from the investigators. Use the FDA field investigators as well as CBER reviewers as sounding boards and advisors from the development phase and beyond commercialization.

14 NEW BIOLOGICS

Finally, there is really nothing terribly new with the production and development of the single-lot therapeutic or alternative source products. The CBER is most concerned that manufacturers can develop a safe, pure, potent, and effective biological product and bring it to market. The same kind of infectious contamination, impurity, and control issues apply to a product whether it is made in animals, cell culture, or plants. The same kind of thoughtful and critical evaluation of the manufacturing process is required. The same attention to details and proving the robustness of the process are important. If it is remembered that the earliest biologic products were made from infectious agents or animals, little has changed as we move into alternative sources.

The same is true for single-patient or single-lot designer products. Much has been made of the fact that the stringency of manufacturing or processing for a single-lot, especially autologous, product should not be held to the same level as a large-scale biological product, yet the same critical analysis of the process is important here too. While infectious disease testing may not be required for each sample of cartilage cells or other tissue used to make an autologous replacement graft, the stringency of controls at the manufacturing plant must be concomitantly heightened to ensure that cross-contamination of samples does not occur. Any loss of control of the starting material causes problems further down the line, and the usual statistical analyses used for multilot production are not highly useful when only a few vials are produced. However, FDA has regulated single-lot products for years. Each unit of blood is considered to be a single lot of product; control is placed on the manufacturing process to assure purity, potency, safety, and efficacy.

The important bottom line is a data-driven critical analysis of the manufacturing process, no matter the source material, to demonstrate sufficient control.

15 THE COST OF NONCOMPLIANCE

One of the most widely publicized events in recent regulatory history was the $100 million dollar fine levied against Abbott Laboratories as part of a consent decree of permanent injunction signed with the Justice Department on behalf of FDA, and the removal from the market of more than 100 Abbott Laboratories IVD products in November 1999 [28]. It is particularly noteworthy that Dr. David Feigal, FDA's director of the CDRH acknowledged that this action was taken despite the absence of any reports of harm to patients. One can justify these actions by presuming that the FDA loves to punish a large manufacturers in order to deter noncompliance by the industry, or one can derive an important lesson in heeding the warning flags that existed and prepare for future inspections that FDA has declared will focus more heavily on the effectiveness of the quality systems of all manufacturers.

When one looks at the Abbott situation retrospectively it appears clear that efforts to implement the expected quality systems were inadequate. There were at least five warning letter concerning in vitro diagnostic devices that carried repetitive messages about failure to fully implement quality systems. In addition, there were 11 other Abbott warning letters between 1992 and 1999 that cite many of the same issues in other Abbott divisions. To avoid FDA warning letters or other regulatory actions, it is always essential to make special efforts to prevent repetition of a previously identified problem and to ensure that the corrective actions extend to the entire manufacturing framework.

The high cost of noncompliance does not just affect final products. Other recent examples of high penalties involve problems with source materials that resulted in enormous cost to suppliers of biologic raw material or users of the unsuitable biologic. A 1995 *Food and Drug Law Journal* article succinctly reviews several cases of liability for bulk suppliers under the common law and speculates on the effect of the Biomaterials Access Assurance Act [29]. Although we are not qualified to posit whether such liability and litigation extends directly to more familiar biologic products, one should be aware of the escalating costs of insurance due to such costly episodes as those listed in Table 1.

TABLE 1 Biologics Noncompliance Scenarios and the Associated Costs

Undetected problem or error	Cost to company
1. Source plasma containing small amounts of anti-C(G) erroneously shipped to a fractionator whose specifications precluded it; no FDA violation involved, same material accepted by other fractionators, and risk to final product very remote.	Supplier had to pay $2,000,000 for cost of plasma pool discarded.
2. Recombinant biologic product contained non-FDA-approved human plasma in tissue culture media used in manufacture of the product.	Final product manufacturer had to go to great expense to research and document the equivalent safety of the foreign plasma that had been used and product approval was delayed until resolved.
3. Nucleic acid test (NAT) lab performing tests on small pools of samples expected only viral marker antibody negative samples to be forwarded for testing.	$1,000 fine levied by the NAT lab every time an antibody-positive sample was shipped in error.
4. New variant CJD precludes use of donor with history of residence in U.K. for 6 months or more between 1980 and 1996.	All prior donations from donor who retrospectively provides this information must be recalled and any fractionated products destroyed. If plasma collector shipped the material in error, it will bear cost (fractionator dependent). This could cost millions of dollars, depending on the time frames and numbers of products affected.
5. German Health Authorities may deny import certificate to plasma collectors who make a single error in interdicting shipment of a viral marker reactive unit or properly notifying fractionator of a sero-converting donor.	Export shipments cease the day of inspection deficiency, and the wait for reinspection may be a year or longer. All plasma in inventory must find another market. Cost could exceed $1,000,000 per instance.
6. BioWhittaker supplied Abbott with human neonatal kidney cells and FDA inspection revealed numerous deviations for cGMP.	FDA warned public (Jan. 25, 1999), and Abbott received warning letter that included reference to failure to ensure supplier-corrected problems. Because FDA found insufficient process validation at Abbott it was told to "review and redesign . . . the manufacturing process."

TABLE 1 (Continued)

Abbott Laboratories' sanctions revisited: warning signs/preventative measures		
1993	Diagnostic product deficiency citations at Lake County and North Chicago plants	Oct. 19, 1993, WL cites validation, SOP gaps and CAPA weakness.
1994	March 28, 1994 WL	Same issues as Oct. 1993 + incomplete quality system audits.
February 1997	VAI (voluntary action) FDA inspection outcome	
July August 1998	Sponsor/monitor operations inspected: warning letter received Nov. 1998	Nine unreported recalls/market corrections (mc); 2 more mc triggered by inspection; no official log of CAPA; several failed lots not entered in CAPA; long delays in entering problems in CAPA.
1999	March, May, June, July 1999 inspections revealed continuing failure to correct multiple problems; March 17 warning letter; FDA talk paper (July 14) re: serious Abbokinase problems.	CAPA still deficient: "Failure to establish and maintain procedures for implementing corrective and preventive action"; failed components reused; inadequate records of investigation of nonconformances (NCM); multiple failed lots "expected" and no statistical analysis of failures; computer system database defects not reported to users through identified since 1997. July inspection resulted in warning letter
November 1999	Consent decree, $100 M fine, suspend sales for 125 products within 30 days	100% FDA testing for biologics lot release reinstituted. Dear colleague letter: "FDA is talking this action because of (Abbott's) long-standing failure to comply with FDA's Good Manufacturing Practices (GMP) regulation, now called the Quality System (QS) regulation, and Abbott's repeated failure to fulfill commitments to correct

Abbott Laboratories' sanctions revisited: warning signs/preventative measures (Continued)

		deficiencies in its manufacturing operations for its in vitro diagnostic (IVD) devices . . . inspections showed non-compliance with GMP/QS regulation requirements involving process validation, corrective and preventative action, and production and process controls, among others." Regarding process validation: FDA is especially concerned about Abbott's long-standing noncompliance with these accepted manufacturing principles because they represent the minimum requirements for manufacturing quality [30].
	Stock price plummets and revenue loss estimated at $250 M for 2000.	Any failure to meet deadlines for correction of manufacturing processes will result in fines of $15,000 per day process, up to a total of $10 M. Required to retain independent experts to evaluate and audit at least twice a year for 4 years or longer and report to FDA. Abbott within 60 days required to submit a " master compliance plan, proposed validation plan and proposed validation procedure . . . to bring processes into compliance with QSR within 365 days."

16 CONCLUSION

Biological and biotechnology products present unique developmental and compliance challenges. Because of the relative delicacy of the protein products, great importance is placed on demonstrating control of the manufacturing process. This has been significantly complicated in the last several years by two developments: the imposition of an entirely new drug cGMP paradigm on a relatively mature industry without adequate time for adjustment and the development of biotechnology products by parsimonious and relatively inexperienced new companies. Both require the same resolution, though at an expense neither appears to be able to afford—a data-rich demonstration of complete control of the manufacturing process. However, by adopting and using the design control and quality systems concepts, the ease and control of process development, comparability testing, and final approval, especially when working with FDA, should made the process faster and easier.

17 REGULATORY REALITY CHECK

With this chapter providing an instructive backdrop, several FDA observations (FD-483s) documented during recent inspections of FDA-regulated facilities are presented below, followed by a strategy for resolution and follow-up corrective and preventative actions.

17.1 FD-483 and Warning Citation

Media fill studies do not adequately support sterile product expectations in that

Worse-case scenario did not specify maximum time for freezer door to be open or maximum number of trays to be loaded into freeze drier.

All personnel/shifts were not represented and pacing demands on operator for manual operations was not challenged.

Failed or invalidated media fills were not reported as deviations and investigation is not documented according to firm's SOP.

Environmental monitoring was not performed during media fills to establish that conditions were representative with respect to bioburden.

Recently added fill size was outside prior tested limits.

An appropriate response from the firm might go something like the following:

> We understand the potential significance of the observations made and have addressed corrective and preventive action with assign-

ment of a project team. This team (composed of an engineering consultant skilled in sterility validation, the quality unit director, the filling supervisor, and the operations manager) will provide the FDA with a comprehensive plan to address all identified deficiencies within 30 days of inspection close-out.

All in-plant lots are being held in quarantine until a final decision is made concerning any potential impact on the quality of products prepared to date. However, a review of historical data collected under current conditions indicates that we have not had a single failure in sterility testing performed on 100% of the 2918 lots manufactured in the past fifteen (15) months [including at least ten (10) lots of the new fill size]. These tests were conducted in full accordance with the requirements of 21 CFR 610.12 and have been validated with respect to sensitivity in identifying minimal levels of contamination. In addition, all product complaints received for product processed at this facility were reviewed and these reports also confirm that there has been no documented instance of sterility failure detected.

Note that the firm makes a point of establishing the integrity and quality of material filled to date in an effort to prevent the necessity of a recall. In response to the FDA's concern that the maximum time for the freezer door to be left open, along with the maximum number of trays that could be loaded into the freeze dryer, the firm should requalify the freezer to accommodate minimum and maximum load configurations, as well as worst-case temperature-excursion scenarios. Additionally, an alarm should be placed on all freezers containing sterile product and stability samples. In response to the second item, related to the issue of all personnel shifts and personnel not being represented, the firm should include in its validation protocol an assessment of operating demands placed on an array of personnel involved with manual processes. In response to the third item, concerning the handling of all failed or invalidated media fills, the firm should retrain applicable personnel to the SOP that addresses handling of failed or invalidated media fills. Additionally, the firm should make a concerted effort to review all failed and invalidated media fills over the last 24 months to ensure the appropriate release of material occurred. With regard to the agency's concern about environmental monitoring not being performed during media fills, the firm should include in its validation protocol an environmental monitoring component to assess bioburden during static and dynamic conditions. The firm should also correlate and compare this environmental monitoring data with data collected for the

> overall operation. Finally, with regard to adding a fill size that had not been tested previously, the firm should include in its validation protocol all fill sizes intended to be used for stability, clinical trials, and commercialization.

One way this company could have prevented receiving this citation would have been to have the production manager most knowledgeable about the process (along with quality assurance) routinely approve all validation protocols. For this kind of operation, sterility documentation will be among the first that the FDA will review. As such, all sterility parameters and operating ranges must be scientifically justified and adequately documented. The thoroughness of the approach should be optimized with regard to every parameter. Justification for the selection of criteria should be put into writing, as well as the rationale for any deliberate omissions based on an understanding of the equipment involved, other potential variables, and their effect on outcomes.

17.2 FD-483 Citation

The firm failed to establish and maintain procedures for implementing corrective and preventive action (CAPA), as evidenced by some events not included, long delays in entry of events, no assurance that affected products were quarantined, lack of central log for all CAPAs, and no tracking or trending reports to management.

17.2.1 Response

Top management is investigating the apparent lapses in quality systems that might have contributed to these deficiencies through the execution of a comprehensive internal audit by both internal quality assurance (QA) and a third-party specialist. Management is also assessing the resources that are currently available or that may be needed to ensure that there is rapid and complete correction of this issue. The CEO will act on recommendations from a senior management working group within 2 weeks. The director of the QA unit will personally chair the task force responsible for developing the needed procedures within 30 days and will report progress weekly to the CEO until full correction is confirmed. Beginning last week and going forward indefinitely, all biweekly senior management meetings will have a standing agenda item for reviewing the data evolving from the new CAPA program, and each department head will provide written comment to respond to any undesirable trends in product or process under their purview. Both the QA unit director and the CEO will sign off on all summary reports (to be provided at least quarterly).

We would like the record to reflect, however, that the delays in database entry of the event information or in the creation of formal records of product quarantine did not reflect a delay in appropriate investigation and action to ensure only safe product was available and prompt correction of precipitating defects. In addition, the informal records of actions related to these events are being gathered from all departments, and the QA director will direct assembly of a complete, consolidated record and determine if any requirements remain unmet. Within 2 weeks the QA director will report the status and conclusions of this project. All personnel with appropriate training for participation in such a project will be relieved of their current duties and directed to focus on these issues. Additional training will be provided to all personnel related to the company's procedure for identifying and implementing corrective and preventative action.

The product inventory hold that was voluntarily instituted as a result of this finding will remain in effect until the QA director has assured the CEO that there are no outstanding concerns related to product integrity and the handling of manufacturing deviations and implementation of corrective and preventative actions. However, we view this as a strictly precautionary measure since no evidence has been revealed to indicate any threat to the consumers from the firm's product.

17.2.2 Prevention

One way this company could have prevented receiving this citation would have been to routinely include as part of its internal audit program an assessment of all manufacturing deviations and follow-up on corrective actions. Handling of manufacturing deviations along with the installation of corrective and preventative measures is a quality system. (For an extensive discussion of handling manufacturing and laboratory deviations see Chap. 13, "Handling Laboratory and Manufacturing Deviation.") An additional preventative measure would have been for the firm to proactively retain a third-party auditor to perform a GAP analysis of the firm's quality systems. It is important to be aware of any organizational changes and how they might impact personnel's understanding of their roles and responsibilities. Finally, periodic quality reviews by a cross-functional team (QA, engineering, manufacturing, quality control) would consistently bring to light deficiencies of this nature.

18 WORDS OF WISDOM

Never make a commitment in a license application, SOP, or response to inspection if the company is not prepared to ensure implementation and monitor the ongoing effectiveness.

Comprehensive programs to identify compliance issues and thoroughly address corrective and preventative action are worth the investment. The investigator's view of the severity of an issue will be significantly reduced if the firm has already identified the problem and developed a formalized plan for correction. The plan must prioritize appropriately and represent realistic time frames for meeting its commitments.

Periodic third-party reviews are an invaluable tool for identifying "blind spots" within an organization. Do not hesitate to augment the company's internal audit efforts with outside experts who are familiar with similar company operations.

APPENDIX A

Critical Compliance Guideposts for Biotechnology Companies to Facilitate a Self-Check of the Adequacy of All Significant Processes and Systems [31]

Review process and documentation for conformance to BLA.

Assess GMP vs. CPGM 7356.002 and 7356.002A.

Enlist objective professional to conduct mock audit based on biotechnology inspection guide and CPGM 7341.001.

1. Components
   Defined specifications, supplier evaluation, established controls, animal sources free from adventitious agents (mycoplasma, BSE, etc.); acceptance documented in batch record.
   a. Cell bank control: storage, identification, handling; characterization and FDA approval for all new cell banks.
   b. Media, buffers: criteria, holding times and conditions, validation, records.
   c. Containers/closures: Specifications and SOPs for receipt, handling, sampling, storage, validation, reconciliation of final associated equipment maintained and requalified as scheduled (211.80–211.94).
2. Manufacturing (211.100–211.115)
   a. Aseptic start point clearly defined; connections and transfers verified aseptic; cleaning and sanitation defined, validated, recorded; process and holding time limits and storage conditions validated; sterile filtration adequacy evaluated and

integrity tested postfill per manufacturer's instructions; training (including gowning); class 100 wherever product/components exposed; environmental monitoring (including gowning), bioburden alert/action levels and identification of contaminants.

b. Endotoxin levels: Microbial load minimized; potential contamination and cross-contamination prevented.
c. Fermentation/bioreactors: Closure, connections, optimal and documented.
d. Validated processes for disruption and harvest; purification; viral inactivation and removal; lyophilization. Routine revalidation, DOP testing of HEPAs, media fills, pyrogen/depyrogen verification, growth promotion, quality test method verification, etc.

3. Validation

   a. Process changes: controlled by protocol and validation documented.
   b. Computers: comply with 21 CFR part 11; no overwrites of data permitted.
   c. Shipping: conditions, containers, methods, effectiveness of temperature controls (211.142–211.150).

4. Testing and laboratory controls: pass/fail criteria and justification if released; sampling methods; reference standards ongoing evaluation; defined retest conditions; investigation and review SOPs; training of analysts and supervisors.
5. Environmental controls/monitoring: programs includes viable and nonviable particles, surface-viable particles and personnel (filling areas); schedules and action/product disposition; media fills cover all shifts and operators, all package sizes and worst-case assessments.
6. Cross-contamination prevention: segregation by space or time; cleaning and personnel precautions.
7. Nonconforming product: investigation; rework records include investigation/corrective action (no blends to reduce adulterants unless CBER approved reduction method); no microbial positive rework; complaints, recalls, errors, and accidents; adverse experience reports all tracked and reported as required (Identity–610.14; purity–610.13).
8. Changes to be reported: meet 21 CFR 601.12 requirements.
9. GMP

a. Equipment SOPs, periodic revalidation, status tagged, column life span specified; cleaning processes validated; calibration schedules and SOPs adequate and include limits for accuracy, precision, and remedial actions (211.63–211.72; 600.11).
b. Buildings: conditions and practices controlled throughout; segregation of raw material, quarantined rejected, in-process and released products; designed to prevent mix-ups and allow cleaning and maintenance; pressure differentials, temperature and humidity limits conform; spore-forming materials isolated; water systems criteria defined and monitored; CIP, SIP, HVAC, and HEPA all validated and periodically reassessed (211.42–211.58; 600.11).
c. Quality control: QC unit has formal processes and controls, is responsible to approve/reject all components, packaging, drug products, labeling, etc (211.160–211.176).
d. Personnel/training: adequate number with background, training and experience; trained to SOP; qualified instructors; competency assessed. (211.22–211.34; 600.10).
e. Waste processing: handling spills and disposal appropriate; prevents cross-contamination.
f. Labeling: meets 21 CFR 201, 610 and 660; all wording approved by CBER.

10. Records: master production records complete and current; batch records reflect accurate process for every lot initiated and get reviewed by QC unit before release; distribution records permit accurate and rapid recall; stability protocols and retention samples for final products and components conform to SOP and support labeled dating period (211.180–211.198).
11. Lot release: conforms to 21 CFR 610.2(a) unless exempt as specified biotechnology products or approved surveillance program.

APPENDIX B

The evolution of the aggressively ambitious and relatively underfunded biotechnology industry has also led to increasing demands for alternative manufacturing arrangements. The design, development, and financing of cGMP quality manufacturing plants is now often left initially to contract manufacturing firms, but the control and oversight of the entire process remains the responsibility of the final licenseholder and commercial distributor. The execution of this control and oversight is important to consider early in the relationship as well as at the time of licensure.

Contract Manufacturing Responsibilities Checklist

License applicant (LA)	Contract manufacturer (CM)
Application integrity and all regulatory interaction with authorities	FDA registration + inspection accepted; cooperation with inspections and prompt correction of deficiencies assured
Qualifications of the CM and establishing the personnel qualifications for work under contract	Access to batch records for review by LA
Entire process GMPs and validation protocols; defining who will validate analytical methods	Adherence to approved SOPs; supervision and control by CM for the LA product
Contract signed prestart with name and locations(s), detailed list of CM functions	Notifications of all proposed changes, including suppliers or their specs
Description: product to be shipped and how	Notice of errors, accidents, and deviations
Description: operations at CM	Notice of adverse inspection findings
Contract process SOP + segregation; summary of systems and equipment validation	Notice of new product manufacture
How CM will be periodically assessed by LA, including readiness for FDA inspection, corrective/preventive action expectations	Description of quality program and change control SOPs; commitment to GMP, including record retention that meets FDA requirements
Mechanisms for identifying, reporting, and resolving problems	Commitment to participate in problem solving

Source: Ref. [32].

APPENDIX C

Resources

Guidance for industry	
Container Closure Systems for Packaging Human Drugs and Biologics (CMC Documentation):	May 1999
For the Submission of Chemistry, Manufacturing and Controls and Establishment Description Information for Human Plasma-Derived Biological Products, Animal Plasma or Serum-Derived Products:	February 1999
For the Submission of Chemistry, Manufacturing and Controls and Establishment Description Information for Human Blood and Blood Components Intended for Transfusion or for Further Manufacture and For the Completion of the Form FDA 356h "Application to Market a New Drug, Biologic or an Antibiotic Drug for Human Use":	May 1999
For the Submission of Chemistry, Manufacturing, and Controls Information for a Theraputic Recombinant DNA-Derived Product or a Monoclonal Antibody for *In Vivo* Use (CMC):	August 1996
For the Submission of Chemistry, Manufacturing, and Controls Information for a Theraputic Recombinant DNA-Derived Product or a Monoclonal Antibody for *In Vivo* Use (CMC):	November 1994
Content and Format of Chemistry, Manufacturing and Controls Information and Establishment Description Information for a Vaccine or Related Product:	January 1999
Content and Format of Chemistry, Manufacturing and Controls Information and Establishment Description for a Biological *In Vitro* Diagnostic Product:	March 1999
On the Content and Format of Chemistry, Manufacturing and Controls Information and Establishment Description Information for an Allergenic Extract or Allergen Patch Test:	April 1999
IND Meetings for Human Drugs and Biologics (CMC Information)—Draft Guidance:	February 2000
INDs for Phase 2 and 3 Studies of Drugs, Including Specified Therapeutic Biotechnology-Derived Products (CMC Content and Format)—Draft Guidance:	February 1999
Guidance for the Submission of Chemistry, Manufacturing, and Controls Information and Establishment Description for Autologous Somatic Cell Therapy Products:	January 1997

Note: These are available from: Office of Communication, Training and Manufacturers Assistance (HFM-40), 1401 Rockville Pike, Rockville, MD 20852-1448. (tel.) 301–827–1800; or 1–800–835–4709; (Internet) http://www.fda.gov/cber/guidelines.htm.

BIORESEARCH MONITORING COMPLIANCE PROGRAMS

Their purpose is to ensure quality and integrity of data and protect rights and welfare of subjects.

CPGM: Program
7348.808: Nonclinical laboratories
7348.809: Institutional review boards
7348.810: Sponsors, monitors/CROs. Sponsors (312.50 59) must: train staff, monitor frequently, retrain personnel as needed, recognize and deal with noncompliance quickly to protect study. Monitors must: Verify and check investigator functions, including: qualifications and resources, product storage and handling, protocol adherence, informed consent and subject documentation, records, reports, accuracy and completeness of CRFs, and reporting to sponsor.
7348:811: Clinical investigators. Investigators (312.60–69) must: conduct study to plan; protect rights, safety; welfare of subjects; obtain informed consent; record drug disposition; file prompt safety reports to sponsor; assure IRB review of protocol and changes.
Inspection program coordinated by ORA/OE. A small percentage of firms are routinely inspected plus selected IND applicants whose submission or history raises questions.
21 CFR: Biological products regulations
Part 312: INDs
Part 314: NDAs
Part 25: Environmental assessments
Part 201, 202: Labeling and advertising
Part 210, 211: CGMPs (FD&C Act)
Part 600–680: Biologics (PHS Act)
Part 800: In vitro diagnostics/medical devices
Part 1270: Tissue regulations

REFERENCES

1. Compliance Program Guidance Manual: CPGM—Drugs and Biologics. Chap. 41 CP 7341.002. Inspection of Tissue Establishments. Nov. 1999.
2. Compliance Program Guidance Manual: CPGM—Drugs and Biologics. Chap. 41 CP 7342.001. Inspection of Licensed/Unlicensed Blood Banks. Oct. 1999.
3. Compliance Program Guidance Manual: CPGM—Drugs and Biologics. Chap. 41 CP 7382.830. Inspection of Medical Device Manufacturers. July 1998.
4. Compliance Program Guidance Manual: CPGM—Drugs and Biologics. Chap. 41 CP 7356.002. Drug Process Inspections. Dec. 1990.
5. Compilation of Documents Relating to Inspection of Source Plasma Establishments. Annapolis MD: ABRA, 1999.

6. 21 CFR Parts 606 and 640. annotated by P. A. Hoppe. Abbott Park, IL: Abbott Quality Institute, Sept. 1996.
7. Blood Bank Regulations: A to Z. 3d edition. McCurdy, K., Gregory, K. AABB Press, Bethesda MD: 2003.
8. Title 21, Code of Federal Regulations, Part 600.3. U.S. Government Printing Office, Washington, D.C.: 2003.
9. Guidance for Industry: Revised Precautionary Measures to Reduce the Possible Risk of Transmission of CJD and new variant CJD by Blood and Blood Products. DHHS/FDA/CBER, Bethesda, MD: Nov. 23, 1999.
10. Precautionary Measures to Reduce the Possible Risk of Transmission of Zoonoses by Blood and Blood Products from Xenotransplantation Product Recipients and Their Contacts. DHHS/FDA/CBER, Bethesda, MD: Dec. 1999.
11. Blood Products Advisory Committee. Notice of meeting, Monday, March 6, 2000, Bethesda, MD. FR FDA 03/06/00 N 65 FR 11785, 65 (44): 11785, March 16, 2000; BPAC proceedings available on www.gov.fda.cber.
12. American Blood Resources Association stats.
13. Revision of Title 21, Code of Federal Regulations, Part 601.22. Fed Reg 64: 56441, Oct. 20, 1999.
14. FDA. Regulatory Procedures Manual: Import Operations/Action. U.S. Government Printing Office, Washington, D.C.: Aug. 1, 1997, Chap. 9.
15. Title 21, Code of Federal Regulations, Parts 210, 211. U.S. Government Printing Office, Washington, D.C.: 2003.
16. Title 21, Code of Federal Regulations, Parts 600 to 799. U.S. Government Printing Office, Washington, D.C.: 2003.
17. American National Standards, ANSI/ISO/ASQC Q10011-1-1994. Milwaukee, WI, July 18, 1994.
18. Committee for Proprietary Medicinal Products, Commission of the European Communities, Directorate General III. Industry-Pharmaceuticals 001329. Brussels, Belgium, March 16, 1994.
19. GMP for Blood Banks. Blood Transfusion Council of the Netherlands Red Cross. Amsterdam, Netherlands, 1993.
20. FDA. Guidance Concerning Demonstration of Comparability of Human Biological Products, Including Therapeutic Biotechnology-Derived Products. April, 1996; www.fda.gov/CBER
21. Public Health Service Act. Biological products. 42 USC 262.
22. Food, Drug and Cosmetic Act. 21 USC 374.
23. Fed Reg FDA N 63 FR 64999. FDA Plan For Statutory Compliance; Notice—Team Biologics. U.S. Government Printing Office, Washington, D.C.: Nov. 24, 1998.
24. Calise opinion (SDNY). 1962. quoted in CBER. Compilation of Judicial Action, 1979.
25. German Drug Law (Arznelmittelgesetz, AMG). rev. version. Art. 72a, paragraph 1, no. 2, Dec. 11, 1998.
26. Guide to Inspections of Source Plasma Establishments in the USA. Regierungspraesidium, Kassel, Germany: Jan. 2000.

27. G. Bobrowicz, The compliance costs of hasty process development. BioPharm, 12(8), 35–38: Aug. 1999.
28. FDA says Abbott Labs kits out of compliance. Wall St J, Sept. 29, 1999.
29. Baker, F. D. Effects of products liability on bulk suppliers of biomaterials. FDLJ 50: 455–460, 1995.
30. Abbott consent decree; Dear Colleague letter; Q&As, warning letters; www.fda.gov/CBER.letters.htm
31. Compliance Program Guidance Manual: CPGM—Drugs and Biologics. Chap. 41—Therapeutic Products—7341.001. March 1999.
32. Guidance for Industry (draft). Cooperative Manufacturing Arrangements for Licensed Biologics. CBER, Bethesda, MD: Aug. 1999.

Index